PLANT FACTORY BASICS, APPLICATIONS AND ADVANCES

PLANT FACTORY BASICS, APPLICATIONS AND ADVANCES

Edited by

TOYOKI KOZAI

GENHUA NIU

JOSEPH MASABNI

ELSEVIER

ACADEMIC PRESS

An imprint of Elsevier

Academic Press is an imprint of Elsevier
125 London Wall, London EC2Y 5AS, United Kingdom
525 B Street, Suite 1650, San Diego, CA 92101, United States
50 Hampshire Street, 5th Floor, Cambridge, MA 02139, United States
The Boulevard, Langford Lane, Kidlington, Oxford OX5 1GB, United Kingdom

Notices

Knowledge and best practice in this field are constantly changing. As new research and experience broaden our understanding, changes in research methods, professional practices, or medical treatment may become necessary.

Practitioners and researchers must always rely on their own experience and knowledge in evaluating and using any information, methods, compounds, or experiments described herein. In using such information or methods they should be mindful of their own safety and the safety of others, including parties for whom they have a professional responsibility.

To the fullest extent of the law, neither the Publisher nor the authors, contributors, or editors, assume any liability for any injury and/or damage to persons or property as a matter of products liability, negligence or otherwise, or from any use or operation of any methods, products, instructions, or ideas contained in the material herein.

Library of Congress Cataloging-in-Publication Data
A catalog record for this book is available from the Library of Congress

British Library Cataloguing-in-Publication Data
A catalogue record for this book is available from the British Library

ISBN: 978-0-323-85152-7

For information on all Academic Press publications visit our
website at https://www.elsevier.com/books-and-journals

Publisher: Megan R. Ball
Acquisitions Editor: Nancy J. Maragioglio
Editorial Project Manager: Lena Sparks
Production Project Manager: Joy Christel Neumarin Honest Thangiah
Cover Designer: Christian Bilbow

Typeset by TNQ Technologies

Contents

13. How to integrate and to optimize productivity

Kaz Uraisami

14. Emerging economics and profitability of PFALs

Simone Valle de Souza, H. Christopher Peterson, and Joseph Seong

15. Business model and cost performance of mini-plant factory in downtown

Na Lu, Masao Kikuchi, Volkmar Keuter, and Michiko Takagaki

16. Indoor production of tomatoes

Mike Zelkind, Tisha Livingston, and Victor Verlage

IV

Advanced research in PFALs and indoor farms

17. Toward an optimal spectrum for photosynthesis and plant morphology in LED-based crop cultivation

Shuyang Zhen, Paul Kusuma, and Bruce Bugbee

Contributors

Yumiko Amagai Center for Environment, Health and Field Sciences, Chiba University, Kashiwa, Chiba, Japan

Bruce Bugbee Utah State University, Logan, UT, United States

Hiroshi Ezura Faculty of Life and Environmental Sciences, University of Tsukuba, Tsukuba, Ibaraki, Japan; Tsukuba-Plant Innovation Research Center, University of Tsukuba, Tsukuba, Ibaraki, Japan

Kazuhiro Fujiwara Graduate School of Agricultural and Life Sciences, The University of Tokyo, Tokyo, Japan

Celina Gómez Environmental Horticulture Department, University of Florida, Gainesville, FL, United States

Eri Hayashi Japan Plant Factory Association, Kashiwa, Chiba, Japan

Ricardo Hernández Department of Horticultural Sciences, College of Agriculture and Life Sciences, North Carolina State University, Raleigh, NC, United States

Kyoko Hiwasa-Tanase Faculty of Life and Environmental Sciences, University of Tsukuba, Tsukuba, Ibaraki, Japan; Tsukuba-Plant Innovation Research Center, University of Tsukuba, Tsukuba, Ibaraki, Japan

Yasuomi Ibaraki Yamaguchi Universiy, Yamaguchi City, Japan

Harumi Ikei Center for Environment, Health and Field Sciences, Chiba University, Chiba, Japan

Murat Kacira Department of Biosystems Engineering, The University of Arizona, Tucson, AZ, United States

Katashi Kai Shinnippou Ltd., Shizuoka, Japan

Kazuhisa Kato Graduate School of Agricultural Science, Tohoku University, Sendai, Miyagi, Japan

Nathan Kelly Department of Horticulture, Michigan State University, East Lansing, MI, United States

Volkmar Keuter Department Photonics and Environment, Fraunhofer Institute for Environmental, Safety, and Energy Technologies UMSICHT, Oberhausen, Germany

Masao Kikuchi Center for Environment, Health and Field Sciences, Chiba University, Kashiwa, Chiba, Japan

Toyoki Kozai Japan Plant Factory Association, Kashiwa, Chiba, Japan

Paul Kusuma Utah State University, Logan, UT, United States

Tisha Livingston 80 Acres Farms, Hamilton, OH, United States

Na Lu Center for Environment, Health and Field Sciences, Chiba University, Kashiwa, Chiba, Japan

Toru Maruo Japan Plant Factory Association, Kashiwa, Chiba, Japan

Joseph Masabni Texas A&M AgriLife Extension Service, Dallas, TX, United States

Genhua Niu Texas A&M AgriLife Research, Texas A&M University, Dallas, TX, United States

Yujin Park College of Integrative Sciences and Arts, Arizona State University, Mesa, AZ, United States

P. Morgan Pattison Solid State Lighting Services, Inc., Johnson City, TN, United States

H. Christopher Peterson Michigan State University, East Lansing, MI, United States

Erik S. Runkle Department of Horticulture, Michigan State University, East Lansing, MI, United States

Joseph Seong Michigan State University, East Lansing, MI, United States

Yutaka Shinohara Japan Plant Factory Association, Kashiwa, Chiba, Japan

Michiko Takagaki Center for Environment, Health and Field Sciences, Chiba University, Kashiwa, Chiba, Japan

Kaz Uraisami Marginal LLC, Tokyo, Japan

Simone Valle de Souza Michigan State University, East Lansing, MI, United States

Viktorija Vaštakaitė-Kairienė Institute of Horticulture, Lithuanian Research Centre for Agriculture and Forestry, Babtai, Lithuania

Victor Verlage 80 Acres Farms, Hamilton, OH, United States

Mike Zelkind 80 Acres Farms, Hamilton, OH, United States

Ying Zhang Department of Agricultural and Biological Engineering, University of Florida, Gainesville, FL, United States

Shuyang Zhen Texas A&M University, College Station, TX, United States

Preface

Plant factory with artificial lighting (PFAL) or indoor vertical farming, a technology-based innovative approach to sustainable food production, is capturing the interests of people with all levels of experience and expertise around the globe. Many universities have created and expanded their research and teaching programs in controlled environment agriculture. The number of plant factories, large and small with varying degrees of productivity and profitability, is also increasing. In the past five years, several technical books on plant factories have been published, including those authored by the editors of this book. However, PFAL technologies are evolving rapidly with the advances in many related fields such as lighting technology, photobiology, plant nutrition, breeding, and environmental control engineering. PFAL applications are expanding to a wider range of vegetable, ornamental, and medicinal crops. All these new developments have encouraged us to write another book that discusses the latest advancements and new topics and covers the basics of PFAL.

This book consists of four parts: introduction, basics, applications, and advanced research. Part I summarizes the role and characteristics of PFALs and its potential contribution to the United Nation's sustainable development goals and beyond. Part II covers the technical basics from lighting terminology, light-emitting diode performance and efficacy, to cultural methods such as hydroponics and aquaponics. Part III discusses aspects of PFAL application such as optimization of crop productivity, economics and profitability, and business models. Part IV introduces the latest advances in research in PFAL.

The target audience for this book are researchers, engineers, students, educators, businesspeople, and policymakers in the field of horticulture, food production, and urban agriculture. We hope this book will provide a good reference for PFAL technology and science, inspire new business opportunities, and advance technological developments that improve the productivity and profitability of PFALs.

We are deeply indebted to all the authors for their timely contribution. We specially thank Ms. Tokuko Takano for her tireless assistance.

Genhua Niu,
Toyoki Kozai,
and Joseph Masabni

PART I

Introduction

1

Introduction: why plant factories with artificial lighting are necessary

Toyoki Kozai[1], Genhua Niu[2] and Joseph Masabni[3]

[1]Japan Plant Factory Association, Kashiwa, Chiba, Japan; [2]Texas A&M AgriLife Research, Texas A&M University, Dallas, TX, United States; [3]Texas A&M AgriLife Extension Service, Dallas, TX, United States

1.1 Introduction

Today, we face global as well as local issues in areas such as climate systems, biodiversity, land utilization, and shortages of phosphorus (P) and nitrogen (N) reserves, all of which are interrelated and affect the sustainability of our planet (Rockstrom and Klum, 2015, Fig. 1.1). Excessive use of P and N as chemical fertilizers in agriculture and excessive wastewater emissions in urban areas have caused and continue to cause eutrophication in rivers and lakes, thus resulting in environmental pollution in surrounding areas and, at the same time, causing a shortage of phosphate ore reserves in mines for future use in agriculture.

According to Rockstrom and Klum (2015), biodiversity and P/N flows are already beyond the high-risk zone with regard to planetary boundaries of the Earth's systems, and land utilization and climate systems are also approaching the high-risk zone. Since major countermeasures to address these issues are also interrelated, they need to be planned and implemented concurrently, step by step, and from various aspects on different scales.

Solving the above four issues concurrently to improve the resilience of the planet's sustainability through "transformation of the economic and social systems" with respect to food, energy, urban, and production and consumption systems (Fig. 1.2) is also in line with the principles of earth stewardship of the global commons (Chapin et al., 2011; The World in 2050, 2019).

We must construct and manage a sustainable food system bearing in mind a future that will be characterized by the following: (1) a significant increase in the urban population (about eight billion or 80% of the world's population by 2050) in tandem with increasing demand for higher food quality and security; (2) increases in social, political, economic,

Plant Factory Basics, Applications and Advances
https://doi.org/10.1016/B978-0-323-85152-7.00015-X

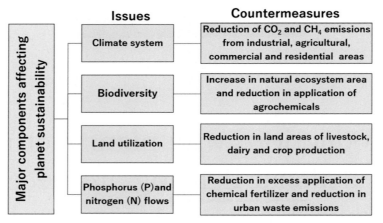

FIGURE 1.1   Planetary boundaries in the zone of increasing risk (climate change and land use change) and beyond the zone of high risk (biodiversity and P/N flows). *Adopted from Rockstrom, J., Klum, M., 2015. Big World Small Planet. Yale University Press, 206.*

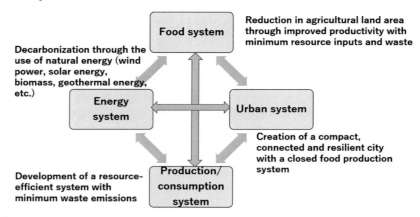

FIGURE 1.2   Social and economic systems to be transformed to improve the resilience of sustainability of our planet. Four interrelated global as well as local issues to be solved under global commons stewardship. *Adopted from Chapin III, F.S., S.T.A. Pickett, Power, M.E., Jackson, R.B., Carter, D.M., Duke, C., 2011). Earth stewardship: a strategy for social—ecological transformation to reverse planetary degradation. J. Environ. Stud. Sci. 1, 44—53.*

and pandemic risks in areas such as international transportation and the trade of safe, secure food; (3) a contraction in the agricultural population due to the aging of farmers and growers, and decreases in arable land area, availability of water for irrigation, and P and K (potassium) fertilizer; and (4) climate change events such as floods, droughts, typhoons/hurricanes, and high temperatures, resulting in crop damage.

Consequently, there is a pressing need for sustainable intensification of agricultural production on a global scale; in other words, agricultural processes or systems where yields are increased without adverse environmental impact and without the conversion of additional nonagricultural land (Royal Society, 2009). It is becoming increasingly clear that the so-called Digital Revolution (DR) is becoming a key driving force in sustainable intensification and system transformation with respect to a sustainable food system. A host of recent technologies is also advancing the DR such as virtual and augmented reality, 3D printing, artificial intelligence (AI), the Internet of Things (IoT), the fifth generation wireless network, cloud computing, robots, smartphones with superior cameras, global positioning systems, bioinformatics, and various kinds of application software for smartphones and their peripheral devices. The DR will radically alter all dimensions of global and regional society and economies and will therefore change the interpretation of the sustainability paradigm itself (The World in 2050, 2019) particularly since the outbreak of the Covid-19 (coronavirus) pandemic in 2020.

1.2 Global land use and environmental impact of livestock

1.2.1 Global land use

As shown in Table 1.1, 29% of the Earth's surface is covered by land, and about 71% of that land area is habitable. Approximately half of all habitable land is used for agriculture. About 77% percent of the global agricultural land area is used for livestock, including pastures for grazing and land for growing animal feed. Only about 37% of habitable land is covered by forests, 11% by shrubs and grasslands, and 1% by fresh water. The remaining 1% or 1.5 million km^2 is comprised of urban and built-up areas which include cities, towns, villages, roads, and other human infrastructure (Richie, 2019). In 2050, 80% of the global population (nearly eight billion people) will live in urban areas, which account for only 1% of the habitable land area on Earth.

The expansion of agricultural land has been one of humanity's largest impacts on the environment, and one of the greatest pressures on biodiversity (Richie, 2019). Furthermore, agricultural land/soil is often degraded by extreme weather and adverse land management by humans. The processes of agricultural land degradation include erosion, organic matter

TABLE 1.1 Global land use for food production (Richie, 2019; Richie and Roser, 2020; Our World in Data https://ourworldindata.org/) (All the percentages are approximate values).

Category	Use by % and area		Other uses (%)
	(%)	10^6 km^2	
Earth's surface (514 10^6 km^2)	29% land	149	71% ocean
Land surface	71% habitable	104	10% glaciers, 19% barren land (deserts, dry salt flats, beaches, sand dunes, and exposed rocks)
Habitable land	50% agriculture	51	37% forest, 11% shrub, 1% freshwater, and 1% urban/built-up land
Agricultural land	77% livestock (meat and dairy)	40	23% crops excluding feed crops

decline, salinization, soil diversity loss, contamination, flooding, landslides, and sealing (Xydis et al., 2020).

1.2.2 Environmental impact of livestock and countermeasures

To produce 18% of calories and 37% of proteins needed in our diet from meat and dairy, 77% of agricultural land is needed for use by livestock and 23% for crops excluding feed crops. However, it should also be noted that 57% of the land used for feed production is not suitable for food production (Mottet et al., 2017) mainly due to unfavorable soil and climate. In the United States, 56% of water drawn from rivers or wells is used for livestock (Poore and Nemeeek, 2018). Moreover, 58% of greenhouse gases (GHGs) and 57% and 56% of water and air pollution, respectively, are attributable to the livestock industry.

From 1970 to 2016, global meat production increased nearly fourfold, with a production of nearly 320 million tons annually, and this growth is set to continue to increase in tandem with global population and economic growth unless people make considerable dietary changes. Dry matter feed required to produce 1 kg of edible parts of beef, lamb/mutton, pork, and poultry is 25, 15, 6.4, and 3.3 kg, respectively, and (Mottet et al., 2017) feed required to produce 1 kg of eggs and milk is 2.3 and 0.7 kg, respectively (Alexander et al., 2016). This conversion efficiency from feed to edible meat can be improved by better livestock management. For example, to produce 1 kg of boneless meat requires 2.8 kg human-edible feed in ruminant systems and 3.2 kg in monogastric systems (Mottet et al., 2017). The three major feed materials of six billion tons of global feed (dry matter) in 2010 were grass and leaves (46%), followed by crop residues such as straws, sugarcane (19%), and human-edible grains (13%) (Mottet et al., 2017).

Reducing meat and dairy dietary intake, thus reducing the land area and feed consumption required for livestock, is the most effective way of reducing the environmental impact on Earth (Carrington, 2018). Improving feed-meat/milk/egg conversion efficiency is another effective way of reducing the environmental impact on Earth.

A simple, direct countermeasure for reducing the land area for livestock is the substitution of meat with plant-based alternatives. Plant-derived meat or plant meat is one such alternative and is similar to meat in appearance and taste. In terms of calorie production and protein for humans, it requires less land area, freshwater withdrawals, and P/N fertilizer, and causes lower emissions of GHGs and water and air pollutants. Therefore, a significant percentage of farmland can be restored to forests and natural habitats. In addition to plant-derived foods, insect-derived foods will become another important option for providing protein as food for humans and feed for livestock and fish (van Huis, 2013).

1.3 Scope and organization of this book

In view of the statistical facts and the opinion of qualified researchers on trends in global land use and the environmental impact of agriculture and changes in agricultural and urban population described earlier, it is clear that we have to change the conventional food chain system and our dietary habits to make them more sustainable.

One of the options for making the food chain system more sustainable is the concept of the vertical farm (Despommier, 2010) and/or plant factory (Kozai et al., 2020). In fact, in the previous decade, a significant number of papers on vertical farms and plant factories have been published by researchers of various backgrounds in educational, research, and business.

1.3.1 Objective

The plant factory with artificial lighting (called PFAL hereafter) discussed in this book is one type of vertical farm as explained in Chapter 2. The objective of this book is to demonstrate the usefulness of PFALs in contributing to solve concurrently a number of issues relating to food, the environment, natural resources, and quality of life. In brief, the PFAL is considered to contribute to the following: (1) solving global and local issues relating to food, urban, energy, and production/consumption systems concurrently (Fig. 1.2); (2) achieving the Sustainable Development Goals (SDGs); and (3) adopting ESG (environmental, social, and corporate governance) criteria through sustainable intensification of plant production (SIPP) (The Royal Society, 2009, Fig. 1.3).

Bearing in mind the SIPP shown in Fig. 1.3, this book intends to stimulate and inspire readers to discover new research areas and business opportunities in regard to PFALs, based on an understanding of basic sciences, creative viewpoints, and innovative methodology. This book is neither a text that aims to cover all aspects of PFAL research and business nor a comprehensive literature review of PFALs. What it does cover are the benefits of PFALs, the challenges of the next generation of PFALs, and selected topics not discussed in detail in previous books on PFALs (Kozai et al., 2016, 2020; Kozai, 2018).

1.3.2 Organization of this book

This book consists of four sections: Introduction, Basics, Applications, and Advanced research. In Part 1, Chapter 2 examines the differences in meaning between the PFAL and

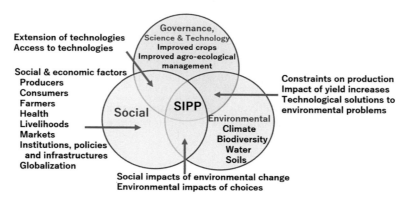

FIGURE 1.3 Sustainable intensification of plant production (SIPP) in relation to ESG (environmental, social, and corporate governance). *Adopted from The Royal Society. 2009. Reaping the Benefits: Science and the Sustainable Intensification of Global Agriculture. The Royal Society, London, 71.*

vertical farm or indoor farm, while Chapter 3 describes the role and characteristics of PFALs, and Chapter 4 discusses how PFALs can contribute to achieving the SDGs.

In Part 2, Chapters 5 and 6 explain basic technical terms and SI (System International) units and their relationship in regard to light and light sources. Chapter 7 describes the current and potential efficacy of light emitting diodes and Chapter 8 discusses resource use efficiency. Chapters 9 and 10 discuss hydroponics and aquaponics as essential components of indoor vertical farms, and Chapter 11 explores plant responses to environments.

In Part 3, Chapters 12 and 13 propose a definition of productivity and methods of optimizing the productivity of PFALs. Chapter 14 discusses the economics and profitability of PFALs, and Chapter 15 discusses their cost performance and presents some business models.

In Part 4, Chapters 16—23 discuss in detail a range of topics including the following: the optimal spectra of light for indoor plant growth, effects of light on plant nutrition and secondary metabolites, transplant production, environmental control, molecular breeding of tomato lines, production of strawberries, production of tomatoes, phenotyping, and a human-centered perspective on urban agriculture.

1.3.3 Benefits and challenges

The benefits of PFALs and challenges in achieving more sustainable PFALs are summarized below as the introductory part of this book. These topics are discussed in the chapters that follow.

1.3.3.1 *Benefits of PFALs*

(1) The land area usage of a PFAL with 10 tiers is reduced to about 1/20th that of conventional open field vegetable production, and the amount of water required for irrigation per kg of produce is reduced to 1/20th of that of greenhouse production. In addition, food loss and fuel consumption during transportation are reduced due to reduced food mileage.

(2) With PFALs, high-yield and high-quality produce with year-round production is possible everywhere in the world regardless of climate, weather, or soil fertility mainly due to the PFAL's high environmental traceability, controllability, and reproducibility.

(3) Pesticide-free, insect-free, and other contaminant-free PFAL-grown leaf vegetables with an extremely low population density of microorganisms are clean enough to be served as fresh salad without washing, and the produce shelf life is about twofold that of greenhouse-grown vegetables.

(4) Since the environment in the cultivation room is more or less the same everywhere in the world, the exchange of information via the Internet among PFAL users is easy and beneficial.

1.3.3.2 Challenges in achieving more sustainable PFALs

(1) The cost of electricity, which accounts for around 20% of production costs in Japan as of 2020, must be further reduced by generating all electricity from natural energy.

(2) The development of software for more efficient lighting systems, optimal environment control, user-friendly universally designed PFALs, online and onsite training courses for beginners and skilled workers, human centered PFALs, and modularized, scalable, and networked PFALs is essential.

(3) The integration of PFALs with other biological and engineering systems with appropriate use of advanced technologies such as AI and the IoT is necessary, taking into consideration the life cycle assessment of various types of PFALs (Kikuchi et al., 2018).

(4) The breeding of cultivars of leaf, fruit and head vegetables, and medicinal and flowering plants suited to PFALs is essential.

More detailed discussion regarding the challenges is provided in Chapter 3.

1.4 Conclusion

The PFAL is an emerging, yet still immature, technology. As a result, proven benefits are limited at present. Nevertheless, the number of PFALs that are generating economic profits has been increasing, particularly since around 2015. On the other hand, there are still a host of challenges to be tackled and solved to actualize the full potential and opportunities that PFALs offer in contributing to the achievement of the SDGs and urban sustainability.

To make full use of PFAL potential in urban areas, citizen science applying interdisciplinary and transdisciplinary approaches is essential (Grandison, 2020) as a means of raising awareness about PFALs, having them accepted, widely utilized, and continuously improved by all stakeholders including consumers, growers/farmers, PFAL employees, city planners, policy makers and business persons, and incorporating them into the everyday life of people everywhere.

References

Alexander, P., Brown, C., Arneth, A., Finnigan, J., 2016. Human appropriation of land for food: the role of diet. Global Environ. Change 41, 88–98. ISSN 0959-3780.

Carrington, D., 2018. Avoiding Meat and Dairy Is 'single Biggest Way' to Reduce Your Impact on Earth. The Guardian. https://www.theguardian.com/environment/2018/may/31/avoiding-meat-and-dairy-is-single-biggest-way-to-reduce-your-impact-on-earth.

Chapin III, F.S., Pickett, S.T.A., Power, M.E., Jackson, R.B., Carter, D.M., Duke, C., 2011. Earth stewardship: a strategy for social–ecological transformation to reverse planetary degradation. J. Environ. Stud. Sci. 1, 44–53.

Despommier, D., 2010. The Vertical Farm: Feeding the World in the 21st Century. St Martin's Press, p. 105.

Grandison, T., 2020. Citizen science: the American experience. Trends Sci. 25 (4), 50–53.

Kikuchi, Y., Kanematsu, Y., Yoshikawa, N., Okubo, T., Takagaki, M., 2018. Environmental and resource use technology options: a case study in Japan. J. Clean. Prod. 186, 703–717.

Kozai, T. (Ed.), 2018. Smart Plant Factory: The Next Generation Indoor Vertical Farms. Springer, p. 436.

Kozai, T., Fujiwara, K., Runkle, E. (eds.), 2016. LED Lighting for Urban Agriculture. Springer, p. 454.

Kozai, T., Niu, G., Takagaki, M. (Eds.), 2020. Plant Factory: An Indoor Vertical Farming for Efficient Quality Food Production. Academic Press, p. 487.

Mottet, A., de Haan, C., Falcucci, A., Tempio, G., Opio, C., Gerber, P., 2017. Livestock: on our plates or eating at our table? A new analysis of the feed/food debate. Glob. Food Sec. 14, 1−8. https://doi.org/10.1016/j.gfs.2017.01.001.

Poore, J., Nemeeek, T., 2018. Reducing food's environmental impacts through producers and consumers. Science 360, 987−992.

Richie, H., 2019. Half of the World's Habitable Land Is Used for Agriculture. https://ourworldindata.org/global-land-for-agriculture.

Ritchie, H., Roser, M., 2020. Land Use. Published online at. OurWorldInData.org. Retrieved from. https://ourworldindata.org/land-use.

Rockstrom, J., Klum, M., 2015. Big World Small Planet. Yale University Press, p. 206.

The Royal Society, 2009. Reaping the Benefits: Science and the Sustainable Intensification of Global Agriculture. The Royal Society, London, p. 71.

The World in 2050, 2019. The Digital Revolution and Sustainable Development: Opportunities and Challenges. Report prepared by the World in 2050 initiative. International Institute for Applied Systems Analysis (IIASA), Laxenburg, Austria. www.twi2050.org. https://iiasa.ac.at/web/home/research/twi/Report2019.html.

van Huis, A., 2013. Potential of insects as food and feed in assuring food security. Annu. Rev. Entomol. 58, 563−583.

Xydis, G.A., Liaros, S., Avgoustaki, D., 2020. Small scale plant factories with artificial lighting and wind energy microgeneration: a multiple revenue stream approach. J. Clean. Prod. 255 https://doi.org/10.1016/j.jclepro.2020.120227.

Terms related to PFALs

Toyoki Kozai

Japan Plant Factory Association, Kashiwa, Chiba, Japan

2.1 Introduction

This chapter describes mainly the similarities and differences in meaning between "vertical farm," "plant factory," and "plant factory with artificial lighting (PFAL)." The terms "indoor farm" and "indoor vertical farm" are sometimes used interchangeably for vertical farm. All of the above terms are recently being increasingly used both as general terms as well as technical terms in research papers.

Nevertheless, there are no standard or clear definitions of these terms which distinguishes one from the other, so they are often used interchangeably, which occasionally leads to misunderstanding or confusion. Thus, establishing clear definitions of these terms and their relationships is warranted to avoid confusion. The author intends for this chapter to be a starting point in stimulating discussion of these terms to clarify their respective meanings for easier communication in the future.

The author believes that the definition of a vertical farm should take into consideration its structure, role, and function including its future integration with other biological facilities such as wastewater and plant residue processing facilities.

2.2 Vertical farm

2.2.1 Vertical farm versus vertical farming

Waldron (2019) pointed out the important difference in concepts between "a vertical farm" as a noun and "vertical farming" as an activity (or gerund). The discussion in this chapter takes up only the term "vertical farm," although the term and concept of vertical farming is important in the design and management of vertical farms.

11

2.2.1.1 Vertical farms as an agricultural production system

2.2.1.1.1 Vertical farms in a broad sense

According to Al-Kodmany (2018), the term "vertical farm" was coined by G. E. Bailey (1915) who authored a book on geology entitled *Vertical Farming* describing the history and processes of the physical structure and substances of the agricultural land. The term had been popularized in an earlier book entitled *The Vertical Farm: Feeding the World in the 21st Century* by Dickson Despommier (2010), who described vertical farming as "the mass cultivation of plant and animal life for commercial purposes in skyscrapers." Since then, the term vertical farm has been used worldwide, particularly in North America and Europe. Banerjee and Adenauer (2014) referred to vertical farming as "a system of commercial farming whereby plants, animals, fungi, and other life forms are cultivated." In a review of vertical farming, Kalantari et al. (2017) also included aquaculture, livestock production, and waste management. It is only natural and logical that plants, animals, and other life forms would be produced in vertical farms since the term "farm" refers to an area of land and buildings for growing crops and rearing animals (Oxford Dictionary of English).

2.2.1.1.2 Key factors for classifying vertical farms

Fig. 2.1 shows the classification of vertical farms based on various key factors. The vertical farm can be further subclassified under the respective key factors.

Waldron (2019) proposed the following factors for considering the classification of vertical farms: (1) scale of production, (2) density/volume of production (high density vertical cultivation), (3) environmental controls (building integrated controlled environment farm and controlled environment agriculture), (4) layout (i.e., racking systems), (5) building type/structure (sky farming), and (6) location (urban farms).

2.2.1.1.3 Classification of vertical farms by produce type

Fig. 2.2 shows a classification of vertical farms by produce types, one of the key factors shown in Fig. 2.1. Vertical farms are classified into the following: (1) plant farms (vertical farms for plant production); (2) insect farms (Madau et al., 2020; Specht et al., 2019); (3) fish/shellfish farms (aquaculture system); (4) animal farms (chicken house, pig house, cow house, etc.); and (5) microorganism farms. In Fig. 2.2, two-spotted cricket (*Teleogryllus occipitalis*), field cricket (*Teleogryllus*), and black soldier fly (*Hermetia illucens*) are considered as feed for fish, shellfish, and livestock and as protein source (food) for humans among others.

The plant farm is subdivided into PFALs, which are defined in Section 3, and plant farms other than PFALs which are called "plant factory" (common noun) in this book. In Fig. 2.2, the word "vertical" is added as a prefix to each farm when it has an upright structure or occupies a tall building with many stories. For example, "vertical plant farm" in place of "plant farm."

The classification shown in Fig. 2.2 is based on the understanding that R&D and the industry of all types of vertical farms will continue to grow in the coming years. Therefore, clear definitions of basic terms common to all types of vertical farms will become increasingly important to avoid possible confusion. This will be especially so when plant farms, for example, are integrated with other types of vertical farms. Vertical farms can be classified based on various key factors other than produce type, as shown in Fig. 2.1.

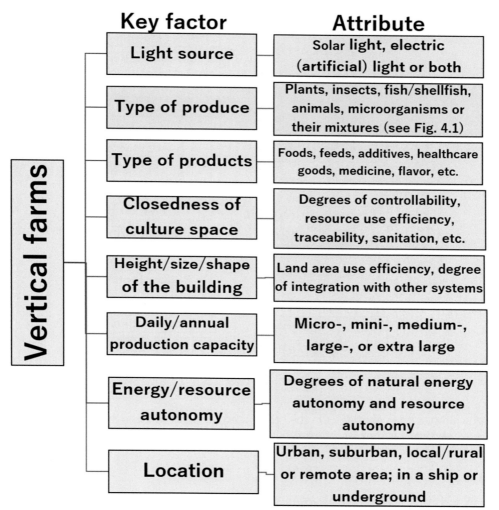

FIGURE 2.1 A classification of vertical farms by key factors.

2.2.1.2 Vertical farms in a narrow sense—a plant production system

In his review of vertical farms, Al-Kodmany (2018) simply described the various types of indoor plant production systems. Sharath Kumar et al. (2020) described the vertical farm as "a multitier indoor plant production system ... independent of solar light and other outdoor conditions." This description also seems to be reasonable since, as of 2020, most existing facilities or buildings referred to as vertical farms exclusively cultivate plants as produce. In fact, many horticultural researchers in the United States and Europe understand vertical farming to be the production of crops (or plants) grown in completely enclosed environments with electric lamps as the sole source of light (i.e., no sunlight, based on the author's

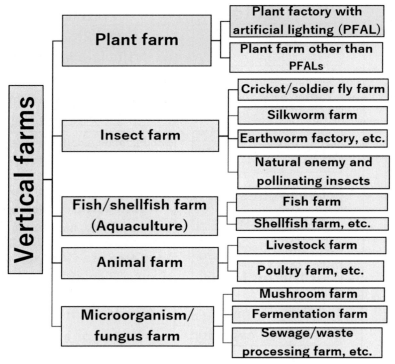

FIGURE 2.2   A classification of vertical farms by types of produce.

communication with E. Runkle). On the other hand, it might become difficult to refer to plant farming as vertical farming as other types of vertical farms such as vertical fish farms and vertical insect farms develop.

2.2.1.3 Type of light source

The description in the previous section on vertical farms raises another point for discussion: the light source. Rooftop greenhouses with or without supplemental electric lamp lighting and indoor plant production systems using both solar and lamp lighting are included in the vertical farms in many papers (e.g., Kalantari et al., 2017). In some cases, greenhouses built on the ground are included in indoor farms. Vertical farms graphically illustrated in Despommier (2010) include rooftop greenhouses and plant cultivation systems with full or partial use of solar light transmitted through glass windows on exterior walls. In fact, some existing vertical farms use both artificial and natural (solar) light (e.g., Gerner et al., 2011), although most existing vertical farms use light from electric lamps as the sole light source.

2.2.1.4 Photosynthetically and physiologically active radiation

The wavelength of solar radiation ranges from around 300 to 2500 nm at the peak wavelength of about 500 nm (end of blue-green). The solar radiation ranging from 800 to 2500 nm accounts for around 50% of total solar radiation energy and is called thermal radiation (i.e., physiologically nonactive radiation for plants and animals).

Photosynthetically active or photosynthetic radiation ranges from about 400 to 700 nm, and physiologically active radiation for plants ranges from about 300 to 800 nm, including ultraviolet (UV) radiation A and B (about 300—400 nm), photosynthetically active radiation, and far-red radiation (700—800 nm). Visible light for humans ranges between 380 and 740 nm. LED lamps used in PFALs/vertical farms emit physiologically active radiation with a very small amount of near-infrared and infrared radiation.

Radiation energy required for the photosynthetic growth of plants is much greater than that required for the photo-physiological reaction of plants such as photomorphogenesis, since photosynthesis is an energy conversion process, while photomorphogenesis is a signal conversion process. This is why electricity consumption for lighting is much greater in PFALs than in livestock/poultry houses and why the design and management of the lighting environment for plant production is considerably different from the design and management for animal production. The differences among photosynthetic radiation, photosynthetic photons, physiologically active radiation, and photons are described in Chapters 5, 6, 11 and 17.

2.2.1.5 The words "indoor," "closed," "factory," and 'farm'

The word "indoor" means "in a building or under cover," and the word "closed" means "surrounded by (side) walls," so that it is clear that all indoor farms, indoor vertical farms, and closed farms refer to farms that are closed or, at least, under cover. A factory is also understood to be a closed structure. On the other hand, the term "vertical farm" itself does not imply that it is closed. This is why the term "indoor vertical farm" is often used to indicate a closed vertical farm.

From a traditional point of view, the word "farm," particularly among older people, evokes images of rural landscape and perhaps even experiences from childhood. Many people thus associate the word "farm" with nature or natural, or not artificial. In this context, the word "farm" has a connotation that differs qualitatively from "factory" and "artificial."

2.2.1.6 The word "vertical"

The word "vertical" in vertical farm means "upright," and thus "vertical farm" refers to a high-rise or tall building (with many stories) or a skyscraper with plants and/or livestock growing inside, which is often associated with the words urban, modern, and/or high tech (Despommier, 2010). On the other hand, the term vertical farm is recently being used also to mean a type of farm built underground as well as a large-scale but flat (or single-story) farm. A greenhouse utilizing solar light only, however, cannot be a tall multistory structure, so it would be difficult to call it a vertical farm even if it were large in scale.

Interestingly, large windowless poultry houses for the mass production of eggs or chickens with or without a raised floor, which have been commercialized for nearly a half century, are seldom referred to as vertical farms.

In most PFALs, the cultivation beds are placed horizontally on each tier of the cultivation rack with the vertical distance between the tiers ranging from 0.5 to 1.5 m. In a limited number of PFALs, however, the cultivation beds are arranged vertically or inclined, hung or stood upright. The word "vertical" in reference to vertical cultivation beds is, of course, unrelated to the word "vertical" in vertical farm.

2.2.2 The term "plant factory" and its classification

2.2.2.1 *The term "plant factory"*

The term "plant factory" was probably used in 1974 for the first time in a technical committee report on the concept and design of a plant factory published by the Japan Electronics and Information Technology Industries Association (JEITIA, 1974; Takakura, 2020). The committee proposed a design concept of a plant factory with respect to the following: (1) the automatic transportation of plants from the seeding area to the shipping area, (2) automation for labor saving, (3) environmental control for the enhancement of growth and labor saving, (4) efficient utilization of light, electric/fuel energy, and land, and (5) an overall structural design encompassing logistics and sales. In 1976, the technical committee published its final report in which it presented a basic plant factory design 40 m wide, 40 m long, and 10 m high for growing leafy vegetables, fruit vegetables, and other crops under solar light (JEITIA, 1976).

The term "plant factory" was popularized, to some extent, by a paperback book by M. Takatsuji entitled *Plant factory—to new home gardening from cultivation without soil* (Takatsuji, 1979). The author used the term plant factory to mean a facility enabling scheduled and stable production of high-quality plants with the use of artificial and/or natural (or solar) light. Since then, the term plant factory gradually came into use as a general term.

2.2.2.2 *PFAL and PFSL (plant factory with solar light)*

From 2010 to 2014, Japan's Ministry of Agriculture, Forestry, and Fisheries provided subsidies for research, training, and business concerning plant factories. The term "plant factory with solar light (PFSL)" was coined by the Ministry to mean a Dutch-style large-scale greenhouse with environmental control and automated handling units. According to this definition, a PFSL with or without supplemental lighting was considered to be one type of Dutch-style greenhouse. Since then, both the terms PFAL and PFSL have been used in Japan and, to some extent, in China, Taiwan, and Korea. In these countries, the term plant factory includes both PFSLs and PFALs. This is why the term PFAL has been increasingly used since 2010 to avoid the confusion between PFALs and PFSLs (Fig. 2.3).

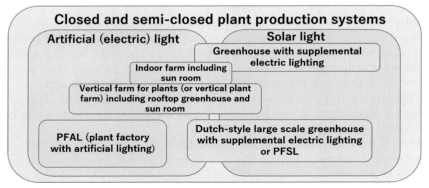

FIGURE 2.3 Diagram showing the differences in light sources of various closed and semiclosed plant production systems.

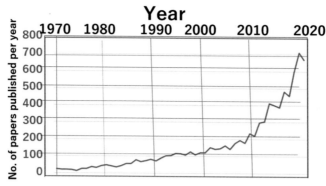

FIGURE 2.4   Yearly trend in No. of papers on plant factory published in journals during 1970–2019 (Total: 8894). https://app.dimensions.ai/discover/publication?search_mode=content&search_text=plant%20factory&search_type=kws&search_field=text_search&order=date. *By courtesy of E. Hayashi and N. Hiramatsu.*

On the other hand, from around 2015, PFAL R&D and business have been more popular than their PFSL counterparts in Japan. Therefore, as of 2020, the term plant factory more often only refers to a PFAL.

Fig. 2.4 shows a yearly trend of the number of research papers with the title of plant factory/plant factories or PFAL/PFALs (excluding the terms vertical farm/vertical farms and vertical farming) published during 1970–2019. The number in 2020 was sevenfold the number in 2000 and threefold the number in 2010.

2.2.3 Definition of the cultivation room in the PFAL

The PFAL is a special type of plant factory that exclusively uses artificial light to produce any kind of plants such as leaf, fruit, head, and root vegetables and herbs without the use of pesticides and herbicides. A PFAL generally consists of a cultivation room, dressing room, air shower room, precooling room, and shipping room (Kozai et al., 2020). This section discusses the cultivation room only.

Produce from a PFAL can be served fresh in salads without the need to wash with tap water before cooking or eating. Requirements and recommendations of the cultivation room of the PFAL described below apply to requirements and recommendations described in Chapters 12 and 13.

2.2.3.1 Requirements of the cultivation room

2.2.3.1.1 Design factors

(1) All walls and roofs are optically opaque (no solar light is transmitted to the cultivation room).

(2) The cultivation room is almost (but not completely) airtight under normal operation conditions (the number of air exchanges of the cultivation room is around $0.02 \ h^{-1}$).

(3) Ventilation fans are installed for use in emergencies such as extremely high CO_2 concentration, high gas concentrations of volatile organic compounds, fire, etc.

(4) Walls and floors are thermally well insulated to minimize heat energy exchange between the interior and exterior of the cultivation room and to prevent water condensation on the internal surfaces of walls and floors (the heat transmission coefficient of the walls/floor is $0.1-0.5$ W m^{-2} °C^{-1}. The value should be lower in hot and cold regions).

(5) Almost all water vapor transpired from plants is condensed and collected at cooling panels of air conditioners for cooling, and the condensed water is returned to the nutrient solution tank for recycling. Around two-thirds of the lamps are always turned on to ensure that some air conditioners are always turned on for cooling to remove the heat energy generated by the lamps even on cold nights (See Fig. 4.5 in Chapter 4 and Section 8.5 in Chapter 8).

(6) Hydroponic cultivation systems including NFT (nutrient film technique), DFT (deep flow technique), aeroponic, and drip irrigation systems are used.

2.2.3.1.2 Management factors

(1) All personnel entering the cultivation room through the air shower room must change into sanitized clothing and put on a sanitized mask, cap, shoes, and gloves.

(2) No pesticides and herbicides are used.

(3) Only inorganic fertilizer is supplied to the cultivation beds of hydroponic cultivation systems. If organic fertilizer is used, it is first decomposed into inorganic fertilizer using microorganisms before being supplied to cultivation beds.

(4) If the nutrient solution is circulated for recycling, it is first sterilized using filters, UV radiation and/or ozone gas, and/or is filtered through fine substrate layer with special types of microorganisms.

(5) Each of the following environmental factors is maintained independently at time-dependent setpoints: (1) room air temperature, (2) photoperiod, (3) CO_2 concentration in the cultivation room, and (4) electric conductivity and nutrient solution pH.

(6) Sanitary conditions are inspected daily and are maintained.

(7) Measures are taken to prevent the growth and spread of algae in the substrate, nutrient solution, and other wet places in the cultivation room.

2.2.3.2 Recommendations for the cultivation room

(1) Room air pressure is kept slightly higher ($5-10$ Pa) than the atmospheric air pressure outside to prevent dust and insects from entering the cultivation room.

(2) Each of the following environmental factors is maintained independently at time-dependent setpoints: (1) vapor pressure deficit or relative humidity of the room air, (2) spectral photon flux density distribution (SPFDD) just above the canopy surface, (3) PPFD just above the canopy surface, and (4) air current speed above the canopy surface. The spatial distribution of each environmental factor is relatively uniform in the plant canopy.

(3) On the other hands, the SPFDD and PPFD are affected by the density, height, structure, and optical properties of the plant canopy even when photosynthetic photon flux and its spectral photon flux distribution of light emitted from the lamps remain unchanged, due to changes in the spectral reflection and absorption of light by leaves over time. The air current speed over the plant canopy is also affected by the density, height, and structure of the plant canopy.

FIGURE 2.5 A PFAL with red LEDs built in 1998 at Iwata, Shizuoka Prefecture, Japan, for commercial production of leafy vegetables. *Courtesy of Hisakazu Uchiyama.*

(4) Electricity generated by natural energy such as solar energy is partially or exclusively used for operation of the PFAL.

(5) Seeding, transporting, harvesting, packing, shipping, and washing are partly or mostly automated or robotized.

Some of the requirements and recommendations described above can be realized only at a relatively large-scale PFAL (Fig. 2.5).

2.2.4 Commercialization of PFALs in Japan

PFALs using high-pressure sodium lamps were first commercialized in the early 1980s for the production of leaf lettuce plants. In 1998, a PFAL with red LEDs and 10 tiers was built for commercial production of leafy vegetables (Fig. 4.4; Watanabe, 2011). In 2000, two PFALs with fluorescent (FL) lamps each were built near Tokyo for commercial production of leafy vegetables, one with 7 tiers (Fig. 4.5) and the other with 10 tiers. In 2003, a closed transplant production system with 4 tiers and FL lamps was commercialized (Fig. 2.6; Kozai et al., 2004).

Most PFALs built in and after 2015 use white LEDs (broad band LED with blue, green, and red light). As of 2020, over 200 PFALs are in operation commercially, and around 10 PFALs with white LEDs are producing 2000—5000 kg of leafy vegetables daily. Research on PFALs conducted during 1960—2010 by Takakura and his group members is summarized in Takakura (2019) (Fig. 2.7).

FIGURE 2.6 A PFAL with fluorescent lamps built in 2000 at Kashiwa, Chiba Prefecture, for commercial production of leafy vegetables. *Courtesy of Wataru Shirao.*

FIGURE 2.7 Closed transplant production system with fluorescent lamps commercialized in 2003. *Partially adopted from Kozai, T., Chun, C., Ohyama, K., 2004. Closed systems with lamps for commercial production of transplants using minimal resources. Acta Hortic. 630, 239–252.*

2.2.5 The words "factory," "artificial," and "natural"

This section describes the author's understanding, which is neither scientific nor technical, of the words "factory," "artificial," and "natural" in relation to PFALs.

2.2.5.1 The word "factory"

"Factory" means a building or a group of buildings where goods are manufactured or assembled chiefly by machine (the Oxford Dictionary of English). In Japan and some other countries, elderly people tend to have a negative image of factories, which they may associate with chimneys spewing clouds of smoke and discharging environmental pollutants, monotonous jobs, and drab uniforms, etc., which were actual phenomena that characterized many factories until the 1980s.

On the other hand, since around 2000, many young people have a positive image of factories manufacturing goods such as food and confectionery products, computer software, electric appliances, motorcars, soft drinks, and alcoholic beverages. They may associate factories with a clean and comfortable working environment, safe and light work, skilled workers, diversified small quantity production and advanced technology, etc. Accordingly, the term "plant factory" has recently become increasingly accepted in a positive light in Japan and possibly many other countries. The majority of young people tend to like both factory-manufactured products and hand-made or natural products.

2.2.5.2 The word "artificial"

According to the Oxford Dictionary of English, "artificial" means "made or produced by human beings rather than occurring naturally, especially as a copy of something natural." Some people dislike the word artificial when used in association with plants and foods, probably because they think that plants should be grown or produced under solar light or natural (outdoor) conditions. They tend to prefer the term "electric light" to the term "artificial light." On the other hand, many people accept the word artificial when used in association with nonfood words such as artificial satellite, artificial intelligence, and artificial teeth.

2.2.5.3 The word "natural"

According to the Oxford Dictionary of English, "natural" means "existing in or derived from nature, not made or caused by humankind." "Nature" refers to phenomena of the physical world collectively including plants, animals, landscape, and other features and products of the earth as opposed to human-made phenomena or human operations. In this context, electric light is considered to be artificial light even when the electricity is generated by natural energy such as wind power, solar energy, hydraulic power, geothermal energy, or biomass. People who have a positive image of "nature" tend to forget that "volcanic eruption," "earthquake," "tsunami," "hurricane/typhoon," etc., are natural phenomena.

2.2.5.4 4 images of the words "factory," "artificial," and "natural"

Persons who have a positive image of factories and do not have any negative impressions of the word artificial will probably tend to use the term "plant factory," while persons who have a positive image of the word farm and do not have any negative impressions of the words nature or tall building will probably tend to use the term "vertical farm." On the other hand, some elderly people may remind an America classic book *Grapes of Wrath* by John Steinbeck, which evokes the struggles of migrant farmworkers under harsh weather.

People's impressions of words such as factory, artificial, and natural largely depend on their cultural background as well as their own personal background, religion, personality, philosophy, and other factors, so it is difficult to define terms that will be acceptable to all people. On the other hand, since these terms are now used frequently in scientific papers and books, they need to be clearly defined in a scientific context.

2.2.6 Toward sustainable plant production systems

2.2.6.1 Sustainability

Any plant production system to be developed in the coming years needs to be sustainable. Sustainability refers to the ability for such a system to coexist ecologically, environmentally, economically, and socially under specific resource constraints within the surrounding human society and ecosystems for many decades or centuries (Wikipedia, https://en.wikipedia.org. 2020/10/17).

Sustainable development is defined as development that "meets the needs of the present without compromising the ability of future generations to meet their own needs" (United Nations, General Assembly, 1987). However, this concept might appear to be paradoxical since development is considered to be inherently unsustainable. In this context, overcoming the paradox of developing sustainable PFALs and/or sustainable vertical farms will be a major challenge.

2.2.6.2 Changing the existing PFALs to be more sustainable

No technology is perfect, and almost all technologies have strengths and weaknesses. Despite some people's negative image of factories, we can continue to use the term plant factory while making efforts to improve the image of existing types of factories by improving their design, management, and use to be more sustainable so that people working in factories and people living nearby feel more comfortable.

There are so many factories in various industries with so many people working inside throughout the world. Simply changing the word factory to a word that has a nicer ring to it will not improve the factory itself. Likewise, if people have a negative image of farms, we should make efforts to improve the image of existing farms by changing them to be more sustainable.

When the term "closed plant/transplant production system" was first used in 2000 (Kozai et al., 2000, 2004), many people thought that the term "closed" was inappropriate because the word had a negative connotation, which was true at that time. Recently, however, people have come to view this term more positively. Likewise, people's impressions of PFALs can also improve if they are proven to be sustainable plant production systems.

The same applies to light sources. If it is shown that artificial light generated by natural energy and stored in batteries for future use is more ecologically and economically sustainable for year-round stable production of quality plants than uncontrollable, inconsistent natural (or solar) light, people's impression of artificial light will change. Such a change will depend on resource productivity (= yield divided by resource input) and monetary productivity (= (unit economic value x yield) divided by (unit cost of resources x resource consumption)) for particular plant species, the purpose of production, and the availability of resources (See, Chapters 12 and 13).

2.2.7 Conclusion

Vertical farms including PFALs can be useful in our society only when they contribute to achieving a sustainable production system of food, feed, and other biological products, reducing CO_2 emissions and waste such as plant residue and urban waste, and helping cities to become ecosystems rather than parasites (Despommier, 2010).

Among its functions, the PFAL will play an important role as a system for absorbing CO_2 from the air through photosynthesis, producing clean water vapor through stomata of leaves, and converting light energy, water, and inorganic fertilizer to carbohydrates and other essential nutrition elements. These photoautotrophic functions of plants will become more and more important in the development of sustainable cities with various types of vertical farms in the coming decades. Establishing clear definitions of the terms frequently used in the field of vertical farming is essential to promote research, development, and widespread practice of vertical farms.

Acknowledgments

The author wishes to express his gratitude to Professor Erik Runkle and Professor Genhua Niu for their valuable comments and correction of English.

References

Al-Kodmany, K., 2018. The vertical farm: a review of development and implication for the vertical city. Buildings 8 (24), 36. https://doi.org/10.3390/building8020024. www.mdpi.cpm/journal/buildings.
Bailey, G.E., 1915. Vertical Farming. E. J. Du Pont de Nemours Powder Co, p. 69. https://www.biodiversitylibrary.org/item/71044#page/77/mode/1up.

Banerjee, C., Adenaeuer, L., 2014. Up, up and away! The economics of vertical farming. J. Agric. Stud. 2 (1), 40—60. https://doi.org/10.5296/jas.v2i1.4526 (ISSN 2166-0379).

Despommier, D., 2010. The Vertical Farm: Feeding the World in the 21st Century. St Martin's Press, p. 105.

Germer, J., Sauerborn, F., Asch, J., de Boer, J., Schreiber, J., Weber, G., Müller, J., 2011. Skyfarming an ecological innovation to enhance global food security. J. für Verbraucherschutz und Leb. 6 (2), 237—251.

JEITIA (Japan Electronics and Information Technology Industries Association), 1974. Concept of plant factory pilot plant (Interim report on plant factory technical committee). Elect. Ind. Monthly Rep. 16 (10), 18—23 (in Japanese).

JEITIA (Japan Electronics and Information Technology Industries Association), 1976. Final Technical Committee Report on Plant Factory Systems, p. 156 (51-A-96) (in Japanese).

Kalantari, F., Tahir, O.M., Lahijiani, A.M., Kalantarim, S., 2017. A review of vertical farming technologies: a guide for implementation of building integrated agriculture in cities. Adv. Eng. Forum 24, 76—91. 10.4028/www.scientific.net/AEF.24-77.

Kozai, T., Kubota, C., Chun, C., Afreen, F., Ohyama, K., 2000. Necessity and concept of the closed transplant production system. In: Kubota, C., Chun, C. (Eds.), Transplant Production System in the 21st Century. Kluwer Academic Publishers (Springer), pp. 3—19.

Kozai, T., Chun, C., Ohyama, K., 2004. Closed systems with lamps for commercial production of transplants using minimal resources. Acta Hortic. 630, 239—252.

Madau, F.A., Arru, B., Furesi, R., Pulina, R., 2020. Insect farming for feed and food production from a circular business model perspective. Sustainability 12, 5418. https://doi.org/10.3390/su12135418.

SharathKumar, M., Heuvelink, E., Marcelis, L., 2020. Vertical farming: moving from genetic to environmental modification. Trends Plant Sci. https://doi.org/10.1016/j.tplants.2020.05.012.

Specht, K., Zoll, F., Schümann, H., Bela, J., Kachel, J., Robischon, M., 2019. How will we eat and produce in the cities of the future? From edible insects to vertical farming—a study on the perception and acceptability of new approaches. Sustainability 11, 4315. https://doi.org/10.3390/su11164315.

Takatsuji, M., 1979. Plant Factory (Syokubutsu Koujou): From Soilless Culture to New Home Gardening. Blue Backs B-410. Koudansya Inc, ISBN 406118010X, p. 232 (in Japanese).

Takakura, T., 2019. Research exploring greenhouse environment control over the last 50 years. Int. J. Agric. Biol. Eng. 12 (5), 1—7.

Takakura, T., 2020. The birth of plant factory and its history of the subsequent 50 years. Nougyo oyobi Engei (Agriculture and Horticulture) 95 (8), 668—674 (in Japanese).

United Nations General Assembly, 1987. Report of the World Commission on Environment and Development: Our Common Future. Transmitted to the General Assembly as an Annex to Document A/42/427, p. 347.

Waldron, D., 2019. Evolution of vertical farms and the development of a simulation methodology. WIT Trans. Ecol. Environ. 217 https://doi.org/10.2495/SDP180821. Sustainable Development and Planning. X 975.

Watanabe, H., 2011. Light-controlled plant cultivation system in Japan - development of a vegetable factory using LEDs as a light source for plants. In: Acta Horticulturae 907 (ISHS Symposium on Light in Horticulture, Tsukuba, Japan), pp. 37—44.

Role and characteristics of PFALs

Toyoki Kozai

Japan Plant Factory Association, Kashiwa, Chiba, Japan

3.1 Introduction

The primary purposes of this chapter are to clarify the role of PFALs in agriculture and our society in the forthcoming decades, and to compare the environmental characteristics of conventional (or existing) PFALs with those of ideal PFALs. The industry of PFAL with use of LEDs has a history of only 20 years or so, compared with the greenhouse industry's history of around one century and open-field modern agriculture of a few centuries. Thus, most aspects of PFAL research, technology, and business are still immature and at the initial stages of development.

On the other hand, advances in recent technologies such as artificial intelligence (AI), the Internet of Things (IoT), virtual reality (VR), augmented reality (AR), camera-image processing, fifth generation communication network (5G), and LEDs have been remarkable. These advanced technologies are efficiently incorporated into PFALs only when they are designed and managed with the appropriate concept and methodology in line with the fundamental and essential environmental characteristics of PFALs that have an appropriate vision, mission, values, and goals. In fact, there have been many efforts to introduce these advanced technologies into PFALs, although there seem to be only a few successful cases as of 2020. After the spread of Covid-19, the technologies described above have been introduced to unmanned or automated stores and restaurants. Similar technologies can be introduced to develop user-friendly PFALs.

In this chapter, the following PFALs topics are discussed: (1) Vision and mission, (2) Types of plants and products suited for production in PFALs, (3) Essential features of PFALs and reasons behind high resource use efficiency (RUE), (4) Characteristics of conventional PFALs, (5) Characteristics of an ideal PFAL, and (6) Important research and development topics not fully discussed in this book. This chapter discusses issues related to the cultivation room only and does not include issues on other PFAL areas such as the changing room, shipping room, bathroom, and office.

3.2 Vision, mission, values, and goals of PFALs (Kozai, 2019: Kozai et al., 2020a,b)

It is essential to have a clear vision, mission, values, and goals when addressing global and local issues in areas such as food, resources, the environment, and quality of life before designing a PFAL (Fig. 3.1). Achieving the vision, mission, values, and goals also requires decisions on strategies, tactics, and approaches needed in management and operation. This book discusses possible problems and the potential of PFALs based on the vision, mission, values, and goals described below.

3.2.1 Vision: to construct and manage a sustainable production system to produce value-added functional plants

A PFAL needs to be ecologically, environmentally, sociologically, and economically sustainable and beneficial to the local, regional, national, or global society at a time when urban populations are growing and consumer demand for safer, healthier, and more affordable functional foods is increasing on the one hand, and the agricultural population, water supply for irrigation, agricultural land area, and other resources such as ores containing phosphate and potassium for fertilizer are diminishing on the other hand.

A PFAL must be a CO_2 absorbing/fixing plant production system that maximizes CO_2 assimilation by photosynthetic plants and is designed for sustainable production of functional plants including vegetables, herbs/medicinal plants, ornamental plants, and transplants, excluding staple food plants primarily consumed for their calories or energy, such as wheat, maize, and rice, although transplants of these staple crops can be produced in a PFAL. The staple crops include wheat (*Triticum aestivum*), maize (*Zea mays*), and rice (*Oryza*

FIGURE 3.1   PFAL pyramid consisting of the vision, mission, strategy, tactics, and management/operation.

sativa), root and tuber crops such as cassava (*Manihot esculenta*) and potato (*Solanum tuberosum*), legumes such as soybean (*Glycine max*), and fruit crops such as banana (*Musa* spp.).

3.2.2 Mission: to maximize the yields and quality of produce in a resilient way and in a pleasant working environment, with minimum resource consumption and waste generation

Quality criteria for produce include safety, composition, and concentration of functional and nutritional components, taste, color, texture, appearance, shelf life, uniformity, and hardness/softness. Resources consumed during plant production include working hours, electricity for lighting and air conditioning, water for irrigation and cleaning/washing, fertilizer, land area, seeds, CO_2 for plant photosynthesis, and other consumables such as substrate and packaging. Waste generated includes plant residue, wastewater containing fertilizer, waste heat energy generated by lamps, used plastic materials, and other consumables and contaminants, which may cause negative impacts on the environment outside PFALs. Contaminants contained in produce include small dead insects, dust, and other small foreign substances.

3.2.3 Values: the basic theory and methodology of PFAL design and management are applicable and adaptable, with minimum modification, to any type of PFAL in any society, location, and climate

A PFAL is just one component of a local ecosystem in a region and is integrated with other biological and energy conversion systems to improve local environmental, social, and economic sustainability. The basic theory and methodology of PFAL design and management must be universal and applicable in a simple way to any type of PFAL, regardless of physical scale and application, with minimum modification of design and management.

3.2.4 Goals: to contribute to achieving the 17 Sustainable Development Goals and 169 targets by 2030 (Seth et al., 2019)

Countries are expected to achieve the Sustainable Development Goals (SDGs) by 2030 (refer to Chapter 4). After 2030, new targets for sustainability, social welfare, and ethical production will be set. Global Action Programs such as education for sustainable development, and environmental, social, and corporate governance criteria are also moving in the same direction.

In view of the SDGs and emerging technologies that will come to the forefront in the period from 2025 to 2030 (Fig.3.2) (Kozai, 2019a; Kozai et al., 2019b, 2020a), the design of any PFAL should envision an ideal or ultimate PFAL based on a new vision, concept, methodology, and technologies that reflect future SDGs and technologies.

FIGURE 3.2   Roadmap for sustainable PFALs achieving the SDGs and beyond.

3.3 Subgoals of PFALs associated with the SDGs

The adoption of methods and systems suggested in the following subgoals helps PFALs achieve the SDGs.

3.3.1 An energy- and material-autonomous PFAL

It is possible to construct an energy- and material-autonomous PFAL by adopting a combination of the following methods and systems: (a) an efficient lighting system in a PFAL for converting electric energy to light energy (or photosynthetic photons), (b) an efficient photosynthetic conversion system (carbohydrate synthesis and accumulation in plants from CO_2, H_2O, and fertilizer under light) at any plant growth stage, (c) an efficient aerial and rootzone environmental control system, (d) efficient secondary metabolite production and morphogenesis in plants, (e) selection and introduction of cultivars suitable for PFALs, (f) an efficient hydroponic plant cultivation system, (g) integration of various biological and energy/material conversion systems such as aquaculture, mushroom cultivation, fermentation and air-conditioning systems, (h) use of renewable energy such as solar, biomass, wind, hydraulic, and thermal for generating electricity required for lighting, aerial and rootzone environmental control, and the operation of machines, and (i) reducing, reusing, and recycling of waste such as plant residue and wastewater.

3.3.2 Food chain system

The aim is to develop a food chain system for PFALs with minimum resource consumption and loss of produce (Fig. 3.3). Resource consumption and loss of produce (mostly plant residue) during plant production can be reduced significantly through optimal production scheduling and management, aerial and rootzone environmental control, and selection of cultivars.

After shipment from a PFAL, produce is transported to destinations such as local grocery stores, restaurants, convenience stores, schools, community centers, and homes by

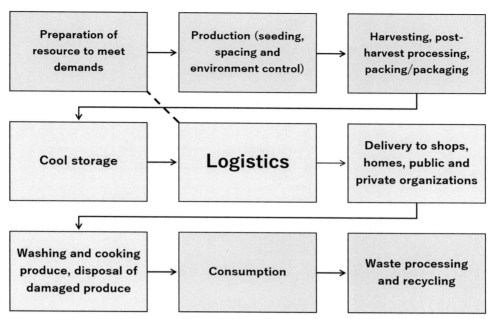

FIGURE 3.3 Resource consumption stemming from cool storage, shipping, transportation, and loss of produce due to damage in a food system chain can be reduced through local production for local consumption of pesticide-free, clean produce with a long shelf life.

minimizing food loss, food quality impairment, food mileage, and time and cost for delivery. A PFAL for local production for local consumption can significantly reduce resource consumption and food loss during transportation.

3.3.3 Conversion of resources into produce and waste

The goal in PFAL design and operation is to maximize production of produce of the highest quality with the minimum use of resources and minimum generation of waste. Fig. 3.4 shows the process of converting resources into produce and waste during the plant production process in a cultivation room.

3.3.4 Multipurpose PFALs

Multipurpose mini- or micro-PFALs can be developed for a wide range of purposes including education, training, self-learning, indoor gardening, small group entertainment, and hobbies (Takagaki, 2020). The software and hardware for mini- or micro-PFALs are the same as for large-scale, commercial PFALs. Users can easily grasp the principle of photosynthesis, the circulation of water and carbon, fertilizer (nitrogen, phosphate, potassium, etc.), plant growth, energy and mass balance, and environmental control. Any actual PFAL can be connected to a virtual PFAL or a PFAL simulator, just like a user-friendly computer game, flight simulator, or car simulator (Kozai, 2018a).

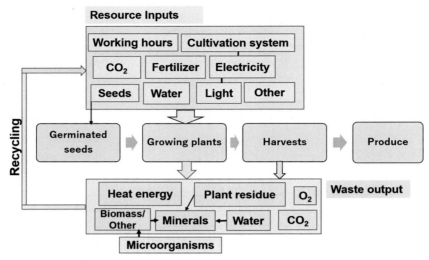

FIGURE 3.4 Conversion of resources into produce and waste during the production process in a cultivation room.

3.3.5 Human resource development

Creating human resource development programs for PFAL designers, engineers, managers, and diverse end users is vital for achieving the SDGs (see Chapter 4 for details). There is a need for software programs that allow users to enjoy learning the principles and concepts of environmental effects on plant growth, efficient energy conversion processes, and the circulation of water, CO_2, and nutrient elements such as nitrogen, phosphate, and potassium in a plant ecosystem, while they grow plants and manage the production process for food and other purposes while considering the resource and monetary productivity (Chapters 12 and 13).

3.3.6 Pleasant work environment

The aim is to create a comfortable, safe, constructive, and inclusive work environment for everyone including the elderly and persons with disabilities, regardless of gender, religion, age, nationality, or personality, to create a friendly and enjoyable community.

3.3.7 Implementation of advanced technologies and models

Advanced technologies and models must be introduced into well-designed PFALs. Advanced technologies include AI, VR, AR, 5G (wireless technology for digital mobile networks), IoT, phenotyping (plant trait measurement), and open databases via the Internet for the development of flexible, user-friendly, sustainable PFALs. Robot and OMICS (genomics, metabolomics, proteomics transcriptomics, etc.) technologies can also be introduced to large-scale PFALs.

Models include mechanistic, multivariate statistic, behavior (or surrogate), and AI models (Kozai, 2018a). Using such technologies, evolutional PFALs can be realized with high resource and monetary productivity (refer to Chapters 12 and 13 for details) and high scalability, traceability, and adaptability.

3.4 Types of plants and products suited for production in PFALs

Functional plants produced in PFALs can be used for various purposes. Table 3.1 shows potential and actual uses of plants produced in PFALs. Only leafy fresh vegetables and various kinds of transplants are currently commercialized widely.

Types of plants shown in Table 3.1 are called functional plants in this book, and exclude staple crops such as maize, wheat, rice, and potatoes. Compared with staple crops, functional plants generally require lower photosynthetic photon flux density (PPFD of $100-200\ \mu mol\ m^{-2}\ s^{-1}$), shorter cultivation periods (a few weeks to a few months) or year-round cultivation, and higher spatial and/or aerial planting density. Besides, monetary value

TABLE 3.1 Main uses of functional plants produced in PFALs. Only leafy vegetables and various types of transplants are currently commercialized widely.

No.	Category	Uses
1	Functional foods	Fresh leafy, fruit, root, and bulbous vegetables
		Processed leafy, fruit, root, and bulbous vegetables
		Dried and freeze dried leafy, fruit, root, and bulbous crops
		Edible flowers and accompaniments/trimmings
		Edible seaweeds (Nori, genus *Porphyra*), spirulina, euglena, and other edible aquatic plants
2	Drink and food additives	Medicinal/herbal components, juices, sweeteners, flavorings, seasonings, and natural coloring pigments
3	Medicine and healthcare goods	Supplements, cosmetics, perfume, aroma, and specialty oils
		Chinese medicine, alternative medicine, and pharmaceuticals including *Cannabis sativa* L. for medical, recreational, and industrial uses
		Potassium (K^+) poor or rich lettuce, Fe^{++} (iron) rich spinach, polyphenol and antioxidant rich vegetables, algae, euglena, spirulina, waxy barley, etc.
4	Disease, virus, and insect-free transplants	Seedlings, micropropagated and grafted transplants, rooted or unrooted cuttings, propagules (stolons, microtubers, minitubers, bulblets), lawn/turf seedling mats, bedding plants, moss, etc.
5	Ornamentals, fruits, and others	Miniature or dwarf ornamentals and fruits, plants for interior decoration, bouquets, pot plants, cut flowers, etc.
6	Functional staple crops	Staple crops containing special functional components showing medicinal and/or health care effects

(economic value per kg of produce) of functional plants is 10—100 times higher than staple crops (Kozai et al., 2020a). PFALs are particularly suitable for growing relatively short plants (30—50 cm in height) with relatively upright leaves, which grow fast under relatively low PPFD and high CO_2 concentration (1000 ppm or higher).

3.5 Land area required for producing fresh vegetables

China ranked the highest in annual vegetable consumption per capita with 328 kg in 2013. Average world annual vegetable consumption per capita was around 135 kg in 2017 according to the Food and Agriculture Organization of the United Nations. Assuming that 5% of 135 kg (0.05×135) or 6.75 kg/person of fresh vegetables are supplied annually by PFALs with 10 tiers in an urban area with a population of 10 million, 67,500,000 kg (6.75×10^7) or 67,500 metric tons of vegetables can be produced annually in PFALs with a total floor area of around 270 ha (67,500 tons ÷ 250 tons/ha).

On the other hand, around 27,000 ha of open fields or around 2700 ha of greenhouses located in a temperate region is necessary to produce the equal amount of 67,500 tons of vegetables annually. As of 2017, there are 47 cities in the whole world with a population of over 10 million, of which three have over 30 million residents, namely Tokyo, Jakarta, and Shanghai.

Martellozzo et al. (2014) wrote that urban agriculture (UA) would require roughly one-third of the total global urban area to meet global vegetable consumption of urban dwellers; the area ratio is higher where urban population density is higher. This estimate does not consider how much urban area may actually be suitable and available for UA. This study suggests that we need to produce a substantial percentage of vegetables in PFALs and greenhouses instead of open fields.

3.6 Essential features of PFALs and reasons behind high resource use efficiency

3.6.1 Resource and waste elements

Types of resource elements supplied to a PFAL and waste elements emitted from the PFALs are shown in Fig.3.5 and 3.6, respectively. In a PFAL, the supply rate (amount supplied per unit time) of each resource element can be accurately measured. Besides, rates of net photosynthesis and water uptake by plants in a cultivation room can be relatively accurately estimated based on CO_2 and water balance equations (Kozai, 2013), because the cultivation room of a PFAL is almost airtight and thermally well insulated.

Similarly, uptake rates of nutrient elements such as N, P, K, Mg (magnesium), and Ca (calcium) by plants from a nutrient solution in a cultivation bed can be estimated based on nutrient ion balance equations, although this type of automatic controller is not available commercially as of 2020 (Kozai et al., 2018a). Furthermore, the emission rate of each waste element can be measured relatively accurately. Moreover, data on production rates of plants, marketable parts of plants, their unit price for sale, and unit cost of each supply element are easily collected daily.

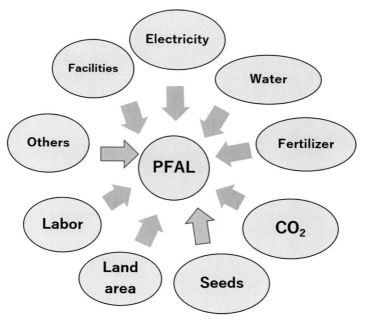

FIGURE 3.5   Resource elements supplied to a PFAL.

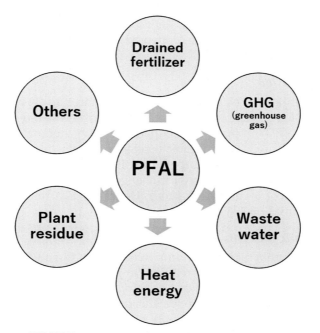

FIGURE 3.6   Waste elements emitted from a PFAL.

3.6.2 Basic characteristics of PFALs

3.6.2.1 Airtightness

A highly airtight and thermally insulated cultivation room guarantees (1) no invasion of viruses, microorganisms, insects, small animals, dust, or the like through air gaps between the interior and exterior of the cultivation room; (2) minimum exchanges of CO_2 and water vapor through air gaps between the interior and exterior of the cultivation room; (3) no influence of weather on the environment inside the cultivation room; and (4) the recycling of CO_2 and water in the cultivation room.

CO_2 respired by plants that accumulates in a cultivation room is reused (assimilated) through plant photosynthesis during a period of light exposure. This improves CO_2 use efficiency to nearly 100% in an airtight cultivation room. CO_2 respired by humans in a cultivation room is also absorbed by plants under light. Nearly 95% of water vapor transpired by plants is condensed and collected by air conditioners during cooling and is recycled as irrigation water. High airtightness means the number of air exchanges in a cultivation room is lower than 0.02 h^{-1} or so.

On the other hand, forced ventilation through a gas filter is required when the following abnormalities are found: (1) Unfavorable gases, such as volatile organic compounds (VOCs) produced by plants and/or structural elements and odors emitted by humans, accumulate in a cultivation room; (2) CO_2 concentration reaches a level higher than around 1500 $\mu mol\ mol^{-1}$ (or ppm) in the presence of workers in a cultivation room; (3) Pathogenic microorganisms and/or viruses are detected in room air; and (4) An extreme environment such as a high air temperature (e.g., 40 °C) is recorded due to a fire and/or malfunction of air conditioners and/or a lighting system. The upper limit of CO_2 concentration in a room varies between 1000 ppm and 5000 ppm, depending on countries, types of building, exposure time, etc.

3.6.2.2 Thermal insulation

High thermal insulation of the cultivation room is required to (1) minimize the impact of fluctuations of the outside air temperature, wind speed/direction, atmospheric pressure, and vapor pressure deficit (VPD) on inside air temperature, CO_2 concentration, and VPD and air movement in a PFAL; (2) reduce or inhibit condensation of water vapor on inside walls, floors, etc., by keeping their surface temperatures higher than the dew point temperature of room air; and (3) minimize heat energy exchange through walls and the floor between the interior and exterior of a cultivation room. Desirable heat transmission coefficients of walls/floors are around 0.1–0.2 W m^{-2} °C^{-1}.

3.6.2.3 Highly hygienic and safe work environment

Clean (or hygienic), semihygienic, and contaminated areas in a PFAL are physically separated from each other. Special caution is required to prevent the invasion, growth, and propagation of (1) bacterial pathogens such as *Escherichia coli* O157: H7 or *Bacillus dysentericus* which cause diseases in humans (hemorrhagic diarrhea), (2) plant pathogens such as *Pythium aphanidermatum which* causes roots to rot; (3) pest and unpleasant insects such as mites (Arachnida) and lake flies (Chironomidae); and (4) small animals such as mice. In conformity to ISO22000, the Japan Plant Factory Industries Association released the following guidelines for the safety of produce: (1) CFU (colony forming units) should be lower than 1000 per g (fresh weight), (2) no use of pesticides during cultivation, (3) *E. coli* negative, and (4) no foreign matter present (JPFIA, 2020).

Once biological contaminants mentioned above are spread in a cultivation room, all the plants inside need to be taken out to the outside for disposal by burning, and the empty cultivation room is disinfected often by a fumigant (smoking). Then, it takes more than 1 month to harvest new produce after disinfection. In this sense, biological contamination poses the highest risk/damage to a PFAL.

3.6.3 Observability, controllability, traceability, predictability, and reproducibility

To achieve high and stable resource and monetary productivity with minimum resource consumption and waste emission throughout the year, high observability, controllability, traceability, predictability, and reproducibility of a cultivation room are indispensable.

High airtightness and thermal insulation of a cultivation room with reliable sensors are required to keep observability high regardless of the weather outside. High observability of a cultivation room with reliable actuators is required to keep controllability high. High degrees of observability and controllability lead to a high degree of traceability of resource consumption, waste emission, plant production rates, the room environment, equipment operation, etc (Fig. 3. 7). High reproducibility and predictability can be achieved by using various types of plant growth and production management models including AI models and of IoT sensors.

High degrees of observability, controllability, traceability, predictability, and reproducibility are special features of a next-generation PFAL that ensures continuous reductions in production cost and waste emission, high RUE, and daily estimation of resource and monetary productivity (Chapters 12 and 13). These measured and estimated variables can be visualized together with other related data on a computer display in its cultivation room and operation room and can be used for further analysis and prediction of production processes.

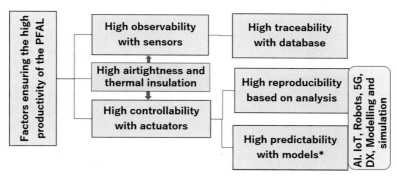

FIGURE 3.7   Models: Mechanistic, multivariate statistical, behavioral (or surrogate), and artificial intelligence (AI) models. IoT stands for Internet of Things. 5G: the fifth generation technology standard for broadband cellular networks. *DX*, digital transformation.

3.6.4 Measurement and estimation (see Chapter 8 for details)

3.6.4.1 Environmental factors

Main environmental factors to be controlled in a PFAL include air temperature, CO_2 concentration, temperature, total ion concentration or electric conductivity, and pH of nutrient solution. Other environmental factors to be controlled or frequently monitored include VPD, air current speed, lighting cycle or photoperiod, PPFD, dissolved oxygen concentration, and flow rates of nutrient solution. Spectral photon flux of lamps and spatial PPFD distribution in cultivation space need to be checked every half year or so.

3.6.4.2 Resource supply and production rates

All resource supply rates can be measured relatively easily, both continuously and online (in real time), for their control and analysis. Time courses of the operating status of all major equipment can also be recorded for analysis. Resources include working hours, electricity for lighting and air conditioning, water for irrigation and cleaning floors and cultivation panels, CO_2 for promoting plant photosynthesis, fertilizer (nutrient solution), seeds, substrates, seed trays, and disinfectants. Daily production rates of produce, wastewater, and waste (plant residue, etc.) can also be measured.

3.6.4.3 Rate variables and RUE estimation

Rates of net photosynthesis, dark respiration, and water uptake or transpiration can be estimated online based on the CO_2 and water balance of a cultivation room. Sensible and latent heat generation and conversion in a cultivation room can be separately estimated based on the heat and water balance equations. The energy conversion process from electrical to chemical energy (carbohydrates in plants) can also be estimated.

The variables measured or estimated above can be used to calculate use efficiencies of electricity (electric energy), light energy (photosynthetic photons), water, CO_2, fertilizer, seeds, and substrates. The coefficient of performance (COP) of air conditioners can be estimated as a function of air temperature difference between the interior and exterior of a cultivation room and the ratio of a cooling load to cooling capacity.

3.6.4.4 High RUEs almost realized

(1) Water and CO_2 use efficiencies are nearly 100% due to their recycled use.
(2) Seed and transplant use efficiencies are nearly 100% (all seeds and transplants are grown to marketable plants).

3.6.4.5 RUEs to be improved

(1) Nutrient elements use efficiencies of conventional PFALs that adopt a nutrient solution recycling system are 0.7–0.8, which need to be improved to nearly 100% hopefully with the use of a one-way (noncirculating) nutrient solution supply system (refer to Section 10).
(2) Substrates (plant-supporting material) and other consumables need to be used at minimum levels or recycled to minimize waste emission.
(3) Cultivation beds need to be always occupied with high-quality plants, and all produce is marketable to achieve a cultivation bed use efficiency of 100%. Spatial density of plants (no. of plants m^{-3} or m^{-2}) in a cultivation room is high at any plant growth stage.

(4) All parts of all plants need to be used, marketed, and/or sold, resulting in no plant residue including physically or physiologically damaged parts of plants.

(5) If only leafy or aerial parts are useable or marketable, the dry weight percentage of roots needs to be minimal as long as the roots do not restrict the growth of aerial parts. The percentage of root weight of whole leafy vegetable plants is often 10%–15% in conventional PFALs.

3.6.4.6 Resource and monetary productivity

Resource and monetary productivity can be estimated and visualized using data on resource supply rates, production rates, resource use efficiencies and unit costs for each resource and waste element, and unit sales prices of produce (refer to Chapters 12 and 13 for details).

Reasons behind potentially high RUEs and resource/monetary productivity are described above (Sections 6.3 and 6.4). These reasons and benefits of PFALs are all related to fundamental characteristics of PFALs described in this chapter. Those benefits are particularly useful to contribute to solving issues concerning food, the environment, resources, and quality of life discussed in Chapter 1.

3.7 Aerial environmental characteristics of conventional PFALs to be improved

Spatial distributions of these environmental factors inside a plant canopy are generally interrelated with each other. These distributions are also interrelated with the architecture (leaf area index, leaf angle, size, shape, etc.) of a plant canopy and physiological status (stomatal conductance, chlorophyll concentration and fluorescence, water potential, etc.) of plants and their changes with time. Accordingly, rates of net photosynthesis, transpiration, and thus growth of plants are interrelated with the environment, canopy architecture, and physiological status.

3.7.1 Light

Under downward lighting conditions, the leaves at the uppermost level of a plant canopy receive more light energy than those at lower levels, particularly in a densely populated plant canopy (e.g., Oikawa, 1977). Then, the PPFD decreases exponentially as the depth increases from the top to the bottom of a plant canopy (Fig. 3.8). Then, the net photosynthetic rate of

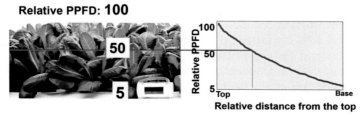

FIGURE 3.8 Scheme showing exponential decrease of PPFD within a densely populated plant canopy. Air current speed shows a similar decrease within the densely populated plant canopy.

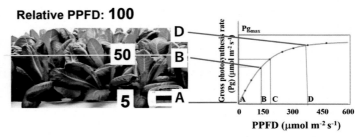

FIGURE 3.9 The exponential decrease of PPFD suppresses the canopy photosynthesis and thus growth. A similar decrease of air current speed within the densely populated plant canopy.

leaves decreases with the increase in the depth of a plant canopy (Fig. 3.9). A similar decrease of air current speed is observed within a densely populated plant canopy (The PPFD in the plant canopy fluctuates if the leaves are fluttered by air currents).

3.7.2 Aerial environmental characteristics and plant response

In a conventional PFAL with downward lighting, the aerial environment inside a plant canopy is significantly different from that just above the plant canopy. Moreover, spatial uneven distributions of environmental factors within a plant canopy are often observed especially when the plant canopy is densely populated. These spatial uneven distributions of environmental factors generally suppress the growth rate of a plant and spatial uniformity of plant growth and chemical components. The environmental factors include PPFD, spectral photon flux density distribution, air temperature, CO_2 concentration, air current speed, and VPD.

If the PPFD is spatially uniform at an appropriate level in a plant canopy regardless of the canopy's depth, the net photosynthetic rate of the whole plant canopy should increase significantly, and the decrease in net photosynthetic capacity of lower leaves due to their senescence may be prevented (Zhang et al., 2015; Joshi et al., 2017). Three-dimensional computer simulation work on the light environment in a plant canopy may be useful to evaluate and optimize a lighting system of PFALs (Saito et al., 2020).

3.7.3 Vertical uneven distribution of spectral photon flux density in a plant canopy

A vertical uneven distribution of spectral photon flux density (SPFD) also exists within a plant canopy in downward lighting conditions using white LEDs emitting blue, green, red, and far-red light. This is because the transmittance of a horizontal single green leaf is around 5% for blue (400–499 nm) and red (600–699 nm) photons, about 20% for green (500–600 nm) photons, and around 40% for far-red (700–800 nm) photons (Taiz and Zeiger, 2006). Thus, percentages of green and far-red photons increase among photons with a wavelength of 400–800 nm with increases in the depth of a plant canopy and the depth of each leaf (Terashima and Saeki, 1985; Sun et al., 1998; Nishio, 2000).

This vertical profile of SPFD inside a plant canopy affects photosynthesis and morphogenesis, and thus the growth and development of a plant canopy (Folta, 2019). In fact, green and far-red photons contribute significantly to plant productivity and RUE, especially in plant production under white LEDs containing a significant percentage of green photons (Smith et al., 2017). It should be noted that percentages of green and far-red photons emitted from white LEDs vary 15%–40% and 0%–5%, respectively, depending on the type and quantity of phosphors covering LED tips (Kozai et al., 2016). Also, as of 2020, most PFALs for commercial production use white LEDs due to their high cost performance (i.e., high performance and low cost).

3.7.4 Profiles of air temperature, CO_2 concentration, and water vapor pressure (Kitaya et al., 1998, 2004; Kitaya, 2016)

In conventional PFALs, horizontal air current speeds and thus gas diffusion coefficients in a plant canopy are significantly higher above and around upper leaves than around lower leaves, which affect the vertical distribution (or profile) of air temperature, CO_2 concentration, and VPD inside a plant canopy (Kitaya, 2016).

Fig. 3.10 shows that the air temperature around leaves during a photoperiod is about 2 °C higher than the air temperature above and in the lower part of an eggplant seedling canopy. The leaf temperature is also 1–2 °C higher than the air temperature. CO_2 concentration is at its lowest at the uppermost part of a plant canopy where the net photosynthetic rate is highest. The VPD is at its highest at the wet substrate surface due to evaporation and at its second highest around leaves due to transpiration.

These profiles are considerably affected by the PPFD, air current speed, and plant canopy architecture. In a tomato seedling canopy, as schematically shown in Fig. 3.11, only hypocotyls (or stems) with two small cotyledons exist in the lower part of the canopy (lower than 30 mm), so that air moves horizontally relatively freely in this space (0–30 mm above the substrate surface).

On the other hand, there is no such empty space in a leaf lettuce canopy shown in Fig. 3.9. Thus, the vertical profile can be more clearly observed in the plant canopy shown in Fig. 3.9 than those of a seedling canopy shown in Fig. 3.10.

Similar uneven spatial distributions of flow speeds, dissolved O_2 concentrations, nutrient element concentrations, and nutrient solutions pH are observed in a hydroponic cultivation bed.

3.7.5 Vapor pressure deficit (VPD)

The VPD in a cultivation room full of plants reaches nearly 0 kPa (relative humidity or RH is 100%) within 10–20 min after all the lamps are turned off at the same time (dark period). This is because air conditioners stop due to a very low cooling load (no heat energy generation from LEDs).

Alternately turning on and off lamps avoids this very low VPD and keeps the VPD and room air temperature at set levels. Air conditioners remove not only sensible heat to lower room air temperatures but also latent heat (water vapor) and dehumidify room air. In this way, the VPD is kept at preferable levels (around 0.3 kPa or relative humidity [RH] of around 80%) all day (Kozai et al., 2019a; Kozai, 2018b).

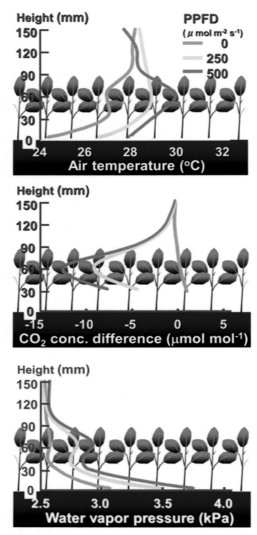

FIGURE 3.10 Vertical distributions of air temperatures, CO_2 concentrations, and water vapor pressure above and within an eggplant (*Solanum melongena* L.) seedling canopy (90 mm high and leaf area index of 6) at PPFD of 0, 250, and 500 mmol/m^{-1}/s, and horizontal air current speed of 0.1 m/s. CO_2 concentration, air temperature, and water vapor pressure were 380 ppm, 28°C, and 2.4 kPa (RH: 65%), respectively, at a height of 150 mm from the substrate surface. *Reproduced from Kitaya, Y., Shibuya, T., Kozai, T. and Kubota, C., 1998. Effects of light intensity and air velocity on air temperature, water vapor pressure, and CO2 concentration inside a plant canopy under artificial lighting conditions. Life Supp. Bios. Sci. 5, 199—203 with permission.*

It should be noted that the VPD decreases (RH increases) with an increase in the total leaf area in a cultivation room, because leaves are a supply source of water vapor to room air. On the other hand, at around 24 °C, the VPD reaches around 1.5 kPa (RH at 50%) or greater when no plants are growing in a cultivation room.

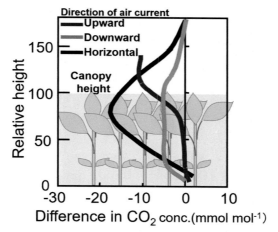

FIGURE 3.11 Vertical distribution of CO_2 concentrations during photoperiod above and within tomato (*Lycopersicon esculentum* Mill.) seedling canopy as affected by air current direction. The reference CO_2 concentration was taken at the air inlet of the system. *Reconstructed from Shibuya T., Tsuruyama J., Kitaya Y., Kiyota M., Enhancement of photosynthesis and growth of tomato seedlings by forced ventilation within the canopy Sci. Hortic. 2006, 218–222 with permission.*

3.7.6 Improving the aerial environment in a plant canopy

3.7.6.1 *Vertical distributions of CO_2 concentrations*

Fig. 3.11 shows the vertical distribution of CO_2 concentration during a photoperiod above and inside the tomato seedling canopy, due to the effects of forced air currents in an upward, downward, or horizontal direction. The figure shows that downward and upward air currents provide relatively uniform vertical CO_2 concentrations in the plant canopy, and that CO_2 diffusion into the stomata of leaves is more pronounced than with horizontal air flows (Shibuya et al., 2006). The CO_2 concentration around the uppermost leaves is about 15 μmol mol^{-1} higher in upward and downward air currents than in horizontal air current. Similar effects of nutrient solution flows in water and nutrient element uptake need to be examined to determine an ideal hydroponic cultivation system.

3.7.6.2 *Promoting net photosynthesis by increasing CO_2 concentration and air current speed*

CO_2 is assimilated (photosynthesized) by leaves mostly through stomata on the abaxial side of leaves only under light, while CO_2 is continuously emitted mostly through stomata under both light and dark (dark respiration) conditions. A very small amount of CO_2 moves through the epidermis of leaves. Dark respiration (or CO_2 emission) increases with an increase in leaf temperature and dry mass of leaves. The net CO_2 exchange rate at leaves or a canopy under light is referred to as the net photosynthetic rate.

Fig. 3.12 shows that the net photosynthetic rate of a tomato seedling canopy increases with an increase in CO_2 concentration and horizontal air current speed over the plant canopy (Kitaya et al., 2004). The net photosynthetic rate generally increases with an increase in CO_2 concentration in a range of 0–1000 μmol $μmol^{-1}$ and air current speed in a range of 0.1–1.0 m s^{-1} at PPFD of around 250 μmol m^{-2} s^{-1}. The optimal air current speed depends on the VPD, LAI, average leaf angle, and PPFD.

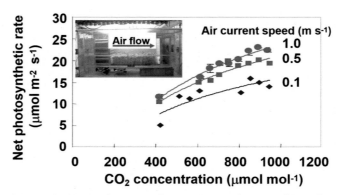

FIGURE 3.12   Net photosynthetic rate of a tomato (*Lycopersicon esculentum* Mill.) seedling canopy as affected by CO_2 concentration and horizontal air current speed at PPFD of 250 mmol/m^{-1}/s and air temperature of 28°C. *Reconstructed from Kitaya, Y., Shibuya, T., Yoshida, M., Kiyota, M., 2004. Effects of air velocity on photosynthesis of plant canopies under elevated CO2 levels in a plant culture system. Adv. Space Res. 34, 1466—1469 with permission.*

3.8 Improving subsystems of a cultivation room

3.8.1 Increasing electric energy use efficiency

In a PFAL, 30%—40% of electric energy is converted to light (or photosynthetic radiation) energy by lamps, 60%—90% of the light energy emitted by lamps is received by leaves, and a portion (around 10% at most) of light energy absorbed by pigments such as chlorophylls in leaves is converted to chemical energy as carbohydrates through photosynthetic activity. As a result, between 1.8% ($100 \times 0.3 \times 0.6 \times 0.1\%$) and 3.6% ($100 \times 0.4 \times 0.9 \times 0.1\%$) of electric energy consumed by lamps is converted to chemical energy as carbohydrates in plants. A portion of carbohydrates is converted to proteins, lipids, and various secondary metabolites. Thus, the percent conversion from electric to chemical energy or electric energy use efficiency is an important parameter to be improved.

3.8.2 Reducing electricity consumption per kg of produce

One of the weaknesses of conventional PFALs with LEDs is that they require 7—9 kWh of electricity, as of 2018, for lighting, air conditioning, etc., to produce 1 kg of fresh leafy vegetables such as leaf lettuce (refer to Chapter 12). This electricity cost accounts for about 20% of total production costs of PFALs in Japan. The following improvements are expected to reduce this electricity consumption per kg of produce to 5—7 kWh:

(1) Increase photosynthetic photon efficacy of LED luminaires by 10% (from 2.5—3.0 µmol J^{-1} in 2020 to 2.8—3.3 µmol J^{-1}). (The term "efficiency" is used when the unit of nominator and denominator is the same, otherwise the term "efficacy" is used). In theory, the maximum efficacy is 3.34 µmol J^{-1} at 400 nm (wavelength), 3.80 µµmol J^{-1} at 455 nm (blue), 4.63 µmol J^{-1}J at 555 nm (green), 5.52 µmol J^{-1} at 660 nm (red), and 5.85 µmol J^{-1} at 700 nm. Thus, the efficacies can be further improved in the near future. See Chapters 5—7 and Chapters 16—19 for further details on efficient use of LEDs.

(2) Increase the ratio of light energy received by the leaves of a plant canopy to light energy emitted from lamps by improving the lighting system and plant canopy architecture.

(3) Optimize SPFD and their spatial and temporal distributions in a plant canopy to achieve higher resource and monetary productivity for various kinds of plants.

(4) Optimize the hydroponic cultivation system and the composition, strength, and supply rate of nutrient solution.

(4) Breed or select cultivars with better phenotypes or plant traits.

(5) Ensure a dynamic environmental control to meet production objectives.

Fortunately, costs have been decreasing year after year for power generation by renewable energy, battery storage, and smart grids.

3.8.3 Increasing harvest index

Electric energy use efficiency can also be improved by increasing the harvest index (HI), which is defined as the weight ratio of marketable part of plants to the whole plants. Production of leafy vegetables and transplants using PFALs is commercialized widely because the HI of leaf vegetables ranges between 0.8 and 0.9 and the HI of transplants is nearly 1.0. However, the HI of fruit vegetables is generally around 0.5. If aerial (or leafy) parts of dwarf root vegetables such as carrots, turnips, and radish are edible, the HI of root crops would improve to nearly 1.0.

The quality and quantity of produce demanded by crop producers, buyers, and/or consumers vary according to the purpose of sale, purchase, or use, but higher quality and lower costs are always required. Such requirements are partially satisfied when all parts (leaves, petioles stems, roots, and flowers/flower buds if any) of the plant can be used, marketed, and/or sold, with virtually no plant residue or loss of plants (i.e., HI = 1.0). The economic or social value per kg of produce is maximized when the quality of all parts of all plants is equally highest. This ideal goal needs to be kept in mind when designing and managing actual PFALs toward next-generation PFALs.

3.8.4 Developing various models for plant production process management

Modeling and simulation work required for PFAL design and management include the following: (1) energy and material balance and energy/material conversion processes using mechanistic models; (2) spatial distributions of environmental factors (PPFD, UV (ultraviolet) and far-red flux density, air current speed, CO_2 concentration, and VPD) above and within a plant canopy and spatial distributions of plant traits (e.g., concentrations of chemical components) within a plant canopy; (3) P = f (G, E, M) model (See Section 3.11.2), or photosynthetic growth, development, and morphogenesis of plants, and secondary metabolite production models; (4) production scheduling, production process control, and financial management; (5) management of risks such as outbreaks of diseases/pest insects/pathogens, sanitation breaches, power outages, machine malfunctions, natural disasters, economic disturbances, and/or changes in consumer preferences; and (6) integration of all the above models into one PFAL management model.

When conducting simulations of points 3, 4, and 5 mentioned above, behavioral (or surrogate) and AI models such as machine learning including deep learning models can be efficiently used in addition to mechanistic and multivariate statistic models. It should be

FIGURE 3.13   Left: The area ratio of total cultivation rack to floor area is about 0.5, and the area ratio of total cultivation bed area to floor area of cultivation room is about 5, in a typical conventional PFAL with 10 tiers for commercial production. Right: The area ratio of total cultivation rack to floor area of cultivation room is about 0.8, and the area ratio of total cultivation bed area to the floor area is nearly 10 (= 0.8 × 4 × 3 = 0.96) in a three-stacked cultivation system modules (CSMs) each with 4 tiers with total of 12 (=3 × 4) tiers for commercial production (to be built in 2022). The area for automatic push-in/pull-out of cultivation panels is located at right side of each CSM. The plants are harvested after transported to a separate space. *Photo: by courtesy of PlantX Corp.*

noted that, in modeling, the relationships among variables in those models are often nonlinear and time dependent and show hysteresis effects in some cases.

3.8.5 Improving rack and cultivation bed area ratios in conventional PFALs

In most PFALs, nearly half of the floor area of a cultivation room is occupied by multi-tiered cultivation racks, and the other half is used for plant management and transportation of plants and supplies as well as maintenance of cultivation systems (Fig. 3.13). For example, in a PFAL with 10 tiers and a rack area ratio (ratio of rack area to floor area) of 0.5, the cultivation bed area ratio (ratio of total cultivation bed area to floor area) is equivalent to 5.0 (0.5 × 10). Thus, in this cultivation room, the ceiling height and distance between tiers and the rack area ratio, which is 1.0 at the maximum (i.e., no walkway on the floor), are the keys to improving production capacity per floor area. This figure can be doubled with the use of Cultivation System Modules (CSMs, see Section 3.11.12) shown in Fig. 3. 13, Right.

3.9 Environmental characteristics of ideal PFALs

3.9.1 Environmental uniformity of ideal PFALs

This section describes the environmental characteristics of ideal PFALs for maximizing resource productivity, that is, having the highest yield and quality with minimum resource inputs and emissions of environmental pollutants (revised from Kozai, 2019a). At a glance, the ideas for maximizing resource productivity presented below may seem unrealistic and impractical. However, it is important to understand the characteristics of ideal PFALs before settling for less due to difficulties at present.

3.9.1.1 *Strategy for optimal environmental control*

During the vegetative growth stage of plants, aerial and rootzone environmental factors other than photosynthetic photons need to be controlled primarily so that photosynthetic photons are converted to chemical energy of carbohydrates in leaves and translocated to

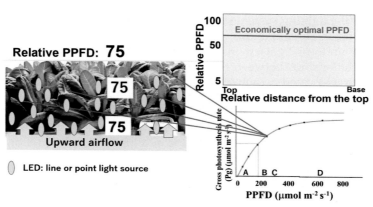

FIGURE 3.14 Omni-directional lighting and enhanced air movement within a densely populated plant canopy will promote the photosynthesis, growth, and their spatial uniformities. Even distributions of photosynthetic photons and air currents to all parts of all leaves maximize photosynthesis and thus plant growth.

appropriate parts of plants at the highest efficiency and lowest costs as intended by PFAL managers. This is because electricity cost for photosynthetic lighting is the highest of all environmental control costs.

3.9.1.2 Hypothetical idea of a lighting system

A hypothetical idea of an ideal lighting system is that all photosynthetic photons (or light energy) emitted by lamps are distributed and/or received equally by all parts of all leaves in a canopy (Fig. 3.14) although a small fraction of the photons are unavoidably absorbed by petioles, stems, etc., of plants. Air currents are also distributed uniformly to all parts of all leaves in a canopy.

This hypothetical idea is based on the PPFD−Pg (gross photosynthetic rate) curve shown in the lower right of Fig. 3.14. The physiological efficiency of Pg over PPFD is highest at PPFD near zero, but the absolute value of Pg is too low; the economic efficiency is highest and absolute value of Pg is appropriate between points B and C; and lowest at point D or greater in Fig. 3.14.

Under ideal LED lighting conditions, photosynthetic photons emitted by LEDs, P_{flux}, are distributed equally to all parts of all leaves with total leaf area of A_{leaf}. The photosynthetic photons received by a unit leaf area are expressed as $r_p^*(P_{flux}/A_{leaf})$ µmol m^{-2} s^{-1} where r_p is a ratio of photosynthetic photons received by leaves to that emitted by the LEDs. By increasing the r_p value and minimizing the spatial variation of (P_{flux}/A_{leaf}), the canopy photosynthesis is maximized.

Zhang et al. (2015) and Joshi et al. (2017) indicated that upward supplemental lighting from below the canopy delayed senescence of the outer leaves and increased the fresh weight of marketable leaf lettuce plants by around 20%. This upward lighting system could be further improved in the near future.

As a general rule, the ideal lighting conditions described above can be realized by arranging a number of small light sources such as LEDs around and inside the plant canopy to provide light from above, below, the sides, and inside the canopy and/or by applying projection mapping technology to use laser lighting.

With the use of the lighting system shown in Fig. 3.14, the plants' morphology and metabolism change considerably, and this positively or negatively affects the economic value of the produce. A uniform light environment in a plant canopy has the following effects: (1) Geometrical relationships between the source (photosynthesizing parts) and sink (accumulating parts of translocated carbohydrates) of plants are changed; (2) All leaves of a plant canopy relatively equally act as producers of carbohydrates; (3) Senescence of lower leaves due to low PPFD is suppressed; and (4) Phytohormone balances in individual plants are changed.

Little is known how plants respond to this spatially uniform light environment in a plant canopy, and this research area is a big challenge. Besides, an optimal lighting system depends on plant canopy architecture (e.g., leafy vegetables, head vegetables, fruit vegetables, etc.) and the purpose of production (served fresh in its original shape, supplied fresh but shredded, supplied as processed produce, etc.).

3.9.1.3 Lighting for head vegetables (see also Chapter 23)

In the case of head vegetables such as cabbage, Chinese cabbage, and head lettuce plants, lighting with green LEDs or white LEDs emitting a considerable percentage (e.g., 40%) of green photons over photosynthetic photons may be beneficial because green photons penetrate inside their heads (Fig. 3.15). Green photons inside the heads enhance the biosynthesis of chlorophyll and other secondary metabolites (Saengtharatip et al., 2020). Green LEDs can be inserted into the base of head vegetables 1–3 days before or after harvest for greening the inside of the head (Fig. 3.15. Lower right). This type of lighting technology can be applied when the greening and subsequent changes in chemical components inside heads are beneficial to improve the value of head vegetables.

3.9.1.4 Spatial uniformity of environmental factors other than photosynthetic photons

Uniform and/or maximum plant growth can be expected by achieving spatial uniformity of all aerial and rootzone environmental factors including spectral distribution of light, air current speed, CO_2 concentration, VPD, and nutrient solution flow speed in the root zone.

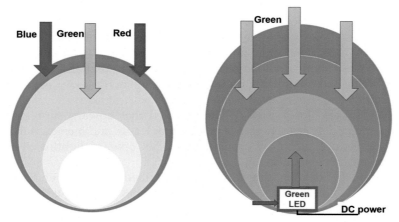

FIGURE 3.15   Scheme demonstrating that green light penetrates more deeply than blue and red light into the core of head vegetables such as lettuce, cabbage, and Chinese cabbage.

Spatially uniform CO_2 concentration and VPD can be realized by enhancing the air movement in a plant canopy at a speed of around 0.5 m s^{-1} Spatially uniform rootzone environmental factors are more difficult to achieve than aerial ones but can be realized by enhancing the movement of nutrient solution at a speed of around 0.1 m s^{-1}. Namely, spatial uniformity of environmental factors is always associated with the movement of air and nutrient solution.

Spatially uniform environmental factors in a plant canopy are expected to enhance spatial uniformness of growth, development, and secondary metabolite production in a plant canopy and plant growth. However, little is known on how plants respond to these spatially uniform environment factors in a plant canopy.

3.10 Ideal hydroponic cultivation systems for PFALs

3.10.1 Conventional and ideal hydroponic cultivation systems

Type 1 in Fig. 3.16 shows a scheme of a typical hydroponic cultivation system, which is often referred to as an NFT (nutrient film technique) or DFT (deep flow technique) system (Lu and Shimamura, 2016). In Type 1, the nutrient solution stored in a nutrient solution tank is pumped up and supplied from one end of the cultivation bed and then flows down slowly to the opposite side of the cultivation bed. The nutrient solution drained from the other end of the cultivation bed flows back in a longitudinal direction to the nutrient solution tank through a slightly inclined pipe or a gutter outside the cultivation bed. A sterilization device with a filtering unit is installed in the middle of a return pipe in most cases.

Clean water mixed with stock solution is added to the nutrient solution tank to replace the nutrient solution absorbed by plants. A portion of the nutrient solution is sometimes drained to the outside when there is an imbalance in ion composition and/or an accumulation of organic acids and/or microorganisms. This type of hydroponic cultivation system is characterized by its recycling use of nutrient solution. Refer to Chapter 9 for realistic and practical hydroponic cultivation systems and to Chapter 10 for aquaponics.

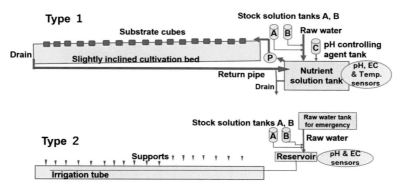

FIGURE 3.16 Hydroponic cultivation systems. Type 1: A conventional system, Type 2: An ideal system.

3.10.2 Design principles of an ideal hydroponic cultivation system

Three important design principles of an ideal hydroponic system are to (1) minimize the overall size, volume, piping/tubing length, and weight of the construction components; (2) minimize the total volume of nutrient solution in a hydroponic system at full production capacity; and (3) supply nutrient solution with optimized composition and strength at an optimum rate (Type 2, Fig. 3.16).

In Type 2, the nutrient solution flows in a one-way (noncirculatory) direction. The cultivation bed is horizontal and without a drain (except for an emergency drainpipe), and the nutrient solution is supplied from small holes along thin irrigation tubes (a few mm in diameter), which is similar to the tubes used in drip irrigation.

This system can significantly reduce the costs of construction components and of a hydroponic cultivation system. The required structural strength of a cultivation rack with multiple cultivation beds containing nutrient solutions can also be reduced considerably.

By installing a highly reliable intelligent controller in Type 2, controllability can be maintained at a high level to optimize the rootzone environment in the cultivation beds. All nutrient element ions are supplied with the water into cultivation beds either intermittently or continuously, and most of nutrient solution ions are absorbed by plants within several hours or so, resulting in nutrient element and water use efficiencies of nearly 100%. Ions that are unnecessary for plants or are minimally absorbed (Cl^-, Na^+, etc.) by plants are not supplied.

To realize a Type 2 hydroponic system, the desirable daily uptake rate of each nutrient element ion by plants must be estimated in advance by using plant growth and environment models. A method to avoid the accumulation of undesirable organic acids emitted by plant roots and sediments of chemical compounds in the cultivation beds needs to be developed.

The ultimate goal of a hydroponic system is to control supply rates, nutrient composition, and nutrient solution strength to meet the exact needs of individual plants based on multiobjective functions of plant cultivation. In this regard, one challenge is to develop such a hydroponic cultivation system with smart software and sensors.

3.10.3 Major characteristics of an ideal hydroponic cultivation system

Major characteristics of an ideal hydroponic cultivation system for PFALs are summarized in Table 3.2.

3.11 Challenges to realize sustainable PFALs

Challenges to realize ecologically, environmentally, economically, and socially sustainable PFALs that are not discussed in this book but would become more and more important as core components in sustainable urban area development include the following:

3.11.1 Breeding of cultivars suited to PFALs

Almost all cultivars (or cultivated varieties) grown in PFALs as of 2021 have been bred for cultivation in open fields or greenhouses. However, the desired genotypes and phenotypes

TABLE 3.2 Major characteristics of an ideal PFAL hydroponic cultivation system.

No.	Factor	Description
1	No circulation of nutrient solution	A return nutrient solution reservoir tank is unnecessary since nutrient solution is delivered via a thin one-way (noncirculating) irrigation tube. Thus, piping to return the drained nutrient solution to its nutrient solution tank is also unnecessary.
2	Recycling of water used in washing or cleaning	Water used for washing/cleaning floors, cultivation panels, etc., and nutrient solution unexpectedly drained from the hydroponic cultivation beds are also recycled after proper filtering and sterilization.
3	Variable nutrient solution supply rates	Nutrients and water are supplied at optimal rates according to the plant species, growth stage, purpose of production, etc., so that no nutrient solution is drained.
4	Variable nutrient solution composition	Nutrient solution composition needs to be adjusted to avoid an imbalance of ions that results in an accumulation of unabsorbed ions in nutrient solution.
5	Zero substrate volume per plant	The use of substrates and time required for handling substrates are none or minimal, which simplifies a cultivation system.
6	Minimum nutrient solution volume per plant	Artificial controllability of nutrient solution is improved, a hydroponic cultivation system is simplified, and initial investment and operating costs are reduced.
7	No propagation and growth of algae and fungi	Propagation and growth of fungi and of unsanitary, unpleasant insects are suppressed due to the absence or low population density of algae in the cultivation room. To maintain such an environment, substrate surfaces, and cultivation beds, floors, and walls must be kept dry.
8	Elimination or control of root Exudates in cultivation beds	Plants exude, from their roots, significant amounts of exudates including allelopathic organic acids such as formic acids, saccharides such as glucose, amino acids such as glycine, and low molecular organics such as citric acid. These exudates often accumulate in cultivation beds and affect plant growth negatively (allelopathy) or positively (feeding beneficial microorganisms around roots) and need to be eliminated or controlled. Root rotting pathogens such as *Pythium* sometimes enter and propagate in the cultivation beds.

(or plant traits) for production in PFALs are different from those for production in open fields and greenhouses. Plant phenotypes suitable for PFALs include the following:

(1) Rapid growth at high CO_2 concentration (1000 µmol mol^{-1} or higher), with low PPFD (100–200 µmol m^{-2} s^{-1}, and in a long photoperiod (20–24 hours a day), without any physiological disorders such as tipburn, intumescence, and necrosis, and producing desired functional chemical components at high concentrations under controlled environments (see Chapter 20 for breeding a new dwarf tomato cultivar for PFAL).

(2) Plant architecture including leaf angle and shape is well adapted to artificial lighting and for easier mechanical or manual harvesting.

(3) Cultivars whose phenotypes such as morphology, color, secondary metabolite production, etc., can be easily manipulated by environmental control, so that multiple products can be derived from one genotype (Folta, 2019).

(4) Cultivars with no resistance to environmental stress and pest insects/viruses because they are grown free from environmental stress and pathogens/pest insects.

(5) Dwarf, compact, and determinate type fruit vegetables such as tomato plants shorter than 0.5 m with short internodes.

(6) Fruiting vegetables with traits of self-pollination or parthenocarpy (fruiting without fertilization of ovules) to avoid the need for artificial pollination by insects, mechanical devices, or humans.

(7) Cultivars suitable for rapid breeding (Watson and Ghosh, 2018).

It should be noted that cultivars with the genetic traits mentioned above can be efficiently bred in PFALs. Furthermore, PFALs can be used as an efficient tool for rapid breeding of indoor, greenhouse, and outdoor plants as well as seed propagation and transplant propagation.

3.11.2 Production of seedlings with uniform phenotype

Production of seedlings with a uniform phenotype is an essential starting point of plant production in a PFAL. Spatial and temporal uniformity of germination time after sowing and unfolding of cotyledonary leaves are the key to uniform seedling production. The seedling or plant phenotype (P) is a function of genotype (G), environment (E), and management (M), or P = f (G, E, M). More specifically, P is a function of seeds' genotype, physiological status of seeds, presowing treatment (priming, coating, etc.) of seeds, chemical, physical, and biological properties of a substrate (or a seed mat), seeds' microenvironment (substrate surface temperature and wetness, contact area of seeds on a substrate, composition and strength of nutrient solution, and SPFD), and human or machine seeding operation (Hayashi et al., 2020). Finding the function P = f (G, E, M) for various types of seeds is a challenge to realize a next-generation PFAL.

3.11.3 Phenotyping

In any commercial PFAL, it is impossible for managers or workers to observe all plants with the naked eye at all times. Plants are elongating their stems, expanding, and moving their leaves even during the dark period. Thus, continuous plant phenotyping is becoming more and more important to determine the optimal environmental factors in a PFAL. Plant phenotypes (or traits) include seed germination (Hayashi et al., 2020), leaf area, leaf angle, number of leaves, plant height, fresh weight, chlorophyll concentration, chlorophyll fluorescence (Moriyuki and Fukuda, 2016), leaf temperature, and chemical composition affecting the quality of produce.

On the other hand, plant phenotypes in a PFAL can be measured relatively easily by using various types of small cameras such as visible, near-infrared, infrared, and/or night-vision cameras with zooming, rotating, mobile, and microscopic functions (Kozai, 2018a). Cost performances of these cameras, related image processing software, and environmental sensors have been improving every year. Time lapse image data on plant phenotypes together with environmental and management data can be used for developing AI and other models.

3.11.4 Power generation using natural energy

Energy autonomy is a key element of sustainable PFALs, because electric energy is indispensable for lighting, air conditioning (mainly cooling), and pumping the nutrient solution. Electricity can be generated using natural energy such as wind power, solar energy, biomass, etc. Besides, the cost performance of batteries for electricity storage has been improving year by year (Uraisami, 2018). Even so, electricity consumption per kg of produce needs to be reduced further.

3.11.5 Life cycle assessment

From the viewpoint of life cycle assessment (LCA), reductions in GHG (greenhouse gas) emission and consumption of N−P−K fertilizer and water per kg of produce are critical to improve the sustainability of PFALs (Kikuchi and Kanematsu, 2019). LCA is a useful tool to quantify environmental impacts and potential impacts based on a product's life cycle from raw material acquisition through production, use, end-of-life treatment, recycling, and final disposal (Kikuchi and Kanematsu, 2019).

Nicholson et al. (2020) compared economic and environmental performances of supply chains for leaf lettuce produced in open fields, greenhouses, and PFALs. The authors included costs for land area, packaging, and transportation in the calculation of production costs and LCA. CO_2 emission in the supply chains was also considered in the LCA analysis.

This type of LCA research in combination with economic analysis of profitability is becoming more and more important for choosing an optimal combination of plant production system, production planning and management, and location or site to produce a particular crop under given social, economic, and climate conditions. For more information on the profitability of PFALs, refer to Chapters 13 and 14.

3.11.6 Construction and operation of PFALs using renewable materials

For PFAL construction, using reinforced, nonflammable wooden, or plant-derived structural elements, which do not emit VOCs, is preferable to help minimize the use of metals, cement, and fossil fuel−derived plastics, which emit CO_2, during production of structural elements and during construction and maintenance of PFALs.

The sustainability of PFALs can also be improved by (1) using plant-derived consumables for cultivation trays, substrates, packaging, etc.; (2) reducing wasted potassium and phosphate fertilizers to preserve their finite natural resources (rock phosphate and potash ore) (Kikuchi and Kanematsu, 2019); (3) reducing wastewater; (4) using no fossil fuels; and (5) minimizing the traffic or flow lines of workers and produce.

3.11.7 Design, construction, and management of human-centered PFALs

As an antidote to the increasing stress in people's lives today due to urbanization and other factors, the practice of 'nature therapy' has demonstrated that an environment rich in plant life can contribute to the relief of physical and mental anxiety (Miyazaki et al., 2011, Chapter 22). Interaction between humans and plants in urban horticulture even in PFALs is believed to contribute to good health and well-being of people (Lu et al., 2020). Therefore, human-centered PFALs based on nature therapy may become an important objective of PFALs for improving quality of stakeholder's life and quality of produce (Lu et al., 2020).

3.11.8 Circadian rhythm

Circadian rhythms (or biological clocks), as characterized by an endogenous period of nearly 24 h, are ubiquitous in almost all living organisms including plants. Use of this characteristic of plants is expected to improve photosynthesis and growth, and thus PFAL productivity (Dood et al., 2005; Higashi et al., 2015). Or, the DNA related to circadian rhythm can be genetically deactivated to achieve production under a 24-hour photoperiod or for lighting at nighttime or irregular time.

3.11.9 Mycorrhizal/symbiotic fungus and biostimulants

Using fungi, including arbuscular mycorrhizae for symbiosis, in a hydroponic cultivation system is a challenge to realize in next-generation PFALs as appropriate use of renewable organic fertilizer is indispensable for sustainable production of plants. Development of a sustainable system to convert organic fertilizer to inorganic fertilizer holds the key as plants basically uptake inorganic mineral ions as nutrient elements (Kozai et al., 2019a). Beneficial fungi can be used to suppress the growth of pathogenic fungi and decompose unfavorable organic matter. Ensuring compatibility of their symbiosis with strict sanitation control is a challenge. See Chapter 9 for a discussion on use of organic fertilizers in hydroponics. Root exudates such as organic acids, polysaccharides, and low molecular substances accumulated in cultivation beds are known to affect plant growth, but little is known about their role in the ecosystem of a cultivation bed.

3.11.10 Seed grain production of staple crops

Interest in production of staple crops in large-scale PFALs has been growing since around 2020 by some large private companies, national institutes, and universities. Foods from staple crops form the basis of a traditional diet containing relatively high percentage starch and some protein, serving as a major source of energy and nutrients. Staple crops include cereal crops such as maize, wheat, rice, and potatoes.

Until recently, commercial production of staple crops in a PFAL did not attract attention except for basic research on space farming (Wheeler et al. (1993); Wheeler (2006)), because (1) produce can often be stored for 1 year or so; (2) physical and physiological damage due to long distance bulk transportation by ship is limited; (3) international trade is common to balance the supply and demand; and (4) economic value per kg of staple foods is much lower compared with that of specialty or functional foods.

However, yields of those crops in open fields have been vulnerable recently to climate change causing droughts, floods, and high/low temperatures as well as a plague of locusts in Africa and Asia. A shortage of water for irrigation, soil deterioration, decreases in arable land area and/or agricultural workers are also serious in some regions. In addition, international trade has been declining in 2020 due to the Covid-19 pandemic and international political and economic frictions. Besides, some oil-producing countries are interested in commercial production of staple crops in PFALs with solar panels for power generation as the next emerging industry. Thus, ensuring staple foods under the above situations is becoming a political and economic issue in regard to food security in many countries.

It might be practical, in the near future, to produce a portion of high quality seed grains of rice, maize and wheat, and microtubers of potatoes for security reasons in PFALs using electricity generated by natural energy.

3.11.11 Laser lamps as a substitute for LEDs

A solid state laser is widely used in such products as laser pointers used for presentations and barcode scanners. Laser is an acronym for "light amplification by stimulated emission of radiation." A laser differs from other light sources in that it emits coherent light, which can (1) be focused to a tight spot (high intensity); (2) the beam remains narrow over great distances (directionality); (3) emit light with a very narrow spectrum (monochromatic); and (4) produce pulses of light with durations as short as a femtosecond (10^{-15} s). Little is known on how plants respond to the light emitted from a laser lamp, although research on plant growth under laser light has been conducted to some extent (Yamazaki et al., 2002; Murase, 2015; Ooi et al., 2016). No commercial PFALs that use laser lamps are present as of 2021. An advantage of lasers over LED lamps is their higher conversion efficiency from electric to light energy (65%–75% at wavelength of 900–999 nm) (Nogawa et al., 2020).

3.11.12 Development of a cultivation system module and its application

The concept, methodology, design principles, and potential application of a CSM are described in Kozai (2018). A CSM becomes smarter by implementing ideas described in this book. However, this scalable and evolutionary CSM with e-learning software for sustainable production of plants is not actualized yet. Developing a standard and/or open-source CSM will accelerate the achievement of the SDGs and the realization of sustainable societies.

3.12 Conclusion

With a history of only about 20 years, the technological and business development of PFALs with LED lamps is still at an early stage. Advanced technologies such as AI and IoT have just been recently introduced to commercial PFALs. As of 2021, no commercial PFALs using 5G, VR, AR, and/or MR (mixed reality) seem to come into being. Such advanced technologies are most efficiently used in well-designed PFAL hardware, software, and firmware units. It is time for all stakeholders to develop sustainable PFALs to contribute to solving issues concerning food, the environment, resources, and quality of life. In this chapter, basic concepts and methodologies were discussed for designing and managing sustainable PFALs for the next generations.

Acknowledgment

The author would like to thank Mr. Masaaki Tamura, Japan Plant Factory Association, for his valuable comments on the original manuscript.

References

Dood, A.N., Salathia, N., Halt, A., Kevel, E., Toth, R., Nagy, F., Hibberd, J.M., Millar, A.J., Webb, A.R., 2005. Plant circadian clocks increase photosynthesis, growth, survival and competitive advantage. Science 309, 630–633. https://doi.org/10.1126/science.1115581.

Folta, K.M., 2019. Breeding new varieties for controlled environments. Plant Biol. 21 (Suppl. 1), 6–12.

Hayashi, E., Amagai, Y., Maruo, T., Kozai, T., 2020. Phenotypic analysis of germination time of individual seeds affected by microenvironment and management factors for cohort research in plant factory. Agronomy 10, 1680. https://doi.org/10.3390/agronomy10111680.

Higashi, T., Nishikawa, S., Okamura, N., Fukuda, H., 2015. Evaluation of growth under non-24 h period lighting conditions in *Lactuca sativa* L. Environ. Control Biol. 53 (1), 7–12.

Joshi, J., Zhang, G., Shen, S., Supaibulwatana, K., Watanabe, C., Yamori, W., 2017. A combination of downward lighting and supplemental upward lighting improves plant growth in a closed plant factory with artificial lighting. Hortscience 52 (6), 831–835. https://doi.org/10.21273/HORTSCI11822-17.

JPFIA (Japan Plant Factory Industries Association), 2020, p. 9 (in Japanese). https://jpfia.org/en/. https://jpfia.org/guideline/#3. Guidelines for health and safety.

Kikuchi, Y., Kanematsu, Y., 2019. Life cycle assessment. In: Kozai, et al. (Eds.), Plant Factory: An Indoor Vertical Farming System for Efficient Quality Food Production, second ed. Academic Press, pp. 383–395.

Kitaya, Y., 2016. Air current around single leaves and plant canopies and its effect on transpiration, photosynthesis, and plant organ temperatures. In: Kozai, T., et al. (Eds.), LED Lighting for Urban Agriculture. Springer, pp. 177–187.

Kitaya, Y., Shibuya, T., Kozai, T., Kubota, C., 1998. Effects of light intensity and air velocity on air temperature, water vapor pressure, and CO_2 concentration inside a plant canopy under artificial lighting conditions. Life Supp. Bios. Sci. 5, 199–203.

Kitaya, Y., Shibuya, T., Yoshida, M., Kiyota, M., 2004. Effects of air velocity on photosynthesis of plant canopies under elevated CO_2 levels in a plant culture system. Adv. Space Res. 34, 1466–1469.

Kozai, T., 2013. Resource use efficiency of closed plant production system with artificial light: concept, estimation and application to plant factory. Proc. Japan Acad., Ser. B 89 (10), 447–461.

Kozai, T. (Ed.), 2018a. *Smart Plant Factory: The Next Generation Indoor Vertical Farms*. Springer, p. 456. ISBA 978-98-13-1064-5.

Kozai, T., 2018b. Benefits, problems and challenges of plant factories with artificial lighting (PFALs): a short review. Acta Hortic. 25–30 (GreenSys 2017, Beijing).

Kozai, T., 2019a. Towards sustainable plant factories with artificial lighting (PFALs) for achieving SDGs. Int. J. Agric. Biol. Eng. 12 (5), 28–37.

Kozai, T., Fujiwara, K., Runkle, E. (Eds.), 2016. LED Lighting for Urban Agriculture. Springer, p. 454. https://doi.org/10.1007/978-981-10-1848-0_1.

Kozai, T., Amagai, Y., Hayashi, E., 2019b. Towards sustainable plant factories with artificial lighting (PFALs). In: Marcelis, L., Heuvelink, E. (Eds.), Achieving Sustainable Greenhouse Cultivation. Burleigh Dodds Science Publishing Ltd, pp. 177–202.

Kozai, T., Niu, G., Takagaki, M. (Eds.), 2020a. Plant Factory: An Indoor Vertical Farming System for Efficient Quality Food Production, second ed. Academic Press, p. 487. ISBN: 978-0-12-816692-8.

Kozai, T., Hayashi, E., Amagai, Y., 2020b. Plant factories with artificial lighting (PFAL) towards sustainable plant Production. Acta Horticult. 1273, 251–259 (IHC 2018, Istanbul, Turkey).

Lu, N., Shimamura, S., 2016. Protocols, issues and potential improvements of current cultivation systems (Chapter 3). In: Kozai, T. (Ed.), LED Lighting for Urban Agriculture. Springer Nature Singapore Pte Ltd., pp. 31–49

Lu, N., Song, T., Kuronuma, H., Ikei, Y., Miyazaki, Y., Takagaki, M., 2020. The Possibility of sustainable urban horticulture based on nature therapy. Sustainability 12, 5058. https://doi.org/10.3390/su12125058.

Martellozzo, F., Landry, J.-S., Plouffe, D., Seufert, V., Rowhani, P., Ramankutty, N., 2014. Urban agriculture: a global analysis of the space constraint to meet urban vegetable demand. Environ. Res. Lett. 9, 064025 (8 pages).

Miyazaki, Y., Park, B.J., Lee, J., 2011. Nature therapy. In: Osaki, M., Braimoh, A., Nakagami, K. (Eds.), Designing Our Future: Local Perspectives on Bioproduction, Ecosystems and Humanity. United Nations University Press, New York, pp. 407—412.

Moriyuki, S., Fukuda, H., 2016. High-throughput growth prediction for *Lactuca sativa* L. Front. Plant Sci. 7, 394. https://doi.org/10.3389/fpls.2016.00394.

Murase, H., 2015. The latest development of laser application research in plant factory. Agric. Agric. Scie. Proc. 3, 4—8.

Nicholson, C.F., Harbick, K., Gomez, M.I., Mattson, N.S., 2020. An economic and environmental comparison of conventional and controlled environment agriculture (CEA) supply chains for leaf lettuce to US cities. In: Aktas, E., Bourlakis, M. (Eds.), Food Supply Chains in Cities, pp. 33—68. https://doi.org/10.1007/978-3-030-3406-5-0_2.

Nishio, J.N., 2000. Why are higher plants green? Evolution of the higher plant photosynthetic pigment complement. Plant Cell Environ. 23, 539—548.

Nogawa, R., Yoshitaka, K., Kaifuchi, Y., Kawakami, T., Yamagata, Y., Yamagichi, M., 2020. High frequency and high power broad area laser diodes. Fujikura Tech. Rev. 133, 22—25. https://www.fujikura.co.jp/rd/gihou/backnumber/pages/icsFiles/afieldfile/2020/06/25/133_R6.pdf.

Oikawa, T., 1977. Light regime in relation to plant population geometry III. Ecological implications of a square-planted population from the viewpoint of utilization efficiency of solar energy. Bot. Mag. Tokyo 90, 301—311.

Ooi, A., Wong, A., Khee Ng, T., Marondedze, C., Gehring, G., Ooi, B.S., 2016. Growth and development of *Arabidopsis thaliana* under single wavelength red and blue laser light. Sci. Rep. 6, 33885. https://doi.org/10.1038/srep33885.

Saengtharatip, S., Goto, N., Kozai, T., Yamori, W., 2020. Green light penetrates inside crisp head lettuce leading to chlorophyll and ascorbic acid content enhancement. Acta Hortic. 1273, 261—270. https://doi.org/10.17660/ActaHortic.2020.1273.35.

Saito, K., Ishigami, Y., Goto, E., 2020. Evaluation of the light environment of a plant factory with artificial light using an optical simulation. Agronomy 10, 1663. https://doi.org/10.3390/agronomy101111663.

Seth, N., Barrado, C.D., Lalaguna, P.D., 2019. Sustainable Development Goals — Main Contributions and Challenges. United Nations Institute for Training and Research (UNITAR), p. 196.

Shibuya, T., Tsuruyama, J., Kitaya, Y., Kiyota, M., 2006. Enhancement of photosynthesis and growth of tomato seedlings by forced ventilation within the canopy. Sci. Hortic. 109, 218—222.

Smith, H.L., McAusland, L., Murchie, E.H., 2017. Don't ignore he green light: exploring diverse roles in plant processes. J. Exp. Bot. 68 (9), 2009—2110.

Sun, J., Nishio, J.N., Vogelmann, T.C., 1998. Greenlight drives CO2 fixation deep withing leaves. Plant Cell Physiol. 39 (10), 1020—1026.

Taiz, L., Zeiger, E. (Eds.), 2006. Plant Physiology, fourth ed. Sinauer Associates, Inc., p. 764

Takagaki, M., 2020. Micro- and mini-PFALs for improving the quality of life in urban areas (Chapter 6. In: Kozai, T., Niu, G., Takagaki, M. (Eds.), Plant Factory: An Indoor Vertical Farming System for Efficient Quality Food Production. Academic Press, London, pp. 117—128.

Terashima, I., Saeki, T., 1985. A new model for leaf photosynthesis incorporating the gradient of light environment and of photosynthetic properties of chloroplasts within a leaf. Ann. Bot. 56, 489—499.

Uraisami, K., 2018. Renewable energy makes plant factory "smart". In: Kozai, T. (Ed.), Smart Plant Factory: The Next Generation Indoor Vertical Farms). Springer, pp. 119—123.

Watson, A., Ghosh, S., 2018. Speed breeding is a powerful tool to accelerate crop research and breeding. Nature Plants 4, 23—27.

Wheeler, R.M., 2006. Potato and human exploration of space: some observations from NASA-sponsored controlled environment studies. Potato Res. 49, 67—90, 0.1007/s11540-006-9003-4.

Wheeler, R.M., Kenneth, A.C., Sager, J.C., William, M.K., 1993. Gas exchange characteristics of wheat stands grown in a closed controlled environment. Crop Sci. 33, 161—168.

Yamazaki, A., Tsuchiya, H., Miyazima, H., Honma, H., Kan, H., 2002. In: Dorias, M. (Ed.), (Proc. 4th IS on Artificial Light). Acta Hort 580, 177—181.

Zhang, G., Shen, S., Takagaki, M., Kozai, T., Yamori, W., 2015. Supplemental upward lighting from underneath to obtain higher marketable lettuce (*Lactuca sativa*) leaf fresh weight by retarding senescence of outer leaves. Front. Plant Sci. 16, 1—9.

Contribution of PFALs to the sustainable development goals and beyond

Toyoki Kozai

Japan Plant Factory Association, Kashiwa, Chiba, Japan

4.1 Introduction

In 2015, more than 190 world leaders committed to achieving 17 sustainable development goals (SDGs) with 169 targets by the year 2030 as a universal call to action to end poverty, protect the planet, and ensure that all people can enjoy peace and prosperity (Sharma and Sobti, 2018; Seth and Suazo, 2019). In short, the 17 SDGs are aiming for: (1) No poverty, (2) No hunger, (3) Good health and well-being, (4) Quality education, (5) Gender equality, (6) Clean water and sanitation, (7) Affordable and clean energy, (8) Decent work and economic growth, (9) Industry, innovation, and infrastructure, (10) Reduced inequalities, (11) Sustainable cities and communities, (12) Responsible consumption and production, (13) Climate action, (14) Life below water, (15) Life on land, (16) Peace, justice, and strong institutions, and (17) Partnerships for the goals.

Under SDG 4 (Quality education), the United Nations Educational, Scientific and Cultural Organization (UNESCO) is responsible for improving access to quality education for sustainable development (ESD) at all levels and in all social contexts, to transform society by reorienting education and helping people develop knowledge, skills, values, and behaviors needed for sustainable development.

In the investment industry and business operations, the SDGs are discussed in relation to environmental, social, and corporate governance (ESG) criteria. These criteria are a set of standards in a company's operations that socially conscious investors use to consider potential investments. Environmental criteria consider how a company performs as a steward of nature. Social criteria examine how a company manages relationships with employees, suppliers, customers, and the communities where it operates. Governance deals with a

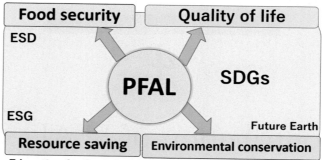

FIGURE 4.1 The PFAL can contribute to achieving the SDGs by solving the four global and local issues in an environment characterized by (1) a declining agricultural population with aging farmers, (2) growing urban populations, (3) increasing demands for higher quality food and lifestyle, and (4) changing global and local climates. ESD and ESG denote, respectively, education for sustainable development, and environment, society, and corporate governance.

company's leadership, executive remuneration, audits, internal controls, and shareholder rights (Chen, 2020). The outbreak of COVID-19, a pandemic which began in early 2020, will continue to affect all aspects of the SDGs, ESD, and ESG, and poses a tremendous challenge for all countries in finding workable solutions (OECD, 2020).

This chapter focuses on how the plant factory with artificial lighting (PFAL) can contribute to achieving the SDGs and related action programs and why the PFAL can be one appropriate and efficient means for achieving the SDGs in today's environment characterized by (1) a declining agricultural population with aging farmers, (2) growing urban populations, (3) increasing demands for higher quality food and lifestyle, and (4) changing global and local climates (Fig. 4.1).

Prior to commencing the discussion of the potential contribution of PFAL to the SDGs, the chapter will present the structure, characteristics, and necessity of the PFAL in Sections 4.2 and 4.3. Therefore, readers already familiar with PFALs can skip these sections. To maintain the independence of this chapter, the description of PFALs in Sections 4.2 and 4.3 partially overlaps with that in Chapters 2 and 8.

4.2 What is a PFAL?

4.2.1 Main components and environmental control factors

A PFAL is a closed plant production system, and its cultivation room consists of a thermally well-insulated, fairly airtight, and optically opaque warehouse-like structure, multitier cultivation units with light emitting diode (LED) lighting and hydroponic cultivation units, air conditioners with circulation fans, a CO_2 supply unit, a nutrient solution supply unit, and an environmental control unit (Fig. 4.2; see Chapter 8 for more details).

Environmental control factors include air temperature, photosynthetic photon flux density (PPFD), CO_2 concentration of the room air, pH, and total ion concentration of the nutrient

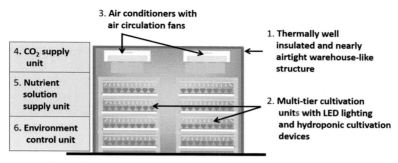

FIGURE 4.2 Six main components of the PFAL cultivation room.

solution. The air current speed, vapor pressure deficit (VPD) or relative humidity (RH) of the room air, and temperature and dissolved O_2 concentration of the nutrient solution as well as the flow speed of the nutrient solution are indirectly controlled within their upper and lower limits.

Since the cultivation room is highly airtight and thermally insulated, all the resource inputs and outputs to and from there can be relatively easily measured and controlled automatically. The resource input and output are not directly influenced by the weather, pest insects, and small animals outside, so the environment in the PFAL can be controlled as desired, and the use efficiency of each resource element can be estimated online based on the measured values of corresponding resource input and output (see Chapter 8 for more details). Thus, the use efficiency of each resource element and the overall resource use efficiency or productivity (or the ratio of yield to cost) can be continuously improved based on the data analyses of the measured inputs and outputs (Kozai et al., 2019a).

4.2.2 Why are airtightness and thermal insulation required?

High airtightness of the cultivation room is required to prevent (1) the invasion of viruses, microorganisms, insects with or without pathogens, small animals, dust, or the like through air gaps, and (2) the loss of CO_2 and water vapor gases through air gaps to the outside. This airtightness also enhances the recycling use of CO_2 and water in the cultivation room, resulting in high CO_2 and water use efficiencies.

Both high airtightness and thermal insulation are required to prevent (1) the impact of weather on the aerial environment of the cultivation room and (2) water condensation on the inner surfaces of walls, roof, and floor (fungi, algae, and some insects propagate and/or grow quickly on wet surfaces), and to ensure that (3) the cooling load for air conditioners or heat pumps is determined only by electricity consumption of lamps, pumps, fans, etc., regardless of weather.

4.2.3 Heating and cooling loads and electricity consumption of equipment

No heating is required even on cold winter nights when more than half of the lamps are turned on, because the heat generated by the lamps is more than adequate for keeping the air temperature of the thermally well-insulated, airtight cultivation room at around 25°C even during outside temperatures as low as −40°C.

On the other hand, the cooling of the cultivation room using air conditioners is required throughout the day, and all year round when most of the lamps are turned on. Electricity consumption for cooling the thermally well-insulated, airtight cultivation room accounts for around 20% of all electricity consumption (lighting, cooling, pumping, etc.) in warmer regions and about 15% in cooler regions (Chapter 8; Kozai, 2013; Kozai et al., 2019a). It is also important to notice that no extra electricity is required to dehumidify the room air or to keep the VPD at around 0.5 kPa or RH of 75% at 25°C if the lamps are turned on. The room air is naturally dehumidified when cooled by air conditioners to keep the room temperature at a set point.

4.2.4 Reduction in maximum electricity consumption by alternating lighting

About 60%−70% of LEDs are alternately turned on all day in the cultivation room in most PFALs. In this case, air conditioners operate all day, even on cold winter nights, to remove heat energy generated by LEDs (Chapter 8). Under such conditions, the electricity consumption for cooling and dehumidification of the room air accounts for 15%−20% of all electricity consumption in the cultivation room, while lighting accounts for 75%−80%, and pumping, air circulation, etc., for about 5% on average annually in regions with a temperate climate.

By alternating the lighting as mentioned above, the electricity consumption for lighting and cooling remains relatively consistent throughout the day. Thus, their maximum values are reduced by about 30% compared to those for all LEDs turned on for $16 \, h \, d^{-1}$ and off for $8 \, h \, d^{-1}$ at the same time, although the total amount of daily electricity consumption for lighting and cooling is the same as those for all LEDs turned on for $16 \, h \, d^{-1}$ and off for $8 \, h \, d^{-1}$ at the same time.

With the same daily cooling load, the daily cost of electricity for cooling is lower at night than in daytime due to higher coefficient of performance (COP) of air conditioners at lower outdoor air temperatures at night. Then, the cooling capacity of air conditioners and the capacity of power supply units can also be reduced by about 30%, resulting in a reduction in the initial cost for air conditioners and power supply units.

Besides, VPD (or RH) is kept relatively constant all day due to continuous lighting of two-thirds of LEDs and continuous cooling as well as dehumidification all day. Under highly airtight, thermally insulated, clean cultivation room conditions, about 95% of water vapor transpired from leaves in the cultivation room is condensed and collected at cooling panels of air conditioners, and then returned to the nutrient solution tank (Kozai, 2013; Kozai et al., 2019a).

4.2.5 Trends in electricity costs

The total electricity costs for the cultivation room with LEDs account for about 20% of the total production costs (with depreciation costs of about 30%, labor costs about 20%, logistics costs about 5%, and costs for other consumables such as fertilizer, water, seeds, CO_2, etc., of about 25%) in Japan where the cost of electricity is relatively high compared to electricity costs in other countries (Ijichi, 2018). When fluorescent lamps were used as the light source prior to the introduction of LEDs, lighting accounted for about 30% of costs.

It should be noted that labor costs, the electricity cost per kWh, and the reliability of the electricity supply vary according to factors such as time, country, region, and season. It must also be noted that the cost of solar panels, wind power generation, and batteries have recently been decreasing relatively rapidly (Kozai, 2018a). In addition, electricity consumption per kg of fresh produce has been decreasing, although around 10 kWh of electricity was required to produce 1 kg of fresh leafy vegetables such as leaf lettuce until 2018.

4.2.6 Mini- and medium-sized PFALs

Fig. 4.3 shows the exterior view of two mini-PFALs without a sanitation room for changing clothes, shoes, etc. As shown in the photo, mini-PFALs look like small windowless warehouses. The external placement of the air conditioning unit (lower left) may have to be changed depending on the climate where the PFAL is located and its purpose of use. The unit can also be placed on the roof or at a height of around 1.5 m rather than on the ground. The air emitted from the outer unit (i.e., condenser) is around 40°C during cooling operation, so the warm air can be used for drying/warming anything by placing it with caution around 1 m from the outer unit.

A mini-PFAL can be placed either inside or outside a building. When a mini-PFAL is placed outdoors under sunlight, a hood or a shade over it may be necessary to avoid heat from strong solar radiation, strong wind/sandstorm, and/or heavy rain/snowfall. When the PFAL is used to produce pesticide- and insect-free transplants, it is covered with an insect screen or placed in an insect-screen net house. In this case, the sanitary room described below is not necessarily required.

A small sanitary room or front room (e.g., 1.5 × 1.5 × 2.5 m) for handwashing and changing from everyday clothing to disinfected or clean clothing needs to be attached to the cultivation room of the mini-PFAL when it is used for the production of vegetables to be served fresh without washing. In this case, the produce is packed in clean bags and then sealed in a clean space such as the cultivation room. If possible, the bags are placed in a heat-insulating box before taking it out from the sanitary room.

Fig. 4.4 shows an exterior view of a medium-sized PFAL with 10 tiers and a daily production capacity of 250 kg (leaf lettuce). The total floor area is about 400 m² and the floor area of the cultivation room is about 300 m². In addition to the changing room area (3 m by 3 m) in

1.8 m wide, 1.8 m deep, 2.4 m high   3 m wide, 2 m deep, 3 m high, 4 tiers

FIGURE 4.3   Exterior view of mini-PFALs for personal or family use.

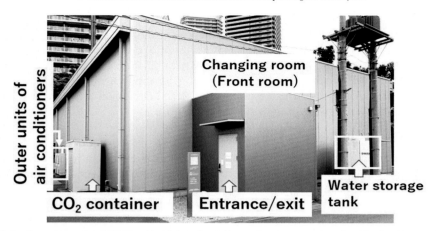

FIGURE 4.4 A medium-sized PFAL with 10 tiers for daily production capacity of 250 kg (leaf lettuce). Total floor area: 404 m^2, Floor area of the cultivation room: 338 m^2. In addition to the changing room area (3 m by 3 m) in the photo, there is another changing room on the opposite side of the PFAL. The annual production per total floor area is (216 kg m^{-2} y^{-1} (= 250 kg d^{-1} x 350 d y^{-1})/404 m^2). *The photo was taken at the Kashiwa-no-ha campus of Chiba University in 2020.*

the photo, there is another changing room on the opposite side of the PFAL. The annual production per total floor area is 216 kg m^{-2} y^{-1} (= 250 kg d^{-1} x 350 d y^{-1})/404 m^2.

4.2.7 Hygiene management of hydroponic cultivation system

Population density of microorganisms is often more than 1000 times higher in the cultivation beds than in the room air because the microorganisms are fed by dead and alive roots, algae, and other organic substances in the cultivation beds (algae grow photosynthetically only under the presence of light, water, and nutrients). Thus, falling down of nutrient solution droplets from the roots to the plant leaves needs to be strictly avoided to keep them clean. Microorganism colony-forming units (CFUs) are generally controlled below 300 per gram of produce fresh weight. No pesticide, insecticide, and bactericide are used during cultivation, so harvests can be used as fresh salad without washing before serving.

The nutrient solution is circulated to minimize amount of drainage in most hydroponic systems. Although sterilization and/or filtering units are placed in the middle of the nutrient solution recirculation pipe, the population density in the cultivation beds is not decreased significantly because a significant portion of the microorganisms and algae tend to stay on the inner surface of the cultivation beds and the roots of plants.

Hygiene management pertains to controlling population densities of unsanitary and/or unpleasant insects, algae, small animals, etc. The invasion of small flying insects, dust, or other foreign substances can be efficiently blocked by keeping the room air pressure slightly (3-5 Pa) higher than outdoor air pressure.

In most hydroponic cultivation systems, inorganic fertilizers only are used to minimize the population density of microorganisms (organic fertilizer is used after its decomposition to inorganic fertilizer through the use of microorganisms in a separate unit). It is desirable to measure and control each of major ion concentrations, although the measurements of PO_4, NH_4^+, etc., using chemical sensors are difficult at present.

4.3 Why is the PFAL necessary?

4.3.1 Use of solar radiation is not free of charge!

A greenhouse is often installed with environmental control units to improve the annual yield and quality of produce and the environment of workers. To use 'free' solar energy, extra initial costs are required for shading screen, thermal screen, insect screen, natural or forced ventilators, units for heating/evaporative-cooling, supplemental lighting, etc. Consumption goods necessary in the greenhouse but unnecessary in the PFAL include oil or natural gas for heating, pesticide/herbicide, etc.

The main reason for these extra initial operation costs aimed at taking advantage of the 'free' solar energy is that the solar radiation (waveband: about 350–2500 nm at sea level) and outside air temperature are often too high or too low depending on the time of day, season, weather, and geographical location. The ever-changing solar radiation and outside air temperature make it difficult to control the greenhouse environment optimally, resulting in variations in the yield and quality of produce.

In addition, only about 50% of solar radiation is photosynthetically active (wavelength: 400–700 nm). Radiation with wavelengths of 700–800 nm is physiologically active (has photomorphological effects) but is photosynthetically inactive, and with 800–2500 nm simply increases the air and leaf temperatures inside the greenhouse.

Therefore, it is reasonable to assume that the cost performance of the PFAL can be higher than that of the greenhouse for selected plant species for a particular purpose of production. This hypothesis has recently been proved for leafy vegetables by researchers and business people in many countries. However, as of 2021, commercial production of fruit vegetables, root vegetables, head vegetables, and medicinal plants is profitable only in a limited number of PFALs. In view of these conditions, there have been efforts to step up PFAL research, development, and business to achieve innovations in PFAL technology considering that: (1) recent advances in smart technology are remarkable, and (2) costs of batteries and electricity generated by natural energy such as solar radiation and wind power have been decreasing.

4.3.2 Comparisons of yield, cost, and cost performance

In general, both the average yield (production per land area) and the initial/operation costs per land area of leafy vegetables in the PFAL are more than 100-fold those of the open field and more than 10-fold those of the greenhouse. Thus, the monetary productivities or cost performances (annual sales divided by annual production costs including depreciation costs) of the PFAL, the greenhouse, and the open field are more or less the same if their unit sales prices per kg of produce are the same (Kozai et al., 2019a). The ratios of annual sales to the initial investment on the PFAL, the greenhouse, and the open field are also more or less the same.

4.3.3 Factors affecting cost performance

In reality, the yield at a given location is considerably affected by the skills of the production manager of the PFAL, the greenhouse, and the open field. The yields in the open field and the greenhouse are also affected by the weather and climate, while the yield in the PFAL is not affected by these factors at all. The high yield in the PFAL is mainly due to: (1) year-round production (longer total cultivation periods per year); (2) use of multitiers; (3) shortened cultivation periods or enhanced growth due to an optimal environment throughout the cultivation period; and (4) no reduction in the yield due to pathogenic diseases, insects, or disastrous weather events such as typhoons (Kozai, 2019; Kozai et al., 2019b). The sales price per kg of PFAL-grown, pesticide-free, and high-quality produce is often 10%—50% higher than that of greenhouse- and open-field-grown produce.

4.3.4 Relative resource consumption per kg of fresh produce

Table 4.1 shows roughly estimated relative resource consumption of components per kg of produce of the PFAL compared to consumption by the greenhouse and the open field (Kozai, 2018b, 2019).

The total production cost per kg of produce, C_T, is expressed by the total sum of a product of the consumed resource component 'i' for 1 kg of produce, Q_i, and the unit cost of resource component 'i', C_i. Namely, $C_T = \Sigma(Q_i \times C_i)$ (i = 1—10). The productivity or the cost performance is expressed by Y/C_T where Y is the yield (see Chapters 12 and 13). The resource consumption of the respective components per kg of produce needs to be estimated for the particular product, location, season, etc., when this method is applied.

TABLE 4.1   Rough estimation of the relative resource consumption of components per kg of produce of a PFAL compared to consumption by a heated greenhouse with soil cultivation and an open field with irrigation and agricultural machinery. The values relative to the PFAL are estimated based on the author's knowledge, experience, and presumptions. For the relative values of fossil fuel and pesticides/herbicides, the reference value of 1 is assumed for the open field.

I	Resource component	PFAL	Greenhouse	Open field
1	Land area	1	10	100
2	Water	1	20	30
3	Fertilizer	1	2	2
4	Working hours	1	2	1
5	Electricity	1	0.01	0
6	CO_2	1	0	0
7	Fossil fuel	0	100	1
8	Seeds	1	1.2	1.4
9	Pesticide/herbicide	0	1	1
10	Others	1	1	0.5

(1) **Land area:** Annual yield (production per land area) of the PFAL with 10 tiers is roughly 10-fold and 100-fold that of the greenhouse and the open field, respectively, in a temperate climate area (Kozai et al., 2019a). In other word, land area required for production of 1 ton of plants by the PFAL is about 1/100th compared to open fields, and 1/10th compared to greenhouses. Thus, we need '1 ha of PFAL' only instead of 100 ha of open field for vegetable production.

(2) **Water for irrigation:** About 95% of transpired water vapor from leaves is recycled for irrigation in the PFAL (Chapter 8). This reduction in water consumption is a significant advantage of the PFAL over the greenhouse and open field where the available water is scarce and/or the water quality is not suitable for irrigation. Furthermore, a significant portion of irrigated water is evaporated from the soil surface and/or drained from the soil of greenhouse and the open field, while the amount of such water loss in the PFAL can be minimized. An extra amount of water is required in a greenhouse with an evaporative cooling unit.

(3) **Water for washing/cleaning:** Water consumption for washing the PFAL-grown produce before cooking or service is less than 1/10th compared to greenhouse- and open-field-grown vegetables because PFAL-grown vegetables are clean (free from pesticide, insects, and other foreign substances), and the population density of microorganisms on leaves is 1/1000th to 1/100th of field-grown vegetables.

(4) **Fertilizer:** With use of hydroponic cultivation system in the PFAL, application amount of P/N/K fertilizer per kg of produce is reduced by more than half compared to open field. In the greenhouse and the open field, both organic and inorganic fertilizers can be used. On the other hand, around 50% of fertilizer is often leached/drained away and/or adsorbed to the soil in the greenhouse and the open field.

(5) **Working hours:** The distance traveled on foot and by vehicle per kg of produce of workers is much shorter with the PFAL than with the greenhouse and the open field. Working hours per kg of produce are assumed to be shorter in the open field than in the greenhouse since vehicles are used for the most part in various operations in the open field.

(6) **Electricity:** Electricity is mostly consumed for lighting and air conditioning in the PFAL. In the greenhouse, electricity is used mainly for ventilation, pumping, heating/cooling, and battery-driven vehicles. All machines/vehicles are driven using electricity only both in the PFAL and greenhouse. Battery-driven agricultural machines are used in some cases in the open field.

(7) **CO_2:** In most cases, pure liquid CO_2 contained in a high-pressure cylinder-type container (30—70 kg of CO_2; 0.5—0.8 US$ per kg in Japan) is supplied during the photoperiod in the PFAL to keep its concentration at around 1000 μmol mol^{-1} to promote photosynthesis and thus the growth of plants. Only about 10% of CO_2 is released to the outside in the fairly airtight PFAL, while about 50%—60% of CO_2 is released to the outside through air gaps of a CO_2-enriched greenhouse with all windows closed. CO_2 is free of charge in the open field but its concentration is always around 400 μmol mol^{-1}. CO_2 is a by-product obtained from a chemical plant using fossil fuel. CO_2 is also a by-product of the respiration of living organisms, so CO_2 produced by fungus/microorganisms, fish, animal/livestock, humans, etc., can be used as a CO_2 source for plant photosynthesis if it is collected efficiently.

(8) Fossil fuels (natural gas, gasoline, and heavy oil): No heating is required in the PFAL even on cold winter nights, while fossil fuels are used mostly for heating in the greenhouse on cold days, and for the operation of agricultural machinery in the open field.

(9) Seeds: The ratio of seeds grown to marketable plants is 10%–20% higher in the PFAL than in the greenhouse and open field due to the higher germination ratio (over 98%), higher growth uniformity, and no damage to plants by insects in the PFAL.

(10) Space use efficiency: The ratio of the total cultivation bed area to floor area (m^2 m^{-2}) or ratio of total cultivation space (m^3) to cultivation room space (m^3) (m^3 m^{-3}) can be maximized as far as there are no known negative effects on plant growth and workability, and to minimize the area of empty cultivation beds to maximize PFAL productivity

(11) Food mileage and food loss: Food mileage is minimum when the food is produced at consumption sites (local production for local consumption), and thus simple and lightweight package materials can be used. Long-distance vehicle traffic causing fuel consumption, CO_2 emission, air pollution, and damage of highways is reduced. Food loss during transportation is minimum when food mileage is minimum, especially when plants are produced in a clean and controlled environment in the PFAL. Global loss and waste of field-grown fruit and vegetables are, respectively, 13% in calories and 44% in fresh weight (Lipinski et al., 2013).

4.3.5 Obstacles and risks

Obstacles to popularize PFALs for SDGs, as of 2021, are lack of human resource development program and friendly smartphone-based software for e-learning and online training systems. One hundred-hour on-site or online interactive course with practical and classroom training is desirable to train beginners, and 200-h course to train skilled PFAL managers and workers.

Skills required for PFAL management are considerably different from those for plant production in the greenhouse and open fields. The PFAL manager requires the knowledge and experience of the plant production under sanitary condition, and of environmental control and its effects on plant growth and quality of produce with respect to CO_2 concentration, air current speed, light quality, PPFD, nutrient solution for hydroponic cultivation, etc., because environments, resource consumption, and plant growth are monitored and partially controlled using a smartphone. Also, the PFAL manager requires skills for managing various risks, personnel, production, resource, and logistics, with use of support software.

A fail-safe and alarm system is a must for power breakdown; a fire; malfunction of equipment for water supply, CO_2, etc.; spread of insects and disease; and a theft. A battery backup system is necessary for a water pump to provide a minimum amount of water irrigation during power failure.

The aforementioned software needs to be downloadable via the Internet to a user's smartphone as an application software with free of charge or a minimum charge, preferably supported by a public international organization. Special interest groups organized by PFAL users and their supporters would be helpful to skill up each other.

4.4 Sustainable development goals

4.4.1 Goal 1: end poverty in all its forms everywhere

A PFAL with a hydroponic cultivation system can be built and used efficiently anywhere in the world—in hot, cold, dry, and wet regions with nonfertile soils—to produce high-quality functional plants such as leaf lettuce plants and herbs such as basil all year round, as well as for producing high-quality transplants (seedlings, micropropagated plantlets, and grafted transplants) regardless of weather, using minimum land/floor area and other resources such as water and fertilizer.

Over the past decade there have been various initiatives to commercially produce in PFALs root vegetables such as mini-carrots, mini-turnips, and mini-radishes; fruit vegetables such as strawberries and cherry tomatoes; and some medicinal plants and ornamental plants but only a limited number of PFALs are currently making a profit as of 2021. Staple crops such as wheat, rice, and maize can also be grown in PFALs relatively easily for research, educational, and recreational purposes, but the cost performance (ratio of its economic value of produce to production costs) for these is too low to be economical and practical.

Nonprofit or public-supported microfinancing organizations could lease to farmers and other interested parties mini-PFALs with a floor area of $3-20 \, m^2$ or funds to rent mini-PFALs to start a small personal, family, or group business to help end poverty. The annual yield of leaf lettuce in a mini-PFAL with four tiers is about $85 \, kg \, m^{-2}$ (floor area) y^{-1} and $213 \, kg \, m^{-2} \, y^{-1}$ ($= 85 \, kg \, m^{-2} \, y^{-1} \times 10/4$) in a PFAL with 10 tiers if operated by a skillful professional grower, regardless of the weather; however, the yield would probably be around $10-20 \, kg \, m^{-2}$ if operated by a beginner. Thus, well-organized, remote on-site training is essential for improving skills in PFAL management.

The initial investment cost of a PFAL per unit floor area is about 10 times higher than that of a greenhouse with environment control units, and is around 100 times higher than that of an open field equipped with irrigation systems and standard cultivation machinery. However, the initial investment cost per kg of produce for PFAL, greenhouse, and open field is almost the same because the average annual yield per unit land area of the PFAL is around 10-fold and 100-fold higher than the greenhouse and the open field, respectively.

The basic unit of a scalable PFAL is called a cultivation system module (CSM), with a floor area of around $10-2100 \, m^2$ (Kozai, 2018a). Using multiple CSMs, a PFAL can be scaled up to any size. This scalable characteristic using CSMs without any extra costs for scaling up can play an important role in ending poverty and improving income by increasing production capacity every 1–3 years.

4.4.2 Goal 2: end hunger, achieve food security and improved nutrition, and promote sustainable agriculture

The mini-PFAL operated by electricity generated from renewable energy can be a sustainable plant production system contributing to ending hunger, achieving food security, and improving the nutrition of people. Operation and maintenance manuals and other necessary information on the mini-PFAL can be downloaded from an open database and a self-learning system via the Internet. The mini-PFAL can be used as a small private/family business or for self-consumption at home or by a small community. The establishment of a nonprofit or public organization to support the introduction of mini-PFALs by providing free software and documents could be instrumental in promoting the success and widespread use of mini-PFALs.

PFAL measurements, control, and maintenance can be conducted using a smartphone after downloading the corresponding software. The land area and amount of water for irrigation required for vegetable and/or seedling production are about 1/20th of the land and water required for open fields, and there is no damage to plants due to weather, pest insects, worms, or wild animals.

The reason for the considerable water saving is briefly explained in Fig. 4.5 in Goal 6, and is explained in detail in Section 8.3 in Chapter 8. The PFAL can be locked, if necessary, for security against strong winds and rains and attacks by animals and thieves (Kozai et al., 2019b). A power failure for about 15 h would not damage the plant growth significantly provided a minimum battery unit for data collection and a water pump for irrigation are installed.

A prefabricated PFAL with a floor area of $10-20 \, m^2$ composed of renewable energy—powered CSMs can be built and operated within a few months from the start of on-site work anywhere including a disaster area, thereby providing locally produced and delivered fresh, tasty, nutritious vegetables that are ready to eat (without washing).

After the seeds are sown, sprouts can be harvested within 1 week, microgreens within 2 weeks, baby leaf greens within 3 weeks, and leafy greens (100 g per head) within 6 weeks when CO_2 (carbon dioxide) concentration is kept at around $1000 \, \mu mol \, mol^{-1}$ (or ppm) to promote plant photosynthesis. Once these prefabricated CSMs are developed, the initial investment cost will be considerably reduced, and the PFAL can be scaled up easily by adding CSMs.

To prevent (1) the loss of CO_2 and water vapor in the room air to the outside, (2) the invasion of small insects and/or pathogens, and (3) fluctuations in air temperature and RH or VPD of the room air, the PFAL should not be ventilated to lower the room temperature even when the temperature is lower than the air temperature outside.

4.4.3 Goal 3: ensure healthy lives and promote well-being for all at all ages

Growing plants for food and other purposes under comfortable controlled environment is often enjoyable, provides healing effects, and generally improves the quality of life of people of all ages (Ingrid et al., 2004; Chapter 22 of this book; Haller and Kramer, 2006). Growing plants using a mini-PFAL is neither hard nor dirty work, and there is no need to weed. A mini-PFAL also provides a space for human communication in a family or local community (Kozai et al., 2019a). The mini-PFAL is a new type of home or community garden that connects with the Internet.

By using the software downloaded via social networking sites (SNSs) of the Internet, users can obtain appropriate basic knowledge on environmental set points, species, cultivars, and cultivation know-how. This approach brings to mind the proverb: *Give a man a fish and he will eat for a day. Teach a man how to fish and you feed him for a lifetime.*

To serve clean, fresh vegetables without the need to wash them after harvesting, careful handwashing using disinfectant and wearing disinfected overalls/protection coats, shoes/boots, caps and masks, and daily sanitary control of the cultivation room are essential. To prevent the spread of infectious diseases to humans, persons with a body temperature of higher than 37 deg. C or symptoms such as coughing or diarrhea should not work in the PFAL or handling PFAL produce. The preparation of an educational program on sanitation control for workers and other persons engaged in PFAL cultivation is essential.

4.4.4 Goal 4: ensure inclusive and equitable quality education and promote lifelong learning opportunities for all

Growing plants in a PFAL with the support of an e-learning system will enable people to learn about basic plant physiology, ecology, recycling of water, CO_2, energy, and nutrient elements for growing plants in an integrated, simplified, and understandable form.

The use of virtual PFALs (computer simulation software and peripherals) connected with a corresponding real PFAL will enable people to efficiently learn the fundamentals of PFAL science and technology for growing plants (Kozai, 2018a). A micro-PFAL with a floor area of 0.1—0.5 m^2 can be used for educational purposes or personal use as a hobby (Chapter 6 by Takagaki et al. in Kozai et al., 2020; Chapter 15 in this book). In this way, people can enjoy the computer simulation by virtual PFALs under different environments, management methods, and plant species/cultivars using a virtual PFAL before growing plants using actual PFALs, like a flight simulator that can be used by beginners and professionals alike.

Furthermore, the experience and knowledge gained from using a PFAL help people understand the essence of more complex natural and open-field agricultural ecosystems relatively easily. On the other hand, PFAL managers need more advanced skills and deeper knowledge about controlling the environment and its effect on plant growth than farmers and growers of conventional crops under sunlight.

When growing leafy vegetables under controlled environments in a PFAL, routine cultivation practices such as sowing, transplanting, harvesting, packing, etc., can be experienced every week or every month throughout the year, so that we can learn how to cultivate them well much sooner than growing them under variable environments in the open field or in the greenhouse. This characteristic is one of merits of the PFAL in education.

4.4.5 Goal 5: achieve gender equality and empower all women and girls

Many people—regardless of gender or age and whether or not they have physical or mental handicaps—will be able to derive enjoyment from the light, safe work of growing plants in a comfortable, protected PFAL environment in or near their residential area and at the same time earn money from it. Most children and adults will experience less mental or physical stress when working in a PFAL than when working under uncomfortable conditions in an open field, and they will experience gender equality and empowerment during most working hours.

PFAL-grown plants are provided to and consumed by local consumers of different gender, age, health status, and preferences, so producers of PFAL crops will be expected to produce a variety of crops for diverse consumers with different preferences and backgrounds simply by adjusting the plant environment and the selection of cultivars. Since crops in PFALs are cultivated throughout the year, workers can secure full-time employment in PFALs, while many workers in open fields tend to find it difficult to secure full-time work throughout the year.

4.4.6 Goal 6: ensure availability and sustainable management of water and sanitation for all

As mentioned in Goal 1, about two-thirds of the lamps are always turned on in most PFALs to minimize varation in electricity consumption with time and to avoid high RH in darkness (see also Section 4.4.11 for more details), so that the highly airtight and thermally insulated cultivation room is at all times cooled by air conditioners to remove heat generated by lamps even on cold nights. As the cultivation room cools, nearly 95% of water vapor transpired from the leaves or 75%—95% of water taken up by the plants in the cultivation room is condensed (thus, dehumidified) and collected at the cooling panels (with a temperature of around 5°C) of air conditioners, and is returned to the nutrient solution tank for recycling after a proper water purification/filtering process (see Chapter 8 for further details).

Fig. 4.5 shows how the water balance of the cultivation room is maintained. The net amount of water supply for irrigation is expressed as (W_S − W_V - W_D - W_P), where W_P is the water retained in plants and removed to the outside after harvesting as produce or plant residue. Thus, decreasing the W_V and the W_D, which increases W_C, is essential to save water for irrigation. This large reduction of 75%—95% in net water supply for irrigation is beneficial when the PFAL is built in an arid region and/or where the quality and/or quantity of available water for irrigation is insufficient.

A portion of the nutrient solution in the cultivation beds, W_D, needs to be drained from time to time due to an imbalance of ion concentrations of nutrient elements. The W_D should be minimized, recycled, or reused as far as possible. The entire nutrient solution needs to be replaced with a new nutrient solution in the case of excess propagation of harmful microorganisms, pathogens, and algae. Thus, careful sanitary management of the nutrient solution and hydroponic cultivation system is necessary. Even then, however, the water and nutrient ions need to be recycled.

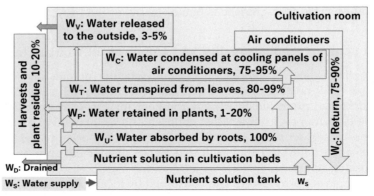

FIGURE 4.5 Scheme showing the water balance in the airtight and thermally insulated cultivation room of the PFAL. Water consumed for cleaning/washing the floor, cultivation panels, and tools is not considered in this scheme (see Chapter 8 for more details).

In addition to the water supply for irrigation, a significant amount of water is consumed for washing the floor, cultivation panels, tools, hands and boots of workers, etc. A large portion of the drained water can be reused after appropriate sanitation processing. Most pathogens (fungi, bacteria, and viruses) are disinfected or inactivated by keeping the nutrient solution at 65–80°C for about 1–10 h (the temperature and the time are species dependent (WHO (World Health Organisation), 2011)). For seed disinfection, 50–55°C for 20–30 min is common. Wiping the empty cultivation beds and panels with a wet rag containing a disinfectant is effective in decreasing the population density of pathogens.

Even when the plants absorb unclean water, they transpire clean water vapor from their leaves. Then, the condensed water at the cooling panels is as clean as distilled or drinking water if the cultivation room is hygienic and the air conditioner and its piping are kept clean through regular maintenance.

It is also noted that the daily amount of transpired water per floor area in the PFAL with 5 and 10 tiers are, respectively, about 5 and 10 times that in the greenhouse with one flat cultivation bed (1 tier). Thus, the PFAL is a kind of plant-based water collection and purification system. This is the reason why PFAL-grown pesticide-free vegetables can be served as fresh salads without washing with tap water. However, the population density of microorganisms in the rootzone and nutrient solution in the cultivation beds is extremely high because organic substances such as dead roots exist and feed the microorganisms in the cultivation beds.

Thus, the dripping of nutrient solution containing microorganisms to the aerial part of the plants should be strictly avoided to keep the aerial part of the plants clean.

4.4.7 Goal 7: ensure access to affordable, reliable, sustainable, and modern energy for all

The energy-autonomous PFAL with a battery backup unit will be commercialized in the near future for use with electricity generated by solar energy, wind power, biomass, or geothermal energy since the cost of electricity generated by such renewable energies has recently become less expensive than that generated by fossil fuels and nuclear power in many regions in the world (Kozai, 2018a), and will continue to decrease in the coming years.

The downsized batteries used for electric cars need to be replaced with new ones when the electricity storage capacity decreases by around 30%. At that time, they are replaced with new ones. This means that a new market for the used downsized batteries will be created and will grow in the coming decades. By connecting these used batteries, an inexpensive battery backup system can be installed in the PFAL and many other electricity-driven systems. The reduction in storage capacity of used batteries by 30% will not cause any problem in those systems.

As of 2019, electricity costs for a PFAL with LEDs accounted for about 20% of the total production cost in Japan (Kozai et al., 2019a). The electricity consumption per kg of produce (7–10 kWh; see Chapters 12 and 13 in this book) will decrease by 15%–30% within several years as a result of the improvement in lighting and other environmental control systems and the introduction of new cultivars expressly bred to suit the PFAL, which grow well under high CO_2 concentration, low PPFD, and pathogen-free conditions.

In the open field, solar light energy is often too abundant or too scarce for plant growth, CO_2 concentration at daytime (about 400 μmol mol^{-1}) is always too low to maximize the photosynthesis and thus plant growth, and the air/soil temperatures are often too low and/or too high, so that the annual average of light energy use efficiency of plants is much lower than that of PFAL-grown plants (Kozai, 2011). Furthermore, risks of plant damage due to drought, strong winds, and/or rain, and the spread of pest insects are high in open fields. The PFAL's protection of plants from these hazards is also beneficial for human physical and mental health.

Dry air in the Earth's atmosphere contains 78.1% nitrogen (by volume), 21% (or 210,000 μmol mol^{-1}) oxygen, 0.04% (or 400 μmol mol^{-1}) CO_2, and small amounts of other gases. The atmospheric air also contains a variable amount of water vapor, on average around 1% at sea level. It should be noted that the low concentration of CO_2, sole carbon source of plants, is one of the key limiting factors in photosynthesis (Taiz and Zeiger, 2006). In the airtight PFAL containing plants, the CO_2 concentration in the room air during the photoperiod decreases to around 100 μmol mol^{-1} within 1 h or so if CO_2 supply is stopped. This is why CO_2 gas needs to be supplied continuously during the photoperiod in the PFAL. In this sense, the PFAL is a CO_2-absorbing system.

4.4.8 Goal 8: promote sustained, inclusive, and sustainable economic growth, full and productive employment, and decent work

The productivity of advanced PFALs will more than double that of existing PFALs by 2030 (Kozai, 2018a, 2019) and will make a substantial contribution to employment and economic growth. It is expected that people who want to start a small business can rent a mini-PFAL and build it on a limited unfertile and/or contaminated land area with a shortage of irrigation water for producing and selling fresh vegetables, medicinal plants, and various seedlings as cash crops. Since some recent PFALs are designed to be scalable, it will be possible to expand their production capacity easily year after year (Kozai, 2018a).

As of 2021, leafy vegetables and seedlings for greenhouse production are produced in most commercial PFALs. However, a few PFALs for commercial production of fruit vegetables such as cherry tomato plants exist (see Chapter 23), and the number of PFALs or commercial production of strawberry and medicinal plants will increase significantly in the 2020s.

4.4.9 Goal 9: build resilient infrastructure, promote inclusive and sustainable industrialization, and foster innovation

PFALs can be reinforced to construct a resilient infrastructure in a local area. The infrastructure is strengthened more through integration with other systems such as wastewater treatment, bio-waste treatment, electric power supply with use of natural energy, CO_2, and heat energy recovery/delivery, all of which promote inclusive and sustainable industrialization and agriculture. Food mileage and food loss are also minimized, and food security is strengthened when fresh vegetables are produced in the consumption area.

An advantage of the PFAL built in residential and industrial areas is that the peak time zone of electricity consumption for lighting and cooling for the PFALs occurs between evening and morning, while the peak time zone in residential and industrial areas often occurs between morning and evening. Furthermore, the plants in the PFAL absorb CO_2, while CO_2 is emitted in the residential and industrial areas. The PFALs produce waste heat (around 40°C) from the outer units of air conditioners for cooling all day, which can be used for drying food and clothing and warming the room air.

The LED lighting system, air conditioners with fans, hydroponic cultivation beds/racks with water pumps, CO_2 and nutrient solution supply units, and environment controllers are major units of the cultivation room (Fig. 4.2). The expected life spans of such systems are 30,000—50,000 h, so they can be used for 5—10 years if maintained properly.

4.4.10 Goal 10: reduce inequality within and among countries

PFALs can be built and operated anywhere, regardless of climate, soil fertility, and water availability. They can be operated commercially in cold (as low as −50°C), hot and dry (as high as 50°C and RH of as low as 10%), and hot and humid areas (around 45°C and RH of 90%), with a slight increase in electricity consumption and costs for cooling in hot areas.

Annual electricity consumption for cooling the thermally insulated, airtight cultivation room accounts for 10%—15% in cold, 20%—25% in hot and dry, and 25% of costs in hot and humid areas. In other words, the COP (coefficient of performance, the ratio of the cooling load to the electricity consumption of air conditioners) of air conditioners ranges from 9 (= (85 + 5)/10) in cold to 3 (= (70 + 5)/25) in hot areas when operated properly. The COP for cooling increases in tandem with the increase in the difference between room air and outside air temperatures.

The electricity consumption for lighting is the same anywhere in the world and accounts for 70%—85% of total electricity consumption (Fig. 4.6). The electricity consumption for pumping nutrient solution and air distribution fans accounts for around 5% of the total annual electricity consumption irrespective of the PFAL's location.

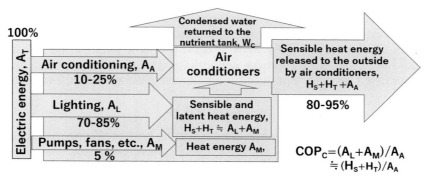

FIGURE 4.6 Scheme showing the flows of electric and heat energy in the airtight, thermally insulated cultivation room of the PFAL. Chemical energy fixed by plants as carbohydrates, $\varDelta C_P$, is less than 5% of the A_L. The COP_C (coefficient of performance for cooling) is the electric energy use efficiency of air conditioners for cooling (see Chapter 8 for more details).

Since most aspects of PFAL technology are universal and largely open to the public, PFALs are suited to local production for local consumption. Thus, PFALs have the potential to help reduce income inequality within and among countries. A small PFAL can be built at a local community center or school.

People living in remote regions often struggle to transport and sell their heavy fresh produce in large cities, resulting in a significant loss of produce during transportation. People living in cold and hot regions struggle to grow plants in greenhouses and open fields as fresh produce for economic profits due to high production costs, low yields, and the long distance to consumers. Under such conditions, they could use PFALs to grow medicinal plants/herbs and vegetables for use as dried products all year round and dry them on-site for sale as ingredients for medicine, cosmetics, seasonings, and food/drink additives before transporting them to buyers every few months when convenient.

4.4.11 Goal 11: make cities and human settlements inclusive, safe, resilient, and sustainable

Integrating a CO_2-absorbing PFAL with a CO_2-emitting biological system such as mushroom and aquaculture facilities, or another CO_2-emitting system such as an office building can enhance the sustainability of cities (Kozai et al., 2019a), making cities and human settlements more inclusive, resilient, sustainable, and safer. In this case, however, a method of introducing CO_2 gas into the cultivation room of PFAL without causing any negative issue needs to be considered.

The reason for lighting the PFAL at night, early morning, and late evening when electricity consumption in office buildings and homes is lower than in the daytime include: (1) instead of turning all LEDS on and off at the same time, lighting two-thirds of the lamps alternately throughout the day can reduce maximum hourly electricity consumption for lighting and cooling by about 30%, resulting in reduced initial investment for lighting and power supply systems and reduced electricity consumption due to increased COP of air conditioners since the air temperature outside is lower at night than in the daytime; (2) RH in the cultivation room can be kept at around 70%−80% throughout day without the use of a dehumidifier due to the dehumidifying function of air conditioners operated for cooling; and (3) plants grow well even at night if the daily light integral (PPFD x photoperiod per day) is high enough. Accordingly, the electric power network/grid can be developed in the region to share the generated electricity optimally with homes, offices, industries, businesses, and PFALs.

4.4.12 Goal 12: ensure sustainable consumption and production patterns

The PFAL is suitable for scheduled production of plants from the viewpoint of both resource availability and demand for produce in the region. Thus, sustainable local production for responsible local consumption can be ensured with relative ease. Local production for local consumption minimizes the loss of produce during transportation, storage, and handling, and prolongs shelf life due to the cleanness (low microorganism population in produce) and less physical damage of produce. Local people tend to try to minimize vegetable and food waste when they know who, how, and where they are produced. In the PFAL, historical and cultural cultivars can be produced without any technical difficulty for local people.

Since the PFAL is thermally well insulated and highly airtight and the plant environment is controlled as desired, use efficiencies of electricity, water, fertilizer, and CO_2 are much higher than those in the greenhouse and open field (see Kozai (2013); and Chapter 8 of this book). In addition, waste heat energy generated mainly by lamps and transferred to the outer units (or condensers) of air conditioners can be efficiently used for drying the plant residue or other wet substances at around 40°C, thereby improving hygienic conditions and mitigating food security risks such as local hazardous weather, pest insects, animals, and human crimes.

4.4.13 Goal 13: take urgent action to combat climate change and its impacts

The PFAL will reduce the negative impact of climate change on plant productivity. The plants growing in the PFAL are almost completely protected against strong wind, heavy rain, flood, drought, hail/snow, extreme temperatures, insects, animals, and thieves because the PFAL is airtight and thermally insulated and all the doors can be locked. CO_2 emissions per kg of PFAL-grown produce due to electricity generation and fertilizer manufacturing have been reduced in recent years and will be further reduced in the near future. The consumption of water and fertilizer per kg of produce in the PFAL has already been significantly reduced compared with consumption in the greenhouse and open field.

On the other hand, the CO_2 emitted during the manufacturing of iron, cement, and aluminum used as structural components of the PFALs needs to be reduced. Likewise, the plant residue produced in PFALs as waste needs to be reduced and/or recycled on-site. Reinforced wooden structural components need to be used more for the PFAL building and cultivation racks (Kikuchi and Kanematsu, 2019). Biodegradable plastics and paper bags also need to be used more in cultivation trays and panels.

4.4.14 Goal 14: conserve and sustainably use the oceans, seas, and marine resources for sustainable development

PFALs constructed under water or underground can be more sustainable than those built above ground. PFALs can also be built on ships for providing fresh vegetables to people on board. Besides, closed-type inland aquaculture systems for producing ocean fish constructed near or in urban areas are becoming more and more resource efficient (in terms of water, electricity, feed, and labor) recently. Thus, this type of aquaculture system using food waste produced in urban areas as feed will contribute to the conservation and sustainable use of the oceans, seas, and marine resources.

On the other hand, in aquaponics, a combination of aquaculture and hydroculture, fish waste can be used as fertilizer for plants, and the plant residue can be used to feed fish (see Chapter 10 in this book). Since both the PFAL and aquaponics are closed systems, they do not interfere with the conservation and sustainable use of oceans, seas, and marine resources. Insect factories for growing crickets (Gryllidae), earthworms, etc., are set to be commercialized for the production of fish and livestock feed.

4.4.15 Goal 15: protect, restore, and promote sustainable use of terrestrial ecosystems, sustainably manage forests, combat desertification, and halt and reverse land degradation and halt biodiversity loss

The PFAL is a powerful system for producing millions or billions of high-quality transplants or seedlings free of disease, insects, and pesticides with high resistance to environmental stresses for use in reforestation, forestation, and desert greening, with minimum use of irrigation water, land area, and labor. Thus, PFALs can play a key role in halting and reversing land degradation and biodiversity loss. In contrast, the open field and greenhouse are generally unsuited to producing millions or billions of high-quality transplants especially in areas for reforestation, forestation, and desert greening (Kozai et al., 2005). Transplants preacclimatized to the climate and the soil can be produced on-site in PFALs where the climate and soil are generally inappropriate for producing high-quality transplants in greenhouses and open fields.

The PFAL would require only about 1%–2% of land area compared to the open field to produce such transplants (Kozai et al., 2019a). Thus, by using PFALs rather than open fields or greenhouses for producing transplants, more land area can be used for conserving natural ecosystems with improved biodiversity, while people can use open spaces for producing fresh vegetables or for the conservation of natural ecosystems. In dry regions in particular, most of the water used for irrigation in transplant production evaporates from the soil surface without being absorbed by the transplants. On the other hand, most of the water used for irrigation in transplant production in the PFAL is absorbed by the transplants. Furthermore, transpired water from plants is mostly condensed, collected, and reused for irrigation (Fig. 4.5).

The transplants produced in the PFAL can be stored at low temperatures and a low PPFD for a few weeks until the transplants can be transplanted under favorable environmental conditions. For details, refer to Kurata and Kozai (1992), Kubota and Chun (2000), and Kozai et al. (2005).

4.4.16 Goal 16: promote peaceful and inclusive societies for sustainable development, provide access to justice for all, and build effective, accountable, and inclusive institutions at all levels

A PFAL composed of multiple CSMs is scalable or expandable with relative ease. The smallest PFAL is a CSM itself. Setting up and operating a CSM with the assistance of a self-learning or educational software tool, anyone can enjoy learning about plant physiology, energy and material conversions, and material circulation processes in the closed ecosystem and productivity management of a PFAL. Users can also learn efficiently how to skillfully grow plants well with minimum resource consumption and waste generation, and how to use or sell the produce. This characteristic of the PFAL connected with a networked platform is beneficial in promoting a peaceful, inclusive society for sustainable development, providing access to justice for all and building effective, accountable, and inclusive institutions at all levels.

4.4.17 Goal 17: strengthen the means of implementation and revitalize the global partnership for sustainable development

Since most aspects of PFAL technology are universal and describable, general information on the PFAL can be shared with people living anywhere in the world using a platform with simultaneous interpretation software via the Internet. The establishment of a global organization for information sharing and capacity building can be expected to strengthen the means for introducing PFALs and to revitalize global as well as local partnerships for sustainable development in society.

It has been difficult to spread an agricultural or horticultural technology for open fields and soil cultivation in the greenhouse to various regions with different climate and soils. However, the PFAL can strengthen the means of implementation and revitalize the global partnership for sustainable development because the PFAL technology is basically universal.

4.5 Learn a simple ecosystem first, then complex ecosystems

Compared to the greenhouse, open field, and natural ecosystems, the PFAL is an extremely simple artificial ecosystem. Nevertheless, the PFAL ecosystem is still often too complex to understand in its entirety, especially the rootzone ecosystem where hundreds of environmental and biological variables are interrelated.

A PFAL simulator to be developed in the near future will consist of mechanistic models for energy and material balances and photosynthetic plant growth, multiple-variable statistic models for development, morphology and secondary metabolite production of plants, and behavior (or surrogate) and artificial intelligence (AI) models for optimizing plant environments from the viewpoint of costs and benefits of production under given amounts of resources and social demands. Using these models, users can investigate the plant—environment interactions, resource conversion, and circulation processes under controlled environments in the PFAL.

The mathematical models used in PFAL management are similar to those used in greenhouse and open field management, although the ranges between upper and lower values of variables are much narrower in the PFAL model than in the greenhouse and the open field models. In addition, certain aspects of the greenhouse and open field models such pesticide application and its effect can be eliminated for simplification of the PFAL model. Thus, the PFAL model is a streamlined educational, self-learning software tool that also introduces beginners to the fundamental principles of greenhouse and open field management, and perhaps even natural ecosystem management (Fig. 4.7).

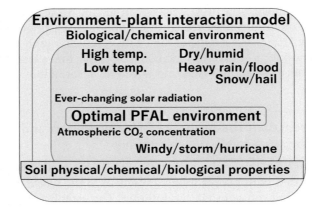

FIGURE 4.7   Learning the functions of an ecosystem starting with the PFAL ecosystem, the simplest one, to more complex ecosystems such greenhouse, open field, and nature ecosystems.

4.6 Conclusion

Achieving the SDGs is a challenge for all people living on this planet, especially after the COVID-19, a recent pandemic. Furthermore, there are so many ideas, tools, methods, systems, and collaborative approaches for contributing to the achievement of the SDGs. The PFAL is just one of them. However, PFALs that are properly applied in the various ways discussed in this chapter have enormous potential in contributing to the SDGs. The descriptions in this chapter largely center on the author's limited experience with and theoretical consideration of PFALs constructed in relatively large cities. Therefore, some adjustments would be necessary for the efficient use of PFALs in a rural or a remote area. Achieving the SDGs by 2030 will help us achieve the next mission of establishing technology and innovative ways of thinking to nourish over 9 billion people and enable them to live meaningful lives on this planet in 2050 with minimum consumption of resources and minimum emission of environmental pollutants (Lee, 2019).

PFAL technology using LEDs emerged around 2000 and is still in its infancy. It is only recently that efforts to apply PFAL technology to achieve the SDGs have started in earnest. Therefore, many promising initiatives to achieve the SDGs through the introduction of PFALs can be expected in the future.

References

Chen, J., 2020. Environmental, Social, and Governance (ESG) Criteria. https://www.investopedia.com/terms/e/environmental-social-and-governance-esg-iteria.asp.

Haller, R.L., Kramer, C.L., 2006. Horticultural Therapy Methods: Making Connections in Health Care, Human Service, and Community Programs (Monograph). Haworth, New York, USA, p. 153.

Ijichi, H., 2018. Chapter 3 Plant Factory Business - Current Status and Perspective of Plant Factory Business. NAPA research report, pp. 58–80 (in Japanese).

Ingrid, S., Mariannne, S., Schlander, S., 2004. Horticultural therapy: the 'healing garden' and gardening in rehabilitation measures at Danderyd hospital rehabilitation clinic, Sweden. Pediatr. Rehabil. 7 (4), 245–260.

5.2.4.1 SI base units

The following names and symbols of the seven SI base units have been fixed since the 11th Meeting of the CGPM. They, respectively, correspond to the seven selected base quantities of time, length, mass, electric current, thermodynamic temperature, amount of substance, and luminous intensity: second [s], meter [m], kilogram [kg], ampere [A], kelvin [K], mole [mol], and candela [cd]. In the Ninth SI Brochure, seven defining constants (Table 5.1) are described and listed before the SI base units (Table 5.2) because the defining constants are characterized as the most fundamental feature of the definition of the entire system of units. The defining constants are chosen such that any unit of the SI can be written either through a defining constant itself or through products or quotients of defining constants. The SI base units are defined using the defining constants.

TABLE 5.1 The seven defining constants of the SI and the seven corresponding units they define.

Defining constant	Symbol	Numerical value	Unit
hyperfine transition frequency of Cs	$\Delta \nu_{Cs}$	9 192 631 770	Hz
speed of light in vacuum	c	299 792 458	m s^{-1}
Planck constant	h	$6.626\ 070\ 15 \times 10^{-34}$	J s
elementary charge	e	$1.602\ 176\ 634 \times 10^{-19}$	C
Boltzmann constant	k	$1.380\ 649 \times 10^{-23}$	J K^{-1}
Avogadro constant	N_A	$6.022\ 140\ 76 \times 10^{23}$	mol^{-1}
luminous efficacy	K_{cd}	683	lm W^{-1}

From Bureau International des Poids et Mesures, 2019. Le Système International D'unités/the International System of Units. https://www. bipm.org/utils/common/pdf/si-brochure/SI-Brochure-9.pdf, 216pp.

TABLE 5.2 SI base units.

Base quantity		Base unit	
Name	Typical symbol	Name	Symbol
time	t	second	s
length	l, x, r, etc.	meter	m
mass	m	kilogram	kg
electric current	I, i	ampere	A
thermodynamic temperature	T	kelvin	K
amount of substance	n	mole	mol
luminous intensity	I_v	candela	cd

From Bureau International des Poids et Mesures, 2019. Le Système International D'unités/the International System of Units. https://www. bipm.org/utils/common/pdf/si-brochure/SI-Brochure-9.pdf, 216pp.

It is noteworthy that candela [cd] is the distinctive unit in the SI base units because it is related to a physical quantity of radiant intensity [W sr^{-1}] through a photobiological/physiological weighting factor: spectral luminous efficacy for photopic vision.

5.2.4.2 *SI derived units*

The SI derived units are defined as products of powers (including negative powers) of the SI base units. That is to say, the SI derived units are units constructed by multiplying, dividing, or powering the SI base units in various combinations. The set of the SI base units and SI derived units is designated as the set of coherent SI units. In this regard, we can find a similar term, a "complete" set of SI units, in the SI Brochure. The complete set of SI units includes both the set of coherent SI units and the multiples and submultiples formed using the SI prefixes (described later).

(1) SI derived units expressed in terms of the SI base units

Examples of the SI derived units expressed in terms of SI base units include: [m^2] (area), [m s^{-1}] (speed, velocity), [A m^{-1}] (magnetic field strength), [mol m^{-3}] (amount of substance concentration), and [cd m^{-2}] (luminance).

(2) SI derived units with special names and symbols

Twenty-two units of the SI derived units are assigned special names and symbols. Together with the seven SI base units, they form the core of the set of coherent SI units. All other SI units of the complete set of SI units are combinations of these 29 units.

Examples of the 22 SI derived units include the following: radian [rad] (plane angle), steradian [sr] (solid angle), hertz [Hz] (frequency), newton [N] (force), pascal [Pa] (pressure), joule [J] (energy, work), watt [W] (power, radiant flux), volt [V] (electric potential difference), siemens [S] (electric conductance), degree Celsius [°C] (Celsius temperature), lumen [lm] (luminous flux), and lux [lx] (illuminance).

It is noteworthy that one radian is the angle subtended at the center of a circle by an arc that is of equal length to the radius (entire plane angle: 360 degrees $= 2\pi$ rad). One steradian is the solid angle subtended at the center of a sphere by an area of the surface that is equal to the squared radius (entire solid angle: 4π sr) (Fig. 5.1). As the units for the plane and solid angles are [rad] $=$ [m m^{-1}] and [sr] $=$ [m^2 m^{-2}], respectively, the unit for both quantities can be expressed as 1. However, unit symbol 1 is not shown explicitly.

The numerical value of a temperature difference or temperature interval is the same when expressed in either degrees Celsius or in kelvin because the relationship of Celsius temperature t [°C] and thermodynamic temperature T [K] is given as t [°C] $= T$ [K] $- 273.15$ [K]. The temperature of 273.15 K is the freezing point of water. The unit of thermodynamic temperature (1 K) is given as $1/273.16$ of the triple point temperature of water (273.16 K). The triple point temperature of a substance is the temperature at which the three phases (gas, liquid, and solid) of that substance in a single-component system coexist in thermodynamic equilibrium.

5.2.4.3 *SI prefixes for expressing decimal multiples and submultiples of SI units*

SI prefixes are the prefixes provided to express decimal multiples and submultiples ranging from 10^{24} to 10^{-24} for use with SI units. In fields related to plant sciences and

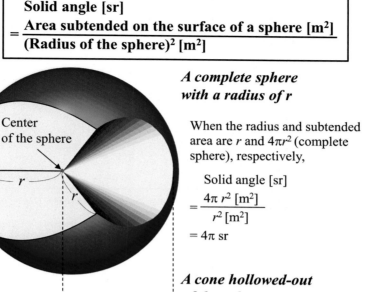

Solid angle [sr]
$$= \frac{\text{Area subtended on the surface of a sphere } [m^2]}{(\text{Radius of the sphere})^2 [m^2]}$$

Center of the sphere

r

r

A complete sphere with a radius of r

When the radius and subtended area are r and $4\pi r^2$ (complete sphere), respectively,

Solid angle [sr]
$$= \frac{4\pi \, r^2 \, [m^2]}{r^2 \, [m^2]}$$
$$= 4\pi \text{ sr}$$

A cone hollowed-out of the sphere

When the radius and subtended area are r and r^2, respectively,

Solid angle [sr]
$$= \frac{r^2 \, [m^2]}{r^2 \, [m^2]}$$
$$= 1 \text{ sr}$$

Subtended Area: r^2

r

FIGURE 5.1 Solid angle and its mathematical explanation.

bioengineering, SI prefixes for expressing decimal multiples and submultiples ranging from 10^{12} to 10^{-12} are expected to be sufficient (Table 5.3).

The SI prefix symbols are printed in upright typeface and are attached to unit symbols with no space between the prefix symbol and the unit symbol. The combination formed by a prefix symbol attached to a unit symbol constitutes a new inseparable unit symbol. Therefore, when raising the combination to a positive or negative power, it is raised as an inseparable unit, i.e., $1.2 \text{ cm}^2 = 1.2 \text{ (cm)}^2 = 1.2 \times (10^{-2} \text{ m})^2 = 1.2 \times 10^{-4} \text{ m}^2$.

The kilogram [kg] is the only coherent SI unit for which the name and symbol include a prefix. Names and symbols for decimal multiples and submultiples of the unit of mass are formed by attaching prefix names and symbols to the unit name "gram" and the unit symbol [g], respectively. For example, 10^{-9} kg is written as microgram [μg] not as nanokilogram [nkg].

TABLE 5.3 SI prefixes for expressing decimal multiples (up to 10^{12}) and submultiples (down to 10^{-12}) of SI units.

Decimal multiples (up to 10^{12})			Submultiples (down to 10^{-12})		
Factor	Name	Symbol	Factor	Name	Symbol
10^1	deca	da	10^{-1}	deci	d
10^2	hecto	h	10^{-2}	centi	c
10^3	kilo	k	10^{-3}	milli	m
10^6	mega	M	10^{-6}	micro	μ
10^9	giga	G	10^{-9}	nano	n
10^{12}	tera	T	10^{-12}	pico	p

Bureau International des Poids et Mesures (2019) provides decimal multiples and submultiples ranging from 10^{24} to 10^{-24} for use with SI units.

From Bureau International des Poids et Mesures, 2019. Le Système International D'unités/the International System of Units. https://www. bipm.org/utils/common/pdf/si-brochure/SI-Brochure-9.pdf, 216pp.

5.2.4.4 Non-SI units that are accepted for use with the SI units

The CGPM has accepted some non-SI units for use with SI units because it is recognized that some non-SI units are widely used and are expected to continue to be used for many years. Non-SI units that are widely used in fields related to plant sciences and bioengineering include the following: minute [min] (time), hour [h] (time), day [d] (time), degree [°] (plane angle), hectare [ha] (area), liter [l] or [L] (volume), and ton [t] (mass). The SI prefixes can be used with liter [l] or [L], but not with other above-listed non-SI units. For example, it is not acceptable to write "15 mh" for "0.015 h". Acceptable alternative options to write "0.015 h" are "54 s" using an SI unit of second [s] and "0.90 min" using a non-SI unit of minute [min].

5.2.5 Writing unit symbols and expressing value of quantities

5.2.5.1 Writing unit symbols

(1) Typeface

Unit symbols are printed in upright type and in lowercase letters unless they are derived from a name, in which case the first letter is a capital letter. An exception is that capital L is allowed as an expression for the liter, to avoid possible confusion between the lowercase letter l (el) and the numeral 1 (one).

(2) Prefix-related matters

An SI prefix symbol, provided to express decimal multiples and submultiples, is part of the unit. It precedes the unit symbol without a space. Compound or mixed prefix symbols are never used. Only one prefix symbol should be used when several unit symbols are combined for expressing a quantity; the prefix should be placed at the beginning of the compound unit symbol.

(3) Nonuse of a period and plural form

Unit symbols are not followed by a period, except at the end of a sentence. They are not abbreviations. One must not use the plural form with a unit symbol.

(4) Products of unit symbols

Multiplication must be indicated by a space or a half-high (centered) dot (•). The author recommends the use of a space.

(5) Quotients of unit symbols

Division is indicated by a horizontal line, by an oblique stroke (/) or by negative exponents. An oblique stroke must not be used more than once in a given expression of compound unit symbols. One can use brackets or negative exponents to avoid more than once usage of an oblique stroke. The author recommends the usage of negative exponents in preferences to the use of an oblique stroke.

5.2.5.2 *Expression of the value of a quality*

(1) Formatting the value of a quantity

The number always precedes the unit. A space is always used to separate the number from the unit. The space between the number and the unit is regarded as a multiplication sign. The sole exceptions to this rule are for the unit symbols for degree, minute, and second for plane angle, [°], ['], and ["], respectively, for which no space is used to separate the number from the unit.

Although not explicitly indicated in the Ninth SI Brochure, a unit symbol following a number of a quantity value need not be put in square brackets or round brackets. It should be written out simply, such as 12 m or 34 kg m^{-2}. However, when a unit symbol follows a number of a quantity symbol, it should be preferred to put the unit symbol in square brackets or round brackets as Q [kJ].

(2) Quantities that are ratios of quantities of the same kind

Quantities that are ratios of quantities of the same kind are simply expressed in numbers because they are quantities with unit [1] (one). They are also expressed with units of quantities of the same kind (m m^{-1}, rad, sr, mol mol^{-1}) to facilitate the understanding of the quantity being expressed.

5.2.5.3 *Frequently found inappropriate writing unit symbols, prefixes, and expressing values of quantities*

Inappropriate writing or expressions, in light of the Ninth SI Brochure, of unit symbols, prefixes, and quantities are frequently found in PFAL-related articles. The author presents some examples in an errata style table (Table 5.4). The author recommends that readers review the "2. International system of units" of the Ninth SI Brochure, which consists of only 5 pages. It provides a full understanding of the rules and style convections for writing unit symbols and expressing values of quantities.

5.3 Photonmetric quantities and their application

5.3.1 Importance of photonmetric quantities

Photonmetric quantities must be used to describe the light environment for plant cultivation because most plant responses to light (strictly speaking, incident photons) are observed

TABLE 5.4   Frequently encountered inappropriate writing unit symbols, prefixes, and expressing values of quantities in PFAL-related articles.

Inappropriate	Appropriate	Section concerned
sec	s	1.4.1
Kg	kg	1.4.1, 1.4.3
lux	lx	1.4.2(2)
hr	h	1.4.5
day	d	1.4.5
μmol m^{-2} s^{-1}	μmol m^{-2} s^{-1}	1.5.1(1)
Wm^{-2}	W m^{-2} or W·m^{-2}	1.5.1(4)
μmol/m^2/s	μmol m^{-2} s^{-1}	1.5.1(5)
12 m	12 m	1.5.2(1)
34°C	34 °C	1.5.2(1)
56 L or 56 l	56 L	1.5.2(1)

as a result of photochemical reactions. To be clear, there are very few situations in which radiometric or photometric quantities must be used to describe the light environment for plant cultivation.

A thermal effect of light or lighting on plant responses is sometimes observed. The thermal effects in terms of energy are usually analyzed based on energy budget equations using measured plant-temperature changes that reflect heat transfer phenomena. When analyzing such thermal effects on plant responses related to light or lighting, one should use radiometric quantities rather than photonmetric quantities because matter is affected by thermal reactions, but not by photochemical reactions.

Although photometric quantities were frequently used until several decades ago to describe light environments for plant cultivation in articles of all types, plant physiologists and bioengineers today agree that those are definitely inappropriate quantities to describe the light environment for plant cultivation. Simply put, photometric quantities are useful for quantifying the perceived brightness for the human eye. Possible situations requiring the use of photometric quantities are descriptions of light environments for workers, not plants, in a PFAL.

Comprehension and appropriate use of photonmetric quantities are necessary for discussion of light environments for plant cultivation. Nevertheless, photonmetric quantities are rarely explained in books and articles. Instead, radiometric and photometric quantities are readily found in many books and articles of the field (e.g., Meyer-Arendt, 1968; Ohno, 1997). Herein, the author simply explains fundamental and important photonmetric quantities to describe light environments using appropriate terms with SI units. For that purpose, equally situated quantities in radiometry, photometry, and photonmetry are explained, respectively. Then they are mutually contradistinguished.

Terminology definitions are mostly quoted from International Electrotechnical Commission (IEC) 60050-845-01 (1987). Several terms associated with photonmetric quantities are also explained using illustrations and graphs. Several parts of this section include partially reconstituted and modified contents of the author's earlier works (Fujiwara, 2013, 2016, 2019).

5.3.2 Fundamental radiometric, photometric, and photonmetric quantities with their SI units

The author selected four radiometric quantities as fundamental ones: radiant intensity, radiant flux, radiant energy, and irradiance. Those quantities are placed along with their SI units in the second-left column. Then photometric and photonmetric quantities with their associated units are put in the right columns so that the relations among the quantities and SI units can be readily understood (Table 5.5). Relational expressions among the quantities in the same law are provided in the leftmost column to indicate quantitative relations. Quantities in the same row are mutually equivalent in a metric sense.

The following definition in the first sentence for each term, except for "photon flux density," is a direct quote from International Electrotechnical Commission (IEC) 60050-845-01 (1987) in which the term definitions are presented clearly. Quantity symbols and equations

TABLE 5.5 Fundamental radiometric, photometric, and photonmetric quantities with their SI units.

Quantitative relations	Radiometric quantities	Photometric quantities	Photonmetric quantities
A	radiant intensity	luminous intensity	photon intensity
	$[\text{W sr}^{-1}]$	$[\text{cd}]$	$[\text{mol s}^{-1}\text{ sr}^{-1}]$
$A \times 1\text{ sr}$	radiant flux	luminous flux	photon flux
$= B$	$[(\text{W sr}^{-1})\text{ sr}]$	$[\text{cd sr}]$	$[(\text{mol s}^{-1}\text{ sr}^{-1})\text{ sr}]$
	$= [\text{W}]$	$= [\text{lm}]$	$= [\text{mol s}^{-1}]$
$A \times 1\text{ sr} \times 1\text{ s}$	radiant energy	quantity of light	photon number
$= B \times 1\text{ s}$	$[\text{W s}]$		$[(\text{mol s}^{-1})\text{ s}]$
	$= [\text{J}]$	$[\text{lm s}]$	$= [\text{mol}]$
$A \times 1\text{ sr} \times 1\text{ m}^{-2}$	irradiance	illuminance	photon flux density
$= B \times 1\text{ m}^{-2}$			(photon irradiance)
		$[\text{lm m}^{-2}]$	$[(\text{mol s}^{-1})\text{ m}^{-2}]$
	$[\text{W m}^{-2}]$	$= [\text{lx}]$	$= [\text{mol m}^{-2}\text{ s}^{-1}]$

A and B, respectively, denote quantity symbols representing quantities for the intensities in the first row and fluxes in the second row.

Modified from Fujiwara, K., 2013. Fundamentals of light on plant cultivation and LED light irradiation technology. Refrigeration 88, 163–168. (in Japanese); Fujiwara, K., 2016. Radiometric, photometric and photonmetric quantities and their units. In: Kozai, T., Fujiwara, K., and Runkle, E. (eds.) LED Lighting for Urban Agriculture, Springer Science+Business Media Singapore, p. 367–376; Fujiwara, K., 2019. Light sources. In: Kozai, T., Niu, G., and Takagaki, M. (eds.) Plant Factory: An Indoor Vertical Farming System for Efficient Quality Food Production, second ed., Academic Press, London, UK, p. 139–151.

were deleted from the original definitions. Then SI units were added. Supplementary explanations have been added for several terms to elucidate their definitions.

5.3.2.1 Fundamental radiometric quantities

Radiant intensity [W sr^{-1}] (of a source, in a given direction): Quotient of the radiant flux [W] leaving the source and propagated in the element of solid angle [sr] (see Fig. 5.1 for its explanation) containing the given direction, by the element of solid angle [sr]. In other words, radiant flux [W] leaving the source per unit solid angle [sr] in the considered direction.

Radiant flux (radiant power) [W] (= [J s^{-1}]): Power [W] emitted, transmitted, or received in the form of radiation. The amount of radiant energy [J] emitted, transmitted, or received per unit of time [s].

Radiant energy [J]: Time integral of the radiant flux [W] over a given duration [s].

Irradiance [W m^{-2}] (at a point of a surface): Quotient of the radiant flux [W] incident on an element of the surface containing the point, by the area [m^2] of that element.

5.3.2.2 Fundamental photometric quantities

Luminous intensity [cd] (of a source, in a given direction): Quotient of the luminous flux [lm] leaving the source and propagated in the element of solid angle [sr] containing the given direction, by the element of solid angle [sr]. In other words, it is the luminous flux [lm] leaving the source per unit of solid angle [sr] in the considered direction. A common error is comparison of the luminous output capability of the light source, especially LEDs, with the luminous intensity [cd]. The luminous flux [lm] must be compared for that purpose.

Luminous flux [lm]: Quantity derived from radiant flux [W] by evaluating the radiation according to its action upon the International Commission on Illumination (CIE) standard photometric observer. The luminous flux is calculated by integrating, from 0 nm to infinity (actually 360−830 nm for photopic vision), the product of spectral radiant flux [W m^{-1}], the CIE standard spectral luminous efficiency function for photopic vision (usually denoted as $V(\lambda)$, with value ranges of 0−1 according to wavelength λ), and the maximum spectral luminous efficacy (of radiation) for photopic vision [lm W^{-1}] (usually denoted as K_m; approximately 683 lm W^{-1} at 555 nm (540 THz)). Ohno (1997) presented explicit and detailed descriptions of the relation between radiometric and photometric quantities.

Quantity of light [lm s]: Time integral of the luminous flux [lm] over a given duration [s].

Illuminance [lx] (at a point of a surface): Quotient of the luminous flux [lm] incident on an element of the surface containing the point, by the area [m^2] of that element. Even in recent years, a few reports in fields related to plant science have used illuminance [lx] as a quantity in cases where photon flux density or photon irradiance should have been used.

5.3.2.3 Fundamental photonmetric quantities

Photon intensity [mol s^{-1} sr^{-1}] (of a source, in a given direction): Quotient of the photon flux [mol s^{-1}] leaving the source and propagated in the element of solid angle [sr] containing the given direction, by the element of solid angle [sr]. In other words, it is the photon flux [mol s^{-1}] leaving the source per unit solid angle [sr] in the considered direction.

Photon flux [mol s^{-1}]: Quotient of the number of photons [mol] emitted, transmitted, or received in an element of time [s], by that element.

Photon number (number of photons) [mol]: Time integral of the photon flux [mol s^{-1}] over a given duration [s].

Photon flux density (photon irradiance) [mol m^{-2} s^{-1}] (at a point of a surface): In International Electrotechnical Commission (IEC) 60050-845-01 (1987), the term "photon flux density" is not listed. The term "photon irradiance" is listed instead. The definition of "photon irradiance" is that the quotient of the photon flux [mol s^{-1}] incident on an element of the surface containing the point, by the area [m^2] of that element. In the fields of horticultural science and related sciences, "photon flux density" has been used widely instead of "photon irradiance."

5.3.3 Electromagnetic radiation at wavelengths of 400—700 nm

Electromagnetic radiation at 400—700 nm wavelengths has been called "photosynthetically active radiation" in the fields of horticultural science and related sciences because the radiation in wavelength range described above was proved with experiments using mainly edible crops (McCree, 1972; Inada, 1976) to drive plant photosynthesis.

The question is what radiometric and photonmetric quantities at wavelengths of 400—700 nm should be called. Probably, adding "photosynthetic" in front of those quantity names is the simplest and most readily comprehensible nomenclature, producing expression such as photosynthetic photon intensity, photosynthetic radiant flux, photosynthetic photon number, and photosynthetic irradiance. Although adding "photosynthetic active" in front of those quantity names might be reasonable, the use of *photosynthetically active photon flux density* cannot be accepted by most horticultural scientists because they have used "photosynthetic photon flux density (PPFD)" exclusively as an established technical term to describe "photon flux density," of which photons have 400—700 nm wavelength. It is noteworthy that, according to standardized IEC terminology, "photosynthetic photon irradiance" is the recommended quantity name for "photosynthetic photon flux density."

In fact, the PPFD is a frequently measured quantity when plants are cultivated. The value is an important element used for evaluation of the light environment. Fig. 5.2 presents a typical spectral response of a quantum sensor and ideal quantum response for measurement of the PPFD.

5.3.4 Spectral distribution and quantitative relations of radiometric, photometric, and photonmetric quantities

5.3.4.1 Spectral distribution and spectral quantity

Spectral distribution is a term used to describe a quantity pertaining to radiation as a function of wavelength or quantity per unit wavelength, arranged in wavelength sequence. A quantity, pertaining to radiation, per unit wavelength is designated as the "spectral quantity." When describing a spectral quantity, nanometer [nm] is often used conventionally as the unit designating "unit wavelength." This practice is contrary to the indication by the Ninth SI Brochure (Bureau International des Poids et Mesures, 2019) that the prefix should be placed at the beginning of a compound unit symbol, as mentioned in Section 5.2.5.1. In

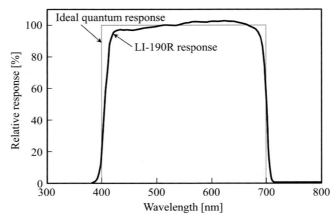

FIGURE 5.2   Typical spectral response of LI-190R quantum sensors and ideal quantum response for the measurement of photosynthetic photon flux density (photosynthetic photon irradiance). *The figure was redrawn from an LI-190R catalogue.*

the following discussion, the author hesitantly uses nanometer [nm] in units for spectral quantities to indicate that the quantities are used per "unit wavelength."

The adjective "spectral" is useful for the 12 terms presented in Table 5.5 and the other terms describing quantities pertaining to radiation when the property or wavelength dependence of the quantity is described, such as spectral radiant flux, spectral photon flux, spectral irradiance, and spectral photon flux density. Those quantities are also examples of spectral quantities.

5.3.4.2 *Relations of radiometric, photometric, and photonmetric spectral quantities*

The relations of spectral irradiance, illuminance, and photon flux density (photon irradiance) for 300−800 nm wavelengths are presented in Fig. 5.3 for spectral irradiance, taking a constant value of 1 W m^{-2} nm^{-1}. Values outside and inside of parentheses presented in the upper left of each graph, respectively, represent the irradiance, illuminance, and photon flux density for wavelengths of 300−800 nm and 400−700 nm. They are the integrals of spectral irradiance, illuminance, and photon flux density in their respective wavelength ranges.

The spectral illuminance curves presented in Fig. 5.3 were drawn using the CIE standard spectral luminous efficiency function for photopic vision ($V(\lambda)$) by the Commission Internationale de l'Éclairage (CIE) Standard (2005), in which the values of the function at 1 nm increments are tabulated. The value of spectral illuminance (for photopic vision) at a wavelength λ [nm], $L_{is,\lambda}$ is calculable as shown below.

$$L_{is,\lambda} = \text{Km} \cdot R_{is,\lambda} \cdot V(\lambda) \tag{5.1}$$

In that equation, $R_{is,\lambda}$ stands for the spectral irradiance at wavelength λ [W m^{-2} nm^{-1}], for irradiance [W m^{-2}] per 1-nm wavelength range [nm], and where K_m is a constant relating to the radiometric quantities and photometric quantities, called the "maximum spectral luminous efficacy of radiation for photopic vision." The value of K_m is given by the 1979 definition

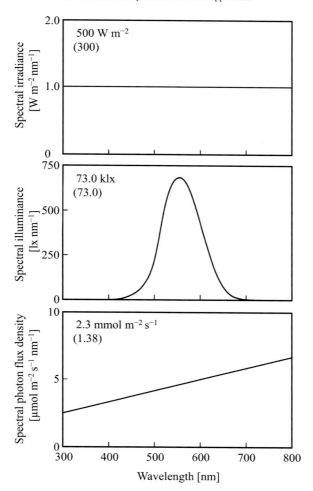

FIGURE 5.3 Spectral irradiance taking a constant value of 1 W m^{-2} nm^{-1} and its corresponding spectral illuminance and photon flux density for wavelengths of 300–800 nm. Values outside and inside of parentheses shown in the upper left of each graph, respectively, represent the irradiance, illuminance, and photon flux density for the respective wavelengths of 300–800 nm and 400–700 nm.

of candela, which defines the spectral luminous efficacy of light at 540×10^{12} Hz (at 555- nm wavelength) as 683 lm W^{-1} (Ohno, 1997).

The curve of spectral photon flux density in Fig. 5.3 was drawn using the theoretical equation presented below. Energy, E [J] that n moles of photons with wavelength λ [nm], is given as the following equation.

$$E = n \cdot N_A \cdot h \cdot \nu = \frac{n \cdot N_A \cdot h \cdot c}{\lambda \times 10^{-9}} \tag{5.2}$$

Therein, h is Planck's constant $(6.626 \times 10^{-34} \, \text{J s})$, c represents the speed of light $(2.998 \times 10^8 \, \text{m s}^{-1})$, and N_A stands for Avogadro's number $(6.022 \times 10^{23} \, \text{mol}^{-1})$.

The relation between $R_{is,\lambda}$ [W m^{-2} nm^{-1}] and the spectral photon flux density at wavelength λ [mol m^{-2} s^{-1} nm^{-1}], which is the photon flux density [mol m^{-2} s^{-1}] per 1-nm wavelength range [nm], is calculable by substituting $R_{is,\lambda}$ into E, and $P_{is,\lambda}$ into n in Eq. (5.2).

$$R_{is,\lambda} = \frac{P_{is,\lambda} \cdot N_A \cdot h \cdot c}{\lambda \times 10^{-9}} \tag{5.3}$$

$$\therefore \quad P_{is,\lambda} = R_{is,\lambda} \frac{\lambda \times 10^{-9}}{N_A \cdot h \cdot c} \tag{5.4}$$

For example, the spectral photon flux density for λ of 480 nm, $P_{is,480}$ in Fig. 5.3 is calculable by substituting $R_{is,480}$ (= 1.00 in Fig. 5.3) into $R_{is,\lambda}$ in Eq. (5.4).

$$P_{is,480} = R_{is,480} \frac{\lambda \times 10^{-9}}{N_A \cdot h \cdot c}$$

$$= 1.00 \times \frac{480 \times 10^{-9}}{(6.022 \times 10^{23}) \times (6.626 \times 10^{-34}) \times (2.998 \times 10^8)}$$

$$= 1.00 \times \frac{480 \times 10^{-9}}{119.6 \times 10^{-3}}$$

$$= 4.012 \times 10^{-6} \, \text{mol m}^{-2} \, \text{s}^{-1} \, \text{nm}^{-1}$$

$$= 4.012 \, \mu\text{mol m}^{-2} \, \text{s}^{-1} \, \text{nm}^{-1}$$

$$\approx 4 \, \mu\text{mol m}^{-2} \, \text{s}^{-1} \, \text{nm}^{-1}$$

The spectral photon flux density at λ of 480 nm is noteworthy, taking a value of approximately 4 μmol m^{-2} s^{-1} nm^{-1}, meaning that the photon flux density within a 1-nm wavelength range of 479.5–480.5 nm is approximately 4 μmol m^{-2} s^{-1}.

Fig. 5.4 portrays quantitative relations among spectral irradiance ($R_{is,\lambda}$), spectral illuminance ($L_{is,\lambda}$), and spectral photon flux density ($P_{is,\lambda}$). The quantitative relations among the three metrics hold for quantities categorized in the same row in Table 5.5, e.g., spectral radiant flux, luminous flux, and photon flux. Incidentally, the photon number in μmol corresponding to energy of 1 J at a wavelength of λ [nm], M_λ [μmol J^{-1}], is calculable using the following equation, which is derived from Eq. (5.2), when wavelength λ [nm] of the photons is designated.

$$M_\lambda = \lambda / 119.6 \tag{5.5}$$

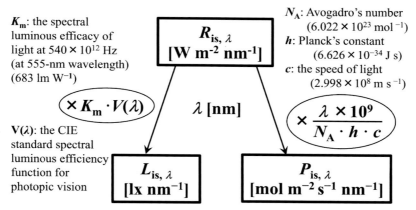

FIGURE 5.4 Relations among spectral irradiance ($R_{is,\lambda}$) [W m^{-2} nm^{-1}], spectral illuminance ($L_{is,\lambda}$) [lx nm^{-1}], Table 5.2 and spectral photon flux density ($P_{is,\lambda}$) [mol m^{-2} s^{-1} nm^{-1}]. The relations among the three metrics hold for the quantities categorized in the same row in Table 5.5.

For instance, when λ equals 660 nm, the photon number per joule is calculated as approximately 5.5 µmol J^{-1}.

5.4 Classification of electromagnetic radiation and light in terms of wavelength

5.4.1 General classification of electromagnetic radiation

The International Electrotechnical Commission (IEC) (2020) defines electromagnetic radiation as the energy that emanates from a source in the form of electromagnetic waves or photons and is transferred through space (IEV ref 705-02-01), noting that the physical concepts of photons and electromagnetic waves are used to describe the same phenomenon of transmission of radiant energy in different ways, depending on the characteristics of interaction of the energy with the physical world. The electromagnetic radiation includes γ radiation, X radiation, ultraviolet radiation, visible radiation, infrared radiation, and radio waves (including microwaves) in increasing order of wavelength (Fig. 5.5).

On the upper slip of Fig. 5.5, the classification of electromagnetic radiation is shown, with the classification from UV-C to the end of visible radiation shown on the lower slip. Some designated boundary wavelengths used to classify the electromagnetic radiation are not standardized among different academic disciplines; they may differ somewhat according to the academic discipline to which they belong. For instance, the wavelength range of ultraviolet radiation differs depending on the academic discipline; the shortwave end is defined as 1 nm or 10 nm; the longwave end is done 360, 380, or 400 nm according to the academic discipline. Regarding color boundaries, the wavelengths designated in the lower slip are based to a large degree on the author's personal views.

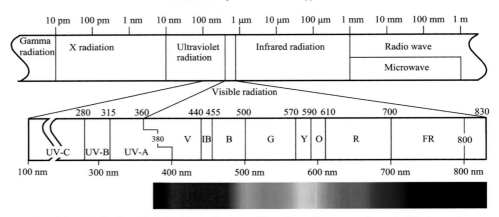

V: violet, IB: indigo blue, B: blue, G: green, Y: yellow, O: orange, R: red, FR: far red

FIGURE 5.5 Classification of electromagnetic radiation and visible radiation in terms of wavelength. The boundary wavelengths designated in the figure are not definite; they differ among academic disciplines. *Modified from Fujiwara, K., 2013. Fundamentals of light on plant cultivation and LED light irradiation technology. Refrigeration 88, 163—168. (in Japanese).*

5.4.2 Definition of light and its wavelength ranges

Light is defined, in physics and its related sciences, as electromagnetic radiation covering ultraviolet radiation, visible radiation, and infrared radiation. In this case, the wavelength range covers 10 nm to 1 m. However, in illuminating engineering, the term "light" refers to visible radiation. The International Electrotechnical Commission (IEC) (2020) defines visible radiation as any optical radiation capable of causing a visual sensation directly (IEV ref 845-01-03). The CIE provides the CIE standard spectral luminous efficiency function for photopic vision for wavelengths of 360—830 nm (Ohno, 1997), together with an introduction of the maximum spectral luminous efficacy of radiation for photopic vision. Based on the CIE standard spectral luminous efficiency function for photopic vision, the author fundamentally includes wavelengths of 360—830 nm as that for light.

5.4.3 Terms to avoid when describing a plant light environment

The terms "light intensity" and "light quality" have been used frequently in scientific articles of all types, and even in academic journal titles. These terms, however, should be replaced with an appropriate term in every case.

Two underlying problems exist in the use of the term "light intensity." First, the term "light intensity" itself refers to no specific quantity despite the author's intention. Therefore, the usage of "light intensity" can only lead the readers of an article to misunderstanding and incomprehension. Secondly, in IEC and CIE terminology, the word "intensity" is used to mean a "flux (radiant flux [W], luminous flux [lm], or photon flux [mol s^{-1}]) leaving the source per unit solid angle [sr] in the direction considered" as described in Section 5.3.2, such as radiant intensity, luminous intensity, and photon intensity. In most cases, however, the term "light intensity" has been used to mean PPFD [mol m^{-2} s^{-1}] or photosynthetic irradiance [W m^{-2}] in the fields of horticultural science and related sciences. In such cases, the term PPFD or photosynthetic irradiance should be used directly instead of "light intensity."

The term "light quality" has been used when referring roughly to a "light spectrum," mainly because of lack of sufficient knowledge of light-related terms. Use of the term "light quality" presents two underlying points of difficulty. First, it provides no information related to what spectral quantity is being featured. Secondly, it does not specify whether the spectral quantity is presented as an absolute value or relative value. The term "light quality" should be replaced with an appropriate concrete expression such as spectral radiant flux distribution, spectral photon flux density distribution, relative spectral irradiance distribution, or relative spectral photon flux distribution.

Finally, the author must address the particularly unfavorable term of "daily light integral," which is often used in the fields of horticultural science and related sciences to designate PPFD integrated over a 24-h period. The term "daily light integral" sounds manifestly strange because no academic field defines "light" as photosynthetically active radiation. It is readily apparent that "daily PPFD integral" is much appropriate than "daily light integral," considering the unit assigned to "daily light integral."

References

Bureau International des Poids et Mesures, 2019. Le Système International D'unités/the International System of Units, 216pp. https://www.bipm.org/utils/common/pdf/si-brochure/SI-Brochure-9.pdf.

Commission Internationale de l'Éclairage (CIE) Standard, 2005. Photonmetry—The CIE System of Physical Photonmetry, CIE S 010/E:2004. CIE Central Bureau, Vienna, Austria.

Fujiwara, K., 2013. Fundamentals of light on plant cultivation and LED light irradiation technology. Refrigeration 88, 163—168 (in Japanese).

Fujiwara, K., 2016. Radiometric, photometric and photonmetric quantities and their units. In: Kozai, T., Fujiwara, K., Runkle, E. (Eds.), LED Lighting for Urban Agriculture. Springer Science+Business Media Singapore, pp. 367—376.

Fujiwara, K., 2019. Light sources. In: Kozai, T., Niu, G., Takagaki, M. (Eds.), Plant Factory: An Indoor Vertical Farming System for Efficient Quality Food Production, second ed. Academic Press, London, UK, pp. 139—151.

Inada, K., 1976. Action spectra for photosynthesis in higher plants. Plant Cell Physiol. 17, 355—365.

International Electrotechnical Commission (IEC), 2020. Electropedia: The World's Online Electrotechnical Vocabulary (International Electrotechnical Vocabulary (IEV) Online). http://www.electropedia.org/iev/iev.nsf/d253fda6386f3a52c1257af700281ce6?OpenForm.

International Electrotechnical Commission (IEC) 60050-845-01, 1987. International Electrotechnical Vocabulary, Chapter 845: Lighting.

McCree, K.J., 1972. The action spectrum, absorptance and quantum yield of photosynthesis in crop plants. Agric. Meteorol. 9, 191—216.

Meyer-Arendt, J., 1968. Radiometry and photometry: units and conversion factors. Appl. Opt. 7 (10), 2081—2084.

Ohno, Y., 1997. NIST Measurement Services: Photometric Calibrations. NIST Special Publication 250-37. Gaithersburg, MD, 66pp.

LED product terminology and performance description of LED luminaires

Kazuhiro Fujiwara

Graduate School of Agricultural and Life Sciences, The University of Tokyo, Tokyo, Japan

6.1 Introduction

Basic information on LEDs that is necessary for the use of LEDs as a light source for plant cultivation includes the following (Fujiwara, 2016b): (1) definitions of LED product terms; (2) light-emitting principle of an LED; (3) LED package configuration types; (4) basic terms for expressing optical, electrical, and radiational characteristics of an LED; (5) optical, electrical, and radiational characteristics in LED operation; (6) lighting methods; (7) radiant flux control methods; and (8) special requirements for LED lamps to cultivate plants.

This chapter first presents some LED-related basic information, particularly addressing the items (1) definitions of LED product terms and (3) LED package configuration types. The definitions of LED product terms are presented simply and clearly because misuse of LED product terms is encountered frequently in articles related to plant factory with artificial lighting research and development. These descriptions contribute to a comprehensive understanding of terminology related to LED products and LED luminaire component formation.

In response to the vibrant demand for LED luminaires to cultivate plants, markets are recently providing widely diverse LED luminaires. This chapter then describes LED luminaire performance and its evaluation items, which are necessary to compare the performance among LED luminaires for greenhouse growers and plant factory managers to select an appropriate or desirable LED luminaire among widely various commercialized ones. Selecting an appropriate or desirable LED luminaire among them is not easy for greenhouse growers and plant factory managers because few manufacturers and suppliers provide the necessary performance data of LED luminaires for plant cultivation. Furthermore, no widely accepted standard exists to describe LED luminaire performance. An LED performance description sheet that has been publicized by the Japan Plant Factory Association is introduced as an example of a well-organized information resource offering easy access via a website with no viewing restriction.

Several sections of this chapter include partially reconstituted, modified, and updated contents of the author's earlier work (Fujiwara, 2016b, 2019a, 2019b).

6.2 LED product terminology

Primary LED product terms must be presented clearly to avoid confusion and misunderstanding. The primarily and frequently used LED product terms are LED chip, LED package, LED module, LED lamp, LED light source, LED luminaire, and LED lighting system. The term "LED control gear," which appears in the explanation of the terms listed above, should also be presented. Most definitions for these terms are direct quotes from IEC 62504 (2014). The International Electrotechnical Commission (IEC) is the world's leading organization that prepares and publishes International Standards for electrical, electronic, and related technologies.

For easy access to check the definitions for LED product terms, the International Electrotechnical Vocabulary (IEV) Online produced by IEC (2020) is expected to be the most useful and recommendable sites.

6.2.1 Primary LED product terms

6.2.1.1 LED

An LED is a solid device embodying a *p-n* junction, emitting incoherent optical radiation when excited by an electric current (IEC, 2014). Under that definition, an LED chip is used with the same meaning. In other words, what can be paraphrased as an LED chip is called an LED. Generally speaking, most people, including relevant specialists, use the term "LED" to refer to an LED package, LED module, LED lamp, and LED light source without regard to their classification. Special attention must be devoted to the selection of an appropriate term when considering usage of the term "LED" alone to refer to a specific type of LED product.

6.2.1.2 LED package

An LED package is a single electrical component encapsulating principally one or more LED dies, possibly including optical elements and thermal, mechanical, and electrical interfaces (IEC, 2014). An LED die, under the definition presented above, is used almost interchangeably with an LED chip. The lamp-type (also called round type or through-hole type), surface-mount device (SMD) type, and chip-on-board (COB)—type LED packages (Figs. 6.1 and 6.2) are major examples of it.

6.2.1.3 LED module

An LED module is an LED light source having no cap, incorporating one or more LED packages on a printed circuit board, and possibly including one or more of the following: electrical, optical, mechanical, and thermal components, interfaces, and control gear (IEC, 2014). A component built in a bulb-type or straight tube—type LED lamp,

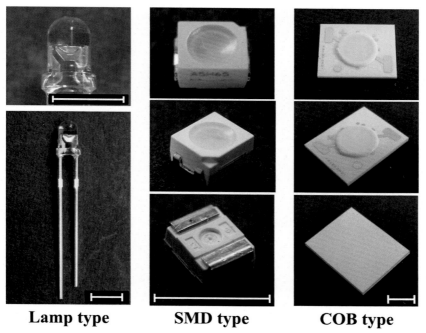

Lamp type   **SMD type**   **COB type**

FIGURE 6.1   Lamp type (NSPB310B; Nichia Corp.), surface-mount device (SMD) type (NESG064; Nichia Crop.), and chip-on-board (COB) type (NTCWT012B—V3; Nichia Corp.) LED packages. White bars represent a scale of approximately 5 mm. *Modified from Fujiwara, K., 2019b. Light sources. In: Kozai, T., Niu, G., and Takagaki, M. (Eds.) Plant Factory — an Indoor Vertical Farming System for Efficient Quality Food Production, second ed., Academic Press, London, U.K., p.139—151.*

with dozens of SMD-type LED packages placed on a printed circuit board, is an example of an LED module. The COB-type LEDs (Figs. 6.1 and 6.2), which are fundamentally dozens of LED chips bonded directly to a substrate to form a single LED module, also constitute an example of an LED module. Incidentally, the COB-type LEDs are categorized as high-power LEDs, which can produce a large luminous flux from a small light-emitting area.

6.2.1.4 LED control gear

An LED control gear is a unit inserted between the electrical supply and one or more LED modules, which serves to supply the LED module(s) with its (their) rated current, and may consist of one or more separate components and may include means for dimming, correcting the power factor and suppressing radio interference, and further control functions (IEC, 2014). Most single LED chips are low-current devices (mostly less than 0.1 A). Generally, LED modules consist of a combination of series and parallel connection strings. Commercial electric power supplies rarely match the voltage or current requirements of most of LED modules. Control gear is necessary for regulating the voltage or current for the LED module(s).

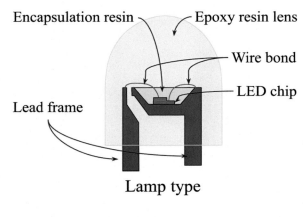

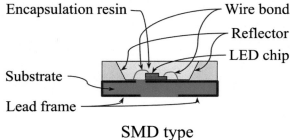

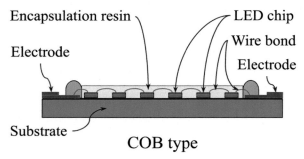

FIGURE 6.2 Schematic cross-sectional diagram of lamp-type, surface-mount device (SMD) type, and chip-on-board (COB)—type LEDs. *Credit:* Original.

6.2.1.5 LED lamp

An LED lamp is an LED light source provided with (a) cap(s) incorporating one or more LED modules and possibly including one or more of the following: electrical, mechanical, and thermal components, interfaces, and control gear (IEC, 2014). A bulb-type LED lamp and tube-type LED lamp are examples of an LED lamp.

6.2.1.6 LED light source

An LED light source is an electrical light source based on LED technology (IEC, 2014). The term "electrical light source" refers to a device emitting light by transforming electric energy into light. An electrical light source comprises one or more electrical lighting devices and lamps, possibly together with components designed to distribute the light, to position and protect the lamps, and to connect the lamps to the electric power supply. According to the definitions presented above, the category of LED light sources can include LED chips, LED packages, LED modules, and LED lamps.

6.2.1.7 LED luminaire

An LED luminaire is a luminaire designed to incorporate one or more LED light sources (IEC, 2014). The term "luminaire" designates an apparatus which distributes, filters, or transforms the light transmitted from one or more lamps and which includes, except the lamps themselves, all the parts necessary for fixing and protecting the lamps and, where necessary, circuit auxiliaries together with the means for connecting them to the electric supply (IEC, 2020).

According to the definition presented above, an LED luminaire is not included in the category of LED light sources because LED luminaires have no LED chip, LED package, LED module, or LED lamp for illumination. However, a luminaire is sometimes used as a term to mean "complete lighting unit," including one or more lamps, reflective surfaces, protective housings, electrical connections, and circuitry. According to the author's impression, more researchers (including the author) and professionals in fields related to illuminating engineering use the term "luminaire" to mean a "complete lighting unit." In such researchers' and professionals' opinions, an LED luminaire is included in the category of LED light sources.

6.2.1.8 LED lighting system

The term "LED lighting system" seems to have no literature-based definition. If the author defines the term, then the definition should be the following: a system composed of one or more LED light sources and a luminaire in the IEC definition with or without an instrument to control the radiant flux or relative spectral radiant flux distribution of emitted light such as a timer, dimmer, or computer-programmed control system.

6.2.2 Illustrated explanation of LED product relations

Based solely on the descriptions given above, one might have difficulty understanding what each LED product name specifically indicates, in addition to the conformational relations among those LED products. Although one can readily find articles with an illustration indicating what each LED product name specifically refers to, one rarely encounters an article with an illustration showing LED product relations. Therefore, the author prepared an illustration describing the conformational relations of LED products (Fig. 6.3). It is noteworthy that an LED luminaire refers to a complete lighting unit and that it is included in the category of LED light sources in this figure, differing from the IEC definition. The definition of an LED lighting system here is based on the author's view.

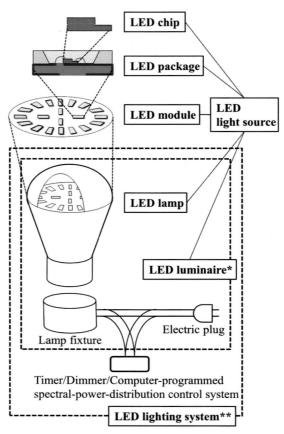

FIGURE 6.3 Schematic illustration displaying LED product relations. * Under the IEC definition, an LED luminaire does not include any LED lamp. ** It is not defined by the IEC; its definition presented here is based on the author's view. *Credit:* Original.

6.3 Performance description of LED luminaires

The American Society of Agricultural and Biological Engineers (ASABE) recognized the plant growth community's need for a standardization committee for developing LED horticultural lighting standard documents. The ASABE formed the new Plant Growth LED Lighting Committee in early 2015 (Jiao, 2015), and in September 2018, the American National Standards Institution (ANSI) and ASABE released a standard designated as "Recommended methods for measurement and testing of LED products for plant growth and development" (ANSI/ASABE S642 SEP2018., 2018). A copy of the standard document can be downloaded for a fee. Both et al. (2017) presented an example of a product label for an LED lamp designed specifically for plant growth applications. The configuration example is well organized with nine item groups.

In 2016, the Plant Growth LED Lighting Research Committee of the Japan Plant Factory Association proposed a tentative standard description sheet to record and compare LED luminaire (complete lighting unit) performance for plant cultivation (Goto et al., 2017). In early 2018, it publicized an improved version via a website with no viewing restriction (Fujiwara, 2019a). The sheet has been improved continually by the committee. The latest version was publicized in late 2019 (Japan Plant Factory Association, 2019). The latest version of the description sheet requires the filling out of 8 items as general information and 18 items in 6 item groups as LED luminaire performance data. The author introduces the description sheet explaining all items listed in the sheet.

This section describes an updated version from previously reported ones (Fujiwara, 2019a, 2019b), reflecting the latest update of the LED luminaire performance description sheet. It is noteworthy that the term "luminaire" used in this section refers to "complete lighting unit," in contrast to the IEC definition of luminaire.

6.3.1 Description sheet to record and compare LED luminaire performance

The standard description sheet which is used to record and compare LED luminaire performance for plant cultivation (Fig. 6.4) consists of two description blocks: a general information block (GI block) and an LED luminaire performance data block (LLPD block). The sheet is intended and designed to allow greenhouse growers and plant factory managers to compare LED luminaire performance for selecting an appropriate LED luminaire for their own facility on well-targeted items. For that intention, the items are limited to a bare minimum. If necessary, refer to Fujiwara (2016a; 2019b) for terms related to radiometric, photometric, and photonmetric quantities that appear on the sheet.

The sheet includes no document that prescribes methods for the measurement and testing of LED luminaires to obtain item values to be filled in the sheet column. Methods necessary for the measurement and testing of LED luminaires should be subjected to well-established standards such as IEC standards and International Commission on Illumination (CIE) standards. ANSI/ASABE S642 SEP2018: "Recommended methods for measurement and testing of LED products for plant growth and development" (ANSI/ASABE S642 SEP2018, 2018) cited above can be an appropriate document for reference when one needs to check a method for the measurement and testing of LED luminaires.

6.3.1.1 General information block

The GI block includes columns declaring general information related to the LED luminaire and the performance description task. The GI block requires that eight items be filled out: (1) manufacturer, (2) trade name, (3) model number, (4) measurer, (5) measurement date, (6) respondent's contact name, (7) date of description, and (8) change record. The manufacturer or sales company might commission measurements of several or all items listed in the sheet to an authorized public or private institution capable of measuring those items and of ensuring the accuracy of the measured values.

Description of LED luminaire performance for plant cultivation

Manufacturer	
Trade name	
Model number	
Measurer	
Measurement date	
Respondent's contact name	
Date of description	
Change record	

Item group	Item	Unit	Write-in column	Measuring instrument /method	Remarks
Measuring condition	Temperature of ambient air	℃			
Electric power	Power-supply current type	−			AC/DC
	Rated voltage	V			Rated voltage for lighting and controlling light output. Separate description of rated voltage for each component is acceptable.
	Rated current	A			Rated current for lighting and controlling light output. Separate description of rated current for each component is acceptable.
	Rated power consumption	W			Rated power consumption for lighting and controlling light output. Separate description of rated power consumption for each component is acceptable.
Spectral characteristics	Spectral photon flux distribution	To be shown separately[1]			Wavelength: 300–800 nm X-axis: wavelength [nm] Y-axis: spectral photon flux [μmol s^{-1} nm^{-1}]
	Spectral radiant flux distribution	To be shown separately[1]			Wavelength: 300–800 nm X-axis: wavelength [nm] Y-axis: spectral radiant flux [W nm^{-1}]
	Photosynthetic photon flux (PPF)	μmol s^{-1}			Wavelength: 400–700 nm
	Photosynthetic radiant flux	W			Wavelength: 400–700 nm
	Luminous flux	lm			
	Color temperature	K			Only for white LEDs
	Color rendering index	−			Only for white LEDs
Spatial distribution characteristics	Spatial distribution curve of photosynthetic photon intensity	To be shown separately[1]			A spatial distribution curve of luminous intensity is acceptable.
Efficacy/Efficiency	Photosynthetic photon number efficacy	μmol J^{-1}			Wavelength: 400–700 nm
	Photosynthetic radiant energy efficiency	J J^{-1}			Wavelength: 400–700 nm
	Luminous efficacy	lm W^{-1}			
Maintainability	Lifetime from the viewpoint of PPF decrease	h			Time to 90% of the initial PPF
	Dust-proof and water-proof characteristics	−			IP code ("IP" followed by a two-digit number)

[1]Graph and measured data files should be submitted.

FIGURE 6.4 Standard description sheet of LED luminaire performance for plant cultivation by the Japan Plant Factory Association's Committee on LED Lighting. *Reproduced from Japan Plant Factory Association, 2019. https:// npoplantfactory.org/information/committee/746/.*

6.3.1.2 LED luminaire performance data block

The LLPD block describes the required values or required figures and the related information including measuring instruments and methods. The LLPD block requires that 18 items of 6 item groups be filled out: (1) measuring condition, (2) electric power, (3) spectral characteristics, (4) spatial distribution, (5) efficiency/efficacy, and (6) maintainability.

(1) Measuring conditions

The most important measuring condition must be the temperature of ambient air of a luminaire to be measured. The relative humidity of ambient air does not exert a marked effect on most items in the LLPD block. However, it should be provided in the remark column for lifetime from the viewpoint of photosynthetic photon flux (PPF) decrease because it exerts a marked effect on the LED luminaire lifetime.

Temperature of ambient air

Items included in the LLPD block other than those in the group item of maintainability must be measured at a constant temperature (e.g., 25°C) because the item values are nonnegligibly affected by temperature.

(2) Electric power

Data for the electric power necessary to drive the LED luminaire are usually stated clearly in the luminaire specification sheet provided by the manufacturer or responsible vendor.

Power-supply current type

Designated current type: alternating current or direct current.

Rated voltage [V]

It is the voltage to be applied to the luminaire for which the manufacturer guarantees normal operation for an assumed operating time under the assumed operating conditions. A separate description of the rated voltage for each component is acceptable.

Rated current [A]

Rated current is either the current that is presumed to be carried when the specified rated voltage is applied to the luminaire or current that is to be applied to the luminaire for which the manufacturer guarantees normal operation for an assumed operating time under the assumed operating conditions. Separate description of the rated current for each component is acceptable.

Rated power consumption [W]

Rated power consumption is the total effective power that is presumed to be consumed when the specified rated voltage is applied to the luminaire. A separate description of the rated power consumption for each component is acceptable.

(3) Spectral characteristics

Spectral photon flux distribution, PPF, photosynthetic radiant flux, and luminous flux of a luminaire are calculable when the spectral radiant flux distribution of the luminaire is provided (Fujiwara (2016a) presents detailed calculations). However, all that information should be provided on the sheet for the customer's convenience. The "flux" used in PPF, photosynthetic radiant flux, and luminous flux refer to the geometrically total flux of a luminaire. It is given by the spatial integration or by the integrated flux over a solid angle of 4π sr (steradian).

Spectral photon flux distribution

Spectral photon flux distribution is a spectral distribution for 300–800 nm wavelengths representing spectral photon flux [μmol s^{-1} nm^{-1}] (y-axis) as a function of wavelength [nm] (x-axis). It should be shown on a separate sheet (Fig. 6.5 presents a sample illustration).

Spectral radiant flux distribution

Spectral radiant flux distribution is a spectral distribution for 300–800 nm wavelengths representing spectral radiant flux [W nm^{-1}] (y-axis) as a function of wavelength [nm] (x-axis). It should be shown on a separate sheet (Fig. 6.6 presents a sample illustration).

Photosynthetic photon flux (PPF) [μmol s^{-1}]

Photosynthetic photon flux is photon flux [μmol s^{-1}] with 400–700 nm wavelength. When numerical data of the spectral photon flux distribution are provided, PPF is calculable using the data.

Photosynthetic radiant flux (PRF) [W]

Photosynthetic radiant flux is a radiant flux [W] with 400–700 nm wavelength. When numerical data of the spectral radiant flux distribution are provided, PRF is calculable using the data.

Luminous flux [lm]

It is quantity derived from radiant flux [W] by evaluating the radiation according to its action upon the CIE standard photometric observer (IEC, 1987).

Color temperature [K]

Color temperature is a temperature of a Planckian radiator (blackbody radiator) with radiation having the same chromaticity as that of a given stimulus (IEC, 1987). The color temperature is also designated technically as the correlated color temperature. The value is necessary only for a luminaire with white LEDs.

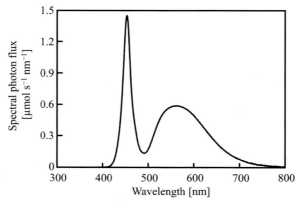

FIGURE 6.5 Sample illustration of the spectral photon flux distribution of light emitted from a luminaire with phosphor-converted white LEDs. *Modified from Fujiwara, K., 2019a. Introduction of a performance description sheet for plant cultivation LED luminaires. Paper Presented at: 2019 International Symposium on Environment Control Technology for Value-Added Plant Production, Beijing, China, Huwan Hotel).*

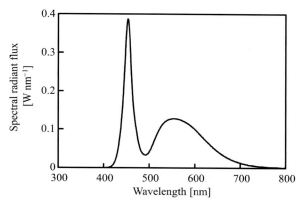

FIGURE 6.6 Sample illustration of spectral radiant flux distribution of light emitted from a luminaire with phosphor-converted white LEDs. *Modified from Fujiwara, K., 2019a. Introduction of a performance description sheet for plant cultivation LED luminaires. Paper Presented at: 2019 International Symposium on Environment Control Technology for Value-Added Plant Production, Beijing, China, Huwan Hotel).*

Color rendering index (CRI)

This is defined as the mean of the CIE 1974 special color rendering indices (CRIs) for a specified set of eight test color samples (Schanda, 2015). A CRI value of the color of light that is close to 100 does not mean that the light has a similar relative spectral distribution of sunlight observed on the ground. For the column of CRI, the CIE 1974 general CRI, Ra, must be filled in. The value is necessary only for a luminaire with white LEDs.

(4) Spatial distribution

Spatial distribution curve of photosynthetic photon intensity

This curve is explainable roughly as a visual representation of photosynthetic photons emitted and diffused using a luminaire. The curve is to be shown on a separate sheet (Fig. 6.7 presents a sample illustration). An illustration of the spatial distribution of the relative photosynthetic photon intensity is acceptable. Most single-LED packages that can be regarded roughly as a point light source show a symmetrical distribution with respect to the optical axis (0-degree axis). A three-dimensional illustration of (relative) photosynthetic photon intensity distribution should be provided for tube-type and panel-type luminaires. When it is difficult to provide a spatial distribution of the photosynthetic photon intensity, a spatial distribution curve of luminous intensity is acceptable.

(5) Efficacy/Efficiency

Photosynthetic photon number efficacy [μmol J^{-1}]

This is defined as the number of photons with 400–700 nm wavelength or the number of photosynthetic photons [μmol] emitted per unit of electric energy [J] input to a luminaire (or lamp). This is one mode of energy–photon conversion efficacy (Fujiwara, 2016b). An ideal maximum value of approximately 5.85 μmol J^{-1} is obtained with 700-nm single-color light emitted using a luminaire with its theoretical maximum energy–photon conversion efficacy. Commercially available phosphor-converted white LEDs

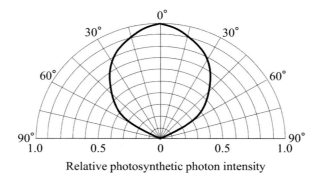

FIGURE 6.7 Sample illustration of spatial distribution of relative photosynthetic photon intensity of an LED luminaire. *Modified from Fujiwara, K., 2019a. Introduction of a performance description sheet for plant cultivation LED luminaires. Paper Presented at: 2019 International Symposium on Environment Control Technology for Value-Added Plant Production, Beijing, China, Huwan Hotel).*

would take a photosynthetic photon number efficacy of roughly 1.2–2.0 μmol J^{-1}. Although the efficacy is designated as "photosynthetic photon efficacy" in ANSI/ASABE S640 JUL2017 (2017), it sounds strange because "photosynthetic photon" is merely a substance name, not a quantity name. It cannot be expressed with units.

Photosynthetic radiant energy efficiency [J J^{-1}]

This is defined as radiant energy with 400–700 nm wavelength or photosynthetic radiant energy [J] emitted per unit of electric energy [J] input to a luminaire (or lamp).

Luminous efficacy [lm W^{-1}]

Strictly speaking, luminous efficacy should be designated as the luminous efficacy of a source (IEC, 1987). It is defined as the luminous flux [lm] obtained per unit of electric power [W] input to a luminaire (or lamp). In practice, it has no important implications for plant cultivation.

(6) Maintainability

Lifetime from the viewpoint of photosynthetic photon flux (PPF) decrease [h]

Time to 90% of the initial PPF. The relative humidity of ambient air during the lifetime test should be provided in the remarks column because it exerts a marked effect on the LED luminaire lifetime.

Dust-proof and water-proof characteristics

An IP code must be filled out. The IP code, defined in IEC 60529 (IEC, 2001), classifies the degrees of protection from solid foreign objects and water provided by enclosures of the electrical equipment with a rated voltage not exceeding 72.5 kV. An IP code consists of the letters "IP" followed by a two-digit number. The first digit is an indication of the degree of the equipment's protection against solid foreign objects. The second digit is that against water. The appropriate sections of IEC 60529 provide detailed information.

6.3.2 Continuous and further improvement for standardization

Because the performance description sheet publicized in late 2019 by the Plant Growth LED Lighting Research Committee of the Japan Plant Factory Association has not yet been perfected, the committee continues to improve the sheet with consideration of suggestions and advice from related specialists and LED luminaire users. Cooperation among different groups seeking to improve standardization related to LED luminaires and/or LED products for plant cultivation and seeking to improve their practical usage can be expected to enhance the maturity and completeness of standardization.

References

ANSI/ASABE S640 JUL2017, 2017. Quantities and Units of Electromagnetic Radiation for Plants (Photosynthetic Organisms). Developed by the ES-311, Electromagnetic Radiation Application for Plants Committee, p. 12.

ANSI/ASABE S642 SEP2018, 2018. Recommended Methods for Measurement and Testing of LED Products for Plant Growth and Development. Developed by the Plant Growth LED Lighting Committee, p. 11.

Both, A.J., Bugbee, B., Kubota, C., Lopez, R.G., Mitchell, C., Runkle, E.S., Wallace, C., 2017. Proposed product label for electric lamps used in the plant sciences. HortTechnology 27, 544–549.

Fujiwara, K., 2016a. Radiometric, photometric and photonmetric quantities and their units. In: Kozai, T., Fujiwara, K., Runkle, E. (Eds.), LED Lighting for Urban Agriculture. Springer Science+Business Media Singapore, pp. 367–376.

Fujiwara, K., 2016b. Basics of LEDs for plant cultivation. In: Kozai, T., Fujiwara, K., Runkle, E. (Eds.), LED Lighting for Urban Agriculture. Springer Science+Business Media Singapore, pp. 377–393.

Fujiwara, K., 2019a. Introduction of a performance description sheet for plant cultivation LED luminaires. In: Paper Presented at: 2019 International Symposium on Environment Control Technology for Value-Added Plant Production. Huwan Hotel, Beijing, China.

Fujiwara, K., 2019b. Light sources. In: Kozai, T., Niu, G., Takagaki, M. (Eds.), Plant Factory — An Indoor Vertical Farming System for Efficient Quality Food Production, second ed. Academic Press, London, U.K., pp. 139–151

Goto, E., Fujiwara, K., Kozai, T., 2017. Proposed standards developed for LED lighting. Urban AG News (16), 73–75.

International Electrotechnical Commission (IEC), 1987. International Electrotechnical Vocabulary, Chapter 845: Lighting.

International Electrotechnical Commission (IEC), 2001. Degrees of Protection provided by Enclosures (IP Code), Edition 2.1, IEC 60529:1989+A1:1999(E).

International Electrotechnical Commission (IEC), 2014. General Lighting - Light Emitting Diode (LED) Products and Related Equipment — Terms and Definitions, Edition 1.0, IEC 62504:2014.

International Electrotechnical Commission (IEC), 2020. Electropedia: The World's Online Electrotechnical Vocabulary (International Electrotechnical Vocabulary (IEV) Online). http://www.electropedia.org/iev/iev.nsf/d253fda6386f3a52c1257af700281ce6?OpenForm.

Japan Plant Factory Association, 2019. Description of LED luminaire performance for plant cultivation. https://npoplantfactory.org/information/committee/746/ (Accessed 28 June 2021).

Jiao, J., 2015. Stakeholders make progress on LED lighting horticulture standards. LEDs Magazine 39–41.

Schanda, J., 2015. CIE color-rendering index. In: Luo, R. (Ed.), Encyclopedia of Color Science and Technology. Springer, Berlin, Heidelberg. https://doi.org/10.1007/978-3-642-27851-8_2-1.

7

Photon efficacy in horticulture: turning LED packages into LED luminaires

Paul Kusuma[1], P. Morgan Pattison[2] and Bruce Bugbee[1]
[1]Utah State University, Logan, UT, United States; [2]Solid State Lighting Services, Inc., Johnson City, TN, United States

7.1 Introduction

The final photon efficacy[1] of a light-emitting diode (LED) luminaire[2] is determined by the inherent efficiency and photon wavelength of the LED package(s)[3], multiplied by four losses associated with the design of the LED luminaire. Depending on the design of the LED luminaire and choice of operating conditions, there can be additional LED package—related losses of current droop and thermal droop; and there will be non-LED package losses related to power supply efficiency and optical efficiency. Here, we describe the typical performance of a range of high-end LED packages with peak wavelengths across the photo-biologically active range of radiation (280—800 nm). We describe how current and thermal droops affect the efficiency of the LED luminaire. Finally, the performances of some state-of-the-art LED luminaires are described.

LED technology is the most efficient lighting technology, with some LED luminaires now providing double the photon efficacy of the highest performing alternative technology, double-ended high-pressure sodium (DE-HPS) luminaires. LED luminaires can achieve a

---

[1] In this chapter, photon efficacy refers to μmol of *all* photons divided by joules of input energy (μmol J^{-1}). The term *photosynthetic* photon efficacy includes only photons with wavelengths between 400 and 700 nm.

[2] Here, the term LED luminaire refers to a "complete lighting unit" including LED packages mounted on a circuit board, power supplies, reflective surfaces, protective housing, electrical connections, and circuitry (see chapter 6).

[3] An LED package refers to an LED chip/die within a housing, which provides a thermal path, enables electrical connections, and affects the angles of photon output. For white LED packages it also includes the phosphor (see chapter 6).

photosynthetic photon efficacy of over 3.4 compared to 1.7 µmol of photons per joule of input energy (µmol J^{-1})[4] for DE-HPS (DesignLights Consortium, 2021). But, these high photon efficacies are only possible under certain operating conditions and spectral combinations. Here, we discuss the changes in efficiency/photon efficacy as LED packages are incorporated into LED luminaires. These include the following: (1) current droop, (2) thermal droop, (3) power supply efficiency, and (4) optical efficiency (Kusuma et al., 2020).

7.2 Typical LED package performance

LED technology is rapidly evolving, with advancements driven by emerging applications for specific types of LEDs. LEDs can be categorized into three groups based on the typical/primary elemental composition of semiconductor materials: (1) GaN/AlGaN for LEDs with a peak wavelength between 220 and 360 nm, (2) InGaN for LEDs with a peak wavelength between 360 and 550 nm, and (3) AlInGaP for LEDs with a peak wavelength between 550 and 1000 nm. Improvements to InGaN LEDs have been enabled by advancements in material quality and device structures that have been driven by their large scale use in general illumination (Feezell and Nakamura, 2018; Tsao et al., 2015). Improvements in AlInGaP LEDs have historically been driven by signage, indicator, and automotive applications but are now being driven primarily by horticultural applications. Development of GaN/AlGaN-based ultraviolet (UV) LEDs has been driven by applications in sterilization and curing.

Table 7.1 and Fig. 7.1 describe the efficiency and photon efficacy of commercially available LED packages with peaks from 280 to 850 nm (the data in Fig. 7.1 are normalized to 1 watt of electrical input). The data presented here and in subsequent figures were developed from data from the following companies: Nichia Corp. for the 280 nm LED package (Nichia, 2020); SeoulViosys Co. for the 310 nm LED package (SeoulViosys, 2020); Lumileds holding BV for the 385 and 405 nm LED packages (Lumileds, 2018), the 850 nm LED package (Lumiled, 2020), and the 3000 and 6500 K LED packages (Lumileds, 2021); and OSRAM Opto Semiconductors GmbH for the 450 nm LED package (Osram, 2020c), 470 nm LED package (Osram, 2020b), 500 nm LED package (Osram, 2020h), 530 nm LED package (Osram, 2020g), 590 nm LED package (Osram, 2020i), 620 nm LED package (Osram, 2020a), 635 nm LED package (Osram, 2020f), 660 nm LED package (Osram, 2020e), and 730 nm LED package (Osram, 2020d). The photon efficacy of all of these LED packages is reported at a junction temperature of 25°C and a nominal drive current specified in Table 7.1. Values here differ from Kusuma et al. (2020) for three reasons: (1) these are reported at the nominal current densities for each type of LED package as opposed to 100 mA mm^{-2}, (2) these values represent typical performance rather than top bin performance, and (3) LEDs have continued to improve (this is most true of the 660 nm red LED package). It is important to recognize that this is an evolving industry and improvements in efficiency will continue.

The LED package performances described here reflect both the fundamental LED material properties as well as commercial development focused on certain wavelengths. For example, green LEDs (with peak wavelengths between 500 and 600 nm) would be commercially viable,

---

[4] The units for photon efficacy are µmol of photons per second divided by watts of input power. Because a watt is a joule per second, the seconds cancel in the numerator and the denominator resulting in µmol J^{-1}.

but LED material challenges in both the InGaN and AlInGaP LED materials systems limit the performance of green LEDs. Among LED scientists, this is referred to as the "green gap." An example of commercial impact is that, based on material properties, an LED at 635 nm should be more efficient than an LED at 620 nm (Table 7.1), but the 620 nm LEDs are more important for general illumination, and thus developmental focus in 620 nm LEDs has led them to surpass 635 nm LEDs.

Fig. 7.1A shows the variation in the power flux (at 1 W electrical input) for direct emitting LED packages with different emission wavelengths and associated spectral bandwidths. Bandwidths are generally defined by the spectral width in nm at half of the maximum output, called

TABLE 7.1 Typical performance of select LED packages. Narrow bandwidth LED packages are described by their approximate peak wavelength and phosphor-converted (PC) white LED packages (bottom) are described by their correlated color temperature (CCT). Percent of theoretical maximum describes the efficiency of the LED packages. For the narrow bandwidth LED packages, efficiency is the power output divided by the power input, while efficiency of the PC white LED packages is the product of the efficiency of the underlying 450 nm blue LED multiplied by the phosphor-conversion efficiency. Both of these parameters must be 100% efficient for a PC white LED package to reach its theoretical maximum performance. The operating conditions to obtain the efficiencies (percent of theoretical maximum) and photon efficacies described here are a 25°C junction temperature and a nominal drive current specified in the table. Choosing LED packages from the highest performance bin can increase the efficiency and photon efficacy values by 5%—10%. Dividing the photon efficacy by the percent of theoretical maximum (as a fraction) provides the theoretical maximum efficacy for that LED package.

Wavelength (nm) or CCT (K)	Nominal drive current (mA)	Photon efficacy (μmol J^{-1})	Percent of theoretical maximum
280 nm	100	0.08	3
310 nm	20	0.03	1
385 nm	500	0.9	28
405 nm	500	1.6	48
450 nm	350	2.8	74
470 nm	350	2.4	62
500 nm	350	2.0	50
530 nm	350	1.3	30
590 nm	350	1.1	21
620 nm	350	3.4	64
635 nm	350	2.5	47
660 nm	700	4.1	74
730 nm	350	3.6	59
850 nm	1000	3.0	42
3000 K	65	2.8	74
6500 K	65	2.9	77

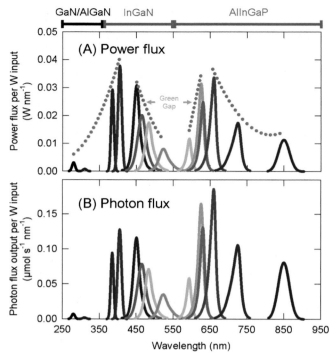

FIGURE 7.1   (A) Power flux of 14 LED packages across the photo-biologically active region of radiation. These fluxes assume 1 watt of electrical input. (B) Data in (A) converted to a photon flux using Planck's equation.

the full width at half maximum. Fig. 7.1 highlights the difference in efficiency for different wavelength LED packages, especially the green gap in LED performance. There are research examples that show it is possible for InGaN LEDs to provide emission across the visible spectrum, but for now, AlInGaP LEDs have a much better performance at emission wavelengths around the red portion of the spectrum (DOE BTO Lighting R&D Program, 2019).

Fig. 7.1B converts the power output in Fig. 7.1A into a photon output using Planck's equation, demonstrating the fact that LEDs with longer peak wavelengths have an advantage in that their photons have less energy. For example, although 450 nm blue and 660 nm red LED packages have similar efficiencies, the photon efficacy of 660 nm red is almost 50% higher than the 450 nm blue.

7.2.1 Binning

Although the LED packages presented in Table 7.1 and Fig. 7.1 are some of the highest performing products currently on the market; these values represent typical, and not maximum, performance of these products. LED package products are generally separated into bins based on the specific performance of that product. The performance characteristics that LED packages are binned by include photon output, forward voltage, and wavelength. The performance differences among individual LED packages are the result of the LED semiconductor crystal growth processes, which require very precise (high) temperature, pressure, and gas flow control in very clean chambers. Even with the best control, there are variations in the semiconductor crystal properties that result in performance variations. The impact of binning is that LED packages from the same manufacturer can have a range of performance levels, so it is necessary for LED luminaire manufacturers to specify not just the color and the package type of the desired LED but also the specific performance bin. By choosing LED packages from a higher performance bin, the efficiencies and photon efficacies in Table 7.1 may vary by 10% depending on the type of LED.

7.2.2 White LED packages

There are two ways to generate white light with LEDs: Phosphor-converted (PC) and color-mixed (CM). PC white LEDs incorporate an optical down-conversion material (typically a phosphor) into an LED package with a blue (450 nm) LED. This material absorbs a portion of the blue photons and then reemits them at longer wavelengths. The relative composition of blue light and the PC light can be engineered to achieve different hues of white light. CM white LEDs are made up of a combination of the narrow-bandwidth LEDs described in Fig. 7.1 and Table 7.1. For example, combining 450, 530, and 620 nm LEDs would provide white light. The process of down-converting photons in a PC white LED introduces conversion efficiency losses. There are two typical phosphors used in generating white light from a blue LED, one with a broad emission that peaks at about 550 nm (green) and another with a peak emission at about 600 nm (red). The green phosphor is more efficient than the red phosphor, 88% compared to 81% photon efficiency at 150°C and 1 W input (DOE BTO Lighting R&D Program, 2019). The photon efficiency of the phosphor material depends on the temperature and the photon intensity. The green and red phosphors have similar efficiencies at low drive currents.

The theoretical maximum photon efficacy of a PC white LED package is limited by its foundational 450 nm LED. Using Planck's equation, this theoretical maximum is equal to 3.76 μmol J^{-1}. This would require both the LED itself and the phosphor-conversion (photons emitted/photons produced) to be 100% efficient. The efficiency (percent of theoretical maximum) of the white LED packages in Table 7.1 assumes the product of these two efficiencies, rather than W_{out}/W_{in}. Because of this limitation, CM white LEDs have a higher potential photon efficacy than PC white LEDs, but for making white light the current green gap in the technology significantly limits CM white LEDs. Therefore, PC white LEDs dominate the market to provide broad spectrum white light. Additionally, the very high demand for PC white LEDs in human lighting has enabled price reductions and performance improvements that enable their effective use for horticulture.

7.2.3 LEDs for horticulture

The lower photon efficacy of narrow bandwidth LED packages within the green gap, especially in comparison to PC white LED packages, limits their application in horticulture. But, green LED packages with peaks between 500 and 600 nm are a useful tool for research purposes. UV-A, UV-B, and UV-C[5] LED packages may find a use in horticulture, as these

---

[5] UV-C is most likely be used for germicidal or sterilization purposes.

wavelengths may have beneficial effects on crop quality, and can readily be incorporated into LED luminaires. However, UV LED packages with a peak below about 380 nm suffer from low efficiency, which can limit their practical effectiveness. Far-red LED packages (730 nm) have the potential to be a valuable addition to LED luminaires for their role in increasing leaf expansion, although caution should be taken as many species increase their stem length under far-red photons. Near infrared LED packages (850 nm) are used in security cameras, which may be installed in plant factories with artificial lighting. Photons at these long wavelengths have the potential to affect plant development, although effects are expected to be minimal.

The most commonly used LED packages in horticulture include PC white, 660 nm red, and 450 nm blue LED packages. Other LED packages, including 730, 405, 385, 280, and 310 nm LED packages, are being considered for incorporation into LED luminaires, but caution should be taken with their inclusion as many responses are intensity and species dependent (see chapter 17).

7.3 Current droop

The incorporation of LED packages into LED luminaires requires a number of operating and design considerations. Driving LED packages lower than the nominal drive current will increase the photon efficacy, while driving them higher than the nominal drive current will decrease it. This is known as current droop. This effect is a direct function of the current density (the current divided by the LED chip area), but the chip area is often not reported by LED package manufacturers. Fig. 7.2 presents the typical current droop as

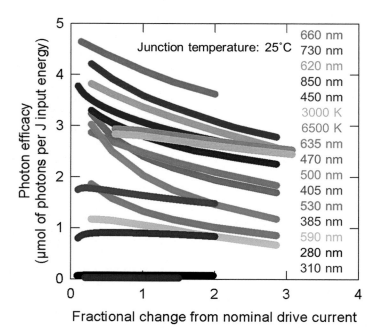

FIGURE 7.2   Current droop of the 16 LED packages described in Table 7.1. Current droop is plotted as a function of fractional change from the nominal drive current. This is because current droop is actually a function of current density and the die size of an LED package is often not reported. The junction temperature is specified at 25°C and the type of each LED package is listed (roughly in order) on the right hand side of the figure.

the fractional change in the drive current, with one being equal to the nominal drive current (note the differences in the nominal drive current between LED packages in Table 7.1). Current droop is a result of a decrease in photon output per ampere input. At higher currents, LED packages also experience an increase in the forward voltage (related to electrical resistance within the LED device and package). This means there is both a relative decrease in the output and an increase in the input at higher current operation (electrical power, $W = V_f A$).

LED luminaire manufacturers often operate the LED packages at a low drive current in order to maximize photon efficacy with additional benefits of (1) reducing thermal management requirements and (2) extending the LED lifetime. The operating drive current for PC white LED packages in LED luminaires is often below 100 mA. However, increasing the drive current increases the output of the LED package. Therefore, a trade-off exists between photon efficacy (along with thermal management and lifetime) and photon output. An LED luminaire with both high photon efficacy and high photon output requires the incorporation of many LED packages. Some luminaires contain more than 5000 LED packages. This means that although the photon output and the photon efficacy of the LED luminaire will both be high, the price of the LED luminaire will also be high. Because very low drive currents can cause excessively low photon outputs per LED package, the economics of the optimal drive current must be considered.

Although a decrease in the drive current generally increases the photon efficacy, extremely low drive currents also cause a decrease in photon efficacy. At very low current, crystal defect and/or impurity-related nonradiative recombination dominates, while at high drive currents, current droop dominates. With InGaN LEDs, Auger recombination causes droop. This can be thought of as the interaction of two excited charge carriers, which leads to an increase in the energy of one of the charge carriers, rather than the desired relaxation of the charge carrier. Relaxation of charge carriers is what causes the emission of a photon. For AlInGaP LEDs, droop is a result of charge carriers escaping from the LED active region and nonradiatively recombining elsewhere in the LED device structure. There is a range of drive currents between these two extremes (defects/impurity losses and droop losses) where luminaire manufacturers can operate the LED packages with optimal photon efficacy (and economics).

7.4 Thermal droop

Efficiency losses in an LED package produce heat (100% efficiency minus LED package efficiency = unwanted heat). This leads to self-heating of the LED package and LED luminaire. LED packages decrease in photon efficacy when operated at high junction temperature, either due to self-heating or ambient conditions.

Junction temperature is the temperature at the LED $p-n$ junction (also known as the active region). The LED junction temperature is measured at a defined temperature probe point with a known thermal resistance between that point and the junction, such that the junction temperature can be determined.

The efficiencies and photon efficacies in Table 7.1, Figs. 7.1, and 7.2 are specified for a junction temperature of 25°C (with the exception of the 280 nm LED package, which is specified

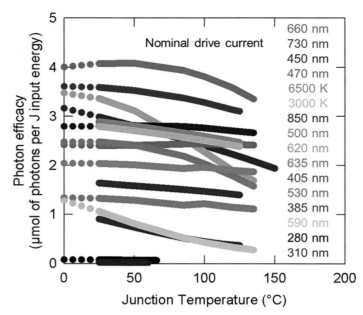

FIGURE 7.3 Thermal droop of the 16 LED packages described in Table 7.1. These data are specified at the nominal drive current described in Table 7.1. Photon efficacy will continue to change as temperature decreases below 25°C and data from LED package manufacturers have been specified down to −40°C, but the junction temperature rarely drops to ambient temperature. A junction temperature of −40°C would require a high amount of energy input for cooling. Therefore, only a dotted line is shown for junction temperatures below 25°C as these values are practically unrealistic.

at an ambient temperature, the 310 nm LED package is specified at a solder joint temperature, and the 850 nm LED package is specified at a case temperature). Typical LED packages, under typical operating conditions, will have junction temperatures between 55 and 100°C. This is despite the fact that typical performance is often defined at 25°C. As with current droop, this decrease in efficiency (and resulting photon efficacy) is caused by a decrease in photon output. Unlike increasing the current, increasing the temperature actually decreases the forward voltage—meaning a reduced input power—but the net effect is still that LED packages have reduced efficiency at higher temperatures.

Fig. 7.3 shows typical thermal droops from increasing junction temperature. It is apparent from this figure that AlInGaP-based LEDs experience a greater degree of thermal droop compared to lnGaN- and GaN/AlGaN-based LEDs.

7.5 Power supply efficiency

LED luminaires require power supplies to convert AC to DC power. Most power supplies can achieve this power conversion with 80%–95% efficiency. As with LED packages, higher performance power supplies are more expensive.

Power supplies can also provide dimming capabilities, either through current modulation or pulse width modulation, which is achieved through pulsing the electrical power through the LED packages at a high frequency.

7.6 LED luminaire optical efficiency

This efficiency describes the number of photons that are emitted from the LED luminaire divided by the number of photons produced by the LED packages. This efficiency can approach 100%. However, if a glass or plastic lens is used to engineer the optical distribution (including making the light diffuse) or protect the LED packages, then there can be optical losses due to reflection and absorption. The LED luminaire optical efficiency does not describe the percentage of emitted photons that reach and are absorbed by the plants. In some cases, it may make sense to use lenses to more optimally distribute the light across the plant and improve yield, but this comes at the cost of optical losses that reduce the efficiency of the LED luminaire. This is another trade-off that horticultural lighting manufacturers and growers must consider. Optical performance and losses can occur through both refractive and reflective optical elements and these elements may degrade over time, especially in the greenhouse environment, which contributes to the depreciation of photon output.

7.7 LED luminaire photon efficacy

Multiplying the LED package performance (photon efficacy) by the four efficiencies described above provides a final estimate of the LED luminaire photon efficacy.

$$\begin{array}{c}\text{LED luminaire} \\ \text{photon efficacy}\end{array} = \begin{array}{c}\text{LED package} \\ \text{photon efficacy}\end{array} \times \begin{array}{c}\text{current} \\ \text{droop}\end{array} \times \begin{array}{c}\text{thermal} \\ \text{droop}\end{array} \times \begin{array}{c}\text{power supply} \\ \text{efficiency}\end{array} \times \begin{array}{c}\text{optical} \\ \text{efficiency}\end{array}$$

Two example cases can be considered:

7.7.1 Case 1

An LED luminaire contains 90% 6500 K PC white LED packages and 10% 660 nm red LED packages (Table 7.1). The white LED packages are operated at the nominal drive current and the 660 nm LED packages are operated at about half of the nominal drive current (350 mA). It is assumed that under these conditions, both types of LED package will operate at a junction temperature of about 85°C. The power supply for the white LED packages is 90% efficient and the power supply for the 660 nm red LED packages is 85% efficient. The LED packages are unprotected, achieving a high optical efficiency.

$$\text{White LEDs: } 2.9 \frac{\mu\text{mol}}{\text{J}} \times 1.0 \times 0.92 \times 0.90 \times 0.99 = 2.4 \frac{\mu\text{mol}}{\text{J}}$$

$$\text{Red LEDs: } 4.1 \frac{\mu\text{mol}}{\text{J}} \times 1.07 \times 0.96 \times 0.85 \times 0.99 = 3.5 \frac{\mu\text{mol}}{\text{J}}$$

One thing that is immediately obvious from this calculation is the fact that decreasing the drive current below the nominal provides an increase in photon efficacy rather than a decrease. The weighted average of the overall LED luminaire would be

$$\text{LED luminaire photon efficacy} = \left(0.9 \times 2.4\frac{\mu\text{mol}}{J}\right) + \left(0.1 \times 3.5\frac{\mu\text{mol}}{J}\right) = 2.5\frac{\mu\text{mol}}{J}$$

7.7.2 Case 2

An LED luminaire contains 90% 660 nm red LED packages and 10% 450 nm blue LED packages. Additionally, by choosing LED packages from a higher performance bin, the performance can be assumed to be 5% better than that indicated in Table 7.1. Both types of LED package are operated at about 100 mA. At this low drive current, heating would be minimal and the junction temperature of the LED packages might be about 50°C. The power supplies for both types of LED packages are 90% efficient and the LED packages are unprotected.

$$\text{Red LEDs: } 4.3\frac{\mu\text{mol}}{J} \times 1.14 \times 0.99 \times 0.90 \times 0.99 = 4.3\frac{\mu\text{mol}}{J}$$

$$\text{Blue LEDs: } 2.9\frac{\mu\text{mol}}{J} \times 1.15 \times 0.99 \times 0.90 \times 0.99 = 2.9\frac{\mu\text{mol}}{J}$$

The overall photon efficacy of the LED luminaire would be

$$\text{LED luminaire photon efficacy} = \left(0.9 \times 4.3\frac{\mu\text{mol}}{J}\right) + \left(0.1 \times 2.9\frac{\mu\text{mol}}{J}\right) = 4.2\frac{\mu\text{mol}}{J}$$

The first case represents typical operating conditions for an LED luminaire, while the second case represents a state-of-the-art LED luminaire—although even this theoretical LED luminaire could be further optimized. Table 7.2 provides the measured photosynthetic photon efficacies of a variety of LED luminaires taken from the DesignLights Consortium (DLC) website (DesignLights Consortium, 2021; https://www.designlights.org/horticultural-lighting/search/). The DLC is a third-party organization that maintains a list of products that qualify for rebates. One of the required qualifications is a photosynthetic photon efficacy of at least 1.9 $\mu\text{mol J}^{-1}$. Following the website link above, LED luminaires can be sorted by photosynthetic photon efficacy in ascending or descending order. The LED luminaires presented in Table 7.2 may represent slightly higher photosynthetic photon efficacies than the average of that product, due to binning and self-selection. Additionally, the current average LED luminaire photosynthetic photon efficacy on the DLC website is about 2.5 $\mu\text{mol J}^{-1}$. It is apparent that the highest photosynthetic photon efficacy LED luminaires contain a high fraction of red photons, and that photon output (PPFD) does not appear to be correlated with photosynthetic photon efficacy.

TABLE 7.2   Photosynthetic photon efficacies of a range of LED luminaires reported by the DesignLights Consortium. The types of LED packages used in the LED luminaire are described in the first column, although some of the LED luminaires with PC-W LED packages may also contain 450 nm blue LED packages. Ratios of blue (%B), green (%G), red (%R), and far-red (%FR) are all described as a percent of the PPFD and not total photon flux density. PC-W: Phosphor-converted white.

LED packages	PPFD	%B	%G	%R	%FR	Photosynthetic photon efficacy (μmol J^{-1})
PC-W	839	10	38	52	8	1.81
PC-W + 660 nm	755	18	41	41	2	2.04
450 nm + 660 nm + 730 nm	700	26	0	74	9	2.06
PC-W + 530 nm + 660 nm + 730 nm	1443	12	23	65	13	2.09
PC-W + 660 nm	138	17	28	55	1	2.10
450 nm + 660 nm + 730 nm	1235	26	0	74	9	2.26
PC-W + 660 nm	1969	18	45	37	2	2.30
PC-W + 660 nm	766	17	30	53	1	2.35
PC-W	64	23	46	31	3	2.40
PC-W + 660 nm	491	24	45	31	2	2.44
PC-W + 660 nm	1593	19	41	40	2	2.49
450 nm + 660 nm	498	17	0	83	0	2.55
PC-W + 660 nm	1639	19	41	40	2	2.56
PC-W + 660 nm + 730 nm	1672	19	38	43	10	2.56
PC-W + 660 nm	781	9	17	74	1	2.57
PC-W + 660 nm	1727	22	41	37	2	2.67
PC-W + 660 nm	277	27	70	73	0	2.72
450 nm + 660 nm	1721	48	0	52	0	2.73
PC-W + 660 nm	1681	8	9	83	1	2.77
PC-W + 660 nm	1796	17	41	42	3	2.77
PC-W + 660 nm	1805	12	23	65	1	2.88
PC-W + 660 nm	506	11	5	84	1	2.90
450 nm + 660 nm	1724	8	0	92	0	2.90
450 nm + 660 nm	93	34	0	66	0	2.94
450 nm + 660 nm	292	5	0	95	0	3.00
PC-W + 660 nm	4971	16	26	58	1	3.18
PC-W + 660 nm	1700	11	5	84	1	3.30
450 nm + 660 nm	2196	5	0	95	0	3.40
450 nm + 660 nm	2195	4	0	96	0	3.51
450 nm + 660 nm	2311	4	0	96	0	3.69

One drawback of Table 7.2 is that LED luminaires are classified by the photosynthetic photon efficacy (400–700 nm) rather than the total photon efficacy. This means that LED luminaires that contain 730 nm far-red LED packages pay a penalty even though these photons are expected to have beneficial effects on the growth and development of many species (Zhen et al., 2021; Zhen and Bugbee, 2020a, 2020b).

7.8 LED luminaire longevity

LED luminaires can operate well beyond 50,000 h, which is longer than most other lighting technologies. LED packages rarely completely fail. They are highly reliable with a predictable degradation over time if the operating conditions (junction temperature and drive current) are known. When LED luminaires do fail, it is typically due to a failure in the power supply, electrical connections, or manufacturing defect (DOE BTO Lighting R&D Program, 2019).

Depreciation of LED packages within an LED luminaire is predicted from the depreciation of the LED packages over time under fixed conditions (junction temperature and drive current). The method of testing LED packages, called LM-80, is approved by the Illuminating Engineering Society (IES) (IESNA Testing Procedures Committee, 2008). LED package manufacturers perform these tests on their products and share the results with their customers (LED luminaire manufacturers). When the LED packages are engineered into an LED luminaire, the drive current is known, and the junction temperature can be measured. The depreciation of the LED packages within the LED luminaire can then be projected according to IES technical memo 21 (TM-21) (IESNA Testing Procedures Committee, 2011). This enables a consistent projection of photon maintenance among different LED luminaire manufacturers. However, TM-21 only allows for projection of up to six times the LED package measurement duration under LM-80. For example, if an LED luminaire manufacturer claims a depreciation level for 60,000 h, then the LED packages must have been LM-80 tested for 10,000 h.

LM-80 and TM-21 measure and project the depreciation of the LED packages. Other factors can increase the rate of depreciation of the LED luminaire. These factors include temperature extremes, humidity, chemical incursion (like sulfur which is a common fungicide), electrical surges (voltage and current), and depreciation of lenses or reflectors in the LED luminaire. For growers, photon output depreciation directly results in reduced plant growth, so it is important to predict depreciation and plan for lighting updates even if LED luminaires are still operational. Output depreciation is not unique to LED lighting technology, as all lighting technologies depreciate, but unlike LED luminaires, the lamps typically fail before the depreciation is a major issue.

While LED packages generally degrade over time, LED luminaires can also catastrophically fail, meaning they stop emitting photons partially or entirely. This is generally due to failures in electrical connections or the power supply. A partial failure could occur when one circuit of LED packages loses its electrical connection, while the remaining circuits are still operational. Power supplies are also sensitive to environmental conditions such as temperature, humidity, chemical incursion, and electrical surges. The wide range of circuit types, operating conditions, and power quality make the lifetime of a power supply less predictable than the LED packages. Quality of manufacturing and component selection plays a large role in power supply reliability. LED luminaire manufacturers can select power supplies that have been tested under wet and hot operating conditions, under thermal cycling to check robustness of solder joints, and/or under vibration tests to check component integrity.

Ultimately, a conservative and practical estimate of the lifetime of an LED luminaire is the duration of the manufacturer warranty. Growers ought to inquire after the LM-80 data and TM-21 analysis for the LED luminaires. They should ask for failure rates and reliability testing, as well as documentation regarding the power supplies that are used. Lighting is a critical component for indoor plant growth, and LED luminaire failures can have significant impacts on production. As such, LED luminaire purchasing decisions, including reliability, are paramount.

7.9 Continued improvement

LED technology has rapidly advanced in the past decade, and the most common LED packages for horticulture use (PC white, 450 nm blue, 660 nm red, and 730 nm far-red) are approaching their theoretical maximum performance. Advancements will continue, but will slow for these most advanced LED packages.

Closing the green gap would increase photon efficacy, especially for human lighting. The theoretical maximum photon efficacy of PC white LED packages is dependent on their underlying higher energy 450 nm blue LEDs. Although green photons are photosynthetic (McCree, 1971), they are not required for plant growth. These LED packages are included in LED luminaires because (1) PC white LED packages are inexpensive and (2) green photons aid in human perception of plant color. As 530 nm green and 590 nm yellow LED packages improve, these can replace PC white LED packages to achieve this human-centric goal.

The theoretical maximum photosynthetic photon efficacy is 5.85 μmol J^{-1} for photons concentrated at 700 nm, but this is not a practical maximum. A plant grown under only 700 nm photons would likely be subject to developmental problems. Several studies have shown that some plants require at least some blue photons for *normal* plant growth (Yorio et al., 2001; Hernandez and Kubota, 2016), although other studies have shown some plants can be grown under pure red (Son and Oh, 2013; Meng et al., 2020). Determining which species can grow normally under pure red may help maximize the theoretical potential in horticulture. Finally, recent studies have provided compelling evidence that the spectral range for photosynthesis should be extended to 750 nm (Zhen et al., 2021; Zhen and Bugbee, 2020a, 2020b). This would allow the use of highly efficient far-red LED packages in selected horticulture applications.

References

DesignLights Consortium, 2021. Horticultural Lighting. https://www.designlights.org/horticultural- lighting/search/ (accessed 9 February 2021).

DOE BTO Lighting R&D Program, 2019. 2019 Lighting R&D Opportunities. https://www.energy.gov/sites/prod/files/2020/01/f70/ssl-rd-opportunities2-jan2020.pdf (accessed 9 February 2021).

Feezell, D., Nakamura, S., 2018. Invention, development, and status of the blue light-emitting diode, the enabler of solid-state lighting. Compt. Rendus Phys. 19 (3), 113–133.

Hernández, R., Kubota, C., 2016. Physiological responses of cucumber seedlings under different blue and red photon flux ratios using LEDs. Environ. Exp. Bot. 121, 66–74.

IESNA Testing Procedures Committee, 2008. LM-80-08: Measuring Lumen Maintenance of LED Light Sources. Illuminating Engineering Society.

IESNA Testing Procedures Committee, 2011. TM-21-11: Projecting Long Term Maintenance of LED Light Sources. Illuminating Engineering Society.

Kusuma, P., Pattison, P.M., Bugbee, B., 2020. From physics to fixtures to food: current and potential LED efficacy. Horticulture research 7 (1), 1–9.

Lumileds, 2018. DS185 LUXEON UV FC Line Product Datasheet. https://www.lumileds.com/wp- content/uploads/files/DS185.pdf (accessed 9 February 2021).

Lumileds, 2020. DS191 LUXEON IR Domed Line Product Datasheet. https://www.lumileds.com/wp- content/uploads/files/DS191-luxeon-ir-domed-line-datasheet-1.pdf (accessed 9 February 2021).

Lumileds, 2021. DS267 LUXEON 3030 HE Product Datasheet. https://www.lumileds.com/wp- content/uploads/DS267-LUXEON-3030-HE-datasheet.pdf (accessed 9 February 2021).

McCree, K.J., 1971. The action spectrum, absorptance and quantum yield of photosynthesis in crop plants. Agric. Meteorol. 9, 191–216.

Meng, Q., Boldt, J., Runkle, E.S., 2020. Blue radiation interacts with green radiation to influence growth and predominantly controls quality attributes of lettuce. J. Am. Soc. Hortic. Sci. 145 (2), 75–87.

Nichia, 2020. PART NO. NCSU434A(T) Datasheet. https://www.nichia.co.jp/specification/products/led/NCSU434A-E.pdf (accessed 9 February 2021).

Osram, 2020a. GA CSSPM1.23 Datasheet. https://www.osram.com/ecat/OSLON%C2%AE%20SSL%20120%20GA%20CSSPM1.23/com/en/class_pim_web_catalog_103489/prd_pim_device_2402546/ (accessed 9 February 2021).

Osram, 2020b. GB CSHPM1.13 Datasheet. https://www.osram.com/ecat/OSLON%C2%AE%20SSL%20150%20GB%20CSHPM1.13/com/en/class_pim_web_catalog_103489/prd_pim_device_2402541/ (accessed 9 February 2021).

Osram, 2020c. GD CSSPM1.14 Datasheet. https://www.osram.com/ecat/OSLON%C2%AE%20SSL%20120%20GD%20CSSPM1.14/com/en/class_pim_web_catalog_103489/prd_pim_device_2402548/ (accessed 9 February 2021).

Osram, 2020d. GF CSSPM1.24 Datasheet. https://www.osram.com/ecat/OSLON%C2%AE%20SSL%20120%20GF%20CSSPM1.24/com/en/class_pim_web_catalog_103489/prd_pim_device_2402545/ (accessed 9 February 2021).

Osram, 2020e. GH CSSRM4.24 Datasheet. https://www.osram.com/ecat/OSLON%C2%AE%20Square%20GH%20CSSRM4.24/com/en/class_pim_web_catalog_103489/prd_pim_device_10285510/ (accessed 9 February 2021).

Osram, 2020f. GR QSSPA1.23 Datasheet. https://www.osram.com/ecat/OSCONIQ%C2%AE%20P%203030%20GR%20QSSPA1.23/com/en/class_pim_web_catalog_103489/prd_pim_device_8882698/ (accessed 9 February 2021).

Osram, 2020g. GT CSSPM1.13 Datasheet. https://www.osram.com/ecat/OSLON%C2%AE%20SSL%20120%20GT%20CSSPM1.13/com/en/class_pim_web_catalog_103489/prd_pim_device_2402547/ (accessed 9 February 2021).

Osram, 2020h. GV QSSPA1.13 Datasheet. https://www.osram.com/ecat/OSCONIQ%C2%AE%20P%203030%20GV%20QSSPA1.13/com/en/class_ppi_web_catalog_103489/prd_pim_device_8931522/ (accessed 9 February 2021).

Osram, 2020i. GY CSHPM1.23 Datasheet. https://www.osram.com/ecat/OSLON%C2%AE%20SSL%20150%20GY%20CSHPM1.23/com/en/class_pim_web_catalog_103489/prd_pim_device_2402540/ (accessed 9 February 2021).

SeoulVioSys, 2020. UV CA3535 Series Datasheet. http://www.seoulviosys.com/en/product/PKG/?sub=13&seq=1> (accessed 9 February 2021).

Son, K.H., Oh, M.M., 2013. Leaf shape, growth, and antioxidant phenolic compounds of two lettuce cultivars grown under various combinations of blue and red light-emitting diodes. Hortscience 48 (8), 988–995.

Tsao, J.Y., Han, J., Haitz, R.H., Pattison, P.M., 2015. The Blue LED Nobel Prize: Historical Context, Current Scientific Understanding, Human Benefit. Annalen der Physik, Leipzig), 527(SAND-2015-4440J).

Yorio, N.C., Goins, G.D., Kagie, H.R., Wheeler, R.M., Sager, J.C., 2001. Improving spinach, radish, and lettuce growth under red light-emitting diodes (LEDs) with blue light supplementation. Hortscience 36 (2), 380–383.

Zhen, S., Bugbee, B., 2020a. Substituting far-red for traditionally defined photosynthetic photons results in equal canopy quantum yield for CO_2 fixation and increased photon capture during long-term studies: implications for redefining PAR. Front. Plant Sci. 11, 1433.

Zhen, S., Bugbee, B., 2020b. Far-red photons have equivalent efficiency to traditional photosynthetic photons: implications for redefining photosynthetically active radiation. Plant Cell Environ. 43 (5), 1259–1272.

Zhen, S., van Iersel, M., Bugbee, B., 2021. Why far-red photons should be included in the definition of photosynthetic photons and the measurement of horticultural fixture efficacy. Front. Plant Sci. 12, 693445.

Balances and use efficiencies of CO_2, water, and energy

Toyoki Kozai

Japan Plant Factory Association, Kashiwa, Chiba, Japan

8.1 Introduction

An important mission of plant factory with artificial lighting (PFAL) is to achieve: (1) highest yield and quality of plants creating economic and/or social values, (2) with minimum consumption of CO_2, water, electricity, land area, etc., resulting in minimum production costs, and (3) with minimum emission of waste such as plant residue, drained water containing fertilizer and used consumables, contributing to environmental conservation and/or reduction in costs for waste processing (Kozai and Niu, 2019).

The PFAL is considered to be suited to achieve the above mission because: (1) since it is a highly airtight and thermally insulated system, all the resource inputs and outputs to and from there can be relatively easily measured and controlled automatically; (2) the resource outputs are not directly influenced by the weather, pest insects, and small animals outside; (3) the environment in the PFAL can be controlled as desired, (4) the use efficiency of each resource element can be estimated online based on the measured values of corresponding resource input and output; and thus (5) the use efficiency of each resource element and the overall resource use efficiency (RUE) or productivity (or the ratio of yield to the cost) can be continuously improved based on the data analyses of the measured inputs and outputs (Kozai et al., 2020).

In this chapter, the balances of CO_2, water, and energy of the cultivation room in the PFAL are described. Then, the use efficiencies of CO_2, water, and electric energy are defined, and the methods of estimating and improving those RUEs are presented. For the fertilizer balance and fertilizer use efficiency under unsteady state conditions, refer to Kozai (2013, 2018) and Kozai and Niu (2019).

8.1.1 Resource inputs

There are two types of resources, Type 1 and Type 2 (Fig. 8.1), supplied to the cultivation room in the PFAL. Type 1 is called essential resources required for plant growth, which consists

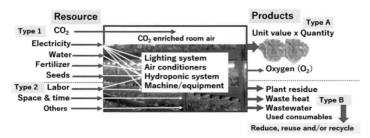

FIGURE 8.1   Resource inputs (Types 1 and 2) and products (Types A and B) in the cultivation room of the plant factory with artificial lighting. Ideally, all the Type 1 resources are converted to the Type A products.

of electricity for lighting, air conditioning, pumping, machine handling, etc., CO_2 for promoting photosynthesis, water for irrigation, fertilizer as inorganic nutrition for plant growth, and seeds.

Type 2 includes space and time for plant cultivation, space and time for human and machine operations, and consumables such as substrate (rockwool cubes, plastic fibrous mats, etc.), water for washing and/or cleaning floor, cultivation beds, etc, and plastic/paper bags except for Type 1 resources. Use efficiencies of Type 2 resources are discussed in Chapter 12 in relation to the productivity of PFAL.

8.1.2 Products

There are two types of resource outputs or products, Type A and Type B, obtained from the cultivation room (Fig. 8.1). Type A includes marketable produce having economic and/or social values, and oxygen (O_2) gas emitted by plants during their photosynthesis. Type B includes: (1) plant residue, (2) waste heat generated by lamps and other equipment, which needs to be removed to the outside by air conditioners (or heat pumps) to keep the room air temperature at its setpoint, (3) used nutrient solution drained from the hydroponic cultivation system, (4) water used for washing/cleaning the floor, cultivation panels, worker's cloth and hands, etc., and (5) used consumables such as plastic/paper bags ad substrate.

In the PFAL, amounts and quality of Type A products need to be maximized, while amounts of Type B products need to be minimized or reduced, recycled and/or reused after proper treatments. The most essential benefit of the relatively airtight and thermally well-insulated cultivation room is that almost all the elements of Types 1, 2, A, and B are relatively accurately and easily measured, and the use efficiency of each resource element can be estimated online based on the energy and/or mass balance of the cultivation room, so that the RUEs can be improved continuously based on the measured supply rates of Types 1 and 2 resources and production rates of Types A and B products.

8.1.3 Plant phenotype as affected by the environment, genome, and management

Phenotype or traits of a plant canopy (**P**) in the cultivation room is determined by the environment (**E**), the genotype (genetic traits) of seeds or transplants (**G**) and the management by human, tool and machine interventions (**M**) (Fig. 8.2). Namely, **P** = Function (**E, G, M**). The

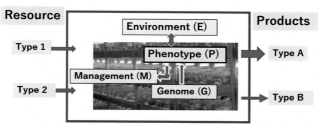

FIGURE 8.2 Phenotypes (or traits, **P**) of plant canopy are affected by the environment (**E**), the genome (genetic traits, **G**), and the management (**M**). Type 1 resource inputs affect the environment, and Type 2 resource inputs affect the management. The phenotype determines the quantity and quality of Types A and B products. The **P** includes plant biomass (fresh weight, leaf area, No. of leaves, leaf shape and size, etc.) and the concentrations of chemical components (chlorophylls a/b, carotene, ascorbic acids, polyphenols, etc.).

Type 1 resources in Fig. 8.1 are mainly used to create the **E**, and the Type 2 resources are mainly used for the **M**. The **G** is expressed or appears as the **P** under the given **E** and **M** (Kozai, 2018).

The **P** includes phytomass or plant biomass (fresh weight, dry weight, leaf area, plant height, etc.), its structure and function (spatial distributions of leaf angle, leaf area, roots, stems, petioles, etc.), and spatial distributions of physiological status and chemical components of plants. Spatial distributions of chemical components in the plant canopy strongly influence color, texture, taste, flavor, mouth feeing of plants, and thus their economic and social values. With this regard, the phenotyping or plant trait measurement in the plant production process plays an essential role in achieving the highest economic and social values of produce with minimum uses of Types 1 and 2 resources and minimum emission of Type B products.

The growth and development of plants and thus the plant phenotype are results of transport of water and solutes in the plants, and biochemistry and metabolism (photosynthesis, translocation in the phloem, respiration and lipid metabolism, assimilation of mineral nutrients, and secondary metabolites and plant defense) in the plants (Taiz and Zeiger, 2006). This chapter discusses the flows and conversions of water (transpiration and fresh weight increase of plants), CO_2 (plant photosynthesis and respiration), and electric, radiative, and heat energy transfers in the plant canopy and/or in the cultivation room.

Fig. 8.3 shows the role of material and energy balance models in the PFAL management. The four balance models in the figure are mutually connected and to be integrated with the models of the environment control, the plant growth and development (Chapter 11),

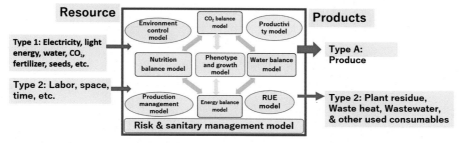

FIGURE 8.3 Role of material and energy balance models in the PFAL management. The balance models are mutually connected and integrated with the models of the environment control, the plant growth and development (Chapter 11), the resource use efficiency (RUE), the productivity (Chapter 12), and the management. The nutrition balance model is not included in this chapter.

the RUEs, the productivity (Chapter 12), and the management (The nutrition balance model, risk and sanitary management model, AI model, database, data collection and network are not described in this chapter).

8.2 CO_2 balance of the cultivation room

8.2.1 CO_2 balance equation

CO_2 balance of the cultivation room is schematically shown in Fig. 8.4 and is expressed by Eq. (8.1) (Wheeler, 1992; Yoshinaga et al., 2000; Yokoi et al., 2005; Li et al., 2012a; Ohyama et al., 2020).

$$C_S = C_P + C_V + C_M + C_A + C_R + C_Q + C_W \tag{8.1}$$

where C_S: CO_2 supply rate; C_P: net photosynthetic (CO_2 fixation) rate of plant canopy; C_V: CO_2 loss rate due to air infiltration/ventilation; C_M: microorganism (in the substrate) respiration rate; C_A: rate of change in CO_2 mass (or weight) in the cultivation room air; C_R: human respiration rate; C_Q: CO_2 absorption/release rate at wall/floor/nutrient solution surfaces. Unit of all the rate variables is kg/h per cultivation room, which can be converted to $kg\ m^{-2}\ h$ (per cultivation bed area or floor area); C_W: $C_A = 0$ when C_{in} remains the same over time; $C_R = 0$ when no workers are in the cultivation room; $C_M = 0$ when no microorganisms exist in the substrate. C_Q is assumed to be $0\ kg\ h^{-1}$ in this chapter because it is negligibly small in most PFALs.

8.2.2 Equation for estimating net photosynthetic rate, C_P

The C_P can be estimated by Eq. (8.2) which is transformed from Eq. (8.1).

$$C_P = C_S - C_V - C_M - C_A - C_R - C_Q - C_W \tag{8.2}$$

In Eq. (8.2), all the variables except for C_A show positive or null values under normal conditions, while the C_A has a negative value when C_{in} (CO_2 concentration inside the cultivation room) decreases during the period of measurement. In Eq. (8.2), The C_s can be measured

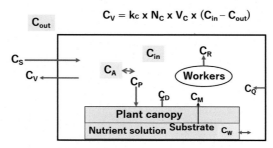

FIGURE 8.4 CO_2 balance of the cultivation room. (See the text and Appendix for the meanings of symbols and units.)

accurately with use of a mass flow meter attached at the outlet of CO$_2$ supply tube. The C$_V$ can be estimated using Eq. (8.3) with the data on the C$_{in}$ and C$_{out}$ as shown in the next sections. The C$_{out}$ (CO$_2$ concentration outside the cultivation room) is often assumed to be constant in a range between 400 and 420 mmol mol^{-1} although it fluctuates to some extent ($\pm$20 mmol mol^{-1} or so) daily, seasonally, and/or locality, and increases gradually year by year (approximately 1 µmol mol^{-1} per year) so that continuous measurement of the C$_{out}$ is recommended to estimate the C$_P$ as accurate as possible. Li et al. (2012a,b) showed hourly changes in the C$_P$ of a PFAL based on Eq. (8.2).

8.2.3 Estimating CO$_2$ loss rate, C$_V$

The C$_V$ can be estimated by:

$$C_V = k_C \times N_C \times V_c \times (C_{in} - C_{out})/10^6 \tag{8.3}$$

where k$_C$: conversion factor from volume to mass of CO$_2$ gas (1.96 kg m^{-3} at 25°C); N$_C$: number of air exchanges per hour of the cultivation room; V$_C$: air volume of the cultivation room (m^3). The "10^6" in Eq. (8.3) is the conversion factor from µmol to mol. The product of the N$_C$ and the V$_c$ is called ventilation rate of the cultivation room (m^3 h^{-1}).

Fig. 8.5 shows the C$_V$ as function of the N$_C$ and the C$_{in}$, assuming that the C$_{out}$ = 400 µmol mol^{-1} and V$_c$ = 5000 m^3. Eq. (8.3) shows that C$_V$ = 0 when (C$_{in}$−C$_{out}$) = 0 regardless of the N$_C$, and that the C$_V$ increases with increasing the N$_C$ and the (C$_{in}$−C$_{out}$). The C$_{in}$ is often kept at around 1000 mmol mol^{-1} or higher under light. For example, when the C$_{in}$ is 1000 mmol mol^{-1}, the C$_V$ is 6 kg h^{-1} at the N$_C$ = 0.4/h (Fig. 8.5), or 144 (= 6 × 24) kg d^{-1} or 52,560 (= 144 × 365) kg y^{-1}. It should be noted that CO$_2$ is a global warming gas when emitted to the atmosphere. On the other hand, it costs 30,000 to 50,000 US$ per year in

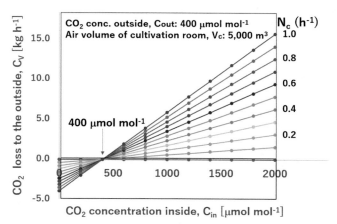

FIGURE 8.5 CO$_2$ loss rate to the outside, C$_V$, as affected by CO$_2$ concentration inside the cultivation room, C$_{in}$, and the number of air exchanges per hour of the cultivation room, N$_C$, which is defined as hourly ventilation rate divided by air volume of the cultivation room, C$_V$. The C$_V$ is expressed as k$_c$ ×V$_C$ × (C$_{in}$−C$_{out}$)/10^6 where k$_c$ is the conversion factor from volume to mass (1.96 kg m^{-3}).

Japan when purchasing 52,560 kg as pure liquid CO_2 for promoting photosynthesis and thus plant growth. It should be also noted that 50%−70% of light-emitting diodes (LEDs) are turned on and thus the air conditioners are always turned on for cooling and dehumidification because the relative humidity reaches nearly 100% soon when all the LEDs are turned off.

It is generally recommended that the N_C be kept at around 0.02 h^{-1}. This low N_C is beneficial to minimize: (1) the loss of CO_2 and water vapor (or latent heat) to the outside and the loss or gain of sensible heat to and from the outside, (2) the invasion of small insects and dusts often with pathogens entering the cultivation room, and (3) the fluctuations of air temperature, vapor pressure deficit (VPD), and the C_{in} in the cultivation room caused by changes in weather outside.

On the other hand, a forced ventilation system with controlled ventilation rates is recommended to be installed to avoid the occasional accumulation of volatile organic compound gases emitted from plants and/or structural elements of the PFAL, bioeffluent gas or odor emitted by workers, which may be harmful or may hinder the comfort, health and safety of workers and plants.

8.2.4 Estimating the number of air exchanges per hour of the cultivation room using CO_2 as tracer gas, N_C

The N_C in Eq. (8.3) can be estimated by Eq. (8.4) using the measured data on C_0 (the C_{in} at time 0 h) and C_T (the C_{out} at time T h) in the cultivation room with absence of plants, as schematically shown in Fig. 8.6. The vertical axis in the figure is logarithmic scale and "ln" in Eq. (8.4) denotes natural logarithm. The C_{in} decreases exponentially over time in the empty cultivation room until it reaches the C_{out}. The gradient of the straight line in Fig. 8.6 represents a negative value of the N_C. The N_C increases, to some extent, with increasing the windspeed and direction outside, so that it is recommended to estimate the N under different weather conditions to get the average N.

$$N_C = -\ln((C_T - C_{out}) / (C_0 - C_{out}))/T \tag{8.4}$$

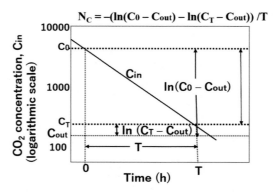

FIGURE 8.6   Method of estimating the N_C of an empty cultivation room, using the data on the C_{in} at time 0 and T. The C_0 and C_T denote, respectively, the C_{in} at time 0 and time T. The C_{out} is often assumed to be 400 μmol mol^{-1}. The T is around 0.5 h when the N_C is around 0.3 h^{-1} and is around 5 h when the N_C is 0.02 h^{-1}. In this method, C_{in} at t = 0 is intentionally increased up to around 3000 μmol mol^{-1} to increase the accuracy of N_C.

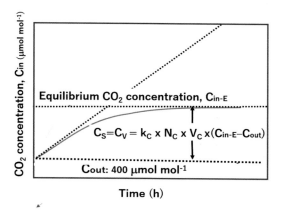

FIGURE 8.7 Time course of CO_2 concentration in the empty cultivation room, C_{in}, under a constant CO_2 supply rate, C_S. At the equilibrium CO_2 concentration of C_{in}-E, $C_S = C_V$. Then, the N can be estimated by Eq. (8.5).

The N_C can also be estimated by measuring the time course of the C_{in} until it reaches an equilibrium CO_2 concentration, $C_{in\text{-}E}$, under a constant CO_2 supply rate, C_S (Fig. 8.7). At the $C_{in\text{-}E}$, $C_V = C_S$. Then, the N_C can be estimated by Eq. (8.5) using the measured data on the C_S, the $C_{in\text{-}E}$ and the C_{out}. The dotted line in Fig. 8.7 shows a virtual time course of the C_{in} in a completely airtight cultivation room, which hourly increase is "$10^6 \times C_S / (k \times V_C)$." This estimation method generally requires several times volume of CO_2 gas and estimation time compared with the method based on Eq. (8.4).

$$N_C = C_S / \left(k_C \times V_c \times (C_{in-E} - C_{out}) / 10^6 \right) \tag{8.5}$$

8.2.5 Estimating the dark respiration rate of plants, C_D

The C_D is respiration in plants regardless of light (i.e., whether light is present or absent) (Taiz and Zeiger, 2006). This term "dark respiration" is used to distinguish it from photorespiration that occurs only in the presence of light. The net photosynthetic rate, C_P, is given as the difference between gross photosynthetic rate, C_G, and the sum of photorespiration rate, C_H, and dark respiration rate, C_D. Thus, under dark, the C_P is equal to -C_D. Once the N_C is determined by Eq. (8.4) or Eq. (8.5), the C_D can be estimated by Eq. (8.6) under conditions of no supply of CO_2 ($C_S = 0$), darkness [photosynthetic photon flux density (PPFD = 0)], and no change in C_{in} and C_{out} with time.

$$C_D = C_V = k_C \times N_C \times V_c \times (C_{in} - C_{out}) / 10^6 \tag{8.6}$$

By repeating the estimation of the C_D at different temperatures, the temperature effect on the C_D can be determined. The C_D is roughly proportional to the fresh weight of plants and increases exponentially with increasing temperature in a range roughly between 5 and 40°C (Taiz and Zeiger, 2006). However, it should be noted that the estimation of the C_D based on Eq. (8.3) or Eq. (8.6) is not so accurate especially when the ($C_{in} - C_{out}$) is smaller than 50 $\mu mol\ mol^{-1}$ under changing the C_{out}.

8.2.6 Estimating the human respiration rate, C_R

The C_R is often assumed to be 0.042 kg (CO$_2$)/h/person (Li et al., 2012a). In case that the height and body mass (weight) of workers vary, the C_R per person can be expressed as below (Parsily and de Jonge, 2017):

$$C_R/person = 0.00276 \times B \times M \times RQ/(0.23 \times RQ + 0.77) \qquad (8.7)$$

where B: DuBois surface area (m^2); M: metabolic rate of physical activity in unit of met; RQ: respiratory quotient (about 0.85, dimensionless). The B is calculated as $0.202 \times H^{0.725} W^{0.425}$ from the height (H) in m and the body mass (W) in kg. For an average-sized adult, B = 1.8 m^2, and met is between 2.5 and 3.5.

8.2.7 Estimating the microorganism (in the substrate) respiration rate, C_M

In case that the C_M is considered not to be negligibly small, the C_M can be estimated as below: (1) One kg of wet substrate with standard moisture content is placed into a chemically inert plastic box; (2) After sealing the box, measure CO$_2$ concentration in the box, C_{in1} and C_{in2}, at time t_1 and t_2 [The ($t_2 - t_1$) would be around 1 h]; (3)Then, substitute the values of the ($C_{in} - C_{out}$) and the ($t_2 - t_1$) into Eq. (8.8):

$$C_M = M_S \times k_C \times V_B \times (C_{in2} - C_{in1})/(t_2 - t_1)/10^6 \qquad (8.8)$$

where M_S is the total weight of the substrate containing water or nutrient solution in the cultivation room; k_C: conversion factor of CO$_2$ from volume to mass; V_B: the air volume of the plastic box (m^3). By repeating the same measurement at different substrate temperatures, the substrate temperature effect on the C_M can be estimated.

By the way, algae is not a microorganism but a photosynthetic (photoautotrophic) organism, which propagates and grows fast only under the presence of light, fertilizer, water, and CO$_2$. To prevent the algae growth, keeping the substrate surface always dry and preventing the penetration of light to the nutrient solution are essential.

8.2.8 Estimating the net photosynthetic rate, C_P

Now, the hourly C_P can be estimated by Eq. (8.2) with measured values of the C_S, the C_{in}, and the C_{out} because the N_C (Eq. 8.4), the C_R (Eq. 8.7), and the C_M (Eq. 8.8) can be estimated separately in advance (Li et al., 2012a,b,c).

The C_P is affected significantly by the mass of plants, leaf area or LAI, and plant physiological status such as stomatal conductance of plants, and the environmental factors such as PPFD, spectral photon distribution, CO$_2$ concentration, air temperature, VPD, air current speed, composition, strength, pH, and temperature of nutrient solution. Thus, online monitoring of the C_P and other plant traits and the environmental factors can contribute to analyze the relationships among the C_P, physiological traits, and environmental factors, and thus to improve the productivity of the PFAL.

8.2.9 CO_2 concentration and ventilation rate for human comfort and health

Recommended upper limit of CO_2 concentration in office, school, and factory, and recommended lower limit of ventilation rate in a building given by a law or as a guideline for keeping human comfort and health vary country by country. The required ventilation rate is determined to: (1) eliminate odors from human bioeffluents, (2) keep CO_2 concentration at a certain level or lower, and (3) optimize the cooling/heating load of the building. If a space is ventilated at a nominal rate of 7.5 L s^{-1} per person, the CO_2 concentration at steady-state will equal 1000 µmol mol^{-1} for assumed values of the CO_2 generation rate per person (see Section 8.2.6 of this chapter) and the standard outdoor CO_2 concentration. There is nothing in the standard declaring 1000 µmol mol^{-1} (or ppm) or any other CO_2 concentration to be a health or comfort-based limit. The 1000 µmol mol^{-1} "limit" simply is based on its association with a nominal ventilation requirement of 7.5 L s^{-1} per person for control of body odor perception (Parsily and de Jonge, 2017). Also, other ventilation standards contain a range of ventilation rate requirements for different types of spaces, which is associated with CO_2 concentrations other than 1000 µmol mol^{-1}.

8.3 Water balance of the cultivation room

Water balance of the cultivation room is schematically shown in Figs. 8.8 and 8.9, and is expressed by Eq. (8.9).

$$W_S + W_V + W_E + W_F + W_P + W_C + W_X + W_A + W_Q + W_W - W_Z = 0 \qquad (8.9)$$

where W_S: water supply rate; W_V: water vapor loss rate; W_E: rate of plants removed as produce or residue; W_F: rate of water drained for reuse or discard; W_P: fresh weight increase rate of plants; W_C: water condensation rate at cooling coils of air conditioners; W_X: change in water mass in the substrate; W_A: change in water mass of the room air; W_Q: water condensation rate at walls and floor; W_W: water supplied for washing/cleaning/disinfecting; W_Y, W_L, and W_B: a portion of W_W for washing/cleaning the floor/walls, cultivation panels/beds, and hands/boots, respectively.

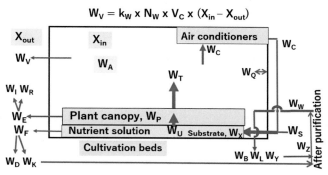

FIGURE 8.8 Water balance of the cultivation room under steady state conditions (See Fig. 8.9 and the Appendix for the meanings of symbols and units.).

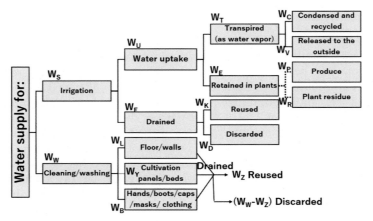

FIGURE 8.9 Water distribution under steady state conditions in the cultivation room. Water mass retained in the cultivation room (W_A and W_X) and W_Q in Fig. 8.8 are neglected.

8.3.1 Water loss due to air infiltration, W_V

W_V can be expressed by Eq. (8.10), which is similar in its structure to Eq. (8.3).

$$W_V = k_W \times N_W \times V_C \times (X_{in} - X_{out}) \tag{8.10}$$

where k_W: conversion factor from volume to mass [1.16 kg m^{-3} at 25°C and relative humidity (RH) of 78%]; N_W: Number of air exchanges of the cultivation room; V_C: air volume of cultivation room; X_{in} and X_{out}: absolute humidity of room air and outside air, respectively.

Fig. 8.10 shows the water vapor loss rate of the cultivation room, W_V, as affected by the absolute humidity of the outside air, X_{out}, and the number of air exchanges per hour, N_W, calculated by using Eq. (8.10). In the graph, when the X_{in} is 0.005 kg kg^{-1} and the N_W is

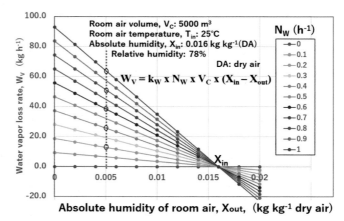

FIGURE 8.10 Water vapor loss rate as affected by absolute humidity of the outside air, X_{out}, and the number of air exchanges of the cultivation room, N_w

0.2, 04, 0.6, 0.8, and 1.0 h^{-1}, the W_V is, respectively, 15, 26, 38, 51, and 64 kg h^{-1}. When the W_V is 64 kg h^{-1}, the latent heat of 156 MJ h^{-1} or 43 kW [= 64 kg h^{-1} × 2442 kJ kg^{-1} (heat of evaporation) at 25°C] is released to the outside. To understand the relationships among the absolute humidity, relative humidity, latent heat and sensible heat of the air, the psychrometric diagram or related equations are useful.

8.3.2 Estimating the N_W using water vapor as tracer gas

Under steady state conditions, the N_W can be estimated using Eq. (8.11) (Li et al., 2012c).

$$N_W = W_V/(k_W \times V_c \times (X_{in} - X_{out})) = (W_S - W_c)/(k_W \times V_c \times (X_{in} - X_{out})) \tag{8.11}$$

The N_W can be estimated in the presence of plants under light in the cultivation room with use of a hydroponic cultivation system, by using Eq. (8.11) with measured values of W_S, W_C, X_{in}, and X_{out}. Thus, this method is advantageous over the estimation method of the N_C using Eq. (8.4) or Eq. (8.5). Li et al. (2012c) showed that the N_W was in good agreement with the N_C. The N_W should be the same as the N_C in Eqs. (8.3) and (8.6) if they are estimated under a similar weather condition.

8.4 Resource use efficiency

8.4.1 Introduction

Use efficiencies of CO_2 (CUE), water (WUE), electric energy for lighting (EUE_L), photosynthetic radiation energy (PrUE), and photosynthetic photons (PUE) are defined in this chapter as the mass, mol, or energy ratio of the resource element fixed, held or assimilated by plants to that of the resource element supplied to the cultivation room. Use efficiency of each nutrient element such as N (nitrogen), P (phosphate), and K (potassium) are similarly defined and estimated (although they are not discussed in the chapter). Then, these efficiencies are dimensionless. Maximum value of the CUE and the WUE is unity (=1.0). However, maximum value of the EUE_L, the PrUE, and the PUE is around 0.1 or lower than 0.1 (Kozai, 2011, 2013).

The term "efficiency" is used in this book when the units of nominator and denominator are the same as above, while the term "efficacy" is used when they are different. The efficacies of space, time, and labor are discussed in Chapter 12 of this book in relation to the productivity.

8.4.2 CO_2 use efficiency without considering C_R and C_M, CUE_W

CUE without considering C_R and C_M, CUE_W, is defined by Eq. (8.12).

$$CUE_W = C_P/C_S = (C_S - C_V)/C_S \tag{8.12}$$

where C_R and C_M are rates of human respiration and microorganisms. The C_P is affected significantly by leaf area per cultivation bed area (LAI: leaf area index), C_{in}, PPFD, air current speed and root zone environmental factors, while the C_V is affected only by the N and the $(C_{in}-C_{out})$ (Yokoi et al., 2005). The CUE_W decreases significantly with increasing the C_V [high N and/or high $(C_{in}-C_{out})$]. At LAI = 1.0, the CUE_W was, respectively, 0.9, 0.7, and 0.2 when the N was 0.02, 0.1, and 1.0 h^{-1} (Yokoi et al., 2005). When the N was around 0.2 h^{-1}, the CUE_W was around 0.4 for the PFAL with floor area of 1300 m^2 and 16–18 layers (or tiers) and maximum daily production rates of 6700 leaf lettuce plants, showing the significant effect of the N on the CUE_W (Ohyama et al., 2020).

The C_R and the C_M are not considered in Eq. (8.12) because no monetary costs are required for the C_R and the C_M. In this case, however, the CUE_W could be greater than 1.0 when the C_R and the C_M account for a significant percentage of the C_P.

8.4.3 CO_2 use efficiency considering C_R and C_M, CUE_C

The CUE_C defined by Eq. (8.13) is always lower than or equal to unity.

$$CUE_C = C_P/(C_S + C_R + C_M) = (C_S - C_V)/(C_S + C_R + C_M) \tag{8.13}$$

The CUE_W and the CUE_C are useful indexes to monitor how CO_2 supplied to the cultivation room is effectively used by plants in the cultivation room. In a PFAL with 10 layers and cultivation room floor area of 338 m^2, producing leaf lettuce plants, the CUE_C was 0.95. The reason for this high CUE_C is that the C_R was comparable to the C_V, and the C_S and the C_R, respectively, accounted for 78% and 22% of total CO_2 supplied in the cultivation room (Li et al., 2012a). In this case, the CUE_E was 0.78.

8.5 Water use efficiency

In crop science, water use efficiency of a crop, WUE_D, is defined as the ratio of carbon assimilation rate or dry biomass production rate to the transpiration rate (Kramer and Boyer, 1995). The increase in dry biomass of tomato plug seedlings grown for two weeks in the greenhouse was 2.7 g per kg of water irrigated, and thus the WUE_D is 0.0027 (=0.0027 kg/1.0 kg) (Shibuya and Kozai, 2001).

8.5.1 Water use efficiency for whole plants and produce, WUE_E and WUE_I

In this chapter, water use efficiency of the cultivation room is defined differently from the above used in crop science. The water use efficiency on whole plant (W_E) basis, WUE_E, and that on produce (W_I) basis, WUE_I, under steady state conditions are, respectively, defined by Eqs. (8.14) and (8.15).

$$WUE_E = W_E/(W_S - W_C) = (W_S - W_C - W_V)/(W_S - W_C) \tag{8.14}$$

$$WUE_I = W_I/(W_S - W_C) = (W_E - W_R)/(W_S - W_C) = (W_S - W_C - W_V - W_R)/(W_S - W_C) \tag{8.15}$$

where W_R is the rate of plants removed as residue to the outside. In the airtight and thermally insulated cultivation room, almost all the transpired water, W_T, is condensed at the cooling panels of air conditioner, W_C, and returned to the nutrient solution tank for reuse. In this case, the W_T is almost equal to the W_C with some time lag, and net water consumption for irrigation is equal to the $(W_S - W_C)$. Thus, the WUE_E and WUE_I increase with decrease in the W_V and increase in the W_C. The WUE_I increases also with decrease in the W_R.

8.5.2 Measured WUE_E

The WUE_E was 0.93 for a small growth chamber (Room air volume: 5 m^3) (Ohyama et al., 2000). For a larger growth chamber (Room air volume of 112 m^3), the maximum WUE_E was 0.95 at $N_W = 0.02$ and LAI $= 1.2$ (Yokoi et al., 2005). In case that the condensed water is not reused, the WUE_E would be around 0.05, just like in the greenhouse or in the open field.

The WUE_E decreases with increasing the W_V or with increases in the N_W, the $(X_{in} - X_{out})$ and the W_C. (see, Eq. 8.14). The WUE_E is greater than 0.9 when the N_W is lower than 0.02 h^{-1} and the LAI on cultivation bed area basis is greater than 0.5. While, the WUE_E is 0.2 at the N_W of 1.0 h^{-1} and the LAI of 1.0 (Yokoi et al., 2005). On the other hand, the WUE_E is about 0.5 when the N is 0.1 h^{-1} and the LAI is 0.5 (Yokoi et al., 2005).

8.5.3 Effect of harvest index, HI, on the WUE_I

Fresh weights of phytomass of whole plants and produce are, respectively, expressed by $(W_E + D_E)$ and $(W_I + D_I)$. The ratio of $D_E/(W_E + D_E)$ and that of $D_I/(W_I + D_I)$ are called dry matter ratio, and it ranges between 0.05 and 0.10 in most leafy vegetables. The ratio of the $(W_I + D_I)$ to the $(W_E + D_E)$ is called harvest index, HI, which ranges between 0.75 and 0.90 in most leafy vegetables and between 0.4 and 0.6 in most fruit vegetables. The difference between the $(W_E + D_E)$ and the $(W_I + D_I)$ is the fresh weight of plant residue (unmarketed parts of plants). Thus, the WUE_I increases with increase in HI.

8.5.4 Water use efficiency considering W_W and W_Z, WUE_W and WUE_P

Eqs. (8.14) and (8.15) can be extended to Eqs. (8.16) and (8.17) by considering the W_W (water supply rate for washing/cleaning) and the W_Z (rate of water recycled for washing and cleaning) where the WUE_W is the WUE_E considering the W_W and the WUE_P is the WUE_I considering the W_W and the W_Z.

$$WUE_W = W_E/(W_S + W_W - W_C - W_Z) = (W_S - W_W - W_V)/(W_S + W_W - W_C - W_Z) \quad (8.16)$$

$$WUE_P = W_P/(W_S - W_W - W_C - W_Z) = (W_S - W_W - W_V - W_R)/(W_S + W_W - W_C - W_Z) \tag{8.17}$$

The W_W is often greater than the $(W_S - W_V)$ in a commercial PFAL. Use of high pressure steamed fine water droplets instead of water droplet spray significantly decreases the W_W.

8.6 The energy balance of the cultivation room

8.6.1 Introduction

Energy balance of an airtight and thermally insulated cultivation room is schematically shown in Fig. 8.11. Electricity consumption for lighting, A_L, air conditioning, A_A, and machinery (pumps, fans, etc.) operation, A_M, accounts, respectively, for around 80%, 15%, and 5% on the yearly average in a fairy airtight and well-insulated cultivation room located in a temperate climate region (Kozai, 2013; Kozai et al., 2020). The A_A decreases with increasing the difference ($T_{in}-T_{out}$).

The air infiltration or ventilation rate needs to be kept minimum or controlled even when the air temperature outside is lower than that inside, to keep the C_{in} at around 1000 µmol mol^{-1}, which is around 600 µmol mol^{-1} higher than C_{out}.(around 400 µmol mol^{-1}) (see Fig. 8.5). The minimum air infiltration is essential to minimize the invasion of small insects and/or pathogens to the cultivation room, the fluctuations of air temperature and VPD and the loss of water vapor to the outside. Those fluctuations and the heat loss/gain in the cultivation room are also caused by low thermal insulation of the walls and floor.

8.6.2 Condensation on the walls and floor

Avoiding the water condensation on the walls and floor in the winter is necessary to prevent the fungi and algae propagation and growth. The condensation can be prevented by keeping the wall/floor surface temperatures higher than the dewpoint of room air. The condensation also occurs around the cultivation beds/racks when their surface temperatures are lower than the dewpoint temperature of room air due to the low temperature of nutrient solution and a sudden change in room air temperature. The dewpoint of room air is, respectively, 17.2, 19.3, and 21.3°C when the room air temperature is 23°C and the relative humidity is 70%, 80%, and 90%, respectively.

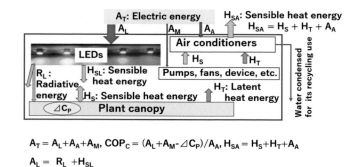

$A_T = A_L+A_A+A_M$, $COP_C = (A_L+A_M-\Delta C_P)/A_A$, $H_{SA} = H_S+H_T+A_A$

$A_L = R_L +H_{SL}$

FIGURE 8.11 Energy balance during photoperiod of the thermally insulated and airtight cultivation room. The radiative energy flux, R_L, consists of thermal and photosynthetic radiation energy fluxes. The ΔC_P denotes the increase in carbohydrates (or chemical energy) in plants. (see the text for details).

8.6.3 Heat exchange through the walls and floor

Heat transmission rate through the walls including roofs between inside and outside the cultivation room under steady state conditions, H_W, is expressed by Eq. (8.18), and that of the floor, H_F, by Eq. (8.19).

$$H_W = K_W \times S_A \times (T_{in} - T_{out}) \tag{8.18}$$

$$H_F = K_F \times F_A \times (T_{in} - T_{soil}) \tag{8.19}$$

where K_W and K_F are the heat transmission coefficient of walls and floor, respectively, S_A and S_F are wall/roof area and floor area, respectively. Both K_W and K_F are around $0.15\ W\ m^{-2}\ K^{-1}$ in the thermally well-insulated cultivation room. In the cultivation room with the total wall area of $1600\ m^2$ (20 m wide, 40 m long, and 5 m high), the H_W is 6 ($=0.15/1000 \times 1600 \times 25$) kW when the ($T_{in}-T_{out}$) is 25°C in Eq. (8.18).

Enthalpy (the sum of sensible and latent heat of moist air) loss due to air infiltration/ventilation is expressed by Eq. (8.20).

$$H_V = k_W \times N_W \times V_C \times (E_{in} - E_{out}) \tag{8.20}$$

where k_W is the conversion factor from volume to mass of the room air; the E_{in} and E_{out} are, respectively, the enthalpy of the air inside and outside the cultivation room. The E_{in} is $65\ kJ\ kg^{-1}$ (DA or dry air) at $T_{in} = 25°C$ and $X_{in} = 0.016\ kg\ kg^{-1}$ (DA) and E_{out} is 8 kJ kg (DA) at 0°C and X_{out} is $0.003\ kg\ kg^{-1}$ (DA). Thus, $H_V = 6612$ [$= 1.16 \times 0.02 \times 5000 \times (65-8)$] kJ $h^{-1} = 1.83$ ($=6612/3600$) kW, and 330,600 [$= 1.16 \times 1.0 \times 5000 \times (65-8)$] kJ $h^{-1} = 92$ kW, respectively, when the N_W is 0.02 and $1.0.h^{-1}$. The thermal insulation is necessary primarily to minimize the heat gain or loss through the air gaps of walls and floor, causing unpredictable fluctuations of room air temperature and VPD.

8.6.4 Cooling load of air conditioners

In the airtight and thermally insulated cultivation room, the steady state cooling load of air conditioners is equal to the sum of A_L and A_M, in case that the heat energy generated by workers and plants in the cultivation room is negligibly small compared with the sum of A_L and A_M.

The heat energy generated by a male worker during walking or standing with light work is about 0.15 kW per person (body mass: 65 kg) or $2.3\ W\ kg^{-1}$ body mass at 25°C room temperature. Plants generate heat energy due to respiration, but its heat energy flux is negligibly small compared with the sum of A_L and A_M. The H_W, H_F, and H_V in Eqs. (8.18)–(8.20) need to be added to the cooling load in case that they are not negligibly small.

8.6.5 Energy conversion at the plant canopy

At a densely populated plant canopy without water stress, around 50% of photosynthetic and thermal radiation absorbed by leaves is converted to latent heat energy, H_T, due to transpiration; the rest is mostly converted to sensible heat energy, and affects the leaf and surrounding air temperatures; a few percent of photosynthetic radiation is converted to chemical energy by photosynthesis, and stored in plants as carbohydrates. The percent of chemical energy fixed by plants depends on LAI, the environmental factors, photosynthetic characteristics of plant canopy, and genetic traits of the plants. Its highest percentage under ideal conditions for a single leaf is around 10%, but its actual percentage of an existing plant canopy is lower than 5% (Kozai, 2011, 2013).

It is noted that the transpiration rate per floor area in the cultivation room with 10 tiers is around 10 times greater than that in the greenhouse in case that the LAI per cultivation area, VPD, and air current speed are similar to each other.

During the transpiration, the temperature of the surrounding air and leaves are lowered by heat of evaporation/transpiration (2442 kJ kg^{-1} at 25°C). The leaf temperature decreases with increasing the transpiration rate and increases with increasing the net radiation flux density. The temperature of upper leaves in the plant canopy is often around 1°C lower than the surrounding air temperature at a moderate transpiration rate, which increases with increasing the VPD, the stomatal conductance and the air current speed (Kitaya, 2016).

8.6.6 Electricity for lighting use efficacy, EUE$_L$ (Kozai, 2011)

The cost of electricity for lighting is the largest single factor among the costs for environmental control. The cost for electricity for air conditioning for cooling increases with increasing the electricity consumption for lighting because over 95% of electric energy consumed for lighting is ultimately converted to sensible or latent heat energy which needs to be removed to the outside by air conditioners. Electric energy consumed by pumps, fans, and vehicles is also converted to the sensible heat energy.

The EUE$_L$ is defined by Eq. (8.21).

$$EUE_L = C_P/A_L \qquad (8.21)$$

In Eq. (8.21), the C$_P$ is expressed in unit of kg h^{-1} (CO$_2$) and electricity consumption for lighting, A$_L$, is expressed in unit of kW or MJ h^{-1}. Thus, unit of the EUE$_L$ is kg kWh^{-1} or kg MJ^{-1}, and the EUE$_L$ is called Electricity for lighting Use Efficacy (not Efficiency). The C$_P$ can also be expressed in unit of kg h^{-1} (dry matter (DM); 1.56 kgCO$_2$/kg DM) (Yoshinaga et al., 2000), mol h^{-1}, MJ h^{-1} (chemical energy; 20 MJ kg^{-1}) (Ohyama et al., 2020). Both the C$_P$ (net photosynthetic rate in Eq. 8.3) and the A$_L$ in Eq. (8.21) can be measured continuously, so that the EUE$_L$ can be estimated online. Energy conversion process in the cultivation room is discussed in detail by Kozai (2011) and Kozai (2013).

8.6.7 Photosynthetic photon flux efficiency

Conversion factor of light source from electric energy to photosynthetic photons is called photosynthetic photon number efficacy, PPnE, in unit of μmol J^{-1}, which is currently around 3 μmol J^{-1} for white LEDs. Then, photosynthetic photon flux efficiency, PPFE, is defined as the ratio of C$_P$ (mol s^{-1}) to photosynthetic photon flux of light source, PPF, in unit of mol

s^{-1}, which is defined as a product of PPnE and electricity consumption of the light source, A_L, in unit of kW or kJ s^{-1}.

$$PPFE = C_P/PPF = C_P/(PPnE \times A_L) \tag{8.22}$$

8.6.8 Photosynthetic photon flux density efficiency

Photosynthetic photon flux, PPF (mol s^{-1}) of LEDs received at the plant canopy surface area or cultivation bed area, S_C (m^2), is called photosynthetic photon flux density, PPFD (µmol m^{-2} s^{-1}). Then, PPFD use efficiency, PPFDE, is defined by Eq. (8.23).

$$PPFDE = C_P/(PPFD \times S_C) \tag{8.23}$$

Similarly, efficiency/efficacy for photosynthetic radiation flux and flux density can be defined.

8.7 Coefficient of performance of air conditioners for cooling

Energy use efficiency of an electric air conditioner is called coefficient of performance (COP), defined as the ratio of heat energy removed to the outside for cooling, H_{SA}, or introduced from the outside for heating, to the electric energy consumed by the air conditioner, A_A (Fig. 8.11; Kozai and Niu, 2019). The COP for cooling, COP_C, is 1.0 lower than the COP for heating, COP_h, as shown in Eq. (8.25).

$$COP_C = (H_S + H_T)/A_A \tag{8.24}$$

$$COP_h = (H_S + H_T + A_A)/A_A = (H_S + H_T)/A_A + A_A/A_A = COP_C + 1 \tag{8.25}$$

The COP_C is expressed as $(A_L + A_M)/A_A$, in case that the cultivation room is completely airtight and thermally insulated.

The COP_C increases with increasing the difference in air temperature between inside and outside the cultivation room ($T_{in} - T_{out}$), and is generally deigned to show its maximum COP_C at around 75% of cooling capacity (i.e., its load factor is 75%) as shown in Fig. 8.12. Thus, adjusting the number of air conditioners in operation to achieve the load factor of around 75% for each air conditioner in operation is important for its energy efficient operation (see Chapter 25 in Kozai et al., 2020 for more details).

In Fig. 8.12, the COP_C of 5.0 means that the electricity consumption of air conditioners for cooling is 20% [$= 100 \times A_A/(A_L + A_M)$] of the sum of electricity consumption for lighting and machine operation in the airtight and thermally insulated cultivation room. In other words, the percentage of A_A over the total electricity consumption [$= A_L + A_A + A_M = 1.2 \times (A_L + A_M)$] is 17% [$= 100 \times A_A/(1.2 \times (A_L + A_M)) = 100 \times 0.2/1.2$].

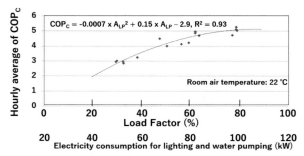

FIGURE 8.12 Hourly average of COP_C as affected by the load factor (percent electricity consumption for lighting and water pumping over its maximum electricity consumption, 120 kW) measured on June 26–27, 2012. The coefficient of performance was about 1.0–1.5 higher in February when the daily average of air temperature outside was 15°C lower than that in July. Maximum electricity consumption of heat pumps: 58.6 kW. *Adopted after revision from Sekiyama, T., Kozai, T., 2015. Issues on cost reduction of heat pumps for air conditioning of horticultural structures. Nougyoudenka (Agric. & Electrif.). 68 (2), 12-16 (in Japanese).*

When the water vapor is condensed at the cooling panels of air conditioners, the H_T (latent heat energy) is converted to the sensible heat energy with a conversion factor of about 2442 kJ kg^{-1} (heat of condensation at 25°C), and the sensible heat energy is removed by the outer units (condensers) of air conditioners to the outside. It is important to notice that no extra electricity is required to dehumidify the room air or to keep the VPD at around 0.5 kPa or relative humidity of 75% at 25°C.

Graamans et al. (2017, 2020) developed mathematical models for computer simulation to compare the transpiration and energy balance in the plant factory and the greenhouse under different design conditions. Zhang and Kacira (2020) developed a simulation model to compare the operational costs and resource-use efficiency of PFALs with different indoor environment and climate conditions. This type of research work is essentially important to develop science-based software tools for designing resource-efficient plant factories.

8.8 Conclusion

As described in this chapter, the balances of CO_2, water, and energy are relatively simple in the airtight and thermally insulated cultivation room, and all the resource inputs and outputs can be measured relatively easily without disturbance of weather outside. Then, use efficiencies of CO_2, water, and electricity can be estimated relatively easily. These features of the PFAL are advantageous over the greenhouse and the open field.

The hydroponic cultivation system used in the PFAL has similar features because it is a kind of closed system, too. Thus, the balance of nutrient elements and use efficiencies of the nutrient elements can be estimated relatively easily, although these advantages are currently not actualized in practice.

The methodology described in this chapter is applied only to the whole cultivation room. In order to apply this methodology to each compartment of the cultivation room, the cultivation room needs to be divided into physically separated compartments or basic cultivation modules (Kozai et al., 2020) This idea can be realized easily for the hydroponic cultivation system because the hydroponic cultivation system can be physically separated easily for each cultivation tier, rack, etc. Further research and development are expected in this field.

It should be noted that the methodology described in this chapter can be applied with no problem during the commercial production for sales. Then, a large-scale experiment can be conducted continuously during the commercial production. This methodology can be applied in parallel with the phenotyping system (Section 8.1.3). Then, plant selection, breeding, and production can be conducted in parallel.

Appendix: List of symbols, their description, equation numbers, and units

Symbol	Description	Eqs. & Figs. No.	Unit
A_A	Electricity consumption for air conditioning (1 kWh $=$ 3.6 MJ)	8.9, 8.10, 8.24, 8.25	kWh
A_L	Electricity consumption for lighting	8.21, 8.22	kWh
A_P	Electricity consumption for pumps, fans, etc.	Fig. 8.10	kWh
B	DuBois surface area related to human respiration	8.7	m^2
C_A	Change in CO_2 mass (weight) in the cultivation room air	8.1, 8.2	kg h^{-1}
C_D	Dark respiration rate under dark	8.6	kg h^{-1}
C_G	Gross photosynthesis rate, $C_G = C_P + (C_h + C_D)$	Section 8.2.5	kg h^{-1}
C_H	Photorespiration rate	Section 8.2.5	kg h^{-1}
C_{in}	CO_2 concentration in the cultivation room (ppm)	8.3, 8.6	μmol mol^{-1}
C_{in1}	CO_2 concentration in the cultivation room at time t_1	8.8	μmol mol^{-1}
C_{in2}	CO_2 concentration in the cultivation room at time t_2	8.8	μmol mol^{-1}
C_{in-E}	Equilibrium CO_2 concentration in the cultivation room	8.5	μmol mol^{-1}
C_M	Microorganism respiration rate	8.1, 8.2, 8.8, 8.13	kg h^{-1}
C_0	C_{in} at time 0	8.4	μmol mol^{-1}
C_{out}	Atmospheric CO_2 concentration (400–410 mmol mol^{-1})	8.3–8.6	μmol mol^{-1}
COP_C	Coefficient of performance for cooling (air conditioners)	8.24, 8.25	—
COP_h	Coefficient of performance for heating (air conditioners)	8.25	
C_P	Net photosynthetic (CO_2 fixation) rate, $C_P = C_G - (C_H + C_D)$	8.1, 8.2, 8.8, 8.12, 8.13, 8.21, 8.22, 8.25	kg h^{-1}
C_Q	CO_2 absorption/release rate at wall/floor surfaces ($\fallingdotseq 0$)	8.1, 8.2	kg h^{-1}
C_R	Human respiration rate ($\fallingdotseq 0.042$ kg h^{-1}/person)	8.1, 8.2, 8.7, 8.13	kg h^{-1}
C_S	CO_2 supply rate	8.1, 8.2, 8.5, 8.12, 8.13	kg h^{-1}
C_T	C_{in} at time T	8.4	μmol mol^{-1}

(*Continued*)

Symbol	Description	Eqs. & Figs. No.	Unit
C_V	CO_2 loss rate due to air infiltration/ventilation	8.1, 8.2, 8.3, 8.6, 8.12, 8.13, 8.4	kg h^{-1}
C_W	Rate of change in CO_2 dissolved as ion in nutrient solution	8.1, 8.2	kg h^{-1}
CUE_C	CO_2 use efficiency considering C_R and C_M	8.13	—
CUE_W	CO_2 use efficiency without considering C_R and C_M	8.12	—
D_E	Dry weight of plants removed as produce or plant residue		kg h^{-1}
D_I	Dry weight of produce removed to the outside	Section 8.2.5	kg h^{-1}
E_{in}	Enthalpy of room air (dry air basis)	8.20	kJ kg^{-1}
E_{out}	Enthalpy of outside air (dry air basis)	8.20	KJ kg^{-1}
EUE_L	Electricity for lighting use efficacy	8.20	kg kWh^{-1}
F_A	Floor area	8.19	m^2
H	Height of a worker	Section 8.2.6	m
H_F	Heat transmission flux through floor	8.19	kWh m^{-2} h^{-1}
HI	Harvest index $((W_I + D_I)/(W_E + D_E))$	Section 8.5.3	—
H_S	Sensible heat energy flux	8.24, 8.25, Fig. 8.11	kWh h^{-1}
H_{SA}	Heat energy released to the outside by air conditioners	Fig. 8.11	kWh h^{-1}
H_{ST}	Sensible heat energy generated by light-emitting diodes (LEDs), fans, and pumps	Fig. 8.11	kWh h^{-1}
H_T	Latent heat energy flux	8.24, 8.25, Fig. 8.11	kWh h^{-1}
H_V	Enthalpy flux due to air infiltration/ventilation	8.20	kWh h^{-1}
H_W	Heat transmission flux through walls and roofs	8.18	kWh m^{-2} h^{-1}
k_C	Conversion factor (1.93) from volume to mass of CO_2 gas	8.3, 8.5, 8.8	kg m^{-3}
k_W	Conversion factor (1.16) from volume to mass of water	8.10, 8.11, 8.20	kg m^{-3}
K_F	Heat transmission coefficient of floor	8.19, 8.20	W m^{-2} K^{-1}
K_W	Heat transmission coefficient of walls and roofs	8.10, 8.18	W m^{-2} K^{-1}
LAI	Leaf area index (leaf area divided by cultivation bed area)	Section 8.4.2	—
M	Metabolic rate	8.7	met
M_S	Weight of the substrate containing water/nutrient solution	8.8	kg
N_C	Number of air exchanges of the cultivation room (tracer gas: CO_2)	8.3–8.6	h^{-1}

Symbol	Description	Eqs. & Figs. No.	Unit
N_W	Number of air exchanges of the cultivation room (tracer gas: water vapor)	8.10, 8.11, 8.20	h^{-1}
PPFD	Photosynthetic photon flux density	8.23	$\mu mol\ m^{-2}\ s^{-1}$
PRFD	Photosynthetic radiation flux density		$J\ m^{-2}\ s^{-1}$
PPFDE	Photosynthetic photon flux density efficacy	8.23	—
PPFE	Photosynthetic photon flux efficacy	8.22, 8.23	—
PPF	Photosynthetic photon flux	8.22	$\mu mol\ s^{-1}$
PPnE	Photosynthetic photon number efficacy	8.22	
R_L	Radiation energy (350−20,000 nm) flux generated by LEDs	Fig. 8.11	$kWh\ h^{-1}$
RQ	Respiratory quotient	8.7	—
S_A	Wall area	8.18	m^2
S_C	Cultivation bed area	8.8, 8.23	m^2
S_F	Floor area	8.19	m^2
T	Time span	8.4	h
t_1	Time of C_{in1} measurement	8.8	h
t_2	Time of C_{in2} measurement	8.8	h
T_{in}	Air temperature inside	8.18, 8.19	°C
T_{out}	Air temperature outside	8.18	°C
V_C	Air volume of the cultivation room	8.3, 8.5, 8.6, 8.10, 8.11, 8.18, 8.20	m^3
V_B	Air volume of the plastic box for measuring C_M	8.8	m^3
W	Body mass of a worker	Section 8.2.6	kg
W_A	Change in water vapor mass (or weight) of the room air	8.9	$kg\ h^{-1}$
W_B	Rate of water used for washing hands, boots, etc.	Figs. 8.8 and 8.9	$kg\ h^{-1}$
W_C	Water condensation rate at the cooling coils of air conditioners for cooling	8.9, 8.11, 8.14, 8.15, 8.16, 8.17	$kg\ h^{-1}$
W_D	Rate of water drained and discarded		$kg\ h^{-1}$
W_E	Rate of plants removed as produce or plant residue	8.9, 8.14, 8.15, 8.16	$kg\ h^{-1}$
W_F	Rate of water drained for reuse or discarded ($W_F = W_K + W_D$)	8.9	$kg\ h^{-1}$
W_I	Rate of plants removed as produce to the outside	8.15	$kg\ h^{-1}$
W_K	Rate of water drained from the cultivation bed for reuse	Figs. 8.8 and 8.9	

(Continued)

Symbol	Description	Eqs. & Figs. No.	Unit
W_L	Rate of water used for washing cultivation panels and beds	Figs. 8.8 and 8.9	kg h^{-1}
W_P	Fresh weight increase (water fixation) rate (= W_U - W_T)	8.9, 8.17	kg h^{-1}
W_Q	Water condensation/evaporation rate at walls and floor	8.9	kg h^{-1}
W_R	Rate of plants removed as residue to the outside	8.15, 8.17	kg h^{-1}
W_S	Water supply rate for irrigation	8.9, 8.11, 8.14, 8.15, 8.16, 8.17	kg h^{-1}
W_T	Transpiration rate of plants	Fig. 8.11	kg h^{-1}
W_U	Water uptake rate of plants	Fig. 8.11	kg h^{-1}
WUE_D	Water use efficiency on dry biomass basis in crop science	Section 8.5	—
WUE_E	Phytomass (plant biomass)-based water use efficiency	8.14	—
WUE_I	Produce-based water use efficiency	8.15	
WUE_P	Produce-based water use efficiency considering W_W and W_Z	8.17	—
WUE_W	Water use efficiency considering recycling use of the drain, W_K and W_Z	8.16	—
W_V	Water vapor loss rate due to air infiltration/ventilation	8.9, 8.10, 8.11, 8.14, 8.16, 8.17	kg h^{-1}
W_W	Water supply rate for washing/cleaning (W_W= W_B + W_L + W_Y)	8.9, 8.16, 8.17	kg h^{-1}
W_X	Change in water mass (or weight) in the substrate	8.9	kg h^{-1}
W_Y	Rate of water used for washing floor, walls, etc.	Figs. 8.8 and 8.9	kg h^{-1}
W_Z	Rate of water recycled for washing and cleaning	8.9, 8.16, 8.17	kg h^{-1}
X_{in}	Absolute humidity of the room air (dry air basis)	8.10, 8.11	kg kg^{-1}
X_{out}	Absolute humidity of the outside air (dry air basis)	8.10, 8.11	kg kg^{-1}

References

Graamans, L., van den Dobbelsteen, A., Meinen, E., Stanghellini, C., 2017. Plant factories: crop transpiration and energy balance. Agric. Syst. 153, 138—147.

Graamans, L., Tenpierik, M., Van den Dobbelsteen, A., Stanghellini, C., 2020. Plant factories: reducing energy demand at high internal heat load through façade design. Appl. Energy 262, 114544.

Kitaya, Y., 2016. Air current around single leaves and plant canopies and its effect on transpiration, photosynthesis, and plant organ temperatures (Chapter 13). In: Kozai, et al. (Eds.), Led Lighting for Urban Agriculture). Springer Nature, pp. 177—187.

Kozai, T., 2011. Improving light energy utilization efficiency for a sustainable plant factory with artificial light. Proc. of Green Light. Shanghai Forum 375—383.

Kozai, T., 2013. Resource use efficiency of closed plant production system with artificial light: concept, estimation and application to plant factory. Proc. Jpn. Acad. Ser. B. 89, 447—461. https://doi.org/10.2183/pjab.89.447.

Kozai (Ed.), 2018. Smart Plant Factory. The Next Generation Indoor Vertical Farms. Springer, p. 456.

Kozai, T., Niu, G., 2019. Plant factory as a resource-efficient closed plant production system. In: Kozai, T., Niu, G., Takagaki, M. (Eds.), Plant Factory: An Indoor Vertical Farming for Efficient Quality Food Production. Academic Press (Elsevier, pp. 93–115.

Kozai, T., Niu, G., Takagaki, M. (Eds.), 2020. Plant Factory: An Indoor Vertical Farming System for Efficient Quality Food Production, second ed. Academic Press, p. 487. ISBN: 978-0-12-816691-8.

Kramer, P.J., Boyer, J.S., 1995. Water Relations of Plants and Soils. Academic Press, pp. 377–404.

Li, M., Kozai, T., Ohyama, K., Shimamura, S., Gonda, K., Sekiyama, T., 2012a. CO_2 balance of a commercial closed system with artificial lighting for producing lettuce plants. Hortscience 47 (9), 1257–1260.

Li, M., Kozai, T., Ohyama, K., Shimamuram, S., Gonda, K., Sekiyama, T., 2012b. Estimation of hourly CO_2 assimilation rate of lettuce plants in a closed system with artificial lighting for commercial production. Eco-Engineering 24 (3), 77–83.

Li, M., Kozai, T., Niu, G., Takagaki, M., 2012c. Estimating the air exchange rate using water vapor as a tracer gas in a semi-closed growth chamber. Biosyst. Eng. 113, 94–101.

Ohyama, K., Yamaguchi, J., Enjouji, A., 2020. Resource utilization efficiencies in a closed system with artificial lighting during continuous lettuce production. Agronomy 10, 723. https://doi.org/10.3390/agronomy10050723.

Oyama, K., Yoshinaga, K., Kozai, T., 2000. Energy and mass balance of a closed type transplant production system (Part 2) water balance. J. Soc. High Technol. Agric. 12 (4), 217–224 (in Japanese with English abstract and figure/table captions).

Persily, A., de Jonge, L., 2017. Carbon dioxide generation rates for building occupants. Indoor Air 27, 868–879. https://doi.org/10.1111/ina.12383.

Sekiyama, T., Kozai, T., 2015. Issues on cost reduction of heat pumps for air conditioning of horticultural structures. Nougyoudenka (Agric. & Electrif.) 68–2, 12–16 (in Japanese).

Shibuya, T., Kozai, T., 2001. Light-use and water use efficiencies of tomato plug sheets in the greenhouse. Environ. Control Biol. 39 (1), 35–41 (in Japanese with English abstract and figure and tables captions).

Taiz, L., Zeiger, E., 2006. Plant Physiology, fourth ed. Sinauer Associates, Inc. Publishers, pp. 253–288.

Wheeler, R.M., 1992. Gas exchange measurements using a large, closed plant growth chamber. Hortscience 27, 777–780.

Yokoi, S., Kozai, T., Hasegawa, T., Chun, C., Kubota, C., 2005. CO_2 and water utilization efficiencies of a closed transplant production system as affected by leaf area index of tomato seedling populations and the number of air exchanges. J. SHITA 17 (4), 182–191 (in Japanese with English abstract and captions).

Yoshinaga, K., Ohyama, K., Kozai, T., 2000. Energy and mass balance of a closed-type transplant production system (part 3) Carbon dioxide balance. J. SHITA 12 (4), 225–231 (in Japanese with English abstract and captions).

Zhang, Y., Kacira, M., 2020. Enhancing resource use efficiency in plant factory. Acta Hortic. 1271, 307–311. https://doi.org/10.17660/ActaHortic.2020.1271.42. ISHS 2020.

Hydroponics

Genhua Niu[1] and Joseph Masabni[2]

[1]Texas A&M AgriLife Research, Texas A&M University, Dallas, TX, United States; [2]Texas A&M AgriLife Extension Service, Dallas, TX, United States

9.1 Introduction

Hydroponics is the practice of growing plants in a nutrient solution with or without a soilless substrate to provide physical support. The word hydroponics comes from the root words "hydro," meaning water, and "ponos," meaning labor, literally "working water." The concept of hydroponics existed thousands of years ago, with the earliest examples of Hanging Gardens of Babylon and the Floating Gardens of China. However, modern hydroponic systems did not thrive until the advent of the greenhouse and plastics industries. Since then, scientists have developed many hydroponic systems for various crops based on locally available resources. Currently used commercial hydroponic systems are the improved versions of these early systems.

In a broad sense, hydroponics can be divided into two types: solution culture and soilless medium culture. Soilless medium culture is sometimes not considered as "true" hydroponics, while solution culture is. Main types of solution culture include nutrient film technique (NFT), deepwater culture or floating raft culture, and aeroponics. Soilless medium culture uses a solid medium (also called substrate) to anchor plant roots while the nutrient solution is provided through sub or top irrigation. Substrate and growing media are often used interchangeably. In this book hereafter, the term substrate is used.

For plant factories, almost all production stages use solution culture, while substrate is often used in the propagation stage. Plant factory is an emerging and innovative system for plant production that commonly uses vertically stacked hydroponic systems in an indoor controlled environment. The basic knowledge of hydroponics and the general understanding of plant nutrition and nutrient management for crop production in plant factories are derived from prior studies on greenhouse and growth chamber hydroponics. That is, the basic concept and principles are the same: to provide essential nutrients, water, and oxygen to the rootzone, and to create a uniform aerial growing environment to maximize plant growth and quality.

9.2 Types of hydroponic systems

9.2.1 Nutrient film technique system

In an NFT system, a very shallow stream (film) of water containing all nutrients required for plant growth circulates between the growing channels (troughs) and the nutrient reservoir (Fig. 9.1). Plant roots are contained in the growing channels and are in contact with the nutrient solution. The growing channels have a slope between 1% and 3% and are up to 15 m in length to ensure uniformity and ease of handling. A properly designed NFT system should ensure adequate supply of water, nutrients, and oxygen to the plant roots. If a channel is too long, the flow of the nutrient solution may be affected, causing localized depressions, and resulting in inadequate oxygen supply for the roots. The key is to provide a thin "film" of nutrient solution so that oxygen is dissolved into the solution. As the roots grow, more are exposed to the air. An NFT system is primarily used for short-term crops with small root systems such as lettuce, herbs, and other leafy greens. The growing channel of an NFT system can be rectangular, square, or circular in shape and can be arranged in a vertical, horizontal, or A-frame design, as long as light distribution across the whole system is uniform for all plants.

9.2.2 Deepwater culture system

In a deepwater hydroponic culture (DWC) system, the majority of plant roots are submerged in a solution containing all essential nutrients, while the crowns are supported by a floating raft, usually a food-grade foam board (Fig. 9.2). DWC is also called raft or floating culture system. DWC is relatively inexpensive to set up and can be built in different sizes even without a nutrient reservoir as long as aeration is provided. Aeration can be achieved by air pumps or venturi tubes to recirculate the nutrient solution within the culture bed or trough. The water depth typically ranges from 15 to 30 cm. Since the amount of nutrient solution per plant is relatively large, there is less fluctuation in nutrient concentrations, pH, and

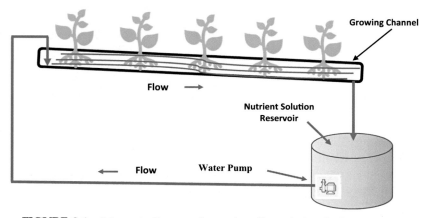

FIGURE 9.1   Schematic diagram of a nutrient film technique hydroponic system.

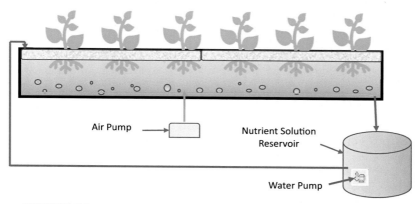

FIGURE 9.2 Schematic diagram of a deepwater culture hydroponic system.

temperature. Therefore, DWC needs less maintenance compared to an NFT system. Small DWC systems are suitable for hobby and small-scale growers.

9.2.3 Aeroponic system

Aeroponics is the practice of growing plants in an air or mist environment without the use of any substrate (Fig. 9.3). That is, the plant roots are suspended in the air and are misted or sprayed periodically with a nutrient solution or aerosol of nutrient solution. Water and nutrient use efficiency in an aeroponic system are higher than those in NFT or DWC systems. The biggest advantage of aeroponics is that roots are exposed to air, thus there is never an issue of insufficient oxygen. Its disadvantages are high initial construction costs, high maintenance of the system, and high level of technical knowledge required.

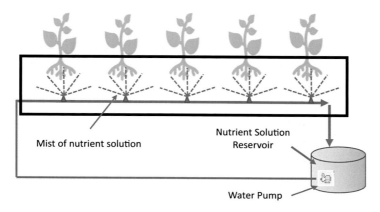

FIGURE 9.3 Schematic diagram of aeroponics.

It is worth mentioning that a hydroponic system can be built in a vertical fashion, commonly known as vertical towers, where the nutrient solution is trickling down. This vertical hydroponic system is marketed as "aeroponics"; however, it is a variant of NFT, and not a true aeroponic system.

9.2.4 Substrate culture system

Substrate culture is the practice of growing plants in a solid soilless substrate. Typical substrates include rockwool, perlite, peat moss, coir (cocopeat or coconut fiber), vermiculite, rice hulls, clay pebbles, etc. A substrate can be a mixture of several ingredients at different ratios depending on crop performance. For example, to achieve best growth, 100% perlite was used for lettuce, while 50% perlite and 50% vermiculite mix was used for komatsuna (Kroggel et al., 2012). The important factors to consider when selecting which substrate to use are the physical and chemical properties, cost, and availability. Different substrate properties require different cultivation and irrigation management.

There are several kinds of substrate culture systems, namely, ebb and flow recirculating system, nonrecirculating bag culture, and container culture. For indoor propagation systems, vertical shelves are employed, and nutrient solution is often supplied through the bottom of the plants. For propagation systems in a greenhouse, both top and subirrigation methods are used; however, subirrigation is preferred to avoid wetting the foliage for disease prevention. Large container- or bag-culture systems are used for fruiting crops such as tomatoes, cucumbers, peppers, and melons. For more information on substrate culture, refer to Raviv et al. (2019).

9.3 Hydroponic systems for plant factories

9.3.1 Hydroponic systems and products for PFAL

The ideal hydroponic system for plant factories, as described in Chapter 3, should be lightweight and simple with minimal pipe length, and no bulky reservoirs to the store nutrient solution because the nutrient solution is not recirculated. That is, all nutrients provided are absorbed by the plants and the water uptake by the plants balances transpiration. In reality, however, it is not easy to estimate accurately the amount of transpiration and the amount of nutrients for each crop at any particular stage. This is because plant physiological and biochemical processes are influenced by the surrounding environment and any change can alter the physiological processes of the plants. Nevertheless, a hydroponic system for use in plant factories should be lightweight and designed and operated in a vertical fashion. NFT or substrate culture systems are better suited for vertical grow racks. As stated earlier, the key is to provide a thin layer (film) of nutrient solution to each culture bed in a multitier grow rack. Aeroponics is also suitable for vertical grow racks. On the other hand, DWC may not be a good choice because of the heavy weight of large amount of nutrient solution, which increases construction costs. Currently, modified NFT and substrate-based hydroponic systems are mainly used in plant factories. Popular substrates used in plant factories are coco, cocopeat, cellulose, foam urethane, peat, and rockwool (Kozai et al., 2019). Selection of substrates depends on multiple factors such as structure of shape of culture beds, physical and chemical properties, price, and availability of the substrates.

Many products related to hydroponic systems are yet to be developed for plant factories. Currently, there are no commercially available turnkey hydroponic products for plant factories. Instead, large commercial operations are designing and developing their own proprietary technologies for various hydroponic components and control devices (Kozai et al., 2019). For example, at AeroFarm, a specially designed substrate, AeroCloth, was developed and patented (Harwood, 2019). Leafy greens and herbs are grown in this reusable cloth substrate and the nutrient solution is provided to the roots through a mist. As interest in plant factories continues to increase, we expect more technologies and innovative products related to hydroponic systems for plant factories will be developed and available commercially and the prices are expected to decrease over time.

9.3.2 Algae and fungus gnats prevention

Algae are plant-like aquatic organisms that need light, water, and nutrients to thrive. Algae appear to be a common problem in many commercial PFALs. Algae absorb light photons, consume nutrients, and make the panels and cultivation beds dirty, which reduces light reflection of the panels, thus decreasing the light use efficiency of the system (Lu and Shimamura, 2018). Therefore, it is important to keep the nutrient solution from exposure to light, which can be achieved by covering any small gaps between floating panels, regular cleaning of the panels, and by keeping the propagation plug surface dry.

Another common problem in commercial PFALs is fungus gnats, especially in the propagation room. Both the larval and adult stages of fungus gnats can cause problems in a PFAL. Fungus gnats are a common pest of indoor plants (PFAL and house plants) where humidity and moisture are high. They are usually first noticed when the harmless adults are seen flying around the plants. Yellow sticky traps can be used to trap the flying pest and as an indicator of the population of the pest. A large population of fungus gnats can cause damage to plants. Their larvae may begin to feed on the roots of the plants, which causes yellowing and wilting. To prevent fungus gnats, it is important to keep the propagation plugs and other materials clean, and by keeping the propagation bed surface dry. When the population of fungus gnats becomes high, it is time to thoroughly clean the system, including all the propagation tools and equipment using an approved cleaning or sanitizing agent.

9.4 Nutrient solution

The fundamental purpose of a hydroponics culture technique is to provide adequate levels of nutrients, water, and oxygen throughout the production period efficiently and uniformly, regardless of the culture system. To achieve this goal, it is of paramount importance to start with a properly designed nutrient formula along with a proper replenishment of nutrients during the crop production cycle. Unlike soil-based crop production, nutrients in all hydroponic culture systems must be provided to plant roots through the nutrient solution.

9.4.1 Essential nutrient elements

Plants require 17 essential nutrient elements for proper plant growth and development. An essential element is defined as one whose absence prevents a plant from completing its life cycle (Arnon and Stout, 1939) or one that has a clear physiological role (Epstein, 1999). Among the 17 elements, 3 of them, hydrogen (H), oxygen (O), and carbon (C), are obtained from water and carbon dioxide; thus, they are not considered as mineral nutrients. Essential elements are usually classified as either macronutrients or micronutrients, based on the amount the plant needs. The essential macronutrients include nitrogen (N), phosphorus (P), potassium (K), calcium (Ca), magnesium (Mg), and sulfur (S). The essential micronutrients are chloride (Cl), iron (Fe), boron (B), manganese (Mn), sodium (Na), zinc (Zn), copper (Cu), and molybdenum (Mo).

In the third edition of *Plant Physiology* book (Taiz and Zeiger, 2002), silicon (Si) and nickel (Ni) were mentioned as a newly added essential macronutrient and micronutrient, respectively. Si is the second most abundant element in the earth crust after oxygen. Si has gained increasing attention in agriculture in recent years due to the increasing evidence in the literature showing its beneficial effects in a range of plants, alleviating diverse forms of abiotic and biotic stresses (Frew et al., 2018; Luyckx et al., 2017). In soil-based cultivation, Si is not lacking in quantity. However, in a hydroponic system, Si needs to be added to the nutrient solution along with the other elements. Early nutrient solution formulas did not include Si. Limited information is available on the beneficial effect of Si under artificial light conditions and hydroponics. As for Ni, it acts as an activator of the enzyme urease (Fabiano et al., 2015), and the understanding of the role of Ni in the nutrition, physiology, and metabolism of a majority of crops is still limited. Thus, Ni is not included in any hydroponic nutrient formula discussed below.

9.4.2 Nutrient solution formulas

Over the years, scientists have empirically developed many formulas for various crops. Early formulas include Hoagland and Arnon (1950), Hewitt (1966), and Steiner (1984). Among these, Hoagland solution is the most widely used and cited formula in scientific research papers and the solution supports the growth of a large variety of crops (Smith et al., 1983). Table 9.1 lists several representative nutrient solution formulas that are used by the hydroponics industry and researchers in comparison with Hoagland's solution. As shown in Table 9.1, the various formulas have different proportions of each element concentration. The "Modified Sonneveld" is a modified formula of Sonneveld and Voogt (2009) and has been used successfully for years for leafy greens and herbs production by the controlled environment agriculture (CEA) group at Cornell University (Mattson and Peters, 2014). The UA CEAE (University of Arizona, Controlled Environment Agriculture Center) formula has been used by another CEA group at the University of Arizona, while Jack's formula is a commercial two-bag preblended mix, one bag with calcium nitrate and the other bag containing all the rest of the essential nutrients (Mattson and Peters, 2014). There are other commercial preblended mixes in concentrated liquid forms that users need to dilute properly before use. For beginners, preblended mixed fertilizers may save the hassle of calculating the amount of

TABLE 9.1 Representative nutrient solution formulas: Hoagland and Arnon, Modified Sonneveld, Jack, and UA CEAE (University of Arizona, Controlled Environment Agriculture Center).

Element	Hoagland and Arnon	Ratio N = 200[a]	Modified Sonneveld	Ratio N = 200	Jack	Ratio N = 200	UA CEAE	Ratio N = 200
Macronutrients (mg L^{-1})								
N	210	200	150	200	150	200	189	200
P	31	30	31	41	39	52	39	41
K	235	223	210	280	162	216	341	361
Ca	160	152	90	120	139	185	170	180
Mg	34	32	24	32	47	63	48	51
S	64	61	32	43	64	85	64	68
Micronutrients (mg L^{-1})								
Fe	2.5	2.4	1.0	1.3	2.3	3.1	2.0	2.1
B	0.50	0.48	0.16	0.21	0.38	0.51	0.28	0.30
Mn	0.50	0.48	0.25	0.33	0.38	0.51	0.55	0.58
Zn	0.05	0.05	0.13	0.17	0.11	0.15	0.33	0.35
Cu	0.02	0.02	0.02	0.03	0.11	0.15	0.05	0.05
Mo	0.01	0.01	0.02	0.03	0.08	0.10	0.05	0.05
TDS[b]	734	698	537	716	601	801	851	901
EC[c]	1.15	1.09	0.84	1.12	0.94	1.25	1.33	1.41

[a]Convert all elements proportionally to N = 200 mg L^{-1}.
[b]TDS: total dissolved salts from essential macronutrients.
[c]Estimated electrical conductivity (EC) from essential macronutrients.
Partially adapted from Mattson, N., Peters, C., 2014. A Recipe for Hydroponic Success. Inside Grower.

fertilizers before mixing them. However, there is little flexibility to modify any individual element concentration during production.

All fertilizer formulas contain all the essential nutrient elements, which are provided from water-soluble fertilizers. Cl and Ni are impurities in water and fertilizer and their quantities are adequate to meet the plant growth requirements. Si is not included in any nutrient formula but adding Si to nutrient solution is recommended (Bugbee, 2004).

9.4.3 Selection of a nutrient solution formula

Selection of a nutrient solution formula should be based primarily on crop type and growth stage. The strength or concentration of the solution should be adjusted based on growing conditions, especially in a greenhouse. When light intensity is strong and air temperature is high, plants transpire more; thus the concentration should be lower compared to conditions with less intense light and lower temperatures. In a plant factory, however, the growing conditions are relatively constant; thus a constant nutrient concentration needs to be applied. Electrical conductivity (EC) is used for continuous monitoring of the strength of nutrient solution. While different formulas have different ratios among the essential elements, all have been successfully used in commercial crop production, as plants can adapt to a wide range of nutrient concentrations. For example, the Hoagland formula has been used for a large variety of species from leafy vegetables to fruit crops, as reported in the literature. However, not one formula is necessarily the optimal one for all plant species and cultivars. Customized formulas may be recommended for a particular crop based on sufficient testing under a special set of conditions of water quality and growing conditions. The Modified Sonneveld and Jack formulas in Table 9.1 are recommended for leafy greens and herbs, while UA CEAE formula is used for fruiting vegetables. To maximize the vegetative growth of leafy greens and herbs, a higher rate of N is preferred. On the other hand, higher rates of P, K, and Ca are required at the reproductive stage of fruiting crops. Even among the same family of vegetables such as leafy greens, some vegetables need higher concentrations. For example, kale and spinach are relatively "heavy feeders" compared to lettuce. The optimal nutrient formula and concentration also vary among cultivars of the same crop, such as lettuce with its different and distinct genetic variability and nutritional needs.

9.4.4 Preparing the nutrient solution

After selection of the appropriate nutrient solution formula, the next step is to make the nutrient solution. Preblended liquid or powder concentrates are available and are easily diluted with water. These products are easy to use but lack the flexibility of modifying their composition. Experienced growers and researchers want to prepare their own nutrient formula by mixing the individual water-soluble mineral components. Fertilizer quality is also important and fertilizers should be 100% water soluble and of high purity and are hence referred to as greenhouse grade or reagent grade. It is a common practice to make a stock solution (usually 50X or 100X). In this case, two tanks are needed to separate the calcium-containing compounds from the phosphate and sulfate-containing compounds to prevent insoluble precipitate formation. For example, tank A contains calcium nitrate and half of the potassium nitrate and iron compounds, while tank B includes all the other compounds (Resh, 2012). In the case of the Modified Sonneveld and UA CEAC formulas, tank A contains calcium and potassium nitrates and iron compounds, while tank B combines all the other minerals. Jack's hydroponic formula consists of one bag with calcium nitrate and another bag for the other compounds (Mattson and Peters, 2014). For detailed steps of preparation, amounts of each compound, and mixing instructions, refer to Resh (2012) and Mattson and Peters (2014).

9.4.5 Differential nutrient uptake rates

Ideally, the nutrient solution in the reservoir should be refilled to provide the necessary nutrients for maximum plant growth without nutrient deficiency or excess. As plants grow, nutrients are absorbed along with water uptake. Many studies have shown that nutrient uptake differs strongly among species and cultivars, not only with respect to the quantity of nutrients absorbed, but also to the nutrient ratios (Niu et al., 2018). Even when starting with an "optimal" formula, the proportion of ions changes over time (Lu and Shimamura, 2018). These results demonstrated that absorption of nutrient elements varied among species and cultivars and the uptake rates of various ions are different. These differential uptake rates result in a situation where some ions are removed quickly; thus their concentrations decreased rapidly, while concentrations of other ions did not decrease or even increase. The quality of source water also has an impact on these nutrients (see Section 9.4.8 below for more details).

For short cycle crops such as leafy greens and herbs, nutrient adjustment in one crop cycle may not be necessary, provided that the nutrient solution reservoir is refilled with half or full strength of the same solution. However, for long cycle crops such as tomato, nutrient adjustment during the production may benefit yield and quality, especially when shifting from vegetative to reproductive growth stages. Ideally, the composition of the refill solution should be matched to the actual concentrations of individual nutrient elements in the reservoir tank to prevent excesses or deficiencies. However, it is difficult to capture the dynamic nature of various minerals with current sensor technologies since individual nutrient monitoring equipment is not available for most nutrients. Analytical instrument like ICP (inductively coupled plasma) is commonly used in commercial laboratories for nutrient solution analysis. However, analysis of solution nutrients using ICP is costly and time-consuming. Turnaround time for a typical commercial ICP lab is usually several weeks. Thus, it is not practical for most researchers and commercial growers to regularly analyze their nutrient solutions. Therefore, EC and pH remain the common parameters for continuous routine monitoring.

Based on the nutrient uptake rate and mobility, essential elements can be divided into three groups as shown in Table 9.2 (Bugbee, 2004). In our hydroponic studies, the concentrations of group I elements dropped to very low levels, while group II and III elements like Ca, Mg, and S concentrations remained at high levels (Niu et al., 2018).

9.4.6 Nutrient solution pH

The pH of the nutrient solution in a hydroponic system is often controlled to a range of 5.5—6.5. As plants grow and selectively absorb certain ions, pH changes over time, especially when the nutrient solution reservoir capacity is small. At low pH (<4.5), the uptake of most nutrient elements becomes slow, while at high pH (>7.5), the uptake of Fe and Mn is slow. Fig. 9.4 shows the effect of soil pH on the availability of essential elements. This graph was

TABLE 9.2 Approximate uptake rates of various essential elements.

Group	Elements	Note
I	NO_3, NH_4, P, K, Mn	Active uptake, fast removal
II	Mg, S, Fe, Zn, Cu, Mo, Cl	Intermediate uptake, intermediate removal
III	Ca, B	Passive uptake, slow removal

developed decades ago based on soil cultivation. Since nutrient uptake varies among species and cultivars and is affected by various growing conditions, it is not surprising that the effect of the nutrient solution pH on the availability of certain nutrient elements does not follow the same pattern for hydroponically grown plants under artificial lighting. In fact, new research finding from The Ohio State University has shown unique plant responses at very low pH on nutrient uptake, growth, and root rot disease incidence of basil (*Ocimum basilicum*) (Gillespie et al., 2020). They observed that K and aluminum (Al) uptake increased with decreasing pH between 5.5 and 4.0, contrary to what is presented in Fig. 9.4 where K uptake should be very low at such a low pH.

So far, there is no relationship developed between solution pH and nutrient uptake for any crop grown hydroponically under artificial lighting. Fig. 9.4 can still be used as a general guide. Thus, to maintain the solution pH at an ideal range, pH adjustments are needed during plant growth by using acids to lower pH, and alkali to increase it. Typical acid products used for lowering pH are nitric, phosphoric, or sulfuric acid, and these acids can be used individually or combined. If additional N is needed to meet the plant requirement, nitric acid is the preferred acid to lower pH. This is because nitrogen requirements for many crops are higher than other macronutrients. Thus, using nitric acid instead of other acid for lowering

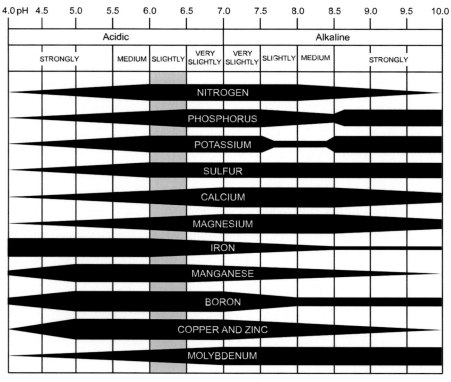

FIGURE 9.4   The effect of soil pH on the availability of plant nutrients. *Source: https://en.wikipedia.org/wiki/PH.*

pH can also supplement nitrogen without adding excessive amounts of other macronutrients (Bugbee, 2004). To raise pH, potassium hydroxide or potassium carbonate is often used.

9.4.7 Dissolved oxygen concentration

The oxygen concentration in the air is 21%. Oxygen is dissolved in water through diffusion and aerating action by wind. Thus, increasing the surface area of nutrient solution exposed to the air increases dissolved oxygen (DO) in the solution. However, solubility of oxygen in water is very low and is affected by water temperature. At 20°C, the maximum DO concentration is only 9 mg L^{-1}. Therefore, the best approach to increase oxygen availability is to have some roots exposed to the air, as in aeroponics. Fig. 9.5 shows the relationship between maximum DO concentration and water temperature.

9.4.8 Water quality

When preparing a nutrient solution, it is necessary to know the initial water quality, unless pure water such as reverse osmosis water or rainwater is used. In a plant factory, the hydroponic system recirculates the nutrient solution. Thus, high-quality water (defined as low salinity and pathogen free) combined with an analysis of the source water are required to understand the ion composition and concentration. Municipal tap water is often the first choice of many plant factories. Groundwater is another common water source. Water quality and mineral composition of municipal tap water can vary greatly with source water and treatment. Ions with significant concentrations include Ca, Mg, SO_4, and bicarbonate (HCO_3). In some cases, Na and Cl concentration can be high. In Texas, USA, the EC of municipal tap water ranges from 0.4 to 1.8 dS m^{-1} in cities across the state. The EC of groundwater is usually higher than that of municipal water, especially in arid and semiarid regions. Even in water-abundant Ohio, USA, groundwater EC can reach 1.6 dS m^{-1} with an Na concentration of 243 mg L^{-1} (Fausey et al., 2009). The dissolved macronutrients, Na and Cl, are often the

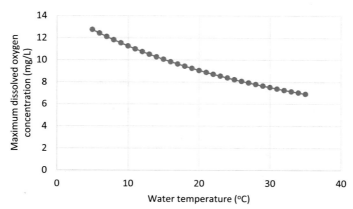

FIGURE 9.5 Relationship between water temperature and maximum dissolved oxygen concentration. *Source: https://dnr.mo.gov/env/esp/wqm/DOSaturationTable.htm.*

major contributors of a high EC. Therefore, when preparing the nutrient solution, elements in the source water need to be taken into consideration.

One approach used by growers to deal with low-quality water in a recirculating hydroponic system is to periodically discharge a partial volume of the nutrient solution to curb the accumulation of unwanted elements. The accumulation of unwanted ions depends on the initial concentration of the unwanted ions in the source water, the capacity of the nutrient reservoirs, the crop, and the duration of production (Niu et al., 2018). In addition to the EC level in the source water, its composition also needs to be considered. The concentration of essential elements (Mg, Ca, etc.) in the source water should be considered when making a nutrient solution to prevent a nutrient imbalance during production. That is, the nutrient formula needs to be adjusted depending on which element and at what concentration they are found in the source water.

9.5 Future perspectives

Hydroponics is essential for plant factories. Current hydroponic systems and knowledge used in plant factories are derived from initial greenhouse and growth chamber studies. It is of no doubt that the hydroponic systems and the control strategies of nutrient solution in the rootzone and of the aerial environmental conditions will be improved over time. Plant responses to manipulations of nutrients under artificial lighting and interaction between light environment and nutrient management on plant quality and yield needs to be investigated. Based on the current and continuing interest in plant factories, we envision the needs for research and development in the following areas related to hydroponics.

Hardware and information technologies for hydroponics in plant factories. There are few products on the market today specifically designed for hydroponic systems in plant factories, such as culture beds and accessories that can be used for vertical grow racks with multitiers, sensors for monitoring individual nutrient elements and controllers to manage the nutrient solution composition, strength and pH at optimal ranges.

Management of nutrient solution composition to enhance vegetable quality. With the ability of precise controls of the environmental conditions in a plant factory, producing quality leafy greens consistently is much easier compared to greenhouse hydroponic crop production. Management of nutrient solution composition, particularly the effects of primary nutrients (N, P, K), secondary macronutrients (Ca, Mg, S), as well as micronutrients, is an important practice in hydroponics to improve vegetable quality (Rouphael et al., 2018). Also, there is a need to increase the nutritional quality, more specifically the content of certain micronutrients since more than 25% of the world population suffers from micronutrient deficiency (Pinstrup-Andersen, 2018). Micronutrient deficiencies are linked to a higher risk of obesity and other dangerous and debilitating diseases (Calton, 2010). Future research is needed to confirm if such deficient nutrients can be added to a nutrient solution and taken up by plants.

Production of functional leafy vegetables. Plant nutritional traits can be enhanced or modified by manipulating the aerial environmental conditions such as light spectrum (refer to Chapter 18), other environmental conditions, or through manipulation of the nutrient solution in the rootzone. For example, low potassium leafy greens would be ideal for people with chronic kidney disease and low nitrate leafy greens can also be produced through manipulation of the nutrient solution composition (Ohashi-Kaneko, 2019) along with controlling light quality.

Organic hydroponics. With a rapid and constant development in hydroponics, urban agriculture, and plant factories, interest in growing organic food crops in hydroponics is also increasing. Regulations on organic labeling differ from country to country around the globe. Despite constant debate regarding the eligibility of soilless growing media and nutrient solution culture to be considered as organic within the food and agriculture community, the US National Organic Standard Board passed a decision in 2017 to allow hydroponically grown food to be labeled as USDA Certified Organic, provided that all products and ingredients used in production and process are listed in the Organic Materials Review Institute website (https://omri.org). In the United States, there are only a few commercial certified organic growers who are using recirculating hydroponic systems, such as NFT, in greenhouses to grow leafy greens using organic fertilizers. The difference between organic and inorganic fertilizers is the form of available nitrogen where the nutrients in organic fertilizers are typically components of complex molecules like proteins (either animal based or plant based) that need to be broken down into mineral elements before they are absorbed by plants. This process transforms organic nitrogen molecules to nitrate, thanks to the activity of microorganisms (nitrifying bacteria) in the system. Depending on whether the organic fertilizer contains microorganisms or not, applying microorganisms in tandem may speed up the breakdown of organic fertilizers. Based on limited research from our group and by other researchers, the drawback of organic hydroponics is lower crop yield. Nevertheless, organically grown leafy greens appear of darker color. Compared to conventional inorganic fertilizers, the actual composition of commercial organic fertilizers is largely unknown. Some users reported that organic fertilizers are smelly and clog the tubing of an NFT system. However, we have used a liquid organic fertilizer in an NFT system without any clogging problems. Apparently, progress has been made in organic fertilizer formulations, but more research and development are needed to provide a balanced and consistent nutrient solution for organic hydroponics.

References

Arnon, D.I., Stout, P.R., 1939. The essentiality of certain elements in minute quantity for plants with special reference to copper. Plant Physiol. 14, 371–375.

Bugbee, C., 2004. Nutrient management in recirculating hydroponic culture. Acta Hortic. 648, 99–112.

Calton, J.B., 2010. Prevalence of micronutrient deficiency in popular diet plans. Sports Nutr. Rev. J. 7, 24. https://doi.org/10.1186/1550-2783-7-24.

Epstein, E., 1999. Silicon. Annu. Rev. Plant Physiol. Plant Mol. Biol. 50, 641–664.

Fabiano, C.C., Tezotto, T., Favarin, J.L., Polacco, J.C., Mazzafera, P., 2015. Essentiality of nickel in plants: a role in plant stresses. Front. Plant Sci. 6, 754. https://doi.org/10.3389/fpls.2015.00754.

Fausey, B., Bauerle, B., Draper, C., Hansen, R., Keener, H., Ling, P., 2009. Research and outreach efforts sustain Ohio hydroponic industry. Acta Hortic. 843, 389–392.

Frew, A., Weston, L.A., Reynolds, O.L., Gurr, G.M., 2018. The role of silicon in plant biology: a paradigm shift in research approach. Ann. Bot. 121, 1265–1273.

Gillespie, D.P., Kubota, C., Miller, S.A., 2020. Effects of low pH of hydroponic nutrient solution on plant growth, nutrient uptake, and root rot disease incidence of basil (*Ocimum basilicum* L.). Hortscience 55 (8), 1251–1258.

Harwood, E., 2019. Selected PFALs in the United States, the Netherlands, and China. In: Kozai, T., Niu, G., Takagaki, M. (Eds.), Plant Factory — an Indoor Vertical Farming System for Efficient Quality Food Production, second ed. Academic Press, San Diego, CA, pp. 419—424.

Hewitt, E.J., 1966. Sand and Water Culture Methods Used in the Study of Plant Nutrition. Technical Communication No. 22. Commonwealth Bureau of Horticulture and Plantation Crops, England.

Hoagland, D.R., Arnon, D.I., 1950. The Water-Culture Method for Growing Plants without Soil. California Agricultural Experiment Station. Circular-347.

Kozai, T., Niu, G., Takagaki, M. (Eds.), 2019. Plant Factory — an Indoor Vertical Farming System for Efficient Quality Food Production, second ed. Academic Press, San Diego, CA, p. 487.

Kroggel, M., Lovichit, W., Kubota, C., Thomson, C., 2012. Greenhouse baby leafy production of lettuce and komatsuna in semi-arid climate: seasonal effects on yield and quality. Acta Hortic. 952, 827—834.

Lu, N., Shimamura, S., 2018. Protocols, issues and potential improvements of current cultivation systems. In: Kozai, T. (Ed.), Smart Plant Factory, pp. 31—49.

Luyckx, M., Hausman, J.F., Lutts, S., Guerriero, G., 2017. Silicon and plants: current knowledge and technological perspectives. Front. Plant Sci. 23 https://doi.org/10.3389/fpls.2017.00411.

Mattson, N., Peters, C., 2014. A Recipe for Hydroponic Success. Inside Grower.

Niu, G., Sun, Y., Masabni, J., 2018. Impact of low and moderate salinity water on plant performance of leafy vegetables in a recirculating NFT system. Horticulturae 4, 6. https://doi.org/10.3390/horticulturae4010006.

Ohashi-Kaneko, K., 2019. Functional components in leafy vegetables. In: Kozai, T., Niu, G., Takagaki, M. (Eds.), Plant Factory - an Indoor Vertical Farming System for Efficient Quality Food Production. Academic Press, San Diego, CA, pp. 235—243.

Pinstrup-Andersen, P., 2018. Is it time to take vertical indoor farming seriously? Global Food Secur. 17, 233—235.

Raviv, M., Lieth, J.H., Bar-Tal, A. (Eds.), 2019. Soilless Culture: Theory and Practice, second ed. Academic Press, San Diego, CA, p. 691.

Resh, H.M., 2012. Hydroponic Food Production. A Definitive Guidebook for the Advanced Home Gardener and the Commercial Hydroponic Grower, seventh ed. CRC Press, Taylor & Francis Group, p. 524.

Rouphael, Y., Kyriacou, M.C., Petropoulos, S.A., De Pascale, S., Colla, G., 2018. Improving vegetable quality in controlled environments. Sci. Hortic. 234, 275—289.

Smith, G.S., Johnston, C.M., Cornforth, I.S., 1983. Comparison of nutrient solutions for growth of plants in sand culture. New Phytol. 94 (4), 537—548.

Sonneveld, C., Voogt, W., 2009. Plant Nutrition of Greenhouse Crops. Springer Verllag, p. 431.

Steiner, A.A., 1984. The universal nutrient solution. In: Proceedings of IWOSC 1984 6th International Congress on Soilless Culture, Wageningen, The Netherlands, Apr 29—May 5, pp. 633—650.

Taiz, L., Zeiger, E., 2002. Plant Physiology, third ed. Sinauer Associates, Inc., Publishers, Sunderland, Massachusetts.

Aquaponics

Joseph Masabni[1] and Genhua Niu[2]

[1]Texas A&M AgriLife Extension Service, Dallas, TX, United States; [2]Texas A&M AgriLife Research, Texas A&M University, Dallas, TX, United States

10.1 Introduction

Aquaponics is evolved from a relatively ancient agriculture practice. The earliest evidence of production that could be qualified as aquaponics is the floating gardens built by the Aztecs of Mexico nearly 600 years ago. The Aztecs called their floating gardens "Chinampas" and demonstrated that fish and plants can be used together to produce these two products sustainably. However, modern aquaponics started in the 1980s in the United States of America and early research was performed primarily by Dr. James Rakocy at the University of Virgin Islands who focused his research to fine tune the aquaponic system (Lennard and Goddek, 2019). Dr. Rakocy's initial approach was to use media beds in his aquaponic system. In later years, Dr. Rakocy focused his research efforts on the floating raft system focused efforts on the floating raft system, also known as deep water culture (DWC). In recent years, research and commercial interest in aquaponics can be found in Europe (Great Britain, Germany, etc.), North America (USA, Canada, and Mexico), Asia (UAE, Singapore, Vietnam, etc.), and Australia, among others.

Aquaponics is the combination of two terms, namely "aqua" from aquaculture and "ponics" from hydroponics. Aquaponics, then, is the culmination of both intensive aquaculture and hydroponics technologies in a recirculating system. Aquaculture is the practice of raising fish in densely stocked tanks. The biggest challenge facing an aquaculture operation that raises fish in densely stocked tanks is the high amount of waste produced. Extensive filtration is required and often the water must be dumped, and new freshwater is added in. On the other hand, hydroponics is the science of growing plants in a soilless environment. In hydroponics, mineral nutrients are provided to plants through inorganic chemicals, which are formulated according to type of crops and growth stage. Most of the fertilizers used in hydroponics are chemical fertilizers. So, there is an energy cost and monetary expense in producing them as they are petroleum derived. In aquaponics, chemical fertilizers are no longer needed because the fish and the bacteria are delivering the nutrients. Additionally, bacterial processing of fish waste and the concomitant plant uptake reduces the need for extensive filtration.

Interest in aquaponics is increasing worldwide, especially in small-scale production systems that provide produce for niche markets in urban areas, such as high-end restaurants, or at farmers markets. However, the development of aquaponics is relatively slow and lacks a planned and aggressive marketing or advertising program by major producers. This is partially because large producers have no interest in switching from hydroponic production technology that they spent years in learning and fine tuning. Milicic et al. (2017) conducted a survey in 16 European countries and concluded that most respondents had positive reaction to aquaponics, even though more than 50% have never heard of it. Survey also indicated that about 17% of respondents were willing to pay a premium price for an aquaponic produce.

In theory, it is feasible from an economic aspect for aquaponics to be profitable and to improve the profit margin of small-size producers (Tokunaga et al., 2015). However, most small size businesses face many hurdles such as lack of consumer acceptance and poor sales of aquaponic fish. Such aquaponic businesses relying only on plant sales cannot compete with hydroponic operations. Only those operations near a large metropolitan city have been able to stay in business. In recent years, many have reported on the economic viability of aquaponics (Bosma et al., 2017; Stadler et al., 2017; Vermeulen and Kamstra, 2013).

In addition, aquaponics lags hydroponics in terms of regulations, food policy, and rate of adoption and scale of production. Food safety regulations are still not clear for both hydroponics and aquaponics, although recent efforts have tried to set standards. Of course, clear and concise food safety regulations will help the consumer public to trust and support the aquaponic business. Continued education is necessary to increase public awareness and ease their fears. For example, consumers fear the concept of pathogen transfer, such as *Escherichia coli*, from fish to plants. Consumers also do not understand the safety of using fish waste as fertilizer in a fruit orchard or a vegetable garden. Therefore, careful research catered to address these concerns and active education of the consumer public is necessary to increase acceptance of aquaponic produce.

The potential of aquaponics to enhance year-round production with minimal use of water or chemical fertilizers is equal to that of hydroponics. Aquaponics is suited for regions with water scarcity, prone to seawater intrusion such as coastal areas, with low rainfall such as arid or semiarid areas, and in urban settings with limited or lack of open spaces for conventional gardening. In urban settings, aquaponics can be and have been built on building rooftops, and in abandoned building, schools, or neglected construction sites.

Looking forward, the integration of aquaponics with plant factory with artificial lighting (PFAL) will most likely improve plant and fish productivity and reduce reliance on chemical fertilizers. The latter will have the most impact in developing countries with little to no access to quality chemical fertilizers. The combination of aquaponics and PFAL has a great potential for space missions or in moon bases where water is scarce and the need for a most efficient and sustainable ecosystem is of highest importance.

Aquaponics also has an educational potential of improving the science base and understanding by middle school and high school students. Aquaponics provides students the opportunity to understand basic and applied sciences, the engineering aspect in terms of design, water flow, air flow, and the environmental and biological aspects in terms of water chemistry, plant and fish biology, and nutrient balance. In summary, this chapter is an introduction to aquaponics and is aimed to overview issues that aquaponics is currently facing and to present a future vision of integrating aquaponics with PFAL.

10.2 The basics

10.2.1 Concept of aquaponics

The concept of aquaponics is simplified in Fig. 10.1. Fish excrete ammonia through their gills or through their urine. Nitrifying bacteria convert this ammonia to nitrate. The plants absorb the nitrate from the water and "clean" water returns to fish. Clean water Fig. 10.1 does not mean potable quality water. In this context, "clean" means that it has less nutrients, some of which (ammonia) can be toxic to the fish.

10.2.2 Aquaponics: a three-legged stool

Although Fig. 10.1 shows only fish and plants, an aquaponic system is a three-legged stool consisting of fish, plants, and bacteria. Fig. 10.2 illustrates how the three parts interact. The fish excrete ammonia through waste and gills. If there is only fish in the system, ammonia concentrations can quickly rise and can kill the fish even at 3 or 5 parts per million. In a pond, river, or anywhere in nature, there exists a bacterium called *Nitrosomonas* that converts ammonia to nitrite. Another bacterium, *Nitrobacter*, converts nitrite to nitrate which is available for uptake by plants. The process of nitrate removal by the plants continues the cycle forward. The plants absorb nitrate and "cleaner" water returns to the fish tanks. The daily production and management activities such as measurements of pH, ammonia, nitrate, nitrite, water temperature, etc., are indirect measurements of the health and viability of the bacteria in a system. For example, if ammonia accumulates and reaches a concentration of 1 mg L^{-1}, *Nitrosomonas*, which is responsible for converting ammonia to nitrite, is either dead or inactive. Similarly, if ammonia concentration is low ($0-0.25 \text{ mg L}^{-1}$) but nitrite is

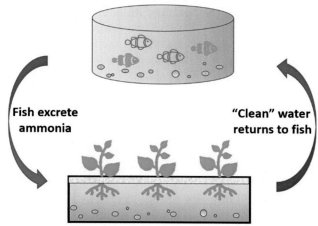

Fish excrete ammonia

"Clean" water returns to fish

Nitrifying bacteria convert ammonia to nitrate, Plants absorb nitrate from water

FIGURE 10.1 Schematic diagram of aquaponics.

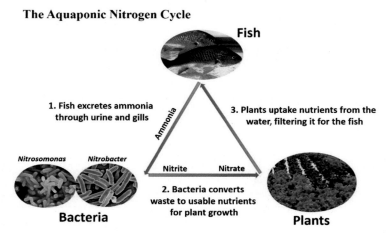

The Aquaponic Nitrogen Cycle

FIGURE 10.2 Schematic diagram of the interaction of fish, bacteria, and plants and the role of the nitrifying bacteria in an aquaponic system. *The Fish photo Credit: Dr. Todd Sink, Aquaculture Extension Specialist, Texas A&M AgriLife Extension.*

high, *Nitrobacter*, which is responsible for converting nitrite to nitrate, is dead or inactive. In brief, the bacteria are the "engine" of an aquaponic system.

Hydrogen ions are released during this process of nitrification, which in turn lowers pH. During the process of converting ammonia to nitrite and then to nitrate, there is a tendency for pH to get lower, or more acidic, with time. An aquaponic system is considered "healthy" when pH tends to get lower with time. Depending on the size of a system and time of year, pH can easily reach 5.5 in a couple of days. If, for any reason, pH stays constant or even goes up (more alkaline), there is something wrong in the system. Frequent and routine pH measurements are a direct indication of the health of an aquaponic system.

Most producers worry about keeping fish and plants alive. Instead, growers should be also concerned in keeping the bacteria alive. Bacteria, *Nitrosomonas* and *Nitrobacter*, are found in nature (in pond water and in and on fish) and do not have to be added at the initial set up of an aquaponic system, if water is from nature. However, addition of bacteria speeds up the process of conversion of ammonia to nitrite to nitrate. We set up and started two aquaponic systems, one inoculated with bacteria, the other system not inoculated. The system inoculated with bacteria was also primed with clear ammonia at 2 mg L^{-1} as a food source for the inoculated bacteria. Then, we measured the levels of ammonia, nitrite, and nitrate daily. In the system that was not inoculated with bacteria, it took about 25 days for the initial detection of nitrate and about 90 days for nitrate levels to reach 40 mg L^{-1} (Fig. 10.3). However, when we inoculated with bacteria at the initial startup, it took only 2 days for initial detection of nitrate, and about 20 days for nitrate to reach 40 mg L^{-1} (Fig. 10.4).

No additional bacteria are needed to add to an aquaponic system if the production and environmental conditions are not stressful to bacteria, fish, or plants. However, in case of accidental stressors (such as power outage, excessively cold or warm water, or cold or warm air temperatures), it is advisable to inoculate the system with additional bacteria. Regular measurements of ammonia, nitrite, and nitrate are the best indicator of the health and activity of the bacteria in a system.

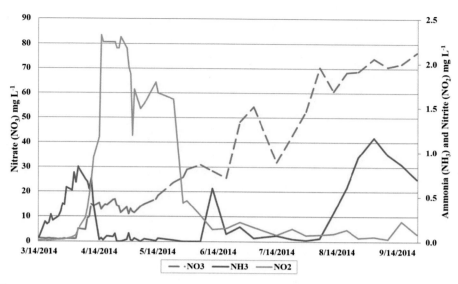

FIGURE 10.3 Levels of ammonia (red), nitrite (blue), and nitrate (green) after initial startup of a new aquaponic system **without** the initial addition of nitrifying bacteria. *Figure courtesy of Dr. John Jifon, Texas A&M AgriLife Research, Weslaco, TX.*

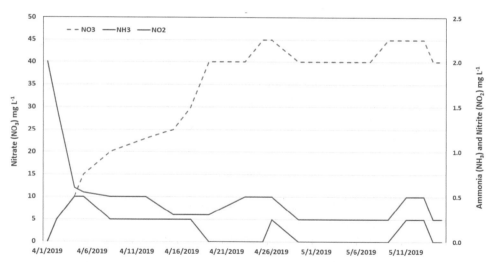

FIGURE 10.4 Levels of ammonia (red), nitrite (blue), and nitrate (green) after initial startup of a new aquaponic system **with** the initial addition of nitrifying bacteria.

10.2.3 Key components

Key components of an aquaponic system are the fish tanks, solid filter, or clarifier for removal of larger settleable solids, a biofilter tank as a secondary stage of waste removal of smaller suspended solids, a degassing tank to vent out excessive gases such as chlorine and chloramine found in the source water, or methane and hydrogen sulfide produced from anaerobic conditions in the filtration system (Fig. 10.5). The degassing tank can also be used to supplement the system with necessary nutrients such as calcium or with pH adjusters when adjusting water pH. A major component of an aquaponic system is the production rafts or grow beds where plants are grown. Of course, water and air pumps are needed.

Other components of an aquaponic system of equal importance are backup parts and pumps. Electricity is needed 24 h per day for the water pump and the air pump. Thus, a backup generator is needed. Backup water pump and air blower also need to be made available and ready to replace old pumps in case of failure. In the case that either the water or air pump fails, a grower should be able to disconnect the old one and attach the new one within 10—20 min. In a situation where an operator must order a new pump and wait a few days for delivery, chances are that the plants may struggle but survive sitting in noncirculating or nonaerated water, but the fish may not survive a few days without oxygenated water, especially in the middle of summer.

In addition to the backup pump and backup generator, a nutritionally complete fish feed is necessary. pH adjusters like calcium carbonate or potassium bicarbonate are also needed to adjust the pH when it drops to levels below 6. Iron chelate is also needed to be supplemented regularly in the system. No matter how well balanced an aquaponic system is, plants may run into a situation where iron can become deficient. A foliar spray of iron chelate can green up the foliage in as little as a couple of days. Other nutrients needed to be supplemented in a system are magnesium and calcium which can be added to the water in the degassing tank.

10.2.4 Biotic and abiotic factors of an aquaponic system

Table 10.1 lists the parameters that play a role in an aquaponic system. Some parameters can be measured readily and should be measured regularly such as pH, water temperature, nitrate, nitrite, and ammonia. Some are not regularly measured such as carbon dioxide or dissolved oxygen. The latter can be measured, but the equipment can be expensive. To avoid the need of measuring dissolved oxygen, a grower needs only to supply sufficient aeration using

TABLE 10.1   List of parameters that play a role in an aquaponic system and how often they should be monitored.

Measured or maintained regularly	Measured or maintained occasionally
pH	Alkalinity
Filtration	Water source
Ammonia	Caron dioxide
Nitrite	Dissolved oxygen
Nitrate	Water flow
Air temperature	Quality and quantity of fish feed
Water temperature	Fish numbers and weight

an air pump. Some parameters affect fish, plants, and bacteria and are also affected by other parameters. For example, air temperature has a big effect on plant growth and on water temperature. In turn, water temperature influences health of fish, plants, and bacteria, and the level of dissolved oxygen. An aquaponic system is highly active and dynamic. All regular measurements of the various parameters and all production activities are meant to reflect the overall health of the system, and more specifically, the health and viability of bacteria.

10.2.5 Fish to plant ratio

The following are the three most common parameters that every aquaponic grower should thoroughly understand: fish stocking density, plant density, and amount of fish feed. In other words, the key to optimizing the system is to determine the ratio between fish waste output, which is directly influenced by fish feed addition, and plant nutrient needs. As mentioned above, fish feed, which is the major source of aquaponic system nutrients, does not contain all the nutrients required for optimal plant growth, and thus external nutrients should be added.

Fish stocking density. Food and Agriculture Organization (FAO) (Small-scale aquaponic food production: Integrated fish and plant farming. Bulletin number 589) recommends 10—20 kg of fish weight per 1000 L of water. For example, assuming the fish tank in a commercial operation has a 9500 L capacity, 9.5—19 kg of fish is required based on the FAO fish stocking recommendations. The FAO recommendation is intended for the final or mature fish size, not for the initial weight when stocking with fingerlings. A grower stocking the system with tilapia fingerlings weighing 50 g each needs 10 kg of fish which is equivalent to 200 fish. Six months later, when the tilapia fish has reached the mature weight of about 454 g, the 200 fish now weigh 91 kg, a much higher stocking density than the system can handle (9.5—19 kg). To avoid excessive fish weights, a grower must regularly cull the fish as the system matures in order to maintain the desired and final stocking density of 10—20 kg per 1000 L. For many aquaponic operations, the FAO recommendation is considered as a "low" fish stocking density and many commercial growers prefer to work with about 60 kg per 1000 L.

Plant density. FAO recommends 20—25 plants per m^2 for leafy greens. For fruiting vegetables, the recommendation is 4—8 plants per m^2. The standard foam board has a size of 0.61×1.22 m or $0.74 \ m^2$. Therefore, the FAO recommendation for leafy greens is equivalent to 15—19 plants per board. This plant density is considered on the low end and is normally adopted by hobbyists or when growing large size lettuce heads. Most commercial operations use boards with 24 holes per board equivalent to a plant density of about 32 plants per m^2.

Fish feed. The FAO recommendation for leafy greens is 40—50 g of feed per m^2 of growing area per day. For fruiting vegetables, the recommendation is 50—80 g per m^2 per day. Now that the three foundational questions have been explained, the following scenario is an example of the calculation process needed to put all the numbers together when designing a new aquaponic system.

- Assuming a producer wants to grow lettuce and wants to harvest 25 heads per week, and the lettuce in that location requires 4 weeks to grow. Therefore, at any time, the producer needs 100 heads of lettuce (25 heads of lettuce per week x 4 weeks per month).

Every week, the producer harvests 25 heads, moves 25 seedlings from the seedling area to the production (or grow out) area, and sows 25 new seeds.

- A good plant density is 25 plants per m². Therefore, the 100 plants need 4 m² of production area (100 plants divided by 25 plants per m²).
- One square meter of production area requires 40–50 g of feed. For 4 m², then, 160–200 g of feed per day is needed.
- Fish are known to consume 1%–2% of their body weight in feed per day. Therefore, the 160–200 g of feed per day is equivalent to 8–16 kg of fish biomass. (160–200 g divided by 1%–2%)
- Finally, at a fish stocking density of 10–20 kg per 1000 L, we need a tank size between 1500 and 6000 L to comfortably house this fish biomass.
- For simplicity and ease of calculations, let us assume we are working with mid-range values. Therefore, 12 kg of fish biomass (mid-range of 8–16 kg) are stocked in a 3780 L tank (mid-range of 1500–6000 L) and are fed 180 g of feed per day (mid-range of 160–200 g).

10.2.6 Crops suitable for aquaponics

The type of plants that are suitable for aquaponics depends on the type of hydroponic system that is coupled with the aquaponic system. For deep water culture (DWC) and nutrient film technique (NFT) aquaponic systems, leafy vegetables and herbs and other small, short-term plants are suitable. For plants with a large root system, such as fruiting vegetables, the ebb and flow aquaponic system is better suited where the solid media also act as physical support for the plants.

10.3 Types of aquaponic systems

As an integration of aquaculture and hydroponics, some growers name their aquaponic systems based on the type of hydroponics, which is described in Chapter 9 of this book, for example, DWC or the raft method, or NFT. Fig. 10.6 shows a typical DWC system for leafy greens production in a high tunnel. Two DWC troughs are seen in the middle, each

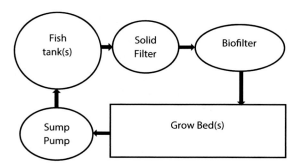

FIGURE 10.5   The key components of a traditional aquaponics system: Fish tank, solid filter, biofilter tank, grow beds, and pumps (Degassing tank is not shown. It is usually located after the biofilter).

FIGURE 10.6 A typical DWC aquaponic system in a high tunnel for leafy greens production.

2.44 m wide and 24.4 m long. Two smaller troughs are visible on the left and right sides used for nursery and seedling production, with widths of 1.22 m each.

Fig. 10.7 shows a close-up photo of red-leaf lettuce in a DWC system using 5-cm-thick foam. Note that the roots appear to be too brown in color due to the adherence of soluble or insoluble solids on the roots. This is an indication that the filtration system is not ideal. Plants will continue to grow and produce in the short term, but the overall system is not "healthy" and will not thrive in the long term.

Fig. 10.8, on the other hand, shows a clean root system of lettuce. The ivory color of the roots with no attached solid waste is a clear indication that the filtration system is clearly removing both settleable and suspended solids.

Another aquaponic system is called flood and drain or ebb and flow (Fig. 10.9). In this system, the rafts are filled with solid media instead of water. Media used in an ebb and flow system include lava rock, Hydroton clay pebbles, or river rock. Also, instead of the water constantly circulating between the fish tanks and the plants, water in an ebb and flow system fills up the trough, reaches a level set by the siphon, and drains back down. The water cycle is in a constant flood and drain instead of a constant circulation as in the DWC system. The ebb and flow water cycle allows the roots to be regularly flooded with nutrient-rich water and drained to allow for roots to get necessary oxygen.

The third type of an aquaponic system is the nutrient film NFT. This system is very closely similar in design to the hydroponics NFT where the plant culture channels have a thin layer of nutrient-rich water running from one end to the other and plant roots are "washed" by the nutrient solution. However, in an aquaponic NFT system, the nutrient solution comes from the bacterial processing of fish waste instead of a nutrient solution tank. For more details on NFT, check Chapter 9.

10.4 Decoupled aquaponic systems

Traditional aquaponics is a coupled system where the nutrient solution (water) is recirculating between aquaculture tank and the plant growing hydroponic system. This design is also called closed loop, single loop, or conventional aquaponics as mentioned above.

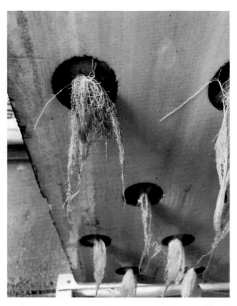

FIGURE 10.7   Lettuce growing in an aquaponic system with poor filtration. Notice the brown roots.

FIGURE 10.8   Clean lettuce roots growing in a well-filtered aquaponic system. Notice the ivory color with no attached solids.

FIGURE 10.9   Ebb and flow aquaponic system filled with hydroton pellets.

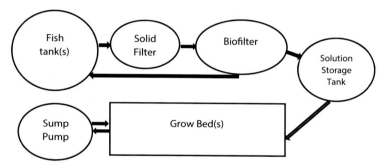

FIGURE 10.10   Schematic diagrams of a decoupled aquaponic system. The red arrow indicates that the flow of the nutrient solution from the storage tank to the grow beds is not permanent. The connecting valve is normally closed and is only opened when a nutrient solution is needed to refill the grow beds.

However, there are challenges because an aquaponic system has three components, plants, fish, and bacteria. Each component has its own set of optimal water quality, temperature, and pH to achieve maximum production. A compromise must be made for all three components to achieve overall conditions, thus reducing productivity and efficiency (Goddek et al., 2019). A decoupled aquaponic system allows the aquaculture and hydroponic system to be operated independently from one another. While there are many variations, a decoupled system typically consists of two independent recirculating units: a recirculating aquaculture system for fish and a hydroponic system for the plants (Fig. 10.10). The two units are connected by a one-way valve that allows flow from the solution storage tank to the grow beds, as needed.

10.5 Aquaponics realities

Aquaponics is fun as a hobby and can be profitable as a commercial operation. It involves raising both fish and growing plants. However, aquaponics is a complex production system that will only be successful if growers realize and observe certain practices. Most aquaponic enthusiasts, even beginning or inexperienced commercial producers, consider aquaponics as a worry-free management model, and that the "business" will be productive or successful with minimum effort. There are some important realities that must be understood when operating an aquaponic system:

- Plants need a robust supply of nutrients which comes from fish waste and from the bacterial processing of that waste. Therefore, growers need not assume that a low fish stocking density can sufficiently supply nutrients for a greenhouse full of plants. In other words, the ratio between fish waste and plant needs of nutrients is key. Fish waste is influenced directly by the number and size of fish and fish feed addition, while plant needs for nutrients depend on the type of crops and growth stage.
- Since nutrients are a result of fish and mineralized wastes, it is also important to know and consider that fish are fed the appropriate feed and the appropriate protein content. Growers need to select a good quality feed appropriate for the fish species used.
- No matter how well an aquaponic system is working, calcium, magnesium, and iron can and will become deficient. Therefore, these three minerals must be kept available to supplement the system when deemed necessary. Calcium and magnesium are sold as a supplement (CaMg+ is an example of a product available in organic or nonorganic formulations). Iron is best applied as a DTPA iron chelate or EDTA chelated iron. Chelated iron is recommended to be applied as a foliar spray on a regular schedule every 2—3 weeks to avoid leaf chlorosis.
- Both bacteria and fish need sufficient dissolved oxygen concentration to survive and function properly. Therefore, a backup generator must be ready to be connected at a moment's notice in case of a power outage. A power outage in the middle of summer can be detrimental to all fish and bacteria if the outage lasts overnight.
- Finally, solid waste removal and constant attention to daily activities are important for proper functioning of an aquaponic system. Poor solid waste removal design and maintenance are the biggest factors contributing to the failure of a new system.

As for challenges of aquaponics, the most important one facing growers is the inability of using many fungicides or insecticides as they may harm the fish health. In the case of a plant disease or insect problem, few pesticides are available for use that can control the disease or manage the insect without harm to the fish. Another challenge is the higher electrical cost because of the constant and uninterrupted operation of water and air pumps. Finally, the initial educational and practical knowledge needed to start or operate an aquaponic system is greater than in hydroponics. In a commercial hydroponic operation, a grower manages the needs of plants only. In an aquaponic operation, however, a grower manages the needs of plants, fish, and bacteria.

10.6 Aquaponics in plant factory

Plant factory is a novel production system with fully controlled environments, usually in stacked layers, in thermally well-insulated warehouses or shipping containers. This type of farming can achieve high productivity while substantially reducing the use of land and water compared to conventional outdoor agriculture. No technology or equipment related to aquaponics has yet been designed just for indoor plant factory. Similarly, technologies and production protocols used in aquaponics were developed originally for greenhouse production. Therefore, when applying aquaponics to plant factories, we are trying to combine two unique technologies.

For aquaponics to be adopted in a plant factory model, an aquaponic system needs to be light weight, and easy to design and operate in a vertical fashion. For example, the aquaponic DWC most popular in commercial greenhouse operations is not suitable and cannot be adapted in a plant factory because the system is too heavy when considering the water weight alone. Additionally, there is an increase in initial construction cost of the fish tanks and filtration equipment, resulting in a low-use efficiency of floor space.

The only link or connection between aquaponics and plant factory is the nutrient solution. In a plant factory, chemical fertilizer is the source of nutrients to the plants. In aquaponics, digested, filtered, and nitrified fish waste is the nutrients for plants. Everything else being equal, if a system is designed to prepare a nutrient solution from an aquaponic source similar in constituents and quality to that of a chemical fertilizer, then it is possible to merge the two technologies—aquaponics and plant factory—in one operation.

As mentioned earlier, the nutrient solution in an aquaponic operation comes from the mineralized fish waste. It would be feasible to design a system where the final nutrient product from an aquaponic operation is filtered to remove dissolvable solids, analyzed for its nutrient content, supplemented with missing amounts of necessary macro- and micronutrients, and finally sent for use in a plant factory. It may be necessary to design and construct a decoupled aquaponic system with a specialized room for filtration, analysis, and supplementation of the aquaponic fish waste into a nutrient solution suitable for a plant factory.

Nitrate production via fish feed is costly compared to chemical fertilizers. Unless chemical fertilizer becomes relatively more expensive, there is no urgent need for alternatives for use in hydroponic operations or plant factories. A typical fish feed with 32% protein costs $534 per metric ton bulk or $15 to $18 for an 18 kg bag when purchased by the bag. When fed to hybrid striped bass, one ton of 32% fish feed will provide 138 kg of nitrate based on protein digestion and turnover rates. To obtain a similar nitrate load from the urea (46-0-0), a common chemical fertilizer, only 285 kg of urea is required and costs only $215 (Dr. Todd Sink, personal communication).

There exists a scenario where all the above-mentioned issues become irrelevant; issues such as the practical feasibility and cost of preparing a complete nutrient solution from an aquaponic source equal to a chemical fertilizer; the additional floor space and cost for the aquaponic components such as fish and filtration tanks; and the additional electrical cost due to the continuous operation of pumps in an aquaponic system. Such a scenario exists in the space program. NASA has been actively researching hydroponic production for use in space. It is conceivable that a well-designed and operational decoupled aquaponic system incorporated with a plant factory can be appealing to moon colonists or astronauts on an 8-month trip to Mars. The aquaponic component will add a diverse and rich diet that includes fish, while the plant factory adds a diverse assortment of plants including microgreens, leafy greens, all kinds of herbs, and even fruiting vegetables.

Incorporating an aquaponic component to a plant factory in the space program has the additional benefit of recycling the plant waste as fish feed. Root or leaf waste from the plant factory can be processed and used to supplement the fish feed diet. Thus, minimal waste is generated. Additional research is needed to prove this concept and this idea is for future research.

References

Bosma, R.H., Lacambra, L., Landstra, Y., Perini, C., Poulie, J., Schwaner, M.J., Yin, Y., 2017. The financial feasibility of producing fish and vegetables through aquaponic. Aquacult. Eng. 78 (b), 146–154. https://doi.org/10.1016/J.AQUAENG.2017.07.002.

Goddek, S., Joyce, A., Wuertz, S., Korner, O., Blaser, I., Reuter, M., Keesman, K.J., 2019. Decoupled aquaponics systems. In: Goddek, S., Joyce, A., Kotzen, B., Burnell, G.M. (Eds.), Aquaponics Food Production Systems. Springer, Cham, pp. 201–229. https://doi.org/10.1007/978-3-030-15943-6_5.

Lennard, W., Goddek, S., 2019. Aquaponics: the basics. In: Goddek, S., Joyce, A., Kotzen, B., Burnell, G.M. (Eds.), Aquaponics Food Production Systems. Springer, Cham, pp. 113–144. https://doi.org/10.1007/978-3-030-15943-6_5.

Milicic, V., Thorarinsdottir, R., Dos Santos, M., Turnsek, M., 2017. Commercial aquaponic approaching the European market: to consumers' perceptions of aquaponic products in Europe. Water 9 (2), 80. https://doi.org/10.3390/w9020080.

Stadler, M.M., Baganz, D., Vermeulen, T., Keesman, K.J., 2017. Circular economy and economic viability of aquaponic systems: comparing urban, rural and peri-urban scenarios under Dutch conditions. Acta Hortic. 1176, 101–114. https://doi.org/10.17660/ActaHortic.2017.1176.14.

Tokunaga, K., Ako, H., Tamaru, C.S., Leung, P., 2015. Economics of small-scale commercial aquaponics in Hawai'i. J. World Aquacult. Soc. 46 (1) https://doi.org/10.1111/jwas.12173.

Vermeulen, T., Kamstra, A., 2013. The need for systems design for robust aquaponic systems in the urban environment. Acta Hortic. 1004, 71–77. https://doi.org/10.17660/ActaHortic.2013.1004.6.

Plant responses to the environment

Ricardo Hernández

Department of Horticultural Sciences, College of Agriculture and Life Sciences, North Carolina State University, Raleigh, NC, United States

11.1 Plant responses to light amount and quality

Light is a key component driving photosynthesis, and adequate light is required for optimal plant growth. Instantaneous light for plant growth is typically quantified in terms of photosynthetic photon flux density (PPFD: $\mu mol\ m^{-2}\ s^{-1}$) in the photosynthetically active radiation (PAR) range between 400 and 700 nm. Plant species, leaf area index (LAI), and growing environment influence the maximum PPFD (light saturation point) of photosynthesis. For example, for crops such as tomato and cucumber, the light saturation point for single leaf measurements has been reported to be close to $1000-1200\ \mu mol\ m^{-2}\ s^{-1}$ compared with $290-400\ \mu mol\ m^{-2}\ s^{-1}$ for lettuce (Kang et al., 2013; Min and Kubota, 2008; Tatsumi, 1969). However, the cumulative light or daily light integral (DLI) is a better predictor of plant growth than the instantaneous PPFD. Yield and DLI are linearly correlated (Fig. 11.1) (Albright et al., 2000; Cockshull et al., 1992; Kubota et al., 2016; Xu and Hernández, 2020). DLI is the cumulative number of photons in the PAR range provided for an entire day (24 h) per area (meter square) with units of $mol\ m^{-2}\ d^{-1}$. Studies have shown that $30-35\ mol\ m^{-2}\ d^{-1}$ is recommended for tomato (Spaargaren, 2001), $20-25\ mol\ m^{-2}\ d^{-1}$ for strawberry fruit production (Kubota, 2020), and $12-17\ mol\ m^{-2}\ d^{-1}$ for lettuce (Albright et al., 2000) and vegetable transplants (Fan et al., 2013). In addition to species-specific DLIs, the optimal DLI also depends on the LAI. Plants with a more developed canopy will have a higher optimal DLI than plants with a less developed canopy (Kubota, 2016). Additional studies are required to standardize optimal DLIs for different plant species under different LAIs and varied environmental conditions.

In a plant factory, the PPFD and photoperiod can be controlled accurately; therefore, the DLI can be provided at any PPFD and photoperiod combination. For example, to induce flowering in a short-day plant while maintaining adequate photosynthesis and growth, a higher PPFD for fewer hours can be provided to deliver the optimal DLI. In contrast, for a long day or day neutral plant, the DLI can be provided at a lower PPFD for a longer photoperiodic time. Although PPFD and DLI are both based on the PAR spectrum (400–700 nm),

Plant Factory Basics, Applications and Advances
https://doi.org/10.1016/B978-0-323-85152-7.00022-7

181

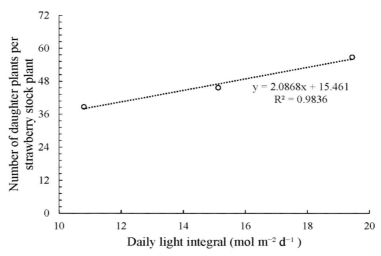

FIGURE 11.1 Production of strawberry daughter plants per strawberry stock plant under increasing daily light integrals (other environmental conditions were the same among light treatments). Note the linear response to the DLI. *Xu and Hernández (2020), unpublished.*

recent research has suggested that far-red (FR) light (700–800 nm), when provided in conjunction with PAR, also has an important contribution to photosynthesis (Zhen and Bugbee, 2020; Zhen et al., 2019). For example, recent research has shown that substituting a small portion of PAR photons with FR photons produced a higher biomass in lettuce. The increase in yield can be explained by the effect of both an increase in leaf area (shade-avoidance response syndrome) and an increase in the canopy net photosynthetic rate (Zhen and Bugbee, 2020). The inclusion of FR light into the light spectrum in plant factories is expected to increase with the improvement of FR diode efficacy ($\mu mol\ J^{-1}$).

Since the adoption of LEDs in horticulture, research reports of the effects of light spectra on plant growth, morphology, and development have been vast and continue to increase (Mitchell et al., 2015). The effect of individual light colors on general plant responses can be summarized for ultraviolet (UV: 280–400 nm), blue (B: 400–500 nm), green (G: 500–600 nm), red (R: 600–700 nm), and FR (700–800 nm) wavelengths.

UV light is perceived by phototropins, cryptochromes, phytochromes, and UV-resistance-locus-8 (UVR-8) photoreceptors (Di et al., 2012; Huché-Thélier et al., 2016). UV light can affect secondary metabolites (flavonoid and anthocyanin contents) and pathogen resistance (JA pathway, flavonoid content) (Huché-Thélier et al., 2016); however, it can also induce similar responses as those to B light (Huché-Thélier et al., 2016).

Responses to B light are mainly controlled by two main receptors: phototropins and cryptochromes. Phototropins perceive mainly UV-A and blue light. Some of the common phototropin-mediated responses are stem bending, leaf positioning, chloroplast accumulation, and stomatal opening (Takemiya et al., 2005). Cryptochromes also absorb light at UV-A and UV-B wavelengths with absorption peaks at 370–450 nm. A reduction in stem length and leaf area, increase in anthocyanin accumulation, and increase in leaf thickness are common responses regulated by cryptochromes (Ahmad and Cashmore, 1996, 1997; Ahmad et al., 2002; Huché-Thélier et al., 2016). Zeitlube (ZTL), flavin-binding Kelch (FKF1), and LOV Kelch proteins (LKP2) form a family of UV-A and blue light photoreceptors and are known to be important for photoperiodic signaling (Imaizumi et al., 2003).

Even though a green light photoreceptor has not been fully characterized, there are exclusive green light plant responses leading to the hypothesis of an unidentified green light photoreceptor (Dhingra et al., 2006; Folta and Childers, 2008; Zhang et al., 2011). In general, green light promotes shade-avoidance syndrome response and reduces B-related responses (Folta and Maruhnich, 2007).

Phytochromes control R- and FR-related responses. Phytochrome exists in two forms: inactive Pr and active Pfr. The transformations between Pr and Pfr are dependent on light quality (or the absence of light) (Kendrick and Kronenberg, 1994). Red and FR light are the two main wavelengths controlling phytochrome responses, but phytochromes also absorb wavelengths in the B and UV spectra (Hernández and Kubota, 2017). An increase in FR light (or a decrease in the R:FR ratio) leads to an increase in phytochrome Pr (inactive form), while an increase in R light (or an increase in R:FR) leads to an increase in phytochrome Pfr (active form), consequently leading to a cellular response. Common plant responses regulated by phytochromes include seed germination, de-etiolation, flowering (Franklin and Quail, 2009; Strasser et al., 2010), fruit quality changes (González et al., 2015; Kalaitzoglou et al., 2019), and shade-avoidance syndrome (SAS) responses (Kendrick and Kronenberg, 1994). SAS responses are elucidated under high plant density and include stem elongation, apical dominance, photoassimilate partitioning, and changes in flowering time (Roig-Villanova and Martínez-Garcí, 2016).

Even though a good understanding of the effect of individual colors is important, plants have more complex responses to different wavelength combinations and ratios. For example, when plants are grown under B-R-only spectra, an increase in the B photon flux ratio increases plant compactness, leaf thickness, and the net photosynthetic rate; however, it also decreases leaf area and fresh mass (Hernández et al., 2016; Hernández and Kubota, 2016; Spalholz et al., 2020). Broad-spectrum white light has become more common in plant factory production systems. In addition, the combination of white diodes (W) with B and R diodes in fixtures has become more common. Research reports have shown inconsistent results when comparing W-B-R diode fixtures with B-R diode fixtures (Lin et al., 2013; Zhen and Bugbee, 2020; Mickens et al., 2019). Dynamic lighting (changes in the spectrum based on the time of day or crop stage) is also possible with new LED technology. Promising research results have been reported for increasing plant growth and quality with dynamic light recipes (Chen et al., 2017; Jishi et al., 2016; Meng and Runkle, 2020; Owen and Lopez, 2015). FR light has been used as a photomorphogenic signal to increase the plant expansion rate during the photoperiod but also as an end-of-day (EOD) signal. EOD FR light can trigger phytochrome-related responses such as stem elongation and leaf area expansion. These responses are triggered by adding FR light after the main photoperiod is over, and the required dosage to maximize the response is low (low photon flux for a short time) (Chia and Kubota, 2010; Eguchi et al., 2016; Kalaitzoglou et al., 2019; Yang et al., 2012). The current research efforts on plant responses to light quality will generate more species- and purpose-specific light recipes with the potential to revolutionize production in plant factories.

11.2 Plant responses to temperature

Plant photosynthesis, growth, and development are linearly correlated with temperature. Plant responses increase with temperature until reaching the optimal/maximal temperature point, and then the responses quickly decline (Fig. 11.2). Since low and maximum temperature thresholds are species specific, it is important to understand the adequate temperature range when producing crops in plant factories. It is also imperative that both shoot and root temperatures be optimized for plant growth. For example, research has shown that maintaining root temperature (through conditioning of the hydroponic solution) at 24°C is optimal to increase the "butterhead" lettuce growth rate (Thompson et al., 1998).

Temperature variation in a 24-h day (night temperature drop) is also a common strategy for the production of several plant species. Plants that originated in tropical regions can grow adequately without day (D) and night temperature (N) fluctuations, while others, such as tomato and cucumbers, grow best under cooler nights (D > N) to reduce the respiration rate (Rn) during the night and reduce the breakdown of carbohydrates. However, the 24-h average day temperature (ADT) is often used as a linear prediction of plant growth (Slack and Hand, 1983) (Fig. 11.3).

The temperature differential (DIF) between day and night can also affect plant morphology. The temperature differential is defined as the difference between day and night temperatures (DIF = Day T − Night T). A positive DIF leads to longer stem extension, and a negative DIF leads to shorter stem extension. DIF techniques have been successfully used to control the plant height of some ornamental crops (Erwin et al., 1989, 1994; Erwin and Heins, 1995). It is important to note that changes in the day and night temperatures that reduce the average 24 h temperature will lead to a reduction in plant growth. The DIF can be a useful tool to control the stem height of transplants produced in plant factories.

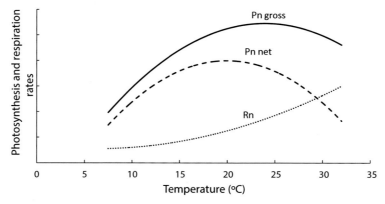

FIGURE 11.2 Effect of temperature on the gross photosynthetic rate (Pn gross), net photosynthetic rate (Pn net), and respiration rate (Rn). Photosynthesis (gross and net) increases with the increase in temperature until the optimal temperature is reached, and then photosynthesis declines with increasing temperature. *After Seginer, I., Shina, G., Albright, L. D., Marsh, L. S., 1991. Optimal temperature setpoints for greenhouse lettuce. J. Agric. Eng. Res. 49, 209—226, reproduced with the permission.*

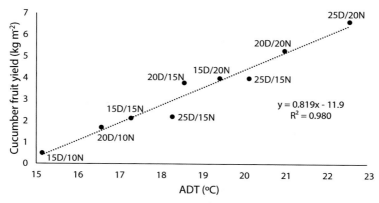

FIGURE 11.3 Impact of the average day temperature (ADT) and different day (D) and night (N) temperatures on cucumber yield (4-week harvest period). *After Slack, G., Hand, D. W., 1983. The effect of day and night temperatures on the growth, development and yield of glasshouse cucumbers. J. Hortic. Sci. 58, 567–573, reproduced with the permission.*

Crop production is commonly divided into several plant growth stages (i.e., germination, young seedlings, vegetative growth, flowering, and fruit development), and the optimal temperature varies depending on the plant growth stage (Hanan, 1997a; Slack and Hand, 1983; White and Warrington, 1988). For example, for tomato, 28–29°C is recommended for germination, 21–26°C day and 14–17°C night is optimal for plant growth, while a 22°C average is suggested for proper flower induction (Adams et al., 2001; Hanan, 1997a).

Optimal temperatures also depend on other environmental variables. For example, increasing PPFD will increase the optimal temperature (Frantz et al., 2004; Pasian and Lieth, 1989). Similarly, an increase in temperature will increase the optimal CO_2 concentration (Both et al., 2017; Frantz et al., 2004; Hanan, 1997a). However, it is imperative to understand the impact of temperature on the different stages of growth (biomass, flowering, dry mass) since increases in the net photosynthetic rate by the increase in temperature may not always lead to a desirable result. For example, for lettuce, it is known that higher temperatures coupled with higher PPFD and CO_2 concentrations can lead to several times higher biomass (Frantz et al., 2004). However, these same conditions will lead to unacceptable levels of tipburn, which reduces plant marketability (Saure, 1998). In another example, warmer temperatures will increase tomato seedling growth and biomass; however, higher temperatures can also delay flowering induction (Adams et al., 2001) and lead to a late harvest.

To predict growth, morphology, and development based on temperature, it is important to understand that such responses are based on plant temperature and not on air temperature. In plant factories, air temperature is commonly measured; in addition, the air temperature is measured in selected locations across the plant factory. However, it is important to understand the relationship between the measured air temperature and the actual plant temperature. Two main challenges typically occur: (1) the measured locations may not be representative of the canopy temperature. Even though the environment in a plant factory can be accurately controlled to desirable set points, microclimates with different environmental conditions can also be formed. For example, on a multilevel racking system, the temperature (and other environmental parameters) inside the rack (under lights) will tend to be different than outside the rack (the hallway). Furthermore, if the growing table in the rack tends to have a large surface area, the temperature in the center of the table may differ from that at the edge of the table. In addition, a temperature gradient is also common between lower-level racks and upper-level racks. Therefore, to make data-driven decisions, it is important to understand the variabilities in the growing system and measure temperature in the most representative location(s). (2) In certain cases, the canopy temperature can be different than the air temperature. For example, a higher radiation load can increase leaf temperature; a higher air flow can decrease leaf temperature; and water stress can increase leaf temperature. Therefore, measuring or accurately predicting leaf temperature is critical for growth optimization in plant factories.

11.3 Plant responses to relative humidity—vapor pressure deficit

Managing humidity inside a plant factory is essential for cultivating a healthy crop. Relative humidity (RH) is often used as the measured variable to estimate the amount of moisture in the air. RH is the amount of water that is in the air over how much water the air can hold (at the same temperature). RH can be a good predictor of moisture in the air and indirectly of the evapotranspiration rate when the temperature is constant. However, if the temperature varies, RH by itself is not a good predictor of evapotranspiration. In other words, it is difficult to estimate evapotranspiration when comparing two environments with different RHs and temperatures. For example, it would be difficult to predict which environment results in greater water evaporation when comparing 20% RH at 20°C to 60% RH at 30°C. Therefore, the vapor pressure deficit (VPD) is the preferred environmental variable to predict evapotranspiration since it is dependent on both temperature and relative humidity. The VPD is the difference between saturation vapor pressure (maximum amount of water that air can hold) minus the actual vapor pressure at a given relative humidity and temperature. Both saturation vapor pressure and actual vapor pressure can be calculated using temperature, RH, and psychometric principles (Hanan, 1997b).

Plant transpiration increases linearly with an increasing VPD, and since plant transpiration is a good predictor of plant growth (Ciolkosz et al., 1998), it is imperative to control and optimize the VPD in the plant factory environment (Ciolkosz et al., 1998; Frantz et al., 2004; Goto and Takakura, 1992). However, to obtain a more reliable estimation of transpiration based on the VPD, it is important to use leaf temperature when calculating the VPD. In general, an optimal VPD depends on crop species/growth stage, and a range of 0.5—1.0 kPa is recommended in research studies (Frantz et al., 2004; Goto and Takakura, 1992; Guichard et al., 2005; Shamshiri et al., 2018; Zhang et al., 2017).

The transpiration rate is affected by several conditions, such as the VPD, temperature, light intensity, water availability, and air velocity. In addition, resistances reduce the flow of water from the plant to the air. The two main sources of resistance (related to the control of environmental conditions) are the leaf boundary layer and stomatal resistance.

The leaf boundary layer is a thin layer of still moist air (viscous fluid) that is adjacent to the leaf surface. Air movement in the environment interacts with this boundary layer.

Inside the boundary layer and close to the leaf surface, the velocity of moving air approaches zero, while at the edge of the boundary layer, the air velocity approaches that of the growing environment (Jones, 2013a). The greater the thickness of the leaf boundary layer, the greater the resistance to gas (H_2O and CO_2) diffusion between the leaf and the environment.

Stomatal resistance is the opposition to diffusion of water out or the diffusion of CO_2 into the leaf by stomata (number, pore size, and pore closure). Stomata resistance can be directly affected by the VPD. Very high VPDs can increase stomatal resistance by the partial or total closure of stomatal guard cells. Other factors affecting stomatal resistance include the light intensity, light spectrum, air movement, leaf temperature, plant anatomy, and water availability, among others (Clavijo-Herrera et al., 2018; Jones, 2013a; Tibbitts and Bottenberg, 1976; Yang et al., 1990).

At the leaf level, stomatal resistance (closed guard cells) can be at least one order of magnitude greater than leaf boundary layer resistance (Jones, 2013a). However, for high-density production systems with low air movement (large LAI in plant factories and closed greenhouses), stomatal control is very minimal, and boundary layer resistance is several times greater than stomatal resistance (Hanan, 1997b; Jarvis, 1985). Therefore, in plant factory production systems, both the leaf boundary layer and stomatal resistances should be considered as a unit, "canopy resistance," and the environment should be optimized to reduce this resistance.

Different crops transfer water from the plant to the air via transpiration at different rates, which are mainly affected by environmental parameters, plant developmental stage, and canopy resistance. For example, Jolliet and Bailey (1992) measured the transpiration rate of young (LAI: 0.56, plant height: 1.0 m) and mature (LAI: 2.94—3.06, plant height: 2.7 m) tomato plants under different radiations (young: 17 W m^{-2}, mature: 129 W m^{-2}) and air temperatures (young: 17.9°C; mature: 20.9°C). Jolliet and Bailey (1992) reported an average transpiration rate of 0.0031 g m^{-2} s^{-1} for young tomato and 0.037—0.041 g m^{-2} s^{-1} for mature tomato. For lettuce, Graamans et al. (2017) modeled the transpiration rate under different light intensities in a plant factory and found that the transpiration rate at 200—400 μmol m^{-2} s^{-1} PPFD was estimated to be approximately 0.032—0.042 g m^{-2} s^{-1}. Similarly, Wheeler et al. (1994) estimated the transpiration rate of 28-day-old lettuce under a DLI of 16—17 mol m^{-2} d^{-1} and found that the canopy transpired approximately 4000 g m^{-2} d^{-1} (24-h day, 0.046 g m^{-2} s^{-1}) at full canopy closure. Based on these numbers, a large amount of water is constantly being introduced in the growing system; therefore, water transpired by plants has to be removed from the air to maintain a desirable humidity.

When plants transpire, they increase the amount of water in the air and increase the energy flux in the system (latent heat), which has to be removed from the system. HVAC systems need to be sized to (1) remove the total energy in the system and (2) remove the humidity in the air (dehumidification). The latent heat flux of mature crops is a large contributor to the total heat gain in a plant factory. For example, Graamans et al. (2017) modeled the distribution of energy fluxes from a lettuce crop in a plant factory by changing the light intensity (PPFD) and the temperature set points. Graamans et al. (2017) found that for a system with 140 μmol m^{-2} s^{-1} light intensity and 25/23°C (day/night) temperatures, the contribution of heat by transpiration (latent heat) was three times higher (81.7 W) than by LED lights (25.9 W).

11.4 Plant responses to air flow

Plant factories and other controlled environmental growing systems are known for their high-density small-footprint production capabilities. A high LAI in plant factories leads to challenges with air movement around the canopy. Increased air movement decreases canopy resistance mainly by reducing the leaf boundary layer (Mason, 1995; Monteith, 1965). Consequently, adequate air movement also increases the net photosynthetic rate (Kitaya et al., 2000) by increasing transpiration and CO_2 diffusion into the stomatal cavity. In addition, adequate air movement at the apical meristem of lettuce enhances transpiration and reduces the incidence of tipburn (Goto and Takakura, 1992), a common disorder of lettuce in environments with high canopy resistance.

However, providing a laminar, uniform velocity and adequate air movement to the canopy continues to be a challenge in plant factories. Studies have shown that air velocity at the leaf level of $0.3 \, \text{m s}^{-1}$ is adequate to reduce the leaf boundary layer and increase the net photosynthetic rate (Downs and Krizek, 1997; Kitaya et al., 2000). However, adequate air velocity depends on the LAI, leaf dimension, leaf angle, and plant height, among other factors (Downs and Krizek, 1997; Jones, 2013a).

Even though the increase in air velocity is known to decrease the leaf boundary layer and consequently increase plant transpiration, increased air velocity can also lead to a reduction in transpiration due to its temperature effect on leaf vapor pressure. If the leaf temperature is greater than the air temperature (high radiation, moderate stomatal closure), then greater air velocity will reduce the leaf temperature by increasing heat loss. At lower leaf temperatures, the water vapor pressure will decrease, and consequently, the VPD between the leaf and the air will also decrease together with the transpiration rate (Downs and Krizek, 1997; Jones, 2013a). However, in plant factories, it is less common to have a greater leaf temperature than air temperature due to the lower amount of infrared radiation.

High air velocities can reduce plant growth mainly through the impact on plant water status and transpiration. High air velocities can induce stomatal closure to prevent excess water loss and consequently reduce the net photosynthetic rate. However, high air velocities can also change plant morphology via thigmomorphogenic (plant shaking) responses, resulting in more compact plants with higher elastic limits in the stem (maximum bending before permanent damage) (Jones, 2013a). However, plant shaking can also lead to a reduction in overall plant growth (Chehab et al., 2008; Mitchell and Myers, 1995).

11.5 Plant responses to atmosphere CO_2

The overall effect of photosynthesis is the removal of water and the production of oxygen for all of the CO_2 that is reduced to carbohydrates (sugars). In other words, CO_2 is a key component to produce photoassimilates in the form of carbohydrates. Just as important as the concentration of CO_2 in the air is the rate at which CO_2 diffuses into the plant (intercellular CO_2) (Jones, 2013b; Lambers et al., 2008). Photosynthesis can be quantified as the diffusion of CO_2 gas from the air to the chloroplast. The rate of diffusion is dependent on several factors, including the difference in CO_2 concentration in the air (C_a) and the CO_2

concentration inside the plant (C_i). The greater the difference between C_a and C_i, the greater the CO_2 diffusion into the plant, and consequently the higher the photosynthetic rate (Jones, 2013b; Lambers et al., 2008). The diffusion of CO_2 from C_a to C_i encounters a series of resistances, including leaf boundary layer resistance and stomatal resistance. The leaf boundary layer is simply explained as a thin air bubble surrounding every leaf in the plant, creating a microclimate that reduces transpiration and CO_2 diffusion (see humidity section). If the boundary layer is large, then C_a will diffuse at a lower rate into C_i, thereby reducing photosynthesis. The main component that reduces the leaf boundary layer resistance is wind velocity (see air movement section). Another resistance that reduces the diffusion of CO_2 into the plant is stomatal resistance (see humidity section) (Jones, 2013b). The main factors influencing stomatal resistance are plant water status, light quality (blue light), and CO_2 concentration in the air (C_a) (Hanan, 1997c; Hernández and Kubota, 2017; Jones, 2013b). It is important for growers to pay close attention to the effect of resistance to CO_2 diffusion from C_a to C_i to maximize the net photosynthetic rate. If the resistances are high, the positive impact of adding CO_2 will be diminished.

The current ambient levels of CO_2 are now close to 410 $\mu mol\ mol^{-1}$ (ppm). This level is acceptable for plant production in plant factories; however, the ambient concentration can drop below 200 $\mu mol\ mol^{-1}$ with a fully developed canopy (Both et al., 2017). Since the difference in C_a and C_i is an important driver of photosynthesis, CO_2 enrichment is often considered a strategy to maintain and/or increase C_a to maintain/increase yield in plant factories.

The impact of enriching CO_2 on the net photosynthetic rate and growth is well documented in research reports. In summary, CO_2 enrichment alone increases the net photosynthetic rate and plant growth; furthermore, enrichment of CO_2 together with increases in light and temperature further improves plant growth (Fig. 11.4). In general, CO_2 enrichment to 700–1000 $\mu mol\ mol^{-1}$ increases the net photosynthetic rate by 30%–50% (Kirschbaum, 2011); however, research has also shown that the increase in the relative growth rate is much lower than that of the net photosynthetic rate (Bunce and Sicher, 2003; Kirschbaum, 2011; Monje and Bugbee, 1998). This phenomenon is known as sink limitations and

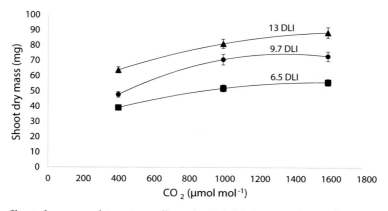

FIGURE 11.4 Shoot dry mass of tomato seedlings (cv Rebelsky) grown for 18 days under increasing CO_2 concentrations and three DLI levels (mol $m^{-2}\ d^{-1}$) (other environmental conditions were the same among treatments). *Huber and Hernández (2020), unpublished.*

downward acclimation. The increased production of photoassimilates by CO_2 enrichment can increase growth only if adequate demand from sink organs (i.e., young leaves, roots, developing flowers/fruits) is present. If the demand for additional carbon in the form of carbohydrates is low, then the increase in the net photosynthetic rate cannot be maintained, and plant signaling will lead to photosynthesis feedback inhibition (Kirschbaum, 2011; Zheng et al., 2019). Therefore, when enriching CO_2 above ambient levels, it is important to maintain adequate sink strength and to provide additional CO_2 during the exponential growth stage to ensure greater yield returns.

In lettuce, several studies have shown a significant increase in fresh and dry mass when the CO_2 concentration was increased to 700 μmol mol^{-1} (Chagvardieff et al., 1994; Pérez-López et al., 2015). In strawberry, CO_2 enrichment resulted in an increase in fruit yield (Enoch et al., 1976) and dry mass (Sun et al., 2012). However, other studies have also shown no yield increase with CO_2 enrichment above ambient levels (Fu et al., 2015; Mortensen, 1994; Park and Lee, 2001).

CO_2 enrichment also alters plant morphology, increasing plant height (epicotyl and hypocotyl) (Khan et al., 2013; Lanoue et al., 2018; Li et al., 2007; Mamatha et al., 2014a) and decreasing stomatal density (Pritchard et al., 1999).

In addition to yield and morphological responses, CO_2 enrichment can also impact plant phytochemicals and nutritional compounds. CO_2 enrichment can increase fructose, glucose, total soluble sugar, starch, ascorbic acid, phenol, and flavonoid contents. However, it can also decrease the contents of nitrogen, chlorophyll, and antioxidant compounds (Dong et al., 2018; Lanoue et al., 2018; Mamatha et al., 2014; McKeehen et al., 1996; Sun et al., 2012).

11.6 Conclusion

In summary, it is important to understand that small changes in one environmental variable will lead to a cascade of changes on other environmental variables and plant responses. Furthermore, the optimization of the multiple environmental factors will always lead to greater benefits (i.e., yield, quality) than optimizing individual components.

In plant factory, even though the technology efficiencies (output per energy input) have improved for temperature, humidity, and CO_2 systems, the accurate control of temperature, relative humidity, and CO_2 concentration has been possible for several decades. However, the accurate control of light intensity and spectrum is relatively a new capability for plant factories. Therefore, research will continue to provide important information on the co-optimization of other environmental factors (T, RH, CO_2) with light intensity and spectrum. In particular, the adoption of precisely controlled FR light and dynamic spectral recipes will continue to increase.

The control and optimization of air movement has been overlooked, some advanced plant factories have air distribution systems capable of delivering a uniform and constant velocity at the canopy level, while most system have more primitive solutions with low air uniformity. Therefore, the possibilities for variable-air-velocity control to reduce crop stage (LAI) specific canopy resistances and to manipulate plant morphology (thigmomorphogenesis) have also been overlooked. The potential advantages of targeted air movement and velocity offer an opportunity for additional research.

References

Adams, S.R., Cockshull, K.E., Cave, C.R.J., 2001. Effect of temperature on the growth and development of tomato fruits. Ann. Bot. 88, 869–877.

Ahmad, M., Cashmore, A.R., 1996. Seeing blue: the discovery of cryptochrome. Plant Mol. Biol. 30, 851–861.

Ahmad, M., Cashmore, A.R., 1997. The blue-light receptor cryptochrome 1 shows functional dependence on phytochrome A of phytochrome B in *Arabidopsis thaliana*. Plant J. 11, 421–427.

Ahmad, M., Grancher, N., Heil, M., Black, R.C., Giovani, B., Galland, P., Lardemer, D., 2002. Action spectrum for cryptochrome-dependent hypocotyl growth inhibition in *Arabidopsis*. Plant Physiol. 129, 774–785.

Albright, D.L., Both, A.J., Chiu, J.A., 2000. Controlling greenhouse light to a consistent daily integral. Transac. ASAE 43, 421–431.

Both, A.J., Frantz, J.M., Bugbee, B., 2017. Carbon dioxide enrichment in controlled environments. In: Lopez, R., Runkle, E. (Eds.), Light Management in Controlled Environments. Meistermedia, Ohio, pp. 82–86.

Bunce, J.A., Sicher, R.C., 2003. Daily irradiance and feedback inhibition of photosynthesis at elevated carbon dioxide concentration in *Brassica oleracea*. Photosynthetica 41, 481–488.

Chagvardieff, P., d'Aletto, T., André, M., 1994. Specific effects of irradiance and CO_2 concentration doublings on productivity and mineral content in lettuce. Adv. Space Res. 14, 269–275.

Chehab, E.W., Eich, E., Braam, J., 2008. Thigmomorphogenesis: a complex plant response to mechano-stimulation. J. Exp. Bot. 60, 43–56.

Chen, X.-l., Yang, Q.-c., Song, W.-p., Wang, L.-c., Guo, W.-z., Xue, X.-z., 2017. Growth and nutritional properties of lettuce affected by different alternating intervals of red and blue LED irradiation. Sci. Hortic. 223, 44–52.

Chia, P.L., Kubota, C., 2010. End-of-day far-red light quality and dose requirements for tomato rootstock hypocotyl elongation. Hortscience 45, 1501–1506.

Ciolkosz, D.E., Albright, L.D., Both, A.J., 1998. Characterizing Evapotranspiration in a Greenhouse Lettuce Crop. International Society for Horticultural Science (ISHS), Leuven, Belgium, pp. 255–262.

Clavijo-Herrera, J., Van Santen, E., Gómez, C., 2018. Growth, water-use efficiency, stomatal conductance, and nitrogen uptake of two lettuce cultivars grown under different percentages of blue and red light. Horticulturae 4, 16.

Cockshull, K.E., Graves, C.J., Cave, C.R.J., 1992. The influence of shading on yield of glasshouse tomatoes. J. Hortic. Sci. 67, 11–24.

Dhingra, A., Bies, D.H., Lehner, K.R., Folta, K.M., 2006. Green light adjusts the plastid transcriptome during early photomorphogenic development. Plant Physiol. 142, 1256–1266.

Di, W., Hu, Q., Yan, Z., Chen, W., Yan, C., Huang, X., Zhang, J., Yang, P., Deng, H., Wang, J., Deng, X., Shi, Y., 2012. Structural basis of ultraviolet-B perception by UVR8. Nature 484, 214–219.

Dong, J., Gruda, N., Lam, S.K., Li, X., Duan, Z., 2018. Effects of elevated CO_2 on nutritional quality of vegetables: a review. Front. Plant Sci. 9, 924.

Downs, R.J., Krizek, D.T., 1997. Air movement. In: Langhans, R.W., Tibbitts, T.W. (Eds.), Plant Growth Chamber Handbook, vol. 99. Iowa agriculture and home economics experiment station special report, Iowa US.

Eguchi, T., Hernández, R., Kubota, C., 2016. End-of-day far-red lighting combined with blue-rich light environment to mitigate intumescence injury of two interspecific tomato. Acta Hortic. 1134, 163–170.

Enoch, H.Z., Rylski, I., Spigelman, M., 1976. CO_2 enrichment of strawberry and cucumber plants grown in unheated greenhouses in Israel. Sci. Hortic. 5, 33–41.

Erwin, J., Velguth, P., Heins, R., 1994. Day/night temperature environment affects cell elongation but not division in *Lilium longiflorum* Thunb. J. Exp. Bot. 45, 1019–1025.

Erwin, J.E., Heins, R.D., 1995. Thermomorphogenic responses in stem and leaf development. Hortscience 30, 940.

Erwin, J.E., Heins, R.D., Karlsson, M.G., 1989. Thermomorphogenesis in lilium longiflorum. Am. J. Bot. 76, 47–52.

Fan, X.-X., Xu, Z.-G., Liu, X.-Y., Tang, C.-M., Wang, L.-W., Han, X.-I., 2013. Effects of light intensity on the growth and leaf development of young tomato plants grown under a combination of red and blue light. Sci. Hortic. 153, 50–55.

Folta, K.M., Childers, K.S., 2008. Light as a growth regulator: controlling plant biology with narrow-bandwidth solid-state lighting systems. Hortscience 43, 504–509.

Folta, K.M., Maruhnich, S.A., 2007. Green light: a signal to slow down or stop. J. Exp. Bot. 58, 3099–3111.

Franklin, K.A., Quail, P.H., 2009. Phytochrome functions in Arabidopsis development. J. Exp. Bot. 61, 11–24.

Frantz, J.M., Ritchie, G., Cometti, N.N., Robinson, J., Bugbee, B., 2004. Exploring the limits of crop productivity: beyond the limits of tipburn in lettuce. J. Am. Soc. Hortic. Sci. 129, 331–338.

Fu, Y., Shao, L., Liu, H., Li, H., Zhao, Z., Ye, P., Chen, P., Liu, H., 2015. Unexpected decrease in yield and antioxidants in vegetable at very high CO_2 levels. Environ. Chem. Lett. 13, 473−479.

González, C.V., Fanzone, M.L., Cortés, L.E., Bottini, R., Lijavetzky, D.C., Ballaré, C.L., Boccalandro, H.E., 2015. Fruit-localized photoreceptors increase phenolic compounds in berry skins of field-grown *Vitis vinifera* L. cv. Malbec. Phytochem. 110, 46−57.

Goto, E., Takakura, T., 1992. Promotion of Ca accumulation in inner leaves by air supply for prevention of lettuce tipburn. Transac. ASAE 35, 647−650.

Graamans, L., van den Dobbelsteen, A., Meinen, E., Stanghellini, C., 2017. Plant factories; crop transpiration and energy balance. Agric. Syst. 153, 138−147.

Guichard, S., Gary, C., Leonardi, C., Bertin, N., 2005. Analysis of growth and water relations of tomato fruits in relation to air vapor pressure deficit and plant fruit load. J. Plant Growth Regul. 24, 201.

Hanan, J.J., 1997a. Greenhouses: Advanced Technology for Protected Horticulture. Taylor & Francis, pp. 167−184.

Hanan, J.J., 1997b. Greenhouses: Advanced Technology for Protected Horticulture. Taylor & Francis, pp. 347−370.

Hanan, J.J., 1997c. Greenhouses: Advanced Technology for Protected Horticulture. Taylor & Francis, pp. 343−344.

Hernández, R., Eguchi, T., Deveci, M., Kubota, C., 2016. Tomato seedling physiological responses under different percentages of blue and red photon flux ratios using LEDs and cool white fluorescent lamps. Sci. Hortic. 213, 270−280.

Hernández, R., Kubota, C., 2016. Physiological responses of cucumber seedlings under different blue and red photon flux ratios using LEDs. Environ. Exp. Bot. 121, 66−74.

Hernández, R., Kubota, C., 2017. Light quality and photomorphogenesis. In: Lopez, R., Runkle, E. (Eds.), Light Management in Controlled Environments. MeisterMedia, Ohio, pp. 29−36.

Huché-Thélier, L., Crespel, L., Gourrierec, J.L., Morel, P., Sakr, S., Leduc, N., 2016. Light signaling and plant responses to blue and UV radiations—perspectives for applications in horticulture. Environ. Exp. Bot. 121, 22−38.

Imaizumi, T., Tran, H.G., Swartz, T.E., Briggs, W.R., Kay, S.A., 2003. FKF1 is essential for photoperiodic-specific light signalling in Arabidopsis. Nature 426, 302−306.

Jarvis, P.P.G., 1985. Coupling of Transpiration to the Atmosphere in Horticultural Crops: The Omega Factor. International Society for Horticultural Science (ISHS), Leuven, Belgium, pp. 187−206.

Jishi, T., Kimura, K., Matsuda, R., Fujiwara, K., 2016. Effects of temporally shifted irradiation of blue and red LED light on cos lettuce growth and morphology. Sci. Hortic. 198, 227−232.

Jolliet, O., Bailey, B.J., 1992. The effect of climate on tomato transpiration in greenhouses. Agric. Forest Meteorol. 58, 43−62.

Jones, H.G., 2013a. Plants and Microclimate: A Quantitative Approach to Environmental Plant Physiology, third ed. Cambridge University Press, Cambridge, pp. 291−295.

Jones, H.G., 2013b. Plants and Microclimate: A Quantitative Approach to Environmental Plant Physiology, third ed. Cambridge University Press, Cambridge, pp. 153−156.

Kalaitzoglou, P., van Ieperen, W., Harbinson, J., van der Meer, M., Martinakos, S., Weerheim, K., Nicole, C.C.S., Marcelis, L.F.M., 2019. Effects of continuous or end-of-day far-red light on tomato plant growth, morphology, light absorption, and fruit production. Front. Plant Sci. 10.

Kang, J.H., KrishnaKumar, S., Atulba, S.L.S., Jeong, B.R., Hwang, S.J., 2013. Light intensity and photoperiod influence the growth and development of hydroponically grown leaf lettuce in a closed-type plant factory system. Hortic. Environ. Biotechnol. 54, 501−509.

Kendrick, R.E., Kronenberg, G.H.M., 1994. Photomorphogenesis in Plants. Kluwer, Dordrecht, The Netherlands, pp. 51−67.

Khan, I., Azam, A., Mahmood, A., 2013. The impact of enhanced atmospheric carbon dioxide on yield, proximate composition, elemental concentration, fatty acid and vitamin C contents of tomato (*Lycopersicon esculentum*). Environ. Monit. Assess. 185, 205−214.

Kirschbaum, M.U.F., 2011. Does enhanced photosynthesis enhance growth? lessons learned from CO_2 enrichment studies. Plant Physiol. 155, 117.

Kitaya, Y., Tsuruyama, J., Kawai, M., Shibuya, T., Kiyota, M., 2000. Effects of air current on transpiration and net photosynthetic rates of plants in closed transplant production system. In: Kubota, C., Chun, C. (Eds.), Transplant Production in the 21st Century. Kluwer Academic, The Netherlands, pp. 83−90.

Kubota, C., 2016. Growth, development, transpiration and translocation as affected by abiotic environmental factors. In: Kozai, T., Niu, G., Takagaki, M. (Eds.), Plant Factory: An Indoor Vertical Farming System for Efficient Quality Food Production. Elsevier, Waltham, MA, US, pp. 154–155.

Kubota, C., 2020. Controlled Environment Berry Production Information, vol. 2021. The Ohio State University, Columbis, OH. https://u.osu.edu/indoorberry/.

Kubota, C., Kroggel, M., Both, A.J., Whalen, M., 2016. Does supplemental lighting make sense for my crop?-empirical evaluations. Acta Hortic. 1134, 403–412.

Lambers, H., Chapin, F.S., Thijs, L.P., 2008. Plant Physiological Ecology, second ed. Springer, New York, pp. 43–45.

Lanoue, J., Leonardos, E.D., Khosla, S., Hao, X., Grodzinski, B., 2018. Effect of elevated CO_2 and spectral quality on whole plant gas exchange patterns in tomatoes. PLoS One 13, e0205861.

Li, J., Zhou, J.-M., Duan, Z.-Q., Du, C.-W., Wang, H.-Y., 2007. Effect of CO_2 enrichment on the growth and nutrient uptake of tomato Seedlings 1 1 Project supported by the National Natural science foundation of China (No. 30230250). Pedosphere 17, 343–351.

Lin, K.H., Huang, M.Y., Huang, W.D., Hsu, M.H., Yang, Z.W., Yang, C.M., 2013. The effects of red, blue, and white light-emitting diodes on the growth, development, and edible quality of hydroponically grown lettuce (*Lactuca sativa* L. var. capitata). Sci. Hortic. 150, 86–91.

Mamatha, H., Rao, N.K., Laxman, R.H., Shivashankara, K.S., Bhatt, R.M., Pavithra, K.C., 2014. Impact of elevated CO_2 on growth, physiology, yield, and quality of tomato (*Lycopersicon esculentum* Mill) cv. Arka Ashish. Photosynthetica 52, 519–528.

Mason, P., 1995. Atmospheric boundary layer flows: their structure and measurement. Bound. Layer Meteorol. 72, 213–214.

McKeehen, J.D., Smart, D.J., Mackowiak, C.L., Wheeler, R.M., Nielsen, S.S., 1996. Effect of CO_2 levels on nutrient content of lettuce and radish. Adv. Space Res. 18, 85–92.

Meng, Q., Runkle, E.S., 2020. Growth responses of red-leaf lettuce to temporal spectral changes. Front. Plant Sci. 11.

Mickens, M.A., Torralba, M., Robinson, S.A., Spencer, L.E., Romeyn, M.W., Massa, G.D., Wheeler, R.M., 2019. Growth of red pak choi under red and blue, supplemented white, and artificial sunlight provided by LEDs. Sci. Hortic. 245, 200–209.

Min, W., Kubota, C., 2008. Effects of electrical conductivity of hydroponic nutrient solution on leaf gas exchange of five greenhouse tomato cultivars. HortTechnology 18, 271–277.

Mitchell, C.A., Burr, J.F., Dzakovich, M.J., Gómez, C., Lopez, R., Hernández, R., Kutoba, C., Currey, C.J., Meng, Q., Runkle, E.S., Bourguet, C.M., Murrow, R.C., Both, A.J., 2015. Horticultural Reviews-Light-Emitting Diodes in Horticulture. Wiley Blackwell, pp. 1–87.

Mitchell, C.A., Myers, P.N., 1995. Mechanical stress regulation of plant growth and development. Hortic. Rev. 17, 1–42.

Monje, O., Bugbee, B., 1998. Adaptation to high CO_2 concentration in an optimal environment: radiation capture, canopy quantum yield and carbon use efficiency. Plant Cell Environ. 21, 315–324.

Monteith, J.L., 1965. Evaporation and environment. Symp. Soc. Exp. Biol. 19, 205–234.

Mortensen, L.M., 1994. Effects of elevated CO_2 concentrations on growth and yield of eight vegetable species in a cool climate. Sci. Hortic. 58, 177–185.

Owen, W.G., Lopez, R.G., 2015. End-of-production supplemental lighting with red and blue light-emitting diodes (LEDs) influences red pigmentation of four lettuce varieties. Hortscience 50, 676–684.

Park, M.H., Lee, Y.B., 2001. Effects of CO_2 concentration, light intensity and nutrient level on growth of leaf lettuce in a plant factory. Acta Hortic. 548, 377–384.

Pasian, C., Lieth, J., 1989. Analysis of the response of net photosynthesis of rose leaves of varying ages to photosynthetically active radiation and temperature. J. Am. Soc. Hortic. Sci. 114 (4), 581–586.

Pérez-López, U., Miranda-Apodaca, J., Lacuesta, M., Mena-Petite, A., Muñoz-Rueda, A., 2015. Growth and nutritional quality improvement in two differently pigmented lettuce cultivars grown under elevated CO_2 and/or salinity. Sci. Hortic. 195, 56–66.

Pritchard, S.G., Rogers, H.H., Prior, S.A., Peterson, C.M., 1999. Elevated CO_2 and plant structure: a review. Global Change Biol. 5, 807–837.

Roig-Villanova, I., Martínez-García, J.F., 2016. Plant responses to vegetation proximity: a whole life avoiding shade. Front. Plant Sci. 7.

Saure, M.C., 1998. Causes of the tipburn disorder in leaves of vegetables. Sci. Hortic. 76, 131–147.

Seginer, I., Shina, G., Albright, L.D., Marsh, L.S., 1991. Optimal temperature setpoints for greenhouse lettuce. J. Agric. Eng. Res. 49, 209–226.

Shamshiri, R.R., Jones, J.W., Thorp, K.R., Ahmad, D., Che Man, H., Taheri, S., 2018. Review of optimum temperature, humidity, and vapour pressure deficit for microclimate evaluation and control in greenhouse cultivation of tomato: a review. Int. Agrophys. 32.

Slack, G., Hand, D.W., 1983. The effect of day and night temperatures on the growth, development and yield of glasshouse cucumbers. J. Hortic. Sci. 58, 567–573.

Spaargaren, I.J.J., 2001. Supplemental Lighting for Greenhouse Crops. Hortilux Schreder, Moster, Netherlands, pp. 1–178.

Spalholz, H., Perkins-Veazie, P., Hernández, R., 2020. Impact of sun-simulated white light and varied blue:red spectrums on the growth, morphology, development, and phytochemical content of green- and red-leaf lettuce at different growth stages. Sci. Hortic. 264, 109195, 1–12.

Strasser, B., Sánchez-Lamas, M., Yanovsky, M.J., Casal, J.J., Cerdán, P.D., 2010. *Arabidopsis thaliana* life without phytochromes. Proc. Natl. Acad. Sci. U. S. A. 107, 4776.

Sun, P., Mantri, N., Lou, H., Hu, Y., Sun, D., Zhu, Y., Dong, T., Lu, H., 2012. Effects of elevated CO_2 and temperature on yield and fruit quality of strawberry (fragaria × ananassa Duch.) at two levels of nitrogen application. PLoS One 7, e41000.

Takemiya, A., Inoue, S.-I., Doi, M., Kinoshita, T., Shimazaki, K.-I., 2005. Phototropins promote plant growth in response to blue light in low light environments. Plant Cell 17, 1120–1127.

Tatsumi, M., 1969. Studies on the photosynthesis of vegetable crops I Photosynthesis of young plant of vegetables in relation to light intensity. Bull. Hortic. Ress. Stn. 8, 127–140 (in Japanese).

Thompson, H.C., Langhans, R.W., Both, A.J.A., Albright, L.D., 1998. Shoot and root temperature effects on lettuce growth in a floating hydroponic system. J. Am. Soc. Hortic. Sci. 123, 361–364.

Tibbitts, T., Bottenberg, G., 1976. Growth of lettuce under controlled humidity levels. J. Am. Soc. Hortic. Sci. 101, 70–73.

Wheeler, R.M., Mackowiak, C.L., Sager, J.C., Yorio, N.C., Knott, W.M., Berry, W.L., 1994. Growth and gas exchange by lettuce stands in a closed, controlled environment. J. Am. Soc. Hortic. Sci. 119, 610.

White, J., Warrington, I., 1988. Temperature and light integral effects on growth and flowering of hybrid geraniums. J. Am. Soc. Hortic. Sci. 113, 354–359.

Xu, X., Hernández, R., 2020. The Effect of light intensity on vegetative propagation efficacy, growth, and morphology of "Albion" strawberry plants in a precision indoor propagation system. Appl. Sci. 10.

Yang, X., Short, T.H., Fox, R.D., Bauerle, W.L., 1990. Transpiration, leaf temperature and stomatal resistance of a greenhouse cucumber crop. Agric. For. Meteorol. 51, 197–209.

Yang, Z.C., Kubota, C., Chia, P.L., Kacira, M., 2012. Effect of end-of-day far-red light from a movable LED fixture on squash rootstock hypocotyl elongation. Sci. Hortic. 136, 81–86.

Zhang, D., Du, Q., Zhang, Z., Jiao, X., Song, X., Li, J., 2017. Vapour pressure deficit control in relation to water transport and water productivity in greenhouse tomato production during summer. Sci. Rep. 7, 43461.

Zhang, T., Maruhnich, S.A., Folta, K.M., 2011. Green light induces shade avoidance symptoms. Plant Physiol. 157, 1528.

Zhen, S., Bugbee, B., 2020. Substituting far-red for traditionally defined photosynthetic photons results in equal canopy quantum yield for CO_2 fixation and increased photon capture during long-term studies: implications for redefining PAR. Front. Plant Sci. 11.

Zhen, S., Haidekker, M., van Iersel, M.W., 2019. Far-red light enhances photochemical efficiency in a wavelength-dependent manner. Physiol. Plant. 167, 21–33.

Zheng, Y., Li, F., Hao, L., Yu, J., Guo, L., Zhou, H., Ma, C., Zhang, X., Xu, M., 2019. Elevated CO_2 concentration induces photosynthetic down-regulation with changes in leaf structure, non-structural carbohydrates and nitrogen content of soybean. BMC Plant Biol. 19, 255.

PART III

Applications

12

Productivity: definition and application

Toyoki Kozai[1], Kaz Uraisami[2], Katashi Kai[3] and Eri Hayashi[1]

[1]Japan Plant Factory Association, Kashiwa, Chiba, Japan; [2]Marginal LLC, Tokyo, Japan;
[3]Shinnippou Ltd., Shizuoka, Japan

12.1 Introduction

Plant factory with artificial lighting (PFAL) is attracting a great amount of attention from all over the world as the novel urban agriculture system that achieves year round high productivity of high quality produce while saving resources and reducing waste to the environment. PFAL can also be built in any location whether shaded, small, or degraded (Kozai et al., 2020). "High quality produce" stands for safe produce containing a reasonable amount of nutrient, medical and/or functional components, while providing good taste, texture, color, appearance, function and/or flavor, and pesticide-free. "Saving resources" means that the necessary irrigation water volume per unit of produce is 1/50th or less compared with field production, and the fertilizer applied is approximately half. Environmental preservation means that the waste per unit of produce is minimal. "Degraded land" means any land with toxic substances such as heavy metal, salt accumulation, or low density of microorganisms.

No matter where located, PFAL reaches a land productivity of 100 times or higher compared with optimal land cultivation in the same region. No matter how cold, hot, or dry the weather of the region is, or with strong winds or frequent pest infestation, PFAL maintains high productivity year round thanks to the precise environmental controls of cultivation space and labor conditions. Excessive consumption of electricity per unit of produce is often regarded as a weakness of PFAL. However, recent functional progress of light-emitting diode (LED) fixtures and lighting systems, optimal environmental controls, newly developed seed lines, and lower costs of electricity due to renewable energy shall alleviate this weakness (Kozai, 2018a).

It must be noted that the PFAL benefits are reached only when the system is properly engineered, procured, and constructed, as well as operated (Kozai, 2018b; Kozai et al., 2019a).

The reality is that PFAL remains with various challenges such as a higher initial investment, uncertainty of optimal cultivation system and environmental controls, needs for more intensive technical and practical training to better manage the cultivation methods that are different from those in an open field or a greenhouse. These challenges are to be solved by improving productivity through proper countermeasures. The issue here is that there is no set of clearly defined "productivity" indexes with appropriate measurement methods to support them. Therefore, hardly any quantitative comparison of various PFALs and measurement of the improvement can be applied.

This chapter focuses on "productivity" and aims to contribute to the following seven purposes: (1) Present various quantitative indexes and the philosophy and definition pertaining to the various productivity aspects of PFAL; (2) Explicitly organize the indexes based on data available from commercially operated PFALs; (3) Compare the productivity of past and present operation of any given PFAL as well as among PFALs in terms of their productivity by using the common indexes; (4) Classify and list factors that affect productivity and contribute to it; (5) Maintain the cycles described here for productivity improvement; (6) Propose other issues to be examined in the future; and (7) Define and provide supplemental explanation and notes of the terminology pertaining to PFAL.

The main theme of this chapter is to compare the productivity of various PFALs based on the productivity indexes using the common calculation formulas. Issues for item (6) described above are introduced in Section 12.7 as issues to be examined in the future.

This chapter is based on the activities of the Japan Plant Factory Association (nonprofit organization) research committee in 2017 and 2018 on "Productivity Improvement of Plant Factory with Artificial Lighting." An article containing essential parts of this chapter is written in Japanese and published by Kozai et al. (2019b).

12.2 Concept of productivity

"Productivity," examined in this chapter, may also be termed as "cost performance," "cost effectiveness," or "cost efficiency" (Wikipedia: https://en.wikipedia.org/wiki/Productivity). Productivity is classified into resource productivity and monetary productivity (Fig. 12.1).

12.2.1 Resource productivity

"Resource Productivity (Pr)" is calculated respectively to such resources as energy, space, and/or time in the definite period of time is expressed as the ratio of amount of produce (S) to the amount of resource supplied or applied (R_{sp}) as follows:

$$P_r = S/R_{sp} \tag{12.1}$$

S is expressed as weight (mass), number, the amount of energy, or mol, etc. R_{sp} is expressed as weight, number, amount of energy, mol, space, or time, etc (For the definition of "mol," see Section 12.8.4). Resource element includes electric energy, light energy, water, CO_2, fertilizer element, cultivation space, etc. The amount of resource supplied (R_{sp}) is expressed as follows:

$$R_{sp} = R_{min}/RUE \tag{12.2}$$

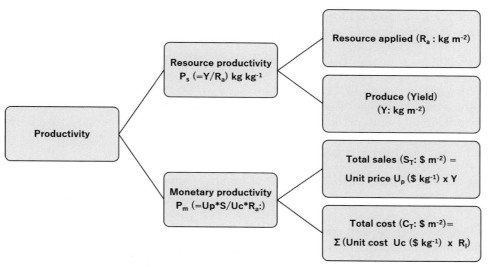

FIGURE 12.1 Definitions of resource and monetary productivities per floor area or cultivation area. R_a is calculated for each resource element, while P_m is calculated for all resources supplied.

where R_{min} is the minimum amount of resources required to produce S ($R_{sp} \geqq R_{min}$) and resource use efficiency (RUE) is the use efficiency of the subject resource ($= R_{min}/R_{sp}$) whose maximum is 1.0 except for light energy use efficiency with a maximum of about 0.1 and electric energy use efficiency of around 0.06 (Kozai, 2013; Kozai et al., 2020).

R_{min} shall be the amount of the subject resource contained in the produce S. For example, the minimum amount of nitrogen to be applied is the aggregate amount of nitrogen contained in the produce. The minimum amount of CO_2 to be applied is the aggregate of carbon (C) and oxygen (O) contained in the carbohydrate in the produce. Use efficiency of water is the amount of water contained in the produce divided by the amount of water irrigated. Light energy use efficiency is the amount of chemical energy fixed in the produce divided by the amount of photosynthetic radiation energy emitted by LEDs. The same is to be applied to the use efficiency of other resources. Formula (12.1) and (12.2) may lead to the formula for P_r as follows:

$$P_r = RUE * S/R_{min} \tag{12.3}$$

This concludes that P_r increases when RUE increases. S/R_{min} is the maximum of P_r at RUE = 1.0.

12.2.2 Monetary productivity

Monetary productivity or monetary term productivity (P_m) is expressed by multiplying the sales unit price (U_p) at the numerator and the purchase unit price or cost (U_c) at the denominator of Formula (12.1) or (12.3):

$$P_m = U_p * S/U_c * R_{sp} = P_r * (U_p / U_c) = (U_p * RUE * S)/(U_c * R_{min}) \tag{12.4}$$

In reality, not all the produce is sold. P_{mr} is the monetary productivity taking into account only the volumetrically sold ratio of produce (r) and is expressed as follows:

$$P_{mr} = (U_p * RUE * r * S)/(U_c * R_{min}) \qquad (12.5)$$

Additionally, in case r and/or RUE are smaller than 1.0, the residue of the produce and/or that of the applied resources shall also be taken into account as the unit cost of produce residue treatment (U_r) and/or as the unit cost of applied resource residue treatment (U_m) thereby amending the formula of P_{mrp} as follows (See "Section 12.8. Supplemental explanation and notes" for further details):

P_{mrp} = Value of produce for sale/(Cost of resources applied

$\qquad$ + Treatment cost of produce not for sale + Treatment cost of resource residue)

Here, with U_{pr} as the unit cost of treatment of produce not for sale (plant residue), U_{rr} as the unit cost of treatment of resource residue, and U_c*R_{sp} as the cost of resources supplied, then the productivity in monetary terms (P_{mrp}) is formula (12.6) as follows:

$$P_{mrp} = (U_p * r * S)/((U_c * R_{min}) + U_{pr} * P_r * (1.0 - r) + U_{rr} * R_{sp} * (1.0 - RUE)) \qquad (12.6)$$

12.2.3 Monetary productivity of all the resources applied

Productivity in monetary terms of all resources applied is expressed as (Value of Produce for sale) divided by (Cost of Resource 1 applied + Cost of Resource 2 applied + Cost of Resource 3 applied + …). In other words, productivity in monetary terms of all the resources applied is equal to value of Produce for sale divided by cost of all the resources applied. The formulas from (12.1) through (12.5) or (12.1) through (12.6) may be applied by considering the total costs of resources applied respectively calculated.

In this chapter, however, Section 12.3 and thereafter examine only formulas (12.1) through (12.3). Formula (12.1) through (12.5) may be applied to the various resources and the result thereof when combined may lead to the total productivity of all resources.

12.3 Productivity indexes, definitions, and formulas

In this chapter, productivity is described in terms of (1) materials, energy, time, and space invested and (2) monetary terms. Three major costs of PFALs in Japan are electricity, labor hours, and depreciation of the investment (Table 12.3). When conducting an international comparison in terms of resource and monetary productivity, keep in mind that some PFAL locations may incur the higher cost of water, fertilizer, seeds, etc., resulting in a substantial impact on production costs. In this case, the subject resources may be regarded as the fourth and/or fifth cost components. As explained in formula (12.4) in Section 12.2, the monetary productivity term is obtained by multiplying the amount of resources applied and the produce amount by the unit cost of the said resource and the unit sale price of the produce, respectively. In this section, we will focus on formula (12.3) below.

Four productivity indexes, a, b, c, and d are proposed in this chapter and three major ones (a-c) are presented in the green boxes in Fig. 12.2A and in Fig. 12.2B below. For a specific production period, the data necessary to calculate the three productivity indexes (a-c) are (i) the total electricity consumption (kWh), (ii) the total man-hours (MH) or hour (h), (iii) the total shelf cultivation area (m^2) and time required from seeding to harvest (h), and (iv) fresh weight of produce for sale (kg). These data are easily obtained at most PFALs and by dividing (i), (ii), and (iii) by (iv) you may obtain the three productivity indexes. Fresh weight of produce for sale may be substituted for number of crops or dry weight.

Fig. 12.2 is presented in two slightly different forms (Figs. 12.2A and 12.2B). In Fig. 12.2A, the hourly productivity of the cultivation space is treated no differently from that of consumable resources. For example, cost of LED fixtures as part of the cultivation space is allocated to production based on the number of light period guaranteed by the manufacturer of LED fixtures. In Fig. 12.2B, the cultivation space is the fixed assets and the cost and productivity is reported through depreciation of the fixed assets. The productivity indexes (a?c) are based on the resources applied and on the resources as converted into monetary terms, while index d represents the monetary terms only. Fig. 12.2A is more applicable to the productivity control of a PFAL as you may increase or decrease resources applied and measure increase or decrease of produce and conclude on productivity. Fig. 12.2B is more applicable to the management or financial control of a PFAL as the productivity indexes are divided into that of operation and that of initial investment.

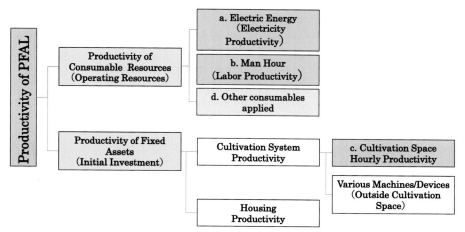

FIGURE 12.2A   Target Productivity Indexes in this Chapter. Three major indexes, a—c, is painted in light green. In this figure, the hourly productivity of the cultivation space is treated no differently from that of consumable resources.

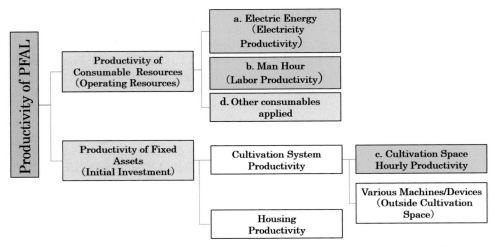

FIGURE 12.2B   Target Productivity Indexes in this Chapter. Three major indexes, a—c, is painted in light green. In this figure, the cultivation space is the fixed assets and the cost and productivity is reported through depreciation of the fixed assets.

In addition to the data on fresh weight of whole plants produced (kg) and amount of resource supplied (kg, kWh, h, etc.), the productivity in monetary terms is also based on the following data:

(i) weighted average of per kg of the crop or unit selling price ($/kg)
(ii) weighted average of per kWh of the electricity cost consumed ($/kWh)
(iii) weighted average of per hour cost of man-hours ($/h)

Then, the productivity in monetary terms is expressed as the ratio of A to B where A is the product of unit selling price, whole plant fresh weight, salable portion ratio, and sold portion ratio, and B is the product of unit production cost and resource supplied. Resource supplied is expressed as a ratio of minimum resource input required (kg, kWh, h, etc.) to RUE.

12.3.1 Definitions

a. Per electric energy index (kg kWh^{-1}) is used for the electricity productivity or electric energy productivity and is defined as "fresh weight (mass) of crop(s) produced for sale per electric energy applied (kWh) inside a cultivation zone or cultivation room." The energy consumption of electric lighting, air-conditioning, and other devices, if measured and recorded separately, is an important factor in the analysis of productivity issues. The time unit used is usually a one year but monthly, quarterly, or per production period may also be used. Electricity productivity for dry weight can also be calculated in case that % dry matter is known.
b. Per hour (kg h^{-1}) or **per man-hour** (kg MH^{-1}) **index** is used for labor productivity or man-hour productivity. Both are defined as "the weight of produce for sale per man-hour (or per working hour) of those laborers whose sole responsibility is cultivation and

related activities (packaging, packing, etc.)." Employees who mostly work outside the cultivation zone, such as a PFAL supervisor, and sales and administration staff are in principle not included as laborers. Labor hours per one year, one quarter, one month, or one cultivation period may be measured per laborer group, individual laborer, or per cultivation crop or variety.

c. **Per cultivation space and time** ($kg\ m^{-2}\ d^{-1}$) **index** is defined as "the weight of produce for sale per cultivation space per day" and may represent cultivation area productivity ($kg\ m^{-2}$) or cultivation cubic space productivity ($kg\ m^{-3}$). For the sake of simplicity, both may be called cultivation space productivity. For more details, refer to the definitions given below and "Section 12.5 Resource productivity and monetary term productivity and the production cost examples."

12.3.2 Cultivation area

The cultivation area (m^2) stands for the aggregate area of grow panels inside a cultivation zone. The aggregate number (width × length of grow panel) is to be multiplied by the number of shelves per rack and by the number of racks inside a cultivation zone. The cultivation area includes trays for seeding, germination, and seedling production. The grow panel shelf area is somewhat larger than the area of grow panels or beds on the shelves. Those shelves may include the area for production and if it is provided with LED lighting, thereon, it shall be treated as part of the cultivation area. Similar to the calculation of the cultivation area productivity, you may also calculate the cultivation shelf area productivity as well as the respective productivity in terms of various produce variety, environment control, and cultivation density.

12.3.3 Cultivation space

Cultivation (cubic) space (m^3) is the aggregate sum of the spaces used for cultivation, seeding, germination, seedling production, and cropping space with LED lighting, and is regarded as the total cultivation space as narrowly defined. The air-conditioning indoor equipment, air circulation fans, and CO_2 tubing, while necessary for environmental control, have no direct impact on labor productivity and shall not be included as cultivation space. Passageways for conveying machines and floor space for the following equipment are not counted as part of cultivation space: floor storage for various materials, cultivation panels and storage, conveying machines, air purifiers, washing machines, nutrient solution tank, pump, sterilizer, transplanting machines, cold storage chambers, precooling rooms, crop preparation facility, packaging equipment, cleaning equipment, power panels, floor stationery air-conditioners, etc.

Cultivation area productivity and cultivation space productivity are mutually converted by applying the measurement data of cultivation racks. The same conversion method is applied to those cultivation systems with vertical, i.e., not horizontal, cultivation panels, to those with movable cultivation panels and pipes.

d. Monetary productivity pertaining to those resources other than those in indexes a—c

Resources or consumables other than a—c are water, CO_2, and fertilizer elements (nitrogen, phosphorus, potash, magnesium, etc.). It is difficult to calculate those productivity indexes as segregated for those consumables. However, assuming that the total cost of these consumables, electricity, labor, and cultivation space and time can be measured, then subtracting indexes a—c from the total and calculating the inverse number provides the productivity number for index d (Refer to Fig. 12.1 and Section 12.8 Supplemental explanation). In this chapter, the total cost excludes those costs categorized as sales and general administration, i.e., transportation, land rent, depreciation of housing, advertisement, sales expenses, etc.

12.4 Cultivation space studied

The labor productivity index shall be measured for the broadly defined cultivation space inside a PFAL (hereinafter cultivation room or hygiene-controlled room). The space of a cultivation room is represented by the product of the floor area and the average ceiling height. The total cultivation space of a PFAL with multiple cultivation rooms is calculated by aggregating those products of the floor area and the average ceiling height of each cultivation room.

Equipment and machines inside the cultivation room shall provide their consumption of materials and electricity, which are part of PFAL productivity. The equipment and machines include floor storage for various materials, cultivation panels and their storage, conveying machines, air purifiers, washing machines, nutrient solution tank·pum·sterilizer, transplant machines, cold insulation chambers, precooling room, crop preparation facility, packaging equipment, cleaning equipment, power panels, floor stationery, air-conditioners, etc.

On the other hand, not included in the calculations of productivity are, in principle, those consumables and electric energy applied by equipment and machines at the nonhygiene-controlled area (i.e., outside of the cultivation room) such as administration offices, dressing rooms, air shower rooms, bathrooms, shipping area, entrance area, etc. The areas in white boxes in Fig. 12.3 are examples of areas not included in the calculations. Excluding the nonhygiene-controlled area is done for the sake of measurement simplicity and not necessarily for the sake of theoretical accuracy. Electric energy applied at the nonhygiene-controlled area may be included and so stated to measure the productivity number and in case possible, to provide the individual number of electric consumption at the nonhygiene-controlled area.

12.5 Case study: resource and monetary productivities and the absolute monetary value

Table 12.1 shows the estimated productivity by resource type and monetary term productivity of PFALs in Japan which report a profitability margin of 10% or more over the crop sales. When the unit of nominator or denominator is monetary, the slant symbol (/) is used instead of a superscript, for example, kg/USD instead of kg USD^{-1}, (see also Section 12.8.3).

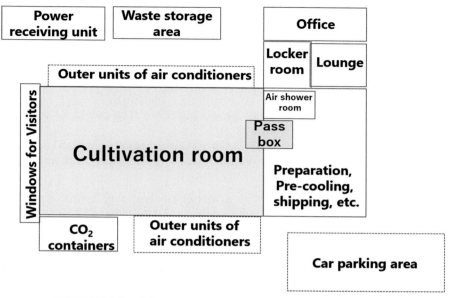

FIGURE 12.3 Cultivation room (Colored area) defined in this article.

TABLE 12.1 Estimated resource productivity and monetary productivity of PFALs in Japan. Numbers and units are cited and converted from the figure of production cost as per resources in Kozai et al. (2019b). Items for sales, general, and administration cost under the accounting rules, such as transportation cost, land lease, housing depreciation, advertisement cost, sales expenses, etc., are not included. Numbers for a, b, and c are calculated by dividing the annual crops produced for sale by annual electricity consumption, annual labor hours, and cultivation area per day (annual total of cultivation area divided by 360). Note that each productivity (a-d) in 2021 has probably been improved by about 20% compared with that shown in this table.

No.	Resources applied	Range of resource productivity	Range of monetary productivity (kg/USD)	Range of production cost (USD/kg)
a	Electric energy	0.11–0.14 kg kWh^{-1}	0.645–0.755	1.09–1.28
b	Labor hours	7.7–10.0 kg h^{-1}	0.591–0.770	1.18–1.63
c	Cultivation area (per day)	0.25–0.33 kg m^{-2} d^{-1}	0.482–0.645	1.28–1.72
d	Other resources	—	0.609–0.827	1.00–1.36
	Total	—	0.166–0.219	**4.55–5.99**

American dollar. One USD and one Euro were equal to 110 JPY and 130 JPY, respectively, in 2019. Note that wage ($/hour) is around 2 times higher in USA and some European countries than in Japan as of 2021

As described above, "other materials" in Table 12.1 include seeds, CO_s, substrate, water, fertilizer, cultivation panels for replacement, packing sacks, transportation boxes, etc., and as described above, does not include sales, general, and administration items. Data in Table 12.1 are obtained by evaluating data of a limited number of PFALs and shall be fine-tuned by evaluating more.

In Table 12.1, electricity cost per kg (the sum of the base rate and the meter rate), labor cost per kg (not received by the laborers but paid by the operator), and cost of cultivation area per unit are:

(1) electricity cost at $1.09/kg to $1.28/kg (i.e., producing 0.11 to 0.14 kg kWh^{-1} for electricity cost of $0.12/kWh to $0.18/kWh) [$1 US = Jpy 110 (Jpy for Japanese Yen) as of 2020].
(2) labor cost at $1.18/kg to $1.63/kg (i.e., producing 5.9–7.0 kg per labor cost of $10 for labor cost of $7.68 to $12.5 per hour)
(3) cultivation area cost at $1.28/kg to $1.72/kg (i.e., producing 0.2 to 0.33 kg m^{-2}d^{-1} per m^2/d cost of $0.32 to $0.56). For cultivation space with multiple cultivation rooms, the average number is applied.

You may recognize that productivity as a matter of monetary terms, e.g., 0.2 kg/USD (US Dollar) is an inverse number of production cost, i.e., 5 USD/kg (=1 USD/0.2 kg). Overall productivity per USD applied (0.136–0.182 kg/USD) is an inverse number of the total cost of a–d as multiplied by 0.166 (=1/5.99) to 0.219 (=1/4.55). It is important that on site laborers recognize, on a daily basis, the productivity indexes in Table 12.1 and the variables affecting them and shall contribute to the continuous improvement of productivity.

It shall be noted that indexes a through d affect one another. For example, an increase in photosynthetic photon flux (PPF) from LED fixtures increases electric energy consumption (a) and may, simultaneously, decrease the necessary cultivation area and labor hours applied. Also, automation system, if installed, shall decrease labor hours thereby improving labor hour productivity. However, the increased cost from the automation system may not necessarily improve the overall monetary productivity of PFALs. Therefore, it is often very difficult to operate PFALs achieving the simple total of the best productivity number of a through d.

A cultivation manager shall optimize individually each index of a through d and eventually optimize the overall productivity of all the resources invested. A research committee (Productivity Improvement of Plant Factory with Artificial Lighting) of JPFA implements research for overall productivity optimization of aggregate resources based on this chapter.

12.6 Factors affecting productivity

Table 12.2 lists factors directly and/or indirectly affecting productivity indexes a through d in Table 12.1. The factors may be broadly categorized into (1) those subject solely to the actions of the cultivation manager and laborers and (2) resulting from (1) in terms of plant growth and amount of resource consumed. These factors may also be categorized into (1) mostly fixed factors upon engineering, procurement, and construction of housing and facilities and not at a later point of time (2) factors altered at a later time such as machine layout and human resource assignment, (3) factors altered daily such as environmental controls. Measurement, analysis, and proposition system for the combined impact of these factors on overall productivity, if developed, shall certainly help contribute to the continuous improvement of productivity.

TABLE 12.2 Factors affecting productivity a through d in Table 12.1.

	Factors group	Factors
Fixed upon design and construction stages	Structure and configuration of cultivation space	Cultivation space capacity/cultivation room capacity, cultivation area/cultivation room area, cultivation shelf area/cultivation room area, height of cultivation shelf, cultivation rack height/ceiling height, thermal insulation performance, airtightness
	Grow (cultivation) system	Structure and function of cultivation and nutrient solution circulation systems, use or no use of substrate, type and layout of substrate, layout and specification of equipment
	Lighting system	Photosynthetic photon number efficacy, maximum photosynthetic photon flux, spectral distribution, angular light distribution curve, lighting fixture layout, optical reflectance of inner walls
	Air-conditioning, hygiene, labor, disaster prevention, safety system	Function and performance of air-conditioner, circulator, sterilization unit, washing machine, insect proofing, disaster/fire/crime prevention, transportation, and work devices
Subject anytime to alteration	Environment control 1 (management)	Time schedule of the set values of environmental factors, control standard of environment, hygiene, and various abnormality detection standard
	Cultivation control 1	Dates for the first and the second transplanting (spacing), seeding and planting density, number of days for cultivation (from seeding to harvesting)
	Employee management and training	Trainings and seminars, technical skill improvement seminars, work efficiency, work standardization, laborer self-detection, mental and body health control, PDCA (Plan-Do-Check-Action) cycle promotion
	Management control	Plan, production, marketing, sales, cost, risk, development, investment, and organization
	Hygiene control 1	Revising hygiene control manual, updating hygiene control devices
Direct impact on productivity	Environmental control 2	Physical and chemical environment of light, air, nutrient solution, and time and spatial variations of environment for micro-organism
	Environmental control 3	Time variation of the amount of consumables applied by respective equipment and use efficiency thereof
	Cultivation control 2	Seed processing (grading by size/weight/color/shape), coating, priming, temperature treatment, etc.), germination under dark or light
	Cultivation control 3	Use efficiency of seeds, seedlings, yield of cropping and shipment
	Cultivation control 4	

(*Continued*)

TABLE 12.2 Factors affecting productivity a through d in Table 12.1.—cont'd

	Factors group	Factors
	Cultivation control 5	Operation rate of cultivation space, occurrence rate of physiological abnormality, compliance rate of hygiene control standard Plant weight, quality, and their dispersion, average and their space and time dispersion as of germination, seedling, and cropping, weight and number of crops produced for sale
	Cultivation control 6	Weight ratio of salable portion over (i) total crop, (ii) roots, (iii) wasted portion other than roots
	Hygiene control 2	Density of microorganism, pathogenic bacterium, and small animals in the produce, nutrient water, air, devices, etc.
	Automation, robotization	Seeding, transplanting, harvesting, quality checking, conveying, packaging, cleaning, sterilization, measurement, recording, analysis, transparency, storage, ordering, purchasing, and book keeping
	Waste water, waste heat, and waste materials	Waste water, waste heat, plant residue, nutrient solution residue, waste substrate, and other wasted consumables
Others	Marketing	Product planning, development, sales promotion, price setting (wholesale or retail and if retail, per gram price variable depending on weight per package), shipment yield, market price, transportation cost, various royalty payment, location, etc.

12.7 Other issues to be examined in the near future (Kozai, 2018b; Kozai et al., 2020)

(1) Contribution to achieving the 17 sustainable development goals (Chapter 4) (Seth et al., 2019)

(2) Conceptual comparison with "carbon footprint," "food mileage," "life cycle assessment," etc.

(3) Introducing other indexes such as comfort, satisfaction, values of work as per labor productivity

(4) Introducing indexes related to waste (waste heat, exhaust gas, etc.) into the productivity formulas. Introducing indexes related to biostimulants, organic acids, biofertilizer, microorganisms and accumulation of inorganic ions in the cultivation beds.

(5) Collecting and organizing actual data of productivity indexes. Recording the any insects, anthropods, small animals found in the cultivation room, and developing the throughgoing countermeasures.

(6) Influence by or impact from the regional considerations in the world, seasons, technology levels, etc., onto the productivity indexes

(7) Development of a software to automatically calculate the productivity indexes

(8) Development of an optimal method for environmental control and production management for improving productivity

(9) Development of automated systems for phenotyping (measurement of plant traits) without destruction or touching

(10) Development of systems applying phenotyping and AI to improve productivity and plant breeding

(11) Productivity of fruit, root, and medical use crops if produced in PFALs

(12) Breeding of high productivity cultivars suitable for PFAL environment [for example, high CO_2 concentration and low photosynthetic photon flux density (PPFD)]

(13) Extending the definition, calculation methods, and software of PFAL to general horticulture applications

(14) Comparing PFAL productivity with greenhouse horticulture and open fields

(15) Developing integrated productivity indexes applicable to mushroom production, aquaculture, fermentation, or biomass utilization

(16) Organizing the relationship with the quality of life of neighbors, the industries in the regions, and energy and food supply autonomy

(17) Developing methods for a versatile expansion of the produce values

12.8 Supplemental explanation and notes

In order to help further understanding of meaning and significance of the various productivity issues described above, supplemental explanation and notes are presented below:

12.8.1 Purposes of keeping cultivation rooms highly airtight and thermally insulated (Kozai et al., 2019a):

(1) prevent the intrusion of midge flies, small animals, viruses, and pathogens through cracks

(2) prevent leakage to the outside of added CO_2 which is maintained at 1000 ppm level, higher than 400 ppm level outside

(3) prevent daily variations of indoor temperature, vapor pressure deficit, and barometric pressure

(4) save electrical use by air-conditioners when outside temperature is higher than inside

(5) prevent condensation on walls and floors by increasing the thermal insulation (floor thermal insulation necessary where average annual temperature is 25°C or lower) as well as minimizing thermal flow of room walls and floors.

On the other hand, the cultivation room should not be completely airtight to avoid accumulation of volatile organic compounds gases emitted from plants, walls, floors, and facilities, which may cause physiological damage to plants and humans. The number of air exchanges per hour of a culture room (hourly air exchange rate divided by the air volume of a culture room) needs to be kept at around 0.02 or higher.

Inside PFALs with LEDs and densely populated plants, the relative humidity tends to increase to around 90% (vapor pressure deficit tends to decrease to 0.5 kPa) at low PPFD, and nearly 100% under dark. In order to keep the relative humidity at around 75%, around 2/3rd of LEDs are turned on. Then, the room air is dehumidified by heat pump (air conditioner) to remove the heat energy generated by LEDs.

When ventilated, the relative humidity tends to be unstable as it is mainly affected by the difference in absolute humidity between inside and outside and the changeable ventilation rate. Besides, when CO_2 is released to the outside, the risk of pathogen/pest insect invasion increases.

12.8.2 Electric power and electric energy (electricity)

kW ($= kJ s^{-1}$) is a unit of electric power. Electric power (kW) multiplied by time (h) is electric energy or electricity in unit of kWh or kJ. One (1) kW equals $1 kJ s^{-1}$, and 1 kWh equals 3.6 MJ (mega-joule) or 3600 kJ. Electricity is measured using an electricity meter (watt meter) and electric power by an electric power meter. Electric power used shall be measured separately for lighting inside a cultivation room from air-conditioning and such other things as nutrient solution circulation, sterilizer·disinfection equipment.

12.8.3 USD/kg and kg kWh^{-1}

In most scientific papers and books, the International System of Units (SI units) is used, which consist of seven base units (s, m, kg, A, K, mol, and cd), derived units [e.g., W ($J s^{-1}$)], and prefixes [kilo (k), milli (m), micro (m), nano (n), etc.]. Since monetary unit (e.g., USD) is not an SI unit, superscript and subscript are not used with the unit. Instead, slant symbol is used (e.g., USD/kg). Derived unit (e.g., $J s^{-1}$) is often expressed as $J s^{-1}$ for convenience as it is difficult to use superscript.

12.8.4 mol and mass

"mol" is a unit for distributive existence such as molecules (CO_2, H_2O), atoms, and photons, and 1 mol is equal to Avogadro's number or 6.02×10^{23}. Space volume of gas and fluid may vary subject to temperature and/or pressure; however, what mol unit stands for does not change and hence the conversion of substance is often reviewed by the unit of mol.

In a dynamic system, the mass does not change whether the object is on the planet surface or in space. The basic unit for mass is kg (kilogram). On the other hand, weight stands for the power that an object receives at any given place on earth due to gravity and the unit is N (newton, $kg m s^{-2}$). In this chapter, the mass unit of kg is used, however as a general practice, is called weight.

12.8.5 Weight of produce for sale and of plant residue

Weight of plant cropped (harvested) is divided into weight of produce for sale and weight of plant residue. Weight of produce for sale (kg) is the weight of a portion of the plant that is salable stated by fresh weight. However, if a portion of produce for sale is dried or of any specific ingredient, weight may be adjusted and stated by applicable weight unit. Productivity improves if the ratio of a portion of produce for sale increases for any given resources applied.

12.8.6 Productivity and cost of production

The inverse number of electric energy productivity (kg kWh^{-1}) is the production cost (kWh kg^{-1}). When the production cost is multiplied by the unit cost of electric energy, the production cost is described in monetary terms [the unit cost (JPY/kWh) $\times$ kWh kg^{-1} = JPY/kg] (the monetary unit can be described in any currency such as US Dollar or Euro). Labor and cultivation systems are the same in terms of the relationship between productivity and production cost. The productivity index used industry wide, "*basic cost,*" is equal to the production cost per unit production yield. Productivity of the aggregate cultivation room differs depending on the size of production rooms, production size, system operation rate, etc. Productivity described above (basic cost) and electricity consumed, labor hours, cultivation area, etc., pertaining to the aggregate cultivation room provide various productivity numbers of the aggregate cultivation room. Production capacity stands for the maximum productivity.

12.8.7 Production cost and its breakdown

Table 12.3 shows the example of percentages and monetary terms of the various cost factors pertaining to the revenues of PFALs producing and shipping leaf lettuce in Japan (Ijichi, 2018).

TABLE 12.3 Percentage and monetary terms of the various cost factors pertaining to the revenues of PFALs producing and shipping leaf lettuce in Japan (Ijichi, 2018). Per crop with sales, general, and administration costs are partially included as production cost. Column A is for percentage over the revenues and Column B in monetary terms (USD). For our information, one US Dollar and one Euro were equal to about 110 JPY and 130 JPY, respectively, as of the end of 2019. Note that wage ($/hour) is around 2 times higher in USA and some Europiean countries than in Japan.

No.	Items	A (%)	B (USD)
1	Depreciation	23.1	0.20
2	Labor	22.4	0.19
3	Electricity	18.7	0.16
4	Transportation	5.7	0.045
5	Major consumables (materials)	5.9	0.054
6	Seeds	2.0	0.017
7	Water, land lease, property tax and others	13.1	0.11
	Subtotal (cost per plant)	90.8	0.776
	Profit	9.2	0.079
8	**Total**	**100.0**	**0.885**

12.8.8 Fixed cost versus. variable cost

Electricity consumption consists of fixed and variable, which is not equal to the basic charge and the volume charge but is the accounting matter of whether the electric energy applied is proportionate to the production volume. Labor consist of fixed operation time and variable operation time, of which the latter increases proportionately to the production volume. Cultivation area consists of fixed and variable as well.

12.8.9 Time span to compare productivity

In order to compare productivity among PFALs, the annual average productivity is to be used in principle. On the other hand, in order to compare productivity on a monthly or seasonal basis of a single PFAL, the average productivity of the subject period or alternatively one production cycle therein should be used.

Net photosynthetic rate (the margin between CO_2 absorbed due to photosynthesis and CO_2 exhausted due to respiration of the plant) responds instantly (within 10 min) to CO_2 concentration, PPFD, etc. Productivity of supplemental CO_2 can be measured within one to two hours, while fresh weight of the plant does not quickly respond to CO_2 concentration, PPFD, etc., causing several to dozens of hours delay. Therefore, the time unit to be applied to productivity measurement shall take into account the time lag of expected responses. Also, when there are plants at various growth stages, the time unit for productivity measurement needs special attention.

12.8.10 Productivity of cultivation system and housing

Initial investment in a PFAL is applied to housing, systems, and other equipment. Depreciation as part of production cost consists of that of housing and that of systems. Number of years for depreciation substantially differs for housing and systems. Cultivation space productivity mostly relates to system productivity and hence, although housing productivity may be defined in the same as system productivity is, as shapes and styles of housing widely vary, we refrain from proposing any measurement method and analysis.

12.8.11 Depreciation of light-emitting diode fixtures

LED fixtures account for a large portion of the systems as initial investment. One may apply not the statutory years for depreciation but the actual light period hours applied. As the manufacturers of LED fixtures provide product guarantee not for a number of years after purchase but for the light period hours and assuming the electric to light energy conversion efficiency declines over the number of light period hours, daily depreciation percentage is regarded as the daily light period hours divided by the guaranteed hours. This depreciation method helps to capture the different depreciation amount of six LED fixtures over $1 m^2$ cultivation area for light period of 16 versus. 19 h or the depreciation amount of five LED fixtures for 19 h. Time lapse degradation of the fixtures may also be taken into account.

12.8.12 Photosynthetic photon number efficacy from light-emitting diode and the light receiving rate of the plants (Kozai et al., 2016)

Number of photosynthetic photons (mol) emitted from LEDs divided by electric energy (J) applied thereto is called "photosynthetic photon number efficacy (μmol J^{-1})." As of 2019, LEDs with efficiencies of about 3.0 μmol J^{-1} or higher are available in the market.

Number of photosynthetic photons received by the plant leaves as organs for photosynthesis over the number of photosynthetic photons emitted by the light fixtures is called the light receiving ratio. After light is received by the plant, it may be absorbed, reflected, or it may penetrate the leaves. Improvement of photosynthetic photon number efficacy of the fixture and/or the light receiving ratio reduces the electric energy consumed without slowing down net photosynthetic rate and hence contributes to electric energy productivity. In case the photosynthetic photon flux (mol s^{-1}) is measurable, the productivity of photosynthetic photons can be estimated.

Productivity of photosynthetic photons is significantly affected by CO_2 concentration, temperature, and vapor pressure deficit of air, composition, and strength of nutrient solution, etc. Accordingly, these environmental factors need to be controlled to maximize the productivity of photosynthetic photons. Because, electricity (or photosynthetic photon) cost is much higher than the costs for controlling other environmental factors.

12.8.13 Reasons for the high efficiency of water use (Kozai, 2011; Kozai et al., 2020)

In the almost airtight and thermally well-insulated cultivation room, when the cooling system operates, water vapor transpires from the plant leaves and then condensates on the cooling plates of air-conditioners, which makes it possible to recover more than 95% of the water transpired as condensation water and to return it back to the nutrient solution tank for irrigation.

The majority of the rest of the 5% of transpired water vapor escapes to the outside through air gaps, air shower room doors, and ventilation fan openings to be used only in emergency. Increase in water mass in plants can be calculated as the difference in weight (mass) between water uptaken by plant roots and transpired water from leaves. The percentage of water retained in plants over irrigated water increases from less than 1% at the seedling stage to around 20% just before harvest with increasing plant fresh weight.

As a result, the net amount of water used for irrigation, i.e., water irrigated minus the sum of water recovered and water retained in plants, is c.a. 5%. Therefore, water use efficiency is defined as the portion of water absorbed and retained by the plant to the portion of water irrigated minus water recovered. In other words, the ratio of water absorbed and retained by the plant to the net amount of water irrigated is 0.95 kg kg^{-1}, virtually equal to the theoretical maximum of 1.0 kg/kg.

It may be noted that, in an open field or in a horticultural facility where no water transpired by the plants is recovered, the water use efficiency is 0.05 kg/kg or lower. It may also be noted that CO_2 applied in the highly airtight cultivation rooms is very efficiently utilized at around 85% efficiency and so is the fertilizer applied in the cultivation rooms of PFALs thanks to a higher use efficiency of the circulated nutrient solution. High use efficiency of the applied resources may not yet guarantee the lower production costs of water, fertilizer, and CO_2 compared to the electricity cost. However, the values for environmental protection and resource conservation will increase in the near future.

12.8.14 Coefficient of performance of air-conditioners (Kozai et al., 2019a)

Air-conditioners (one type of heat pump) are one of the facility equipment in the cultivation rooms. The electric energy efficiency for cooling (ratio of heat energy removed from the cultivation room to the electric energy consumed) is called coefficient of performance (COP). When the room temperature is 23°C and outside temperature is 32°C, COP is around 4. In general, the lower the outside temperature, the higher the COP is. For example, when the outside is 5°C, COP improves to 8. Higher COP of air-conditioners, while maintaining the room temperature constant, enables the electric energy consumed by air-conditioners and eventually of the entire cultivation rooms to decline and hence improve the electric energy productivity.

Highly airtight and highly thermally insulated cultivation rooms maintain nearly zero inflow or outflow of heat energy into and out of cultivation rooms in spite of the heat energy generated by the LED fixtures. The COP of air-conditioners cooling operation is the sum of the electric energy consumed by LED fixtures and other equipment divided by the electric energy consumed by air-conditioners. As the COP moves somewhere between four and eight, the electric energy consumed by lighting and by air-conditioning falls somewhere between 4 to 1 and eight to one.

12.8.15 Cooling load of cultivation room (Kozai, 2013)

In highly airtight and thermally insulated cultivation rooms, turning off all the LED fixtures increases the relative humidity inside to 100% in a short time due to transpiration which converts the sensible heat in the air to latent heat (water vapor). As a solution to avoid physiological abnormality of plants (tip-burn, succulent growth, etc.) and condensation on walls and floors, it is commonly implemented to divide the LED fixtures in the cultivation room to three groups and light them alternately, for example, two of them are on and the one is off. The purpose is to keep the air-conditioning always on, whose cooling operation of the heat from the fixtures simultaneously dehumidifies the room and maintains the relative humidity inside at around 85%.

The cooling load of this operation is approximately equal to the heat energy or the electrical energy consumed by the equipment other than air-conditioners including the LED fixtures. Plants fix the light energy into chemical energy no higher than at about 3% of the electric energy consumed. More than 85% of the heat generated is that of the LED fixtures[4].

On the contrary, when the outside temperature declines, the air-conditioners inside a poorly sealed and not well thermally insulated cultivation room turn off as the room temperature declines simultaneously despite the heat from the fixtures, which causes an increase in relative humidity. This proves that a highly airtight and highly insulated environment is better than purchasing and operating dehumidifiers.

12.8.16 Treatment cost of resource residue

PFAL intends to achieve the theoretical minimum (zero) residue of resources used thanks to RUE of 1 or at the theoretical maximum. Roots or damaged leaves of leafy greens may not be for sale and result in residue (waste). If the medium is not reused, it may also result in residue and the treatment cost. Water used for cleaning the cultivation panels and the floor, if not sterilized and reused, may also causes the treatment cost. External leakage of CO_2 (global warming gas) added to a cultivation room may not be treated and cause the social cost.

It shall be discussed here that the electric energy applied to light LED fixtures lumps is mostly converted into heat energy and whether this heat energy shall be regarded as residue and whether the cooling operation as residue treatment cost. The conclusion is that 95% or more of the electricity consumed by LEDs does not contribute to the plant photosynthesis (i.e., not converted into chemical energy) but is converted into heat energy. To evacuate this heat energy to the outside and to maintain a stable room temperature, air-conditioners are used, which means that electricity consumption by air-conditioners is a treatment cost caused by the unused energy (residue) of the lighting operation. This treatment cost, i.e., electricity consumption by air-conditioners is approximately equal to the electricity consumption for lighting divided by COP of the air-conditioners (four to eight). To be more precise, air fans and nutrient water circulation pumps also generate heat in proportion to their electricity consumption, and the treatment cost of this heat energy (residue) is generated. The discussion here applies to the cost in monetary terms as well. The cooling load of a cultivation room which is almost airtight and thermally well-insulated is equal to the heat energy generated by lights and other equipment such as air circulating fans and pump for nutrient solution circulation because heat energy exchange through walls and floor is negligibly small.

12.8.17 Energy conversion process (Kozai, 2011)

Discussed in more detail, 60% or higher of the electric energy applied to LED lights are converted into heat energy and thermal radiation energy at the LEDs. The remainder 40% is converted into photosynthetic radiation energy, of which 70% reaches plant leaves on average in the cultivation period. About 80% of the 70%, or 56%, is absorbed by leaves and the remainder 14% is reflected or passes through the leaves (Kozai, 2011). In total, 44% of the photosynthetic radiation (44% = 100%−56%) is absorbed by the materials other than the leaves (medium, cultivation panels, racks, ceiling, floor, walls, plant stems, etc.) and converted into heat energy. In reality, only about 10% of the photosynthetic radiation (wavelength of 400−700 nm) energy absorbed by leaves is fixed as chemical energy in the chlorophylls, although the theoretical maximum value is 30%[11]. This percentage varies greatly depending on environmental and plant ecophysiological conditions. Therefore, no more than 5.6% (= 70*80*10/10,000) of the photosynthetic radiation energy generated by LED lumps is converted into chemical energy, which means no more than 2.24% (= 5.6*4/10) is converted. In summary, improving RUE of the electricity and photosynthetic photon flux is vital to improving productivity of PFAL.

12.9 Closing remark

Discussion of indexes, definition, measurement method, sample numbers pertaining to productivity of PFALs and data collection therefor have started only recently. Considering that the number of PFALs is increasing worldwide, comparison of productivity and continuing improvement thereof requires developing an agreement among professionals involved on such issues as the productivity indexes, definitions, and measurement methods, as well as on how to collect and analyze the samples. This chapter hopefully contributes, as one of the first steps, to achieving the six goals proposed in the Introduction.

Acknowledgment

The authors would like to thank the following members of the Committee on Productivity Improvement of Plant Factory with Artificial Lighting who participated in the discussion: Katsunori Shimoshige, Nagamitsu Nozawa, Mariko Hayakumo, Shota Hirose, Yuko Fukui, Hiroshi Morisada, Kosuke Yamada, Hitoshi Wada, Toru Maruo, and Hiroshi Ijichi.

References

Ijichi, H., 2018. NAPA Research Chapter: Chapter 3 Plant Factory Business — Current Status and Perspective of Plant Factory Business, pp. 58—80 (in Japanese). https://www.nomuraholdings.com/jp/company/group/napa/data/20180219_03.pdf.

Kozai, T., 2011. Improving light energy utilization efficiency for a sustainable plant factory with artificial light. Proc. of Green Light. Shanghai Forum 375—383.

Kozai, T., 2013. Resource use efficiency of closed plant production system with artificial light: concept, estimation and application to plant factory. Proc. Jan. Acad. Ser. B 89 (10), 447—461.

Kozai, T., Fujiwara, K., Runkle, E. (Eds.), 2016. LED Lighting for Urban Agriculture. Springer, p. 454.

Kozai, T. (Ed.), 2018a. Smart Plant Factory: The Next Generation Indoor Vertical Farms. Springer, p. 456.

Kozai, T., 2018b. Benefits, problems and challenges of plant factories with artificial lighting (PFALs): a short review. Acta Hortic. 25—30.

Kozai, T., Niu, G., Takagaki, M. (Eds.), 2020. Plant Factory: An Indoor Vertical Farming System for Efficient Quality Food Production, second ed. Academic Press, p. 487.

Kozai, T., Amagai, Y., Hayashi, E., 2019a. Towards sustainable plant factories with artificial lighting (PFALs). In: Marcelis, L., Heuvelink, E. (Eds.), Achieving Sustainable Greenhouse Cultivation. Burleigh Dodds Science, pp. 177—202.

Kozai, T., Uraisami, K., Kai, T., Hayashi, E., 2019b. Indexes, definition, equation and comments on productivity of plant factory with artificial lighting (in Japanese). Nogyo oyobi Engei 94 (8), 661—672.

Seth, N., Barrado, C.M.D., Lalaguna, P.D., 2019. SDGs: Main Contributions and Challenges. United Nations Institute for Training and Research (UNITAR).

13

How to integrate and to optimize productivity

Kaz Uraisami

Marginal LLC, Tokyo, Japan

13.1 Introduction

In the past five years and probably in the next two years before this Book is published, plant factories with artificial lighting (PFALs) have seen and continue to see the various changes inside and outside the academic society and the markets. They include:

(i) PFAL produce is more daily or regular and less special on the family dining table.

(ii) PFAL penetrates into the most price sensitive use of green for precooked commercial operation.

(iii) Field grown green products fail to avoid a drastic fluctuation of the supply and the price to be caused by the harsh climate environment and therefore, loses the advantage to the PFAL green produce.

(iv) New types of the seeds are to be developed and to become available more and best suited for the PFAL production.

(v) The higher light intensity achieves the faster growth speed, which results in the improved space productivity if properly managed by the spacing and the environmental control.

(vi) The initial investment cost of PFALs becomes less without any revolutionary structure or technology thanks to the entry of the Chinese and other suppliers of the systems.

In 2019, NPO Japan Plant Factory Association ("JPFA") identifies and defines three major indexes among others in order to measure the status quo of the technology of PFALs as fully described in Chapter 12. We use the business administration sheet in Chapter 13 to simulate those indexes, which is under the joint copyright of JPFA and Asahi Techno Plant CO., LTD. The indexes for the respective resources in their original unit may not be added to or subtracted from one another, however, when converted into monetary terms, they start contributing to the integrated productivity in the business administration sheet.

Plant Factory Basics, Applications and Advances
https://doi.org/10.1016/B978-0-323-85152-7.00024-0

13.2 Operational efficiency, productivity, and profitability

JPFA organized the research committee for operational efficiency of Plant Factory in spring 2016, and 20 scientists and operation managers shared the academic and business views together on (1) operational efficiency, (2) productivity, and (3) profitability.

Definitions of the three terms are:

(1) Operational efficiency: the result of the amount of direct operational output divided by the direct operational input applied or consumed. An easy example is how many seeds are planted in the cultivation tray by the labor for one hour by one laborer.

You may use the operating efficiency as the index to manage your PFAL. You may compare the index variables of today with those of yesterday, this year with the last year or with five years ago and also, compare them with your peers inside and outside the region.

You learn how to keep it stable or to improve it over the number of years of operation by managing the workers as individuals as well as a team. Improvement may occur in such areas as in the floor plan and the process management to suit the floor plan or the other way around. Needless to say, the process management needs to be supported by the team operation of the workers and the details always matter such as the choice of floor materials or the shoes for the workers.

(2) Productivity: the result of the weight amount of produce divided by the resources applied or consumed in the PFAL.

In case it is denominated in the original unit of the respective resources, it is defined as the resource productivity and three major indexes are described in Section 13.1 Introduction:

  (i) per electric energy (kg/kWh),
  (ii) per man hour (kg/h), and
  (iii) per cultivation space and time (kg/m^2 d)
      When the resource amount is converted into the monetary terms, it is defined as the monetary productivity. The indexes pertaining to all the resources applied or consumed in the PFAL may be summed up and constitute the single index, i.e., the monetary productivity or the integrated productivity.

(3) Profitability: the result of the monetary amount of the produce divided by the monetary amount of all the resources applied or consumed.

The shipment price of the produce, the price per kilogram, which determines the monetary amount of the produce has a substantial influence on the profitability and is affected by many considerations including, but not limited to:

  (i) where the target market resides,
  (ii) whether the target market is for the consumers or for the commercial food or restraint industries,
  (iii) brand recognition of the PFAL produce at the subject market,
  (iv) how much premium the eventual consumers in the target market accept and afford in consideration of clean (hydroponic production), healthy (no synthetic pesticide), convenient ("no-need-to-wash" and longer life in the refrigerators) nature of the produce, and
  (v) an increasing number of the consumers are willing to pay for the contribution to sustainable development goals (the "SDGs") such as the "locally grown" appeals.

(4) Relationship between operational efficiency and productivity:

If all the laborers at the PFAL complete the work within 50% of the time they used to spend, operational efficiency improves 100%, i.e., doubles. However, they do so by the hasty rough handling of the plants and the yield of the produce becomes 50% less, the resource productivity of labor does not improve. In reality, the substantially lower yield causes derivative problems such as the increased cost of waste treatment, extra workload necessary to deal with the uneven or irregular produce, etc. You may also lose the confidence of the consumers and receive the downward pressure on the sales price.

One very tricky thing you should know is that, even when operational efficiency remains the same, productivity of the subject resource, e.g., the man hour labor, may improve or deteriorate. If operational efficiency of man hour labor remains the same but, e.g., the higher photosynthetic photon flux density (PPFD) is applied and the produce weight becomes 10% more without causing the additional workload, resource productivity of the labor improves by 10%. Therefore, you need to watch carefully if and how the shift of operational efficiency affects the other indexes and whether the improvement or the deterioration of resource productivity is active or passive.

In the next subsection, we review and learn how to use the "business administration sheet" and find out how the productivity indexes appear on the sheet.

13.3 Business administration sheet

JPFA holds the international seminar on PFALs every year at Kashiwa-no-ha Kashiwa, Chiba, Japan, since 2018 and in 2020 online. I am in charge of the class unit, "Business Administration on Plant Factory." In this subsection, I guide you to tour where and how the PFAL meets with the business matters. The sheet is under the joint copyright of JPFA and Asahi Techno CO., LTD. in Japan and please contact with Kaz Uraisami at kazuya.uraisami@ gkmarginal.co.jp for further information or for a soft copy.

The business administration sheet consists of the four excel sheets, (1) Elements, (2) Initial investment, (3) Budget and actuals for P/L (profit and loss statement) and Cash flow, and (4) Dashboard. Due to a limited size of book pages and also, for the sake of convenience of explanation, the sample pages are separated into the seven sheets and exhibited as Appendix toward the end of this Chapter as follows:

(1) Elements into the two tables, i.e., Appendix 13.1 Cultivation area elements and Appendix 13.2 Lighting elements, etc.,
(2) Appendix 13.3 Initial investment,
(3) Appendix 13.4 Budget and actuals for P/L and cash flow, Appendix 13.5 Standard operations P/L and cash flow, Appendix 13.6 budget five year operations, and Appendix 13.7 Budget five year P/L and cash flow,
(4) Appendix 13.7 Dashboard

The monetary numbers in the sheet are originally denominated in Japanese Yen and translated into US$ at the exchange rate of US$1 = JPY100.

In the element sheet, Appendix 13.1, we assume that the leafy greens are cultivated in the cultivation tray, 300 × 600 mm in size and made of plastic, and depending on the stage of growth, the spacing is adjusted as follows:

① 300 × 600 mm sponge sheet with 300 holes are used for germination + first stage seedling,
② 300 × 600 mm plastic grow tray with 28 holes for second stage seedling,
③ 300 × 600 mm plastic grow tray with 14 holes for post transplanting first stage cultivation, and
④ 300 × 600 mm plastic grow tray with seven holes for post transplanting second stage cultivation.

You may adjust the numbers, 300, 28, 14, and 7, for more or for less in the cultivation tray and give more or less space to each crop. You may also skip any of ② from ④ if you so decide. Yield stands for the ratio of the number of seeds or seedling moving to the next stage over the number of seeds or seedling at the beginning of the respective stage.

Each shelf in the rack is 1200 × 1200 mm large and accommodates eight of 300 × 600 mm sheets. Each rack then holds six shelves for vertical farming. By assuming how many days each of those four growth stages described above needs, you may calculate how many holes, how many trays are necessary for each step, and how may shelves, how many racks are necessary. You eventually come up with the size of the operation how many plants you grow and subject to the yield assumed, how many you crop in the PFAL.

If you plan to produce baby leaves or micro greens, you prefer each seed hole accepts more than one seed and a sheet shall be a platform for the number of seeds as you choose to plant. You may use the grow trays for (1) germination + first stage seedling and (2) second stage seedling. At the end of second stage seedling will grow c.a. 5 g and you may cut them for cropping and shipping. You may input null to the number column of the cultivation days for (3) transplanting first stage and (4) transplanting second stage and conclude cultivation at the end of second stage seedling. Those numbers of labor man hour for operations related to (3) and (4) are not taken into account.

You move to Appendix 13.2 for the purpose of determining the daily light integral, "DLI." DLI stands for the number of photosynthetically active photons delivered to a specific area over a 24-hour period. Set the PPFD level on the grow tray in the unit of $\mu mol/m^2/second$ ($\mu mol/m^2 s$) and the effective rate how much of them is applied inside the canopy (how much is not leaking to the outside). The sheet shows you how many LED fixtures you need over 1200 × 1200 mm cultivation area on the shelf. You need to assume (i) conversion factor, i.e., how many $\mu mol/s$ of radiation be generated per 1 W of lighting power, (ii) energy conversion efficiency (J/J), i.e., how many photosynthetic photon flux (PPF) is generated per electricity.

In the sheet, the conversion factor and the energy conversion efficiency are assumed to be 4.59 and 0.31, respectively, which, as multiplied, stands for photosynthetic photon efficacy of 1.42 μmol per J, probably an average number of LED fixtures available in the market five years ago and the number has substantially improved by today and further in the next two years. For the smaller purchase cost, the more advanced or the more efficient LED fixtures shall become available in the market and generate higher μmol per J.

For a comprehensive information and review of hydroponics at Plant Factory in Japan, please refer to Toyoki Kozai (2015), Jung Eek Son et al. (2015), Osamu Nunomura et al. (2015).

13.4 How to plan in the "business administration sheet"

(1) First assume the size of housing for the PFAL, number of cultivation trays inside, PPFD on 1 m^2 cultivation trays, the number of LED fixtures necessary, the initial investment in total (then per m^2 cultivation area), fresh weight kilogram per m^2 cultivation area, labor man seconds per each operation, sales price per kg fresh weight produce, etc.

A hypothetical case on the sheet assumes that the housing has a floor area of 1,000 m^2, 3.5 m height and is subject to 10 years lease and PFAL is constructed inside the housing. Then, in order to maximize the value of a PFAL inside the housing, we review and decide what size of facilities and systems fit inside and how many crops of leafy lettuce for how many kilograms are produced daily. All of these numbers fill out the element sheet.

(2) You then determine what you put together as a group of the materials, devices, equipment, etc., and calculate, on the initial investment sheet, how much the aggregate cost is in order to complete the construction of the PFAL. You now complete the initial investment sheet.

(3) After you complete the first two sheets, you work on the budget and actuals for P/L and cash flow. We simulate here how the operation goes in the next five years. As is the case with the actual operations, many elements or indicators need to be planned and assumed including the number of seconds for seeding, transplanting from seeding to germination, from germination to planting, and from planting to shipment, as well as the number of days for germination, seedling, after transplanting, etc.

(4) You simulate in the five-year sheet in the way you think the operation shall improve over the years and get familiar or comfortable with it as if you are virtually operating the PFAL. You are now ready for the case studies:
- Higher PPFD applied
- Longer light hours for the shorter period of cultivation
- Higher or lower yield
- Reflection ratio up or down.

(5) It must be noted that the business administration sheet may not accommodate the following issues regarding the operational efficiency and the productivity:
① Floor zoning for the respective operations in the zones
② Thermal insulation of the housing
③ Ventilation of the air inside the cultivation room
④ Materials for shoes, walls, and floors for the labor efficiency.

For example, spending an additional initial investment for the more air tight housing may improve ③ and the reduced ventilation frequency may reduce the cost for CO_2 supply. In the sample case in the sheet, cost of CO_2 is set at Jpy1 per stock, which is proper when the housing guarantees a low 0.3 ventilation frequency.

④ is extremely important and choosing the most expensive one is not always the best answer. There are the kinds of floor materials not very slippery even when wet and there are some plastic shoes less slippery when the floor is slippery. A round shape of joint of the walls and the floors reduces the time for daily cleaning, etc. "Kaizen" campaigns may help the operation just as much as they do in the Toyota factories.

13.5 Remarks regarding the sheet

Following are the frequently asked questions from those who use the sheet:

(1) Cost of housing: In the sheet, a PFAL housing is assumed to be rent or leased rather than owned and the monthly payment is set at one 120th of the cost of acquisition, i.e., for the 10% cap ratio. The sample sheet assumes that the acquisition cost for 1000 m^2 area housing is Japanese Yen 96 million, c.a., US$1 million and monthly rent is JPY 800,000.

(2) Definition of "target," "standard," and "expected":
   + Target is what is the "ideal" operation for PFALs to achieve.
   + Standard is what is "standard" operation for PFALs to achieve once the operation becomes stable.
   + Expected is what is practical in the first few years for PFALs to achieve.

(3) Consumables: In Appendix 13.4, you find the many cost items including one labeled "consumables," where they add up to $0.272 per crop or $375,192 in total. In Appendix 13.5, the total consumables cost is $154,872 or $.12 per crop. The "$0.272 per crop" cost includes the seeds, the seeding sponge, nutrient hydro, CO_2, packaging materials, transportation cost, waste treatment cost (roots, sponge), etc. The "$0.12 per crop" cost does not include the packaging materials and the transportation costs and transfer them to selling, general and administrative expenses ("SGA") for the accounting reason,

(4) Managers cost: For the sake of simplicity, SGA includes managers cost ($72,000). It is assumed that $72,000 is enough for one senior ($48,000) and one junior ($24,000) managers.

Your questions with respect to the details of the business administration sheet are always very welcome.

13.6 Resource productivity of photosynthetic photon flux density ($\mu mol/m^2$ s)

We acknowledge that:

① The higher PPFD per square meter (m^2) of cultivation tray requires the more LED fixtures to be set over the tray.

② The more LED fixtures are set in the system, the larger amount of the initial investment per m^2 of the cultivation tray becomes.

③ Therefore, the productivity pertaining to PPFD or the electricity consumption declines if the fresh weight produce does not increase in a proratable manner with the increase of the resource applied or consumed.

In fact, ③ is a wrong statement and if the monetary component ratio of the electricity resource is 20% and the consumption of that resource increases 100%, e.g., from 100 to 200 $\mu mol/m^2$ s, as long as the produce increases by 20% or the number of days necessary for cultivation decreases by 20%, the overall, i.e., integrated productivity remains unchanged. If the monetary component ratio of one resource is R% and the consumption of that resource increases X%, the produce needs to increase by the percentage of (X * R) in order to maintain the integrated productivity at the same level. When we increase the light intensity by 100%, from 100 to 200 $\mu mol/m^2$ s over the cultivation tray, and we reduce the number of days we need for cultivation but most often not by 100%. There is the optimal point on the growth curve in terms of PPFD versus. the increase of plant weight and that point is where the resource productivity is the best.

Let's assume that we grow the plants under PPFD of 100 $\mu mol/m^2$ s over the cultivation tray as the optimal lighting intensity and we need to allocate 8 days, 10 days, 10 days, and 10 days for the total of 38 days, respectively, for (i) germination + first stage seedling, (ii) second stage seedling, (iii) first stage transplanting, and (iv) second stage transplanting. If we increase the lighting intensity by 100%—200 $\mu mol/m^2$ s, the number of cultivation days for the respective stages become 50% less, the resource productivity remains the same and the integrated productivity becomes much improved.

Please refer to Tables 13.1 and 13.2 in the next page where the photo period is kept at 16 h for each case and naturally, the fresh weight growth per mol of PPF declines. Then please take one more look to find that, while the resource productivity of PPFD declines, the monetary productivity or the profitability actually improves. The electricity necessary for 1 kg of fresh weight production increases, but the cultivation area necessary for 1 kg decreases and so does the initial investment per 1 kg.

Why does this happen? Should we or should we not target at the optimal point for the respective resource productivity? We now move to the next subsection.

13.7 Optimization on the growth curve

(1) When we pursue the technologies, applications, knowhows, etc., in order to improve the growth curve, we acknowledge that the lighting intensity alone does not improve the growth curve unless accompanied by the proper environmental control such as vapor pressure deficit, density of CO_2, etc. We also recognize that the lighting intensity is not a simple number of $\mu mol/m^2$ s in that the spectrum, the length of photo period, etc., dictates how much PPF the respective kinds of plant absorb efficiently.

(2) The various plant physiological considerations described above improve the growth curve upward, and we shall manage not only the resource by resource productivity but also the overall or integrated productivity in the monetary terms. As we review in Section 13.6, when we optimize the resource productivity of PPFD ($\mu mol/m^2$ s), we do not optimize the integrated productivity in the PFAL.

Why do we fail to? It is actually quite simple or natural as you see on the next Table. The cost, as the denominator of the monetary productivity, is the sum of the all the resources applied or consumed and therefore, if the monetary component ratio of the electricity over all the resources is R% and the consumption of the electricity increases X%, when the produce increases by the percentage of X * R, the integrated productivity does not change, i.e., no improvement or no deterioration. It may not conform with the principle of Pareto analysis and for those resources with the larger R, improvement needs to be only proratably (i.e., R%) larger to improve the integrated productivity and for those with the smaller r, the improvement needs to demonstrate a proratable (again, R%) increase as well. In either case, the productivity of any single resource does not need to demonstrate the same yield improvement in order to improve the integrated productivity as highlighted in Figs. 13.1 and 13.2 below.

TABLE 13.1 Photosynthetic photon flux density versus productivity.

Sensitivity, Initial Investment & Business Planning

Major Productivity Index

Items	Index	(Unit)	$100\,mol\ m^{-2}\ d^{-1}$ Photo Period 16 hours/day	$200\,mol\ m^{-2}\ d^{-1}$ Photo Period 16 hours/day (cultivation days 50% reduced)	$200\,mol\ m^{-2}\ d^{-1}$ Photo Period 16 hours/day (cultivation days 1/3 reduced)	$200\,\mu mol\ m^{-2}\ d^{-1}$ 16 hours/day (cultivation days 1/3 reduced) (yield lowered)
Cultivation Area per m^2	Initial Investment	US$/$m^2$	1,019	1,246	1,246	1,246
	LED fixtures	US$/$m^2$	310	537	537	537
	Other materials	US$/$m^2$	709	709	709	709
per one daily crop	Initial Investment (80g per head)	US$/80g	613	379	528	545
Cultivation Area per m^2 per day	Fresh Weight produced	kg/m^2/d	0.148	0.293	0.210	0.204
	Daily Light Integral (DLI)	$mol\ m^{-2}\ d^{-1}$	5.76	11.52	11.52	11.52
	Cost for DLI	US$/$m^2$/d	438.0	814.0	814.0	814.0
	Electricity consumed by LED fixtures	kWh/m^2/d	1.53	2.84	2.84	2.84
	Cost for Electricity consumed	cents/m^2/d	23.5	43.9	43.9	43.9
	Economic depreciation of LED fixtures	cents/m^2/d	20.3	37.6	37.6	37.6
	Cost for labour	cents/m^2/d	23.3	42.4	31.5	31.2
per mol PPF	Fresh Weight produced	kg/mol	0.026	0.025	0.018	0.018

Yield	80g per head or larger	%	70	70	70	65
	2 heads in 1 package 80g or larger	%	20	20	20	25
Number of Cultivation Days	germination + 1st stage Seedling	days	8	6	6	6
	2nd stage Seedling	days	10	5	7	7
	1st stage Transplanting	days	10	5	7	7
	2nd stage Transplanting	days	10	5	7	7
Annual Revenue		US$ 000	1,516	3,002	2,153	2,086
Gross Profit		US$ 000	450	1,392	648	654
Operating Income		US$ 000	158	884	263	279
EBITDA		US$ 000	452	1,178	628	573
Payback Period		years	7.8	4.4	6.99	7.59
Items	Index	(Unit)	$100\,mol\ m^{-2}\ d^{-1}$ Photo Period 16 hours/day	$200\,mol\ m^{-2}\ d^{-1}$ Photo Period 16 hours/day (cultivation days 50% reduced)	$200\,mol\ m^{-2}\ d^{-1}$ Photo Period 16 hours/day (cultivation days 1/3 reduced)	$200\,\mu mol\ m^{-2}\ d^{-1}$ 16 hours/day (cultivation days 1/3 reduced) (yield lowered)
Initial Investment	per kg daily production	US$/kg	6,816	4,209	5,869	6,058
	per daily crop head (80g)	US$/80g	613	379	528	545
Cultivation Area	Area necessary for 1kg daily production	m²/kg	6.75	3.41	4.75	4.91

Continued

TABLE 13.1 Photosynthetic photon flux density versus productivity.—cont'd

	Daily stock crop production per m² area	crop/m²	1.65	3.26	2.34	2.27
Electricity	Electricity necessary for one stock crop	kWh/crop	1.29	1.21	1.68	1.74
	Lighting & water pumps only	kWh/crop	0.93	0.87	1.21	1.25
	Electricity necessary for 1kg fresh weight	kWh/kg	14.37	13.42	18.70	19.30
	Lighting & water pumps only	kWh/kg	10.34	9.65	13.45	13.88
	Fresh weight per 1kWh electricity consumed	kg/kWh	0.07	0.07	0.05	0.05
	Lighting & water pumps only	kg/kWh	0.10	0.10	0.07	0.07
	Lighting PPF consumed per kg fresh weight	mol/kg	45.73	46.19	64.41	66.48
	Fresh weight per mol Light PPF	kg/mol	0.022	0.022	0.016	0.015
Labour	Man Hour for one crop head	MH/crop	0.014	0.013	0.013	0.014
	MS for man second	MS/crop	50	46	48	49
	Man Hour for 1kg fresh weight	MH/kg	0.154	0.142	0.147	0.150
	MS for man second	MS/kg	554	510	528	540
	Fresh weight per MH	kg/MH	6.49	7.05	6.82	6.66

TABLE 13.2 Depreciation of LED fixtures cost.

LED operations assumed		
LED fixture guaranteed hours	35,000	
LED fixture replacement hours	20,000	
Light hours per day	16	
LED fixtures months before replacement	42	
Economic cost of LED fixtures	(Unit)	Standard operation
One LED fixture for one hour light period	Cent/LED/h	0.43
per kWh electricity consumed by LED fixtures	Cents/kWh	13.28
per mol/m^2 h	Cents/(mol/m^2 h)	2.62
per kg fresh weight	Cents/kg	121.39
per crop head	Cents/crop	10.93

To conclude this subsection, it shall be noted that the improvement is not a resource by resource consideration. When PPFD applied increases from 100 to 200 μmol/m^2 s, the CO_2 density ideal to expedite the photosynthesis corresponding to the new increased lighting intensity changes and needs to increase accordingly. A package of corelated resource factors shall be analyzed as if it is the single resource or the group of resources for the productivity improvement. For PPFD, it includes the CO_2 density and may also does the air ventilation of the cultivation room. The initial investment in the housing interior finishing determines the air ventilation, which then determines the outside leakage of the CO_2 applied, which then the inside density, and eventually the speed of photosynthesis. We shall recognize clearly this sort of the chain reaction grasp the group of corelated resources. Then we may measure the integrated productivity by adding those resources not related to or independent of the subject resource productivity and apply the aggregate monetary amount of the resources as the denominator for the integrated productivity.

13.8 Initial investment as the denominator

As described in the abstract, JPFA proposes three major indexes:

(i) per electric energy (kg/kWh),
(ii) per man hour (kg/h), and
(iii) per cultivation space and time (kg/m^2 d).

These three indexes are not mutually exclusive and collectively exhaustive ("MECE") and of the three, the productivity per cultivation space and time (kg/m^2 d) shall provide the single conclusive view to measure the integrated productivity. The substantial component

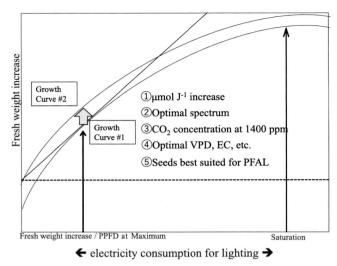

FIGURE 13.1 Growth curve: Fresh weight increase over electricity consumption for lighting.

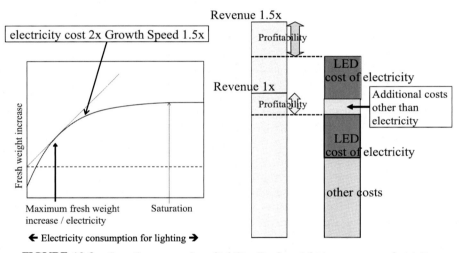

FIGURE 13.2 Growth curve and profitability: Fresh weight increase over electricity.

of this index derives from the amount of depreciation and amortization of the initial investment including LED fixtures.

To monitor the level of lighting intensity of the LED fixtures over the guaranteed period, you may apply the index of (i) per electric energy (kg/kWh) in the different way as follows:

① Convert the energy consumption into the monetary terms, i.e., one kilogram of the fresh weight production over the dollar amount of the electricity consumed per hour (kg/US dollar).

② Calculate the US dollar amount of depreciation of the LED fixtures per hour based on the number of the guaranteed hours.

③ By adding ① and ②, calculate how many kilograms of the fresh weight production is achieved per US$ 1 of the lighting resources applied or, in the other way around, how much resource is consumed per fresh weight produce for the lighting provided.

This approach controls the LED fixtures cost very effectively when you consider increasing or decreasing the photo period per day. You may manage the cost for lighting as the overall productivity.

13.9 Spacing on the tray and productivity

One of the most challenging goals for the environmental control in the PFALs is to unevenly apply the light from the LED fixtures onto the cultivation tray and onto the leaves. You may consider increasing the intensity, increasing the ratio of green light in the spectrum so that the lower leaves get as much lighting as the upper leaves, increasing the reflection ratio of the cultivation space so that the lighting to the corners of the cultivation tray does not leak out but rather reflected back to the corners where the intensity is inevitably weaker than the center of the cultivation tray.

During the earlier stage of plant growth when each seed or seedling has a plenty of space in-between one another, PPFD on the leaf aera remain close to the theoretical number of PPFD on the cultivation tray and as the leaf area index ("LAI"), the total size of the leaves over the size of the cultivation tray, may become larger and beyond 3 to 3.5 times due to over-lapping of the leaves over one another, the lower leaves may not receive the lighting as planned.

Photosynthetic efficacy of respective spectrum is not the major topic for this Chapter, however you may note that Mr. Qingwu (William) Meng reports on July 27, 2017 on URBANAG News that spectrum combination is still a floating target half a century after Dr. Keith McCree created a classic photosynthesis curve (McCree, 1971). Spectrum of two red and one blue combination is now exposed to a challenge that not a leaf by leaf measure-ment but a grow canopy wide measurement is necessary in order to optimize the spectrum for cultivation. Group of researchers at Utah State University (Snowden et al., 2016) finds that a light spectrum with up to 30% green light is generally as good as red and blue light for plant biomass gain and while the upper leaves of a plant absorb most red and blue light, they trans-mit more green light to lower leaves for photosynthesis.

Spacing of the holes for each plant is the key to optimize the productivity in the cultivation tray. As Table 13.3 highlights, you shall plan the fresh weight at the time of cropping. A leaf is two-dimensional and the size of the leaf grows exponentially until the growth saturates as the matter of plant physiology or until the leaves overlap with one another reaching the higher than optimal LAI.

In the Table, the second transplant with 100 plants on the cultivation tray is better if you crop them at the fresh weight of 80 g per crop and the second transplant with 50 plants excels it at the fresh weight of 100 g and grow faster toward 150 g per crop. The second transplant with 100 plants may not grow the plants to reach 150 g per crop as LAI becomes excessive.

TABLE 13.3 Spacing and fresh weight increase.

Spacing small	Seedling 1667 plants/m^2	Transplant #1 333 plants/m^2		Transplant #2 100 plants/m^2	
Cumulative number of days	10	21	42	49	n.a.
Weight of produce (g)	3	28	80	100	n.a.
Spacing large	Seedling 1667 plants/m^2	Transplant #1 333 plants/m^2		Transplant #2 50 plants/m^2	
Cumulative number of days	10	21	34	35	37
Weight of produce (g)	3	28	80	100	150

Spacing small		Weight of produce (g)		
	80	100		
Size occupied (m^2) g/m^2	0.25	0.32		
	321.24	313.45		
Spacing large		Weight of produce (g)		
	80	100		150
Size occupied (m^2) g/m^2	0.30	0.32		0.36
	267.53	313.45		417.79

This spacing consideration confirms that it is not the number of days between seeding and cropping, but the cumulative size of the area occupied by one plant between seeding and cropping that dictates the productivity. You find in Table 13.1 PPFD versus Productivity that one more day in the second transplant stage is worth two more days in the first transplant stage if the number of plants in the second is half of the first stage. Spacing is the key and I welcome any proposals to measure and adjust the spacing to the optimal setting by applying the machine learning or the deep learning technologies.

13.10 Man hour productivity and profitability

Another important productivity index is the "man hour," the fresh weight production over the man hour and as you see in Table 13.4 in the next page, the man hour index is of the cost or cost reduction nature. This is because the total capacity of the PFAL is limited upon construction completion by the size of the housing and the systems and the improvement of the man hour index does not contribute to a substantial acceleration of the growth speed nor to a substantial increase of the fresh weight upon cropping.

However, it shall be noted that, inside the index of the man hour from the earlier stage through the later stage, the quality of the work really matters on the works toward cropping and packing. A poor work at the seeding leads to an uneven growth of the plant and hence

TABLE 13.4 Man hour versus productivity.

Items	Index	(Unit)	Sensitivity, Initial Investment & Business Planning			
			Major Productivity Index			
			10 seconds per crop (seeding, transplanting x 3)	6 seconds per crop (seeding, transplanting x 3)	6 seconds per crop (seeding, transplanting x 3) yield lowered	Labour time total 50% reduced
Cultivation Area per m²	Initial Investment	US$ m⁻²	1,019	1,019	1,019	1,019
	LED fixtures	US$ m⁻²	310	310	310	310
	Other materials	US$ m⁻²	709	709	709	709
per one daily crop	Initial Investment (80g crop)	US$/80g	610	610	633	610
Cultivation Area per m² per day	Fresh Weight produced	kg m⁻² d⁻¹	0.148	0.148	0.144	0.148
	Daily Light Integral (DLI)	mol m⁻² d⁻¹	5.76	5.76	5.76	5.76
	Cost for DLI	US$ m⁻² d⁻¹	438.0	438.0	438.0	438.0
	Electricity consumed by LED fixtures	kWh m⁻² d⁻¹	1.53	1.53	1.53	1.53
	Cost for Electricity consumed	cents m⁻² d⁻¹	23.5	23.5	23.5	23.5
	Economic depreciation of LED fixtures	cents m⁻² d⁻¹	20.3	20.3	20.3	20.3
	Cost for labour	cents m⁻² d⁻¹	23.3	20.9	20.8	14.6

Continued

TABLE 13.4 Man hour versus productivity.—cont'd

per mol PPF	Fresh Weight produced	kg mol-1	0.026	0.026	0.025	0.026
Yield	crop 80g or larger	%	70	70	65	70
	2 crops in 1 package 80g or larger	%	20	20	25	20
Number of Cultivation Days	germination + 1st stage Seedling	days	8	8	8	8
	2nd stage Seedling	days	10	10	10	10
	1st stage Trans-planting	days	10	10	10	10
	2nd stage Trans-planting	days	10	10	10	10
Annual Revenue		US$ 000	1,516	1,516	1,469	1,516
Gross Profit		US$ 000	450	472	427	516
Operating Income		US$ 000	158	180	141	224
EBITDA		US$ 000	452	474	436	518
Payback Period		years	7.8	7.4	7.98	6.68

increases the necessary man hour for cropping, trimming, packing, etc. Reception by the customers depends on the quality of the produce and has a direct impact on the profitability through the sales quantity, the sales price, etc.

While the automation of the operation by the laborers is not the topic of this Chapter, not the saving of the cost or the total cost of ownership of the automation devices but rather the quality of the work the automation guarantees and the resulting quality of the fresh produce shall come first.

13.11 Additional remarks on productivity

As the growing number of PFALs comes into operation in the growing number of regions and countries, the comparative analysis of the respective operating sites shall be of more importance.

(1) Electricity: low in US, China, and Asian countries while high in Japan
(2) Labor or man hour: low in Japan (probably to your surprise), China, and Asian countries while high in US
(3) Water: low in Japan while high in China and the highest in UAE
(4) Food mileage and transportation cost: high per distance in Japan and high due to the longer distance in US

Also, the comparative analysis among the kinds of produce shall contribute to increasing the values of the PFALs.

Plant factory manager shall focus on the integrated productivity and on the overall profitability and have a strong fine-tuned sense to distinguish the results from the causes by knowing how they corelate with one another. PFALs shall benefit from the science of the plant physiology as well as from the science of the management. They set the priority on solving the issues and the problems every day, every month, and every year by applying the Pareto analysis approach.

13.12 How to raise and to secure the fund?

This subsection, the last but the conclusion, briefly discusses how to complete the financial arrangement for the PFALs. The business administration sheet calculates the payback period, i.e., how many years are necessary to recover the investment back. You convince the financiers and yourself, of course, not just talking or convincing but a good planning and management, that you have the enough resources of cash on hand, when financed as proposed, to apply to the engineering, procurement, and construction, and for the working capital in a start-up period of operations before the cash flow turns positive.

The three key indexes and others in the business administration sheet may get better or worse than planned from time to time. Common business wisdom is that:

(1) Set "standard" numbers, not too good or not too bad, for the indexes in the business administration sheet but start with the very practical numbers in the earlier years.
(2) Implement the sensitivity analysis to forecast how much better or worse the numbers fluctuate over the period at 90% probability in the worst case.

(3) Propose the terms and conditions of the financing arrangement to distribute the risk and return to the fund providers for their reasonable satisfaction.

The bankers know the sensitivity analysis very well and ask you many questions. Whether and how you may reduce the number of days for cultivation, how you may reduce the lighting intensity, PPF consumed per 1 kg fresh weight, or the man hour for 1 kg fresh weight, etc. You need to educate them with the peer comparison, with the budget and actuals history, and convince them that you may service the debt in the worst-case scenario.

You agree that the research by JPFA to define the three major and other indexes and to obtain the actual numbers from many PFALs in Japan and in other countries shall bring out the scientific achievement to further improve the PFALs in the world and also the very practical benefits to those who plan the new PFALs and to those who operate and expand the existing ones.

13.13 Conclusion

Operational efficiency, productivity, and profitability are defined and streamlined in the business planning sheet. If the monetary component ratio of one resource is R% and the consumption of that resource increases X%, the produce needs to increase not by the percentage of x by the percentage of (X * R) in order to maintain the integrated productivity at the same level. By converting the resource productivity to the monetary, integrated productivity, you shall optimize the overall efficiency, productivity, and profitability of the PFAL that you manage.

Simulation in the business administration sheet helps you to pursue optimization of the integrated productivity before you make a small mistake or after you find a small one.

Appendix 13.1: Cultivation area elements

Groups	Items		(unit)	Numbers
Cultivation	Seeding per day		Grain	6,000
	Yield: germination/seeding		Stock	90%
	Planting per day		Stock	5,400
	Yield: stocks shipped/stocks picked		Stock	90%
	Shipping per day		Stock	4,860
	Days for Cultivation	Germination + first stage seedling	Day	8
		Second stage seedling	Day	10
		Transplanting first stage	Day	8
		Transplanting second stage	Day	8

Groups	Items		(unit)	Numbers
Tray	Holes for cultivation	Germination + first stage seedling	Hole	48,000
		Second stage Seedling	Hole	56,921
		Transplanting first stage	Hole	43,200
		Transplanting second stage	Hole	43,200
	Size of cultivation tray	Length	mm	600
		Width	mm	300
	Holes per tray	Transplanting second stage	Hole	7
	Cultivation trays	Transplanting first stage	Sheet	3,086
		Transplanting second stage	Sheet	6,172
Cultivation rack	Size of cultivation racks	Length	mm	1,200
		Width	mm	1,260
		Height	mm	2,900
		Size	m^2	1.51
	Trays per shelf in racks	Germination + first stage seedling	Sheet	8
		Second stage seedling	Sheet	8
		Transplanting first stage	Sheet	8
		Transplanting second stage	Sheet	8
	Cultivation shelves	Germination + first stage seedling	Shelf	20
		Second stage seedling	Shelf	255
		Transplanting first stage	Shelf	386
		Transplanting second stage	Shelf	772
		Subtotal	Shelf	1,433
	Shelves per cultivation racks	Germination + first stage seedling	Shelf	6
		Second stage seedling	Shelf	6
		Transplanting first stage	Shelf	6
		Transplanting second stage	Shelf	6

(*Continued*)

Groups	Items		(unit)	Numbers
	Cultivation racks	Germination + first stage seedling	Rack	4
		Second stage seedling	Rack	43
		Transplanting first stage	Rack	65
		Transplanting second stage	Rack	129
		Subtotal	Rack	241
	Space b/w racks	Length	mm	500
		Width	mm	750
	Total bottom area of cultivation racks		m^2	364
	Total area of cultivation shelves		m^2	823
Room area	Picking operation, material storage, refrigerator		m^2	177
	Total size		m^2	1,000

Appendix 13.2: Lighting and other elements

Groups	Items	(unit)	Numbers
Lighting	Photosynthetic photon flux density on tray	µmol/m^2 s	100
	Effective ratio	—	0.85
	Radiation	µmol/s	244,969
	Conversion factor	—	4.59
	Lighting power	W	53,370
	Energy conversion efficiency	—	0.31
	Watt per LED fixture	W	32
	LED fixture per cultivation tray	—	0.47
	LED fixture per cultivation shelf	—	4
	LED fixture total	—	5,784
	Photoperiod	Hours/day	16
Air conditioning	COP (summer time)	—	2.25
	COP (other)	—	3.50
	Weighted average	—	2.56

Groups	Items	(unit)	Numbers
Electricity contracted	Capacity (x1. 1 with allowance)	kW	196.1
Sales	Price per kilogram fresh weight	US$/kg	12
Housing	Monthly rent (one 120th of housing price)	US$	8,000

COP (coefficient of performance) stands for the ratio between energy usage of the compressor and the amount of useful heat extracted from the condenser. Higher COP value2[2] represents higher efficiency.

Appendix 13.3: Initial investment

	Items	(Unit: US$)			
		Unit prices	Pieces	Total	Dprc years
Housing	Interior renovation (water sink included)	per 100 m^2	15,000	10	150,000	50
	Electricity works	per 100 m^2	3,500	10	35,000	15
	Air conditioning	per 100 m^2	6,000	10	60,000	7
	Drain water reuse system			0	7
	CO_2 gas systems	per 100 m^2	1,800	10	18,000	7
Rack	Cultivation racks	per 1.2 × 1.2 m	500	241	120,500	7
	Multipurpose racks	per 100 m^2	1,000	60	60,000	7
	Wind fans for racks	per 100 m^2	200	400	80,000	7
	Water supply and drain systems	per 100 m^2	7,000	10	70,000	7
Lighting	LED	per fixture	85	5,784	491,640	7
	Circuit board	per rack	160	241	38,560	7
	Other materials	per 6 racks	990	40	39,798	7
	Mounting frames for LED	per rack	70	1,446	101,220	7
	Reflection materials	per rack	25	1,446	36,150	7
	Control panel (incl. Dimming device)	per 6 racks	60	40	2,412	7
	Dimming circuit board	per rack	160	241	38,560	7
Cultivation	Cultivation tray	per piece	25	12,000	300,000	7
	Cultivation pool	per pool	100	2,866	286,600	7
	Water pump	per rack	500	241	120,500	7

(Continued)

	Items		(Unit: US$) Unit prices	Pieces	Total	Dprc years
Facilities	Sensors	per set	50,000	1	50,000	7
	Air shower room	per room	20,000	1	20,000	7
	Refrigerating room	per room	10,000	1	10,000	7
	Cropping and shipping board	per set	2,000	1	2,000	7
	Processing board	per board	500	50	25,000	7
	Container for cropping	per rack	10	241	2,410	7
	Packaging machines	per machine	5,000	3	15,000	7
	Storage racks for trays	per set	5,000	2	10,000	7
	Storage racks for materials	per set	5,000	2	10,000	7
	Desks, PCs, lockers	per set	15,000	1	15,000	7
	Total				2,208,350	

2,208,350		Fixed rate	1st year	2 nd year	3rd year	4th year	5th year
50 year depreciation assets straight line	150,000	2.0%	3,000	3,000	3,000	3,000	3,000
15 year depreciation assets straight line	35,000	13.3%	2,333	2,333	2,333	2,333	2,333
7 year depreciation assets straight line	2,023,350	28.4%	289,050	289,050	289,050	289,050	289,050
Total	2,208,350	0	294,383	294,383	294,383	294,383	294,383

Unit cost assumed here may be more than you expect in your home markets, which may decline in the next few years thanks to the development of LED fixtures and other vertical firming systems for less cost and more values.

Appendix 13.4: Standard operations

	Business plan		Standard operation
Production/ sales	Seeding per day	Seed	5,000
	Yield: Germination seeding	%	90.0%
	Transplanting per day	Plant	4,500
	Yield: Stocks shipped as one in one stocks cropped	%	70%
	Yield: Stocks shipped as two in one/stocks cropped	%	20%

	Business plan			Standard operation
	Yield: Stocks not shipped stocks cropped		%	10%
	Weighted yield: Stocks shipped/stocks cropped		%	80%
	Stock packs shipped		Plant	3,600
	Cultivation shelves total—cultivation shelves in operation			12
	Fresh weight per pack		G	90
	Fresh weight shipped per day		Kg	324
	Price per kilo grams		US$	12.0
	Annual sales		US$	1,399,680
	Total number of days in operation		Day	360
Cultivation systems	Number of days necessary	Germination + first stage seedling	Day	8
		Second stage seedling	Day	10
		Transplanting first stage	Day	10
		Transplanting second stage	Day	10
	Number of planting holes necessary	Germination + first stage seedling	Hole	300
		Second stage seedling	Hole	28
		Transplanting first stage	Hole	14
		Transplanting second stage	Hole	7
	Number of cultivation trays necessary	Germination + first stage seedling	Sheet	133
		Second stage seedling	Sheet	1,694
		Transplanting first stage	Sheet	3,214

(Continued)

		Business plan					Standard operation
	Number of cultivation shelves	Transplanting second stage	Sheet				6,429
		Germination + first stage seedling	Shelf				17
		Second stage seedling	Shelf				212
		Transplanting first stage	Shelf				402
		Transplanting second stage	Shelf				804
	Total number of shelves necessary		Shelf				1,434
	Total number of shelves available in plant factory		Shelf				1,446
Lighting	PPFD on trays		$\mu mol/m^2\,s$				100
	Photoperiod		Hours				16
	COP		Numbers				2.56
	Electricity for water pumps and fans		kW				20
	Monthly electricity consumption		kWh	Contract capacity	kWh		139,671
	Annual electricity cost	Capacity payment (US$)	14.45		196		33,996
		Metered payment (US$)	Summer price	0.14	US$/ kWh		222,697
			COP	2.25	Pumps and fans		
			Nonsummer price	0.13	US$/ kWh		
			COP	3.50	Pumps and fans		
Labor	Germination + first stage seedling		Second plant				1
	Second stage seedling		Second plant				3
	Transplanting first stage		Second plant				3
	Transplanting second stage		Second plant				3
	Cropping		Second plant				13

Business plan			Standard operation
Shipping (packing, storing in freezing room)		Second plant	13
Cleaning		Hours per day	8
Total		Total hours	50
Hourly rate		US$	8.50
Number of part-time staff (6 h day)		Headcount	9
Labor cost total		US$	165,240
Welfare and commuting	versus hourly rate	20%	33,048
Labor cost grand total			198,288
Consumables Seeds		Cents/plant	1.5
Seeding sponge		Cents/plant	1.0
Nutrient hydro		Cents/plant	1.0
CO_2		Cents/plant	1.0
Packaging materials		Cents/plant	7.0
Transportation		Cents/plant	10.0
Waste cost (roots, sponge)		Cents/plant	0.7
Others		Cents/plant	5.0
Total		US$/plant	375,192

Appendix 13.5: Standard operations P/L and cash flow

P/L and cash flow			Standard operation
(Unit: US$)			
Gross sales			1,399,680
Fixed cost	Depreciation	Fixed amount	3,000
		Fixed amount	291,383
Maintenance and renovation		3%	66,251
Building monthly rent (US$)		8000 monthly	96,000
Electricity and water (capacity)		Monthly	33,996

(Continued)

P/L and cash flow			Standard operation
	(Unit: US$)		
Subtotal			490,630
Variable cost	Electricity and water (capacity)		222,697
Labor cost			198,288
Consumables			154,872
Subtotal			575,857
Cost over sales	Electricity and water (capacity)		18%
Labor cost			14%
Consumables			11%
Cost per stock	Electricity and water (capacity)	Cents per plant	20
Labor cost	Cents per plant		15
Consumables	Cents per plant		12
Gross profit			333,193
SGA			292,320
Operating profit			40,873
EBITDA			335,256
Cash on hands beginning of year	Cash funded for the project	3,000,000	0
Cash on hands end of year			335,256

EBITDA, Earnings Before Interest, Taxes, Depreciation, and Amortization; *P/L*, profit and loss statement; *SGA*, selling and general administrative expenses.

Appendix 13.6: Budget five years operations

	Business plan		Year 1	Year 3	Year 5
			Plan	Plan	Plan
Production/ sales	Seeding per day	Seed	4,950	4,850	5,000
	Yield: Germination/ seeding	%	85.0%	90.0%	90.0%
	Transplanting per day	Plant	4,208	4,365	4,500

	Business plan		Year 1	Year 3	Year 5
			Plan	Plan	Plan
	Yield: Stocks shipped as one in one/stocks cropped	%	50%	65%	70%
	Yield: Stocks shipped as two in one/stocks cropped	%	25%	20%	20%
	Yield: Stocks not shipped/stocks cropped	%	25%	15%	10%
	Weighted yield: Stocks shipped/ stocks cropped	%	63%	75%	80%
	Stock packs shipped	Plant	2630	3,274	3,600
	Cultivation shelves total—cultivation shelves in operation		17	14	12
	Fresh weight per pack	G	80	80	90
	Fresh weight shipped per day	Kg	210.4	261.9	324.0
	Price per kilo grams	US$	12.0	13.0	13.0
	Annual sales	US$	908,820	1,225,692	1,516,320
	Total number of days in operation	Day	360	360	360
Cultivation systems	Number of days necessary	Germination + first stage seedling	Day	8	8	8
		Second stage seedling	Day	14	12	10
		Transplanting first stage	Day	10	10	10
		Transplanting second stage	Day	10	10	10
	Number of planting holes necessary	Germination + first stage seedling	Hole	300	300	300
		Second stage seedling	Hole	28	28	28

(*Continued*)

	Business plan			Year 1 Plan	Year 3 Plan	Year 5 Plan
	Transplanting first stage	Hole		14	14	14
	Transplanting second stage	Hole		7	7	7
Number of cultivation trays necessary	Germination + first stage seedling	Sheet		132	129	133
	Second stage seedling	Sheet		2,282	1,972	1,694
	Transplanting first stage	Sheet		3,005	3,118	3,214
	Transplanting second stage	Sheet		6,011	6,236	6,429
Number of cultivation shelves necessary	Germination + first stage seedling	Shelf		17	16	17
	Second stage seedling	Shelf		285	247	212
	Transplanting first stage	Shelf		376	390	402
	Transplanting second stage	Shelf		751	780	804
Total number of shelves necessary	Shelf			1,429	1432	1,434
Total number of shelves available in plant factory	Shelf			1,446	1,446	1,446
Lighting	PPFD on trays	$\mu mol/m^2\,s$		100	100	100
	Photoperiod	Hours		16	16	16
	COP	Numbers		2.56	2.56	2.56
	Electricity for water pumps and fans	Kw		20	20	20
	Monthly electricity consumption	kWh	Contract kWh	139,245	139,512	139,679
	Annual electricity cost	Capacity payment(US$)	14.45	capacity 196	33,996	33,996	33,996
			0.14	222,019	222,444	222,710

	Business plan				Year 1	Year 3	Year 5
					Plan	Plan	Plan
	Metered payment (US$)	Summer price		US$/kWh			
		COP	2.25	Pumps&fans			
		Nonsummer price	0.13	US$/kWh			
		COP	3.50	Pumps and fans			
Labor	Germination + first stage seedling	Second per plant			2	1	1
	Second stage seedling	Second per plant			5	4	3
	Transplanting first stage	Second per plant			5	4	3
	Transplanting second stage	Second per plant			5	4	3
	Cropping	Second per plant			20	15	13
	Shipping (packing, storing in f reezing room)	Second per plant			15	15	13
	Cleaning, etc.	Hours per day			8	8	8
	Total	Total hours			63	56	50
	Hourly rate	US$			8.50	8.50	8.50
	Number of part-time staff (6 h/day)	Headcount			11	10	9
	Labor cost total	US$			201,960	183,600	165,240
	Welfare and commuting	versus hourly rate	20%		40,392	36,720	33,048
	Labor cost grand total				242,352	220,320	198,288

(*Continued*)

Business plan			Year 1	Year 3	Year 5
			Plan	Plan	Plan
Consumables	Seeds	Cents per plant	1.5	1.5	1.5
	Seeding sponge	Cents per plant	1.0	1.0	1.0
	Nutrient hydro	Cents per plant	1.5	1.0	1.0
	CO_2	Cents per plant	1.5	1.5	1.0
	Packaging materials	Cents per plant	7.0	7.0	7.0
	Transportation	Cents per plant	10.0	10.0	10.0
	Waste cost (roots, sponge)	Cents per plant	0.7	0.7	0.7
	Others	Cents per plant	5.0	5.0	5.0
	Total	USS per plant	312,908	354,831	375,192

Appendix 13.7: Budget five years P/L and cash flow

P/L and cash flow				Year 1	Year 3	Year 5
		(Unit: US$)		Plan	Plan	Plan
Gross sales				908,820	1,225,692	1,516,320
Fixed cost	Depreciation	Fixed amount		3,000	3,000	3000
		Fixed amount		291,383	291,383	291,383
	Maintenance and renovation	3%		66,251	66,251	66,251
	Building monthly rent (US$)	8,000	Monthly	96,000	96,000	96,000
	Electricity and water (capacity)	Monthly		33,996	33,996	33,996
	Subtotal			490,630	490,630	490,630

P/L and cash flow		(Unit: US$)	Year 1 Plan	Year 3 Plan	Year 5 Plan
Variable cost	Electricity and water (capacity)		222,019	222,444	222,710
	Labor cost		242,352	220,320	198,288
	Consumables		151,971	154,477	154,872
	Subtotal		616,342	597,242	575,870
Cost over sales	Electricity and water (capacity)		28%	21%	17%
	Labor cost		27%	18%	13%
	Consumables		17%	13%	10%
Cost per stock	Electricity and water (capacity)	Cents per plant	27	22	20
	Labor cost	Cents per plant	26	19	15
	Consumables	Cents per plant	16	13	12
Gross profit			−198,152	137,820	449,820
SGA			232,937	272,354	292,320
Operating profit			−431,089	−134,533	157,500
EBITDA			−136,705	159,850	451,883
Cash on hands beginning of year	Cash funded for the project	3,000,000	791,650	679,291	1,282,024
Cash on hands end of year			654,945	839,141	1,733,907

Due to a limited size of the page, only year one, three, and five are shown. Developing the budget for five years or longer lead you to an estimate how many years are necessary to receive a full payback of the investment.

Appendix 13.8: Major productivity indexes

Items	Index	(Unit)	Standard Operation
Cultivation Area per m^2	Initial Investment	US$ m^{-2}	1,019
	LED fixtures	US$ m^{-2}	310
	Other materials	US$ m^{-2}	709
Cultivation Area per m^2 per day	Fresh Weight produced	kg m^{-2} d^{-1}	0.148
	Daily Light Integral (DLI)	mol m^{-2} d^{-1}	5.76
	Cost for DLI	US$ m^{-2}	43.8
	Electricity consumed by LED fixtures	kWh m^{-2} d^{-1}	1.53
	Cost for Electricity consumed	cents m^{-2} d^{-1}	23.5
	Economic depreciation of LED fixtures	cents m^{-2} d^{-1}	20.3
	Cost for labour	cents m^{-2} d^{-1}	23

Appendix 13.9: Other productivity indexes

Items	Index	(Unit)	Standard Operation
Initial Investment	per kg daily production	US$ kg^{-1}	6,816
	per one daily crop head (80g)	US$ per head	613
Cultivation Area	Area necessary for 1kg daily production	m^2 kg^{-1}	6.75
	Daily stock crop production per m^2 area	head m^{-2}	1.65
Electricity	Electricity necessary for one crop head	kWh per head	1.29
	Lighting & water pumps only	kWh per head	0.93
	Electricity necessary for 1kg fresh weight	kWh kg^{-1}	14.37
	Lighting & water pumps only	kWh kg^{-2}	10.34
	Fresh weight per kWh electricity consumed	kg kWh^{-1}	0.07
	Lighting & water pumps only	kg kWh^{-2}	0.10
	Lighting PPF consumed per kg fresh weight	mol kg^{-1}	45.73
	Fresh weight per mol Light PPF	kg mol^{-1}	0.02

Labour	Man Hour for one head	h per head	0.014
	MS for man second	second per head	50
	Man Hour for 1kg fresh weight	h kg^{-1}	0.154
	MS for man second	s kg^{-2}	554
	Fresh weight per MH	kg h^{-1}	6.49

Initial investment per m^2 cultivation area is US\$1019/m^2 and the initial investment per kg daily fresh weight production is US\$6816/kg, the cultivation area necessary for daily 1 kg fresh weight production is 6.69 m^2.

Electricity for lighting and water pumps, not including heat pumps, necessary for 1 kg fresh weight is 10.34 kWh/kg and fresh weight per 1 kWh electricity consumed for lighting and water pumps is 0.10 kg/kWh. These two numbers are inverse to each other and of the pairs of inverse numbers, the former is how small the cost is per production and the latter is how large production is for the cost. The same is the case with (i) lighting PPF consumed per kg fresh weight (45.73 mol/kg) versus fresh weight per 1mol lighting PPF (0.02 kg/mol) and (ii) Man Hour per kg fresh weight (0.154 MH/kg) versus fresh weight per Man Hour (6.49 kg/h).

References

Kozai, T., 2015. Plant production process, floor plan, and Layout of PFAL. In: Chapter 16 of "Plant Factory an Indoor Vertical Farming System for Efficient Quality Food Production". Academic Press.

McCree, K.J., 1971. Significance of enhancement for calculations based on the action spectrum for photosynthesis. Plant Physiol. 49, 704–706.

Nunomura, O., Kozai, T., Shinozaki, K., Oshio, T., 2015. Seeding, Seedling Production and Transplanting (Chapter 18) of the same as above.

Snowden, M.C., Cope, K.R., Bugbee, B., 2016. Sensitivity of seven diverse species to blue and green light: interactions with photon flux. PloS One 11 (10), e0163121. https://doi.org/10.1371/journal.pone.0163121.

Son, J.-E., Kim, H.J., Ahn, T.I., 2015. Hydroponic Systems (Chapter 17) of the same as above.

Emerging economics and profitability of PFALs

Simone Valle de Souza, H. Christopher Peterson and Joseph Seong

Michigan State University, East Lansing, MI, United States

14.1 Introduction

This chapter builds upon and expands the productivity focus of the prior chapters by focusing specifically on profitability as an ultimate objective for PFAL economic success. The Japanese PFAL remains at the heart of the analysis as one, if not the only, fully implemented model of PFAL profitability whose complete range of inputs, costs, and revenues is publicly available anywhere in the world. At the same time, this chapter attempts to broaden the analysis to include emerging profitability issues beyond the Japanese market, especially in the United States.

Productivity of all elements of production is necessary for PFAL profitability but not completely sufficient to assure it. This chapter also analyzes revenue generation, which incorporates price impacts based on consumer demand and willingness to pay for product attributes. Adding the revenue/price component assumes that PFALs are *price makers* and not merely *price takers* given the unique ability of a controlled environment to deliver higher value product attributes than competing growth systems (e.g., conventionally grown or greenhouse).

The first part of this chapter documents what is known and not known about PFAL profitability empirically (especially outside the Japanese industry), while the second part describes emerging issues in the current U.S. PFAL industry. The third part uses the Japanese model to systematically simulate various scenarios and the resulting changes in profitability with implications for PFALs everywhere. Revenue generation and pricing are addressed in the final part.

14.2 The state of knowledge about PFAL profitability

Little exists in the public domain about actual PFAL profitability. The Japanese PFAL model as articulated in Kozai (2018) is perhaps the lone example. For that reason, it is used for simulations later in this chapter. The public domain mostly has academic treatments of this subject and a few industry trade reports. In the academic realm, profitability is analyzed in the context of optimization models.

Profit optimization models in PFAL research, similarly to greenhouse systems research, often focus on specific components of a system, such as production efficiency, aiming to improve grower income by the optimization of crop growth (seeking larger quantity with better quality), reduction of associated costs (i.e., electricity costs) reduction of residues, and the improvement of water efficiency (Ramríez-Arias et al., 2012). An example of a complete system approach model, currently lacking in the PFAL literature, is a multilayered model developed for greenhouse systems (Rodríguez et al., 2003). This model includes three layers: one defining climate control, one for crop control, and the third defining a time scale related to market issues (Rodríguez et al., 2003). Both climate control and crop growth systems influence each other in physical processes involving energy transfer (PAR radiation and heat) and the mass balance (water vapor flux and CO_2 concentration), along with other conditions such as irrigation and fertilizers. Profit, in Ramríez-Arias et al. (2012), is the difference between the income from the sales of the produce and the costs associated with production. Produce quality is defined as a function of sensory attributes, such as color, shape, fruit acidity, and texture. Their three-layered model solves for crop growth at maximum profitability for a range of crop quality and water use efficiency (WUE) levels. Their results, which are specifically for a greenhouse environment, showed profit and WUE increasing as crop quality decreased. Trade-off alternatives, however, demonstrated how benefits from increased quality can outweigh losses in profitability. In this case, a 24% increase in quality generated only a 2% reduction in profitability. A trade-off was also observed in attempts to optimize WUE, which involved compromising both profits and crop quality. Although a good example for modeling PFALs, this model does not consider changes in energy consumption and associated costs in a system that exclusively uses artificial lighting systems and has the ability to create attributes to reach niche market premiums.

The economics of PFALs, specifically, have been sparsely reported, given the lack of publicly available data for PFAL profitability. As expected in the case of emerging industries, any profitability is guarded as proprietary data. Aspects of economic analysis generally explore economic and environmental benefits and social impact of PFALs on surrounding communities. For example, a study case developed for New York City reports specifically on access to healthy, nutritious food; land use and real estate; employment; and environmental sustainability (Goodman and Minner, 2019). Notwithstanding great social economic benefits created by urban farming, the greatest challenge for the development of this industry has been the complexity of CEA profit models (de Nijs, 2017).

PFAL optimization models remain a resource-use–based profit maximization instead of the ideal revenue-based profit optimization model as they take a production perspective versus a market perspective. Often, research on profit optimization targets income improvement through cost reduction/optimization of lighting systems, one of the highest costs in indoor farming. While various environmental factors such as temperature (Hatfield and Prueger, 2015) and carbon dioxide concentration (Kozai, 2018) play an important role, lighting has a profound effect on plant growth (Bian et al., 2015). Lighting is also critical in profitability analysis since light efficiency is the factor that affects cost the most among other inputs (Kozai et al., 2015), most significantly in PFAL operations. Energy accounts for 25%

of total operating expenses in a large indoor facility (the equivalent to $8 per square foot), comparable to 8% of total operating expenses for a greenhouse, primarily used as heating and cooling costs (Agrilyst, 2017).

Lighting optimization models predict increases in yield as a result of changes in lighting, with the objective of at least equating the cost of providing supplemental light with the economic benefit of sales at fixed market prices (Heuvelink and Challa, 1989; Nicole et al., 2016; van Iersel, 2017). A recent study searched for the relationship between lighting conditions and lettuce growth in order to achieve a desired quality of lettuce in a cost-efficient way under controlled environment settings (Kelly et al., 2020). The motivation of this study was to optimize lighting inputs for lettuce growth to maximize yield and minimize costs, but it did not include the economic benefits of the effect of lighting on yield. Another study also tried to find optimal lighting condition, especially supplemental lighting, by conducting cost and return analyses (Kubota et al., 2016). This paper provided specific equations estimating power need for getting the target lighting inputs, but it still focused on profitability from supplemental lighting only, making it a partial model for PFAL profitability.

Other models aim to develop environment perception and data sharing, advanced microclimate control, and energy optimization models (Shamshiri et al., 2018). Nicole et al. (2016) argued that lettuce growth strategies combining crop quality attributes (e.g., color, nutrients, and shelf life) with efficient growth are key for economic viability of plant factories. Optimizing yield meant producing maximum grams of lettuce per mol of light used or having the highest $kg\ m^{-2}\ y^{-1}$. For optimizing quality, they chose to focus on the coloration of the red cultivars and the nitrate content of red and green cultivars. Using a Monte Carlo approach, the model included a photosynthesis model, a carbon pool per plant with assimilate distribution based on the relative sink strength concept, and respiration. Quality was referred to as the point at which the plant grows well "up to market standard," without specific details of the market standard and premium paid for product differentiation.

Some groups of experts have created hypothetical models to predict and examine the economic viability of PFALs. Hypothetical studies can provide useful insights, but they lack the credibility of having been implemented. We found two such models from Germany (Zeidler et al., 2017) and the United States (Mejía, 2020). Vertical Farm 2.0 (Zeidler et al., 2017) is a publicly available report that proposed a modular structure vertical farm system under European context and included some PFAL profitability analysis. Mejía (2020) also proposed a hypothetical model of vertical farming system in California in his thesis paper. He developed a financial plan for an imaginary social enterprise, namely "Vertic Garden," based on two case studies: Vertical Farm 2.0 and Sky Greens, where the latter is a running business model in Singapore using sunlight as a source of light. Although the results of the aforementioned studies provide some good insights for understanding PFAL profitability, they basically bear limited credibility in that their hypothetical approaches do not provide market-validated estimates.

Notwithstanding the importance of examining lighting input of PFAL and understanding the dynamics of plant growth, modeling economic viability of PFAL requires a complete model for profitability, encompassing layers describing the dynamics of plant growth, potential to improve plant growth through environmental control mechanisms, and the ability to increase revenues through value chain innovation and market dynamics. Revenue generating ability will play a crucial role if the producer of plants using a PFAL system could be regarded as a price maker. An integrated market analysis including consumer preference studies would inform about revenue generating ability of a given PFAL. Apart from the potential environmental benefits from reducing water and land use and decreasing environmental footprint by bringing farms closer to the consumer, what really differentiates PFAL from conventional producers is the ability to control quality and further create special attributes. The uniqueness of PFAL products makes it essential to consider the revenue-generating capacity of its attributes when studying PFAL profitability and to regard the PFAL industry as a price maker and not a price taker as in the case of commodity producers.

14.3 The U.S. industry landscape

A description of the U.S. PFAL industry opens an opportunity to draw broader implications for profitability than focusing solely on the Japanese operating and marketing context. Publicly available data identifying and characterizing these farms are scarce and often describe collectively various types of horticultural farms, including PFALs, greenhouses, and any other farm that applies a technology-based approach to food production, while operating within an enclosed growing structure (Lensing, 2018). Although all-inclusive, some of these reports and databases can inform about the size of the specialty horticulture industry in terms of number of players, as well as inform about market volume and prices. For example, in 2014, the USDA Census of horticulture specialties reported a total of 1,603 farms producing "food crops grown under protection," from which 36 farms alone produced $556 million in sales, accounting for 70% of all sales. Between all crops produced in this controlled environment, lettuce represented 5% and herbs would take 15% of total value of production (USDA, 2015). A total of 763 farms produced 5.2 million cwt of all lettuce types, amounting to $55.5 million. The estimated market in wholesale value for leafy greens alone is $20 million and is expected to grow annually by 6.5% (Integrative Economics, 2016). Per state, more than 10% ($6.5 million) originated from Texas, the state with largest sales in value recorded in 2014, followed by New Jersey with $2.5 million and New York with $1.6 million. In 2014, Pennsylvania was the state with the largest number of controlled environment farms (66), followed by Connecticut (57) and Vermont (54). Production of herbs and microgreens reached $71 million in 2014, with more than half of the revenue originating in California ($39 million dollars), followed by Pennsylvania with $4.7 million, and New Jersey and Texas, each having sold $3.7 million.

Market research reports allow us to catch a glimpse of the PFAL industry alone (Newbean Capital et al., 2015; Integrative Economics 2016; Agrilyst 2016, 2017; Newbean Capital 2017; Lensing 2018; Contain 2019; Agritecture Consulting and Autogrow 2019) (Agritecture Consulting and Autogrow, 2019; Agrilyst, 2017) in the United States, albeit lacking a deep analysis on PFAL profitability. According to the report Global CEA Census 2019 (Agritecture Consulting and Autogrow, 2019), vertical farm operators were profitable at a rate of 37%, losing money at 24%, and breaking even at 16%. It is not surprising that a small portion of operators are profitable considering the industry is in an emerging stage. What is more interesting is that 23% chose "I do not know," which perhaps loosely reveals the sensitivity of the information on profitability in this industry. Responses for another question to the same respondent group about intention to increase future production areas are also interesting in that 84% replied "yes." This means there exist quite a few vertical farm operators who are optimistic about markets and technological developments bringing future profits or who simply lack an understanding of current profitability levels.

An earlier market report (Agrilyst, 2017) revealed similar aspects of the current vertical farm industry: a small number of profitable businesses and unclearness in profitability information. In this report, indoor vertical farm operators reported their profitability and revenue per square feet (sq. ft.). Only 27% of the indoor vertical farm operators replied that they were profitable at that time, and the revenue per sq. ft. ranged from $2.13 to $100.00. The report concluded that considering the wide ranges in revenue, operating systems, and farm structure, it is more important to analyze profitability than revenue alone. In one way, these market reports provide a useful snapshot of the industry. For example, the proportion of firms achieving profit in the industry has improved from 2017 to 2019. But on the other hand, proprietary data restrictions do not allow for a detailed analysis of profitability of the PFAL system beyond combinations of most profitable crops, average yield and revenue per sq. ft., and the proportion of costs associated with key production factors.

Returns to capital, in terms of revenue per square feet, are dependent on crop mix choice, with the most profitable indoor farms being those focusing on leafy greens, microgreens, and herbs. Microgreen and herb revenues can vary between $100/sq. ft. and $875/sq. ft. Leafy green farm returns present a wider range of revenue depending on attributes offered. Annual sales of leafy greens can average $22/sq. ft. if wholesale prices are assumed at $1.25/lb., an average price for commodity lettuce in the United States (Integrative Economics, 2016). Should the capacity of this industry to produce a differentiated product be considered, revenues per sq. ft. can reach on average $125 (see prices discussion in Section 14.4). Hydroponically grown leafy greens have on average the highest profit margins (46%) and the lowest operational costs ($20/sq. ft.), with an average revenue of $37/sq. ft. compared to $21/sq. ft. in all other indoor grown crops (Lensing, 2018). Average profits of profitable PFALs in the United States in 2017 were $14.88/sq. ft., with revenues at $51.98/sq. ft., and operational costs at $37.10 (Agrilyst, 2017).

Operational costs are reported to be the biggest challenge of the industry. Labor takes on average 56% of operating costs in PFALs employing eight persons, a rate of 0.00144/sq. ft. The second largest operating cost is energy, which takes another 25% in large farms (10,000 sq. ft. or more) and 12% in smaller farms. In nominal values, a PFAL small farm spends $3.45/sq. ft. and large farms spend on average $8.02/sq. ft., running on an average photoperiod of 16 h (Agrilyst, 2017).

Main constraints for new firms to enter this industry remain the large initial capital and working capital requirements, as well as high operating costs (Agrilyst, 2017) and lack of reliable and skilled labor (Contain, 2019). PFAL capital expenditures in North America vary between $200 and $600 per square feet (Contain, 2019). Operating cost distribution includes the following: 56% in labor; 27% in building and equipment operations (including rent, packaging, energy, etc.); 11% in seeds, nutrients, and growth media; and 6% in total shipping costs (Agrilyst, 2017). Energy expenditures in indoor farms, using an average photoperiod of 16 h, were reported as 12% of OPEX in small farms and 25% of OPEX in large farms (>10,000 sq. ft.), equivalent to $3.45/sq. ft. and $8.02/sq. ft., respectively.

Depending on price premiums in niche markets to succeed, this industry started with PFAL farms strategically positioned to allow control of a large share of the market (Lensing, 2018). The majority of these large PFALs were initially located in the Northeast and Midwest United States, strategically located geographically close to potential markets (Integrative Economics, 2016).

In 2015, the PFAL industry in the United States was comprised of eight commercial-scale PFALs in the United States funded by 2009 or after (Newbean Capital et al., 2015). By September 2016, the total number of farms had grown to 14 (Integrative Economics, 2016), which inspired optimistic reports of the industry showing signs of maturation (Agrilyst, 2017). Indeed, in 2020, announcements of new investments and openings of new PFALs in secondary vertical farm markets such as Texas and Florida, among others, populate the media.

As for 2019, the North American Plant Factory industry, including warehouse farms, plant factories, and rooftop greenhouses, accounted for a total of 63 farms distributed between the East Cost (18), Midwest (18), West Coast (10), and in other parts of the United States plus Canada, another 17 farms (Contain, 2019). This industry was then characterized by an oligopoly of large, mostly plant factory growers focusing on economies of scale and developing their own brands (Fig. 14.1); medium growers focusing on specific niches such as animal fodder and microgreens; and a large number of small growers serving local communities, restaurants, and supermarkets (Contain, 2019).

While the unpredictability of markets and complexity of system operations have been blamed to cause close downs of firms in this industry, undervaluing the potential profitability

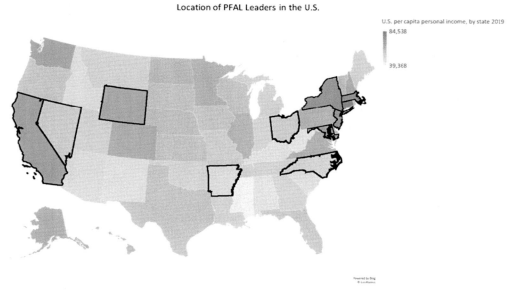

FIGURE 14.1 Distribution of leader PFALs in the United States and per capita income per state. *Source: US Department of Commerce — BEA 2019; Contain, 2019. Automation, AI & the Next Generation of Indoor Agriculture. https://indoor.ag/white-papers/; Companies Websites.*

of a PFAL has, to an extent, contributed to hindering industry growth. Further research is needed in economics of markets, which focuses on the nature of demand and the competitive aspects involved in marketing PFAL products; in other words, the PFAL market-making process (Blin, 1980). A key element in determining revenues is the demand, in terms of quantity and price paid, for potential product attributes created in a PFAL environment.

Generally, market participants take the traditional approach of pursuing increased revenues through economies of scale, facilities and technological improvement, and cropping patterns. This is evidenced through market reports that show 46% of farmers chose to pursue increasing profitability or revenue as a goal, but only 7% considered improving product quality (Agrilyst, 2017). Signals of potential profitability from crop-specific market premiums, on the other hand, have been captured by large farmers who are planning to expand production in herbs and microgreens.

Indeed, economies are accrued with increased size of farm and an opportunity to automate operations, utilizing labor more efficiently, when low-cost capital is available, when there are economies in purchasing materials or when special management skills are available (Dalrymple, 1973). However, improved physical facilities, including size and type of structure, as well as environmental control technology, would not only affect production capacity, but would make it possible to optimize production cycles and product quality, addressing specific needs of market niches that are willing to pay a premium. Price premiums will directly contribute to higher revenues and profits as discussed in the next section.

14.4 Profitability scenarios

Profitability in a PFAL is yet to be defined in a one-size-fits-all framework as physical structure, labor availability, location, lighting systems, and crop mix can be significantly diverse and affect profitability differently. However, a base case for which a complete set of actual production and financial data exists allows examination of profitability for the base case and well-crafted "what if" scenarios for simulating changes in key variables, considering the potential of every activity of the PFAL to generate costs (capital and operating) as well as revenue benefits (Porter, 1979).

Perhaps the only complete base case available in the public domain is a Japanese PFAL (Kozai, 2018) as designed and implemented by the Japanese Plant Factory Association. This case is used for a set of profitability scenarios resulting in estimated Return on Investment (ROI) based on Earnings before Interest and Taxes (EBIT). The ROI equation is simply EBIT/capital. In the base case, capital is relatively simple as the PFAL building is assumed to be rented and capital reduces to equipment costs for the scenarios. This version of ROI is not a particularly strong measure of profitability. It omits interest, taxes, and working capital— estimating each of these creates its own set of uncertainties when searching for profitability in an international context. However, it allows tracking key drivers of profitability and tracing their impact on an often-used key profit measure such as ROI. Setting an appropriately challenging target for this rudimentary ROI allows the analysis to account for these "missing" elements in part if not in their entirety. An ROI target is set at 20% to distinguish between more and less attractive scenarios. Ultimately, the goal should be an internal rate of return on fully articulated cash flows from the life of PFAL facilities that exceeds the cost of capital demanded by capital markets. With only one quarter to one third of existing PFAL firms claiming profitability, a 20% ROI seems a challenging target for this exploratory analysis. As the industry matures, this target can be refined to reflect profitability levels that actually emerge.

In the Japanese PFAL, 241 racks with 6 vertical levels use a hydroponic system in a 1000 m^2 (10,768 sq. ft.) building, which creates a total growing space of 2186 m^2 (23,534 sq. ft.). The farm operates 360 days per year, producing 5000 units of 90-g lettuce heads per day in a growth cycle of 38 days with four growth stages requiring transplanting between each stage. Annual salable biomass is 116,640 kg (257,148 lbs). The system uses LED lights designed to deliver 100 µmol s^{-1} m^{-2}, in a 16-h d^{-1} photoperiod. The prevailing market price for this specialty lettuce is $12/kg, giving total annual sales revenue of $1.4 million dollars.

Table 14.1 provides a summary of key data related to operations and profitability. The base case with 2016 data shows modest profitability with a 1.8% ROI—far from a 20% target level. Transplanting the Japanese PFAL to other countries, particularly the United States, would not be financially feasible with this level of profit.

One of the obvious reasons for modest base case profitability is the high cost of equipment—a common problem across all forms of PFAL technology. High capital costs are one of the distinguishing features of controlled environment agriculture/vertical farming from conventionally grown farming. What if that capital cost could be driven down? Actual events between 2016 and 2020 resulted in a natural experiment on this question. Equipment costs fell approximately 45% over the 4-year period (as described in Chapter 13) from $2.2 million to $1.2 million. The standardized plant design created the incentive for Chinese equipment suppliers to drive out capital costs without hurting performance characteristics. This one change (see last column in Table 14.1) increased the pro forma ROI from 1.8% to 14.3%, a major step toward a 20% target. The decline in capital cost both elevated EBIT directly by cutting annual depreciation (the noncash recognition of capital consumption) and decreased the capital denominator in ROI. As a stand-alone expense, depreciation fell from the largest to the third largest, trailing both electricity and labor after the change.

Based on the observed change in capital cost, nine other scenarios were created using the underlying structure of a Japanese PFAL. Table 14.2 presents a summary of these scenarios focused on changes in profitability resulting from a key underlying change modeled in each scenario. A complete table of results, including all scenarios, is presented in Appendix 14.1, and Tables 14.5 and 14.6.

Changes to scale. The issue of scale is also relevant to the profitability of any PFAL technology. As in most industries, managers are likely to pursue increases in scale to achieve enhanced profits. The Japanese model shows this traditional industrial trend on the upside. Scenario 1 presents results of a half-scale facility, while Scenario 2 presents a doubling of scale (see Table 14.2). In each case, the model considers half or double, respectively, of the current physical structure, costs associated with consumables, a group of costs directly related to number of seeds planted and inputs associated with growth, as well as corresponding required labor and electricity costs.

TABLE 14.1 Operating data, income statement, and potential ROI for the Model Japanese PFAL.

	Base case (2016 data)	Base case (2020 data)
Planted stocks/day	5000	Same
Harvested stocks/day	4500	Same
Total kg sold/day (in packs of 90 g each)	324	Same
Selling price/kg ($)	$12	Same
Annual revenue ('000$)	1400	Same
Annual variable production costs ('000$)	611	Same
Labor	198	Same
Electricity	257	Same
Plant growth consumables	156	Same
Annual fixed, marketing, and admin costs ('000$)	750	616
Building rent	96	Same
Depreciation	294	160
Marketing (pkg/transport), admin, and others	360	Same
EBIT ('000$)	39	173
Equipment cost ('000$)	2208	1200
Crop space m^2	2168	Same
Facility size m^2	997	Same
ROI (based on EBIT/equipment cost)	1.8%	14.3%
Number of employees	9	Same
Growth cycle (days)	38	Same
Salable bio (kg m^{-2} y^{-1})	53.8	Same

Estimated based on Kozai, T., 2018. Smart Plant Factory — The Next Generation Indoor Vertical Farms. Springer, Singapore. https://doi.org/ 10.1007/978-981-13-1065-2.

Half scale essentially results in breakeven profitability with all phases of operation sized to accommodate daily planting of 2500 stocks. 2-X scale pushes ROI to nearly 22%, which is above a 20% target, with operations sized for 10,000 daily plantings.

The fixed costs in the economic structure drive this lowering and heightening of profit. Proportionally, more fixed costs in a facility (e.g., add automation to reduce labor) would leverage profitability even more.

Changes in Operations. Scenarios 3—6 examine four different changes in facility operations. Each of these changes is a rather obvious alternative for a vertical farm manager to increase productivity and profitability:

- decrease the growth cycle (grow plants quicker in a controlled environment)
- increase plant density (grow more plants per unit of space in a controlled environment)
- increase biomass per plant (grow more salable biomass per plant in a controlled environment)

TABLE 14.2 Scenario analysis of changes in Japanese plant factory profitability.

Scenarios	Revenue ('000$)	EBIT ('000$)	ROI	Salable biomass (kg m^{-2} y^{-1})	Driver of ROI change
Base case 2016 capital cost	1400	39	1.8%	53.8	
Base case 2020 capital cost	1400	173	14.3%	53.8	45% decline in equipment cost
Changes to scale					
Scenario 1: half scale	700	−3	−0.5%	53.8	Fixed costs do not halve
Scenario 2: 2X scale	2800	524	21.7%	53.8	Fixed costs do not double
Changes to operations					
Scenario 3: reduce growth cycle from 38 to 34 days	1568	241	19.8%	59.8	Added capacity utilization; input efficiency
Scenario 4: increase plant density (T1 = 17 v 14; T2 = 9 v 7)	1680	306	25.6%	65.1	Revenue benefit exceeds cost
Scenario 5: increase biomass/ plant by 10% (90–99 g)	1540	313	25.9%	59.2	Major boost in revenue Uncertain what mix of inputs would work
Scenario 6: 20% increase in lighting efficiency	1400	201	16.7%	53.8	KWH savings go directly to EBIT
Changes in market context					
Scenario 7: impose U.S. labor costs	1400	21	1.8%	53.8	$12/h (v $8.50) with 50% (v 20%) benefit load
Scenario 8: convert to crop with 1 stage 25-day cycle	2016	464	38.1%	76.8	Redirect facility use to new crop
Scenario 9: increase market price by 5%	1470	243	20.1%	53.8	Adjust price from $12/kg to $12.60 to achieve 20% ROI

- increase the efficient use of a major cost component, such as electricity, in a controlled environment.

These four scenarios are not designed to test the profit impact of all potentially relevant changes in operations. However, they begin to lay out the pattern of such changes on profit.

Scenario 3 reduces the growth cycle from 38 to 34 days, roughly a 10% reduction. The model PFAL has four stages in its growth cycle: 8 days for seedlings, 10 for nursery, 10 for first major transplant (T1), and 10 for second major transplant (T2). The number of stocks per tray moves from 300 to 28 to 14 to 7 across the growth stages. The scenario envisions reducing 1 day per growth stage. Some innovation in operations would be needed to achieve this reduction, namely more light or nutrients, or better genetics. The first two causes could change related cost components, while better genetics might not. Daily planting increases from 5000 to 5600 with the same equipment and one more worker. The scenario allows for the increased cost of the number of plantings but assumes the reduction in 4 days is achieved by efficiency gains in lighting, nutrients, or genetics rather than added costs. The resulting ROI rises from 14.3% in the 2020 base case to just under the 20% target, while annual salable biomass increases by 11% to 59.8 kg m^{-2} y^{-1} (see Table 14.2).

Chapter 13 (Section 13.6) presented a more dramatic version of this scenario where light intensity is increased by 100% (100—200 µmols) resulting in a nearly 50% reduction in the number of cultivation days to 21 (with growth stages falling to 6 + 5+5 + 5 days). To achieve this innovation, an additional $227,000 investment in lighting equipment was needed plus a doubling of the electric consumption and cost for lighting and a near doubling of the labor force. Total output and revenue essentially doubled without changing the physical size of the facility, which resulted in a doubling of the biomass produced per square meter. ROI jumped to 31%, well above the 20% target. Although the profitability implications are attractive, this alternative may be overly optimistic in assuming all other operational components remain unchanged. Scenario 3 (reducing the growth cycle) likely has many other possible variations worth analyzing in greater depth.

Rather than decreasing the growth cycle, Scenario 4 returns to 38 days and increases plant density in growth stages T1 and T2. T1 moves from 14 plants/tray to 17, while T2 moves from 7 to 9, roughly a 20% increase in density for these two stages. As in Scenario 3, the question becomes "can an efficiency innovation be created to allow this increased density?" More precise spacing on the tray may be sufficient, or a slightly taller plant genetically? By increasing the density, daily planting can increase from 5000 to 6000 with the same equipment and one more worker. The scenario allows for the increased cost of the number of plantings but assumes the density increase is achieved by efficiency gains in spacing, lighting, nutrients, or genetics rather than added costs. The resulting ROI rises from 14.3% in the 2020 base case to nearly 26% (well over the 20% target), while annual salable biomass increases to 65.1 kg m^{-2} (see Table 14.2).

Scenario 5 looks at another approach to increasing biomass—change the biomass per plant rather than change growth cycle or density. The scenario envisions a 99-g final pack size rather than 90 g. Efficiency innovation in lighting, nutrients, or genetics is once again essential to implementation. Scenario 5 gets no bump in plantings like the prior two scenarios. The benefit comes entirely from increased pack weight. It is assumed that the increased pack weight is as valuable to the buyer as the lower pack weight. If the increased weight makes the product less desirable or special, this scenario may not work. The resulting ROI rises from 14.3% in the 2020 base case to nearly 26% (well over the 20% target), while annual salable biomass increases to 59.2 kg m^{-2} (see Table 14.2).

Scenario 6 envisions innovation that lowers the cost of a key input without changing the biomass yield. This is a simpler scenario than Scenarios 3—5 in which multiple innovation factors lead to the change in profitability. Electricity is the input targeted. Labor might also be used in a similar scenario. The innovation needed would deliver the light required for growth in a more efficient manner than the current lighting system. KWHs of electricity would be saved and cost lowered. The target is a 20% enhancement in lighting efficiency resulting in a 20% decrease in KWHs consumed. This scenario assumes that this could result from the continuing innovation in lighting systems without necessarily increasing their capital cost. Electricity for lighting is 55% of the facility's electricity use (the remainder being largely for HVAC). The savings in electricity cost of $28,000 go directly to EBIT. The resulting ROI rises from 14.3% in the 2020 base case to nearly 17% (a little less than the 20% target) (see Table 14.2).

The "changes in operations" scenarios 3—6 each results in a significant and desired boost to ROI—three of four meet or exceed a 20% ROI target and the fourth comes close. However, in one way or another, each assumes that a needed innovation can be implemented without cost or, at least, with the value of efficiency gains exceeding the cost of change resulting in increased profitability. If this assumption is incorrect in any of the scenarios, then the profitability gain of that scenario would be diminished. Nonetheless, the scenarios show that innovations to enhance various approaches to biomass production and cost efficiency have desired profitability effects. Such innovations are essential to the future of PFALs and will demand significant changes in production methods and not just minor tweaks (10%—20% improvements were envisioned in each scenario).

Changes in Market Context. The final three scenarios challenge the Japanese model by moving it into some uncharted market territory. Scenario 7 imposes U.S. labor costs. Scenario 8 envisions an entirely different crop. Scenario 9 explores achieving profitability through price enhancement.

Japanese labor costs turn out to be low in comparison to the United States for the model PFAL facility. Hourly wage is $8.50 with only a 20% benefits load. U.S. vertical farms face a significantly higher labor cost. Informal discussions with several U.S. industry firms suggest that an hourly wage of $12 and a benefit load of 50% are reasonable for this scenario. U.S. Bureau of Labor Statistics gives a range of hourly wages for the greenhouse industry from $11.06—17.75 with a median of $12.23 (2019). BLS does not provide information on benefits.

Scenario 7 imposes a $12/h wage with 50% benefit load. Labor costs in the 2020 base case rise from $198,000 to $350,000 in Scenario 7. ROI drops from 14.3% to 1.8% (see Table 14.2). The Japanese model becomes infeasible in a profitability sense at U.S. wage rates. Substantial automation would be needed to substitute capital for labor. This scenario is consistent with the largely held view among U.S. vertical farms that labor costs are their highest single cost.

Partly motivated by Scenario 7, Scenario 8 envisions a Japanese PFAL that grows an entirely different crop from the specialty head lettuce it was designed for. Its physical structure of trays, racks, nutrient delivery, basic growth inputs, and environmental controls is similar to other types of vertical farms and could be used to grow a variety of different crops. Scenario 8 is based on a crop that has one growth stage rather than four and that growth stage only needs 25 days rather than 38. The crop is planted with 14 stocks per tray without need of transplant. The crop uses a similar complement of inputs and also sells finished product for $12/kg. Given these parameters, the converted facility could plant 8500 stocks per day and finish 6120 packs. Revenue would soar to $2.0 million, while EBIT jumps to $464,000 and ROI to 38.1%. Even imposing U.S. labor rates, the ROI only falls to 23% (still above the 20% target). The specifications for this scenario were deliberately selected to generate extraordinary profitability. The prior scenarios build an understanding of the PFAL system that leads to an ability to specify a crop based on profit potential. Scenario 8 highlights the role of crop selection in profitability. The general PFAL structure is quite flexible if built with that in mind. PFALs need to consider not just existing crop options but different genetics designed for indoor control. Selection and creation of "ideal crops" is part of sustaining profitability for PFALs.

Finally, Scenario 9 examines the revenue side of profitability, specifically price/kg. Throughout the first eight scenarios, a wholesale market price of $12/kg was assumed. This price is achieved in Japanese markets for the 90-g pack of specialty lettuce grown by the model PFAL. Profitability hinges critically on a premium price for a specialty product.

Scenario 9 shows that only an additional 5% price increase to $12.60/kg achieves the 20% ROI target without the need for operating or scale changes. Would the Japanese consumer be willing to pay this higher price? The answer depends on whether the added value of PFAL product attributes (quality, consistency, freshness, etc.) versus alternative products is worth a 5% price increase. The 5% is critical to long-term industry success by assuring targeted profitability. Conversely, innovations in operations that merely decrease cost or boost yield without regard to negative effects on product attributes could result in decreased prices, as would overproduction of specialized attributes resulting in commoditization. A 12% price decline to $10.56/kg would reduce a PFAL to breakeven profits. In a third variation on this scenario, a PFAL transplanted to the United States would have to command a 16% price increase to $13.92/kg to achieve a 20% ROI while covering U.S. labor costs. Price and revenue matter in the search for profitability. These components have been too little examined in the public literature to date in favor of the cost and yield components. Price will be further analyzed in the U.S. context in this chapter's final section.

Summary of learning from the scenarios. The Japanese PFAL model shares much of its physical structure (trays, racks, nutrient delivery, lighting system, and environmental control system components) and growing approach (leafy greens, staged growth across a cycle of days, and finished product preparation and packaging) to many other vertical farms being constructed or in use elsewhere in the world. Each vertical farm system is also likely unique in many ways tied to the specific capital, operating, technological, and market circumstances in which it exists. The similarities of physical structure and growing approach suggest that some meaningful, if limited, generalized learning about profitability can be taken from the scenarios presented. Consider the following:

- Significant declines in capital costs, especially equipment (45% in the Japanese case), make profitability increase substantially (ROI rose from 1.8% to 14.3%).
- Scale of operation is critical to profit as well and depends on the proportion of fixed costs in the operating structure. Doubling the size of the Japanese PFAL results in further enhancement of ROI from 14.3% to 22%.
- Innovations to increase biomass per unit of growing surface enhance profitability as well. In the Japanese case, increasing the speed of growth (reducing the growing cycle), increasing plant density, or increasing biomass per plant each results in ROIs in the 20%—25% range. The realization of these profit enhancements hinges on increased productivity in lighting, genetics, precision of environment control, and/or any other component of the growing system. Many innovations across all of these areas are likely needed for long-run farm level and industry profitability.
- Reductions in each and every cost category are also essential to growing profits. Scenario 6 used electricity as a case in point. A 20% enhancement of electric efficiency for lighting and the resulting decrease in KWH use raised ROI to nearly 17%. Notice that the boost in profits from a cost reduction was not as great as that arising from increasing biomass production, which tends to change several system components all at once.
- Finally, profitability hinges on the revenue side as well, particularly on the output of a PFAL commanding a premium price in the marketplace. Profitability does not arise from using the technology to compete with commodity produce. It should be used and priced to reap the benefits of its control over product attributes. Innovating pieces of the system must not only enhance productivity and drive down costs but maintain or

enhance product attributes that buyers will pay for. Every innovation has a revenue impact as well as a cost impact. Both must be evaluated in the decision to innovate.

The specific ROI changes tracked across the Japanese PFAL scenarios are suggestive of the magnitudes of change one might see in other vertical farming systems. More publicly available information (complete sets of operating and profitability data) on other systems is needed to confirm or refute the magnitudes of change seen in the Japanese case. In the end, a need exists to know:

- How sensitive is each alternative system to changes in its components, both cost and revenue?
  - What are the trade-offs in different mixes of inputs?
  - What are the trade-offs in cost and in revenue generation from product attributes?
  - What are the trade-offs between labor and capital?
  - What are the optimal ways to enhance profits from increases in biomass?
- What are the benchmarks for profitability and its components?
- What would need to be done to bring about sharing if information across the global industry is to answer these questions?

Further testing of these models is therefore required to identify trade-offs in different mixes of inputs, costs, and revenue generation from product attributes and identify trade-offs between labor and capital. A single benchmarking for profitability and its components would have to be defined consistently with various production systems, crop mix, and markets.

14.5 Price and revenue implications in the U.S. market

Scenario 9 and its analysis led to significant implications for price and revenue generation on PFAL profitability. Little in depth analysis exists publicly on PFAL pricing. If the bundle of attributes that PFALs can produce is unique given their ability to control attribute production, then PFALs should have some ability to set price, and not merely take price. This section analyzes U.S. market prices to find evidence of the value created by this industry with its potential to increment quality. Market prices for commodity-like lettuce are compared here with those for their premium counterparts. In particular, the analysis focuses on finding market prices that would allow a transplanted Japanese PFAL to the United States to make acceptable profits (20% ROI). Based on the Scenario 9 analysis, the key price point needs to be $13.92/kg or $6.33/lb.

Iceberg and romaine lettuce account for 93% of total volume of sales moved through California and Florida, the two key wholesale markets in the United States (USDA AMS, 2020). Sales in volume average 200 million pounds for iceberg lettuce and 170 million pounds for romaine lettuce outside winter season. Between November and March, these volumes drop to 30 and 14 million pounds, respectively. In total, the 2019 season amounted to 2.5 billion pounds of iceberg lettuce and 2 billion pounds of romaine lettuce being negotiated in these terminals.

Prices of the common iceberg and romaine lettuce vary between regions of the country, as these two types of commonly sold lettuce take turns as the local favorite, implied by relative higher prices. On average, prices at retail are lower for iceberg than romaine lettuce, which

TABLE 14.3 Average daily prices for Romaine and Iceberg lettuce, in specific regions of the United States (pound (lb) = 0.453 kg).

Region of the United States	Romaine lettuce (US$)			Iceberg lettuce (US$)	
	Sold per unit	Sold as 3-count pack	Sold per pound	Sold per unit	Value[a] per pound (lb)
Northeast	1.52	3.26	1.5	1.67	1.41
Southeast	1.47	3.11	1.17	1.35	1.14
Midwest	1.32	2.82	1.09	1.45	1.22
South Central	1.38	2.94	1.48	1.25	1.05
Southwest	1.41	2.43	1.29	1.07	0.90
Northwest	1.47	3.18	1.94	1.41	1.19
Alaska	2.13	3.46	–	2.25	1.89
Hawaii	2.39	3.92	2.19	1.99	1.68
U.S. average	**1.64**	**3.14**	**1.52**	**1.56**	**1.31**

[a]*Value per pound considers a head of iceberg lettuce to weigh 19 oz.*

are sold at $1.31/lb and $1.52/lb, respectively (Table 14.3). In the Southern part of the country, romaine lettuce is sold, per head, for 10%–24% more on average, while the Northeast part of the United States pays 10% more per head for iceberg lettuce. Retail prices are higher in Northern regions than Southern regions, while Alaska and Hawaii pay a premium.

Romaine and iceberg lettuce are the "commodity" leafy green product in U.S. markets. None of the prices in Table 14.3 would support a $6.33/lb price which is the minimum price target needed.

Leafy greens grown in PFALs and high-tech greenhouses are differentiated not only in packaging presentation but can also offer a differentiated flavor and texture, inducing a premium price (Table 14.4). Another valuable attribute is freshness, given the proximity of these farms to the final consumer. On average, wholesale prices of leafy greens sold in small packages carry a premium of 433% in relation to common iceberg and romaine lettuce.

For this analysis (Table 14.4), the lack of formally reported market prices for premium lettuce at the time of this work led to the use of retail prices obtained through an online search of selected chain groceries stores' websites based in the United States (i.e., Wholefoods, Meijer, Kroger, Aldi) in September 2020. Products were selected which were advertised as differentiated produce, ready for consumption, or from farm websites when direct-to-consumer sales were available. Market prices for leafy greens, which include spinach, various types of fresh-cut lettuce, kale, microgreens, and arugula, were taken from produce sold in hard plastic packages varying either between a standard 3.5 and 5 oz, or a medium size of 6–9 oz. Family size options with 10–17 oz were not included as premium, but rather considered commodity-like lettuce for the purpose of this analysis. Basil and herbs in this sample were sold in small packages between 1.2 and 3 oz and microgreens in 2 oz packs. Wholesale prices were estimated using a standard industry gross margin of 40%, an approximation to USDA-

TABLE 14.4   Average package sizes and respective retail and wholesale prices in the United States (1 oz (ounce) = 538 g, 1 lb (pound) = 0.453 kg).

Description	Package size (range)	Retail[a] ($/unit)	Wholesale[b] ($/lb)	Wholesale ($/kg)
			Average price	
Leafy greens	heads[c]	3.09	5.21	11.48
	3.5—5 oz	3.42	7.06	15.58
	6—9 oz	2.87	3.74	8.25
Basil	1.2—3 oz	3.29	21.10	46.52
Herbs[d]	1.2—3 oz	2.25	21.54	47.48
Microgreens[e]	2 oz	3.99	19.15	42.22
Average (basket of high-value greens)			**12.97**	**28.59**

[a]*Retail prices obtained through online search on selected chain groceries stores' websites based in the United States (i.e., Wholefoods, Meijer, Kroger, Aldi), taken in September 2020.*
[b]*Industry gross margin of 40% was used to estimate wholesale prices.*
[c]*"Leafy-greens heads" include living lettuce green butter, butterhead, and watercress greens.*
[d]*"Herbs" include spring onions, cilantro, rosemary, thyme, and wasabi arugula.*
[e]*"Microgreens" include arugula, mustard, kale, and cabbage microgreens.*

published price spread from farm to consumer (USDA ERS, 2019), applied over the cost of goods sold (Clark, 2020).

Prices in Table 14.4 are in the $6.33/lb range for higher value leafy greens and considerably above the range if a PFAL converts to basil, herbs, and microgreens (possible versions of Scenario 8). The 90-g pack of specialty lettuce in Japan is likely not a successful product in the U.S. market unless U.S. consumers would find its specialty uses of similar valuable in their own diets.

Notwithstanding the large price premiums implied by Table 14.4, 73% of indoor farmers in 2017 declared "not being profitable" (Agrilyst, 2017). Determinants of profitability in PFAL also include yields, given the high costs associated with startup and working capital. The PFAL operator is therefore likely to succeed through product differentiation and the pursuit of price premiums paid in niche markets (Lensing, 2018). Large-scale production at startup time is another argument for achieving success and competing with open field or greenhouse producers already in the market (Integrative Economics, 2016). One caveat, in relation to large-scale production aiming to provide fresh product to large urban markets such as NYC and Chicago, is the corresponding large real estate costs. In the long-term, success of PFALs will depend on increased production efficiency and on the selection of a crop mix that is high in revenue generating (Agrilyst, 2017).

Production efficiency can be improved through reduction in the costs of the technology and increased yield per physical space (Integrative Economics, 2016). Both advancements are closely related to plant science discoveries in relation to, and technological developments in, lighting systems used in these PFAL farms. In terms of crop mix selection, plants should be physically short, due to the vertical nature of PFALs, short growth cycles, improving farm's turn over (Agrilyst, 2017), and, finally, taking advantage of PFALs' relatively smaller size as a farm and their flexibility to be established within urban centers, highly perishable produce customized to local niche markets.

The ability to control production yield through flexible growth infrastructure and growth environment, unique to PFALs, is also a key competitive advantage in dealing with market price and consumer preference variability. Using price and sales volume of iceberg and romaine lettuce in the United States (USDA ERS, 2019) as a proxy, large price fluctuations and seasonality are observed in this market. Overall, retail prices have doubled, on a 2012 GDP deflated basis, in both cases, of iceberg and romaine lettuce, in the last 15 years. Large price fluctuations, such as sudden increases by 70% in 1998 followed by a decrease by 40% in 1999, were seen before 2005. Since then, smaller variations, between 10% and 30% up or down, were observed.

14.6 Closing observations on PFAL profitability

The PFAL industry has the unique ability to produce fresh, pesticide free, locally grown high-quality and healthy product to consumers who are seeking these very attributes, especially after the COVID-19 pandemic. Even initial concerns about consumer acceptance of indoor grown produced have proved ill founded (Contain, 2019). Its benefits extend to communities as indoor farms create new jobs and repurpose unused buildings in large urban areas, while addressing the problem of food deserts. It also provides year-round consistent distribution of a fresh product with longer shelf life to local restaurants and supermarkets.

The U.S. PFAL industry is responding accordingly to more accessible technological developments and changes in markets with the proportion of firms achieving profit in the industry reportedly improving from 2017 to 2019. The industry was initially formed by a few growers strategically located to allow control of a large share of the market (Lensing, 2018). In the course of a little more than 5 year, this industry has taken a segmented shape with a few large plant factories producing high-quality leafy green in scale, medium growers focusing on specific niches such as animal fodder and microgreens, and a large number of small growers serving local communities, restaurants, and supermarkets (Contain, 2019). Nevertheless, it remains a combination of different physical and technological structures, which makes it difficult to define a one-size-fits-all profitability model.

More publicly available data are crucial for developing further research in the economics and profitability of this industry, but, for the moment, some important factors driving profitability were discussed here. A well-crafted "what if" scenario analysis suggested that profitability can be significantly enhanced through innovative PFAL technology. However, innovation cannot be solely focused on reduce costs but must have a dual focus on maintaining or enhancing product attributes that buyers will pay for.

For example, new suppliers entering the industry lead to lower investment costs which improved ROI. This chapter showed an example of ROI increasing by 8% once investment costs reduced by 45%. Given its significant contribution to total costs, economies of scale are an important factor. Other reductions in costs, such as a hypothetical 20% increase in electric efficiency for lighting resulting in lower KWH, raised ROI in this analysis by 17% (from 14.4% to 16.7%). Increased labor costs, on the other hand, reduced ROI from 14.3% on a base scenario to 1.8%.

Even so, the boost in profits from a cost reduction was not as great as that arising from increasing biomass production which tends to change several system components all at once. Plant growth and production efficiency led by increased productivity in lighting, genetics, precision of environment control, and/or any other component of the growing system can increase ROI in the 20%–25% range. Finally, a small increase by 5% in market price, without the need for operating or scale changes, was demonstrated to be enough for the Japanese model PFAL to achieve the 20% ROI target, which was stablished here as critical to long-term industry success. A 12% price decline to $10.56/kg would reduce a PFAL to breakeven profits.

These analyses demonstrated that productivity efficiency and resource efficiency, most importantly related to energy use efficiency and labor production efficiency, must be sought after, but it should not be paired with a cost minimizing strategy. Every innovation has a revenue impact as well as a cost impact. Both must be evaluated in the decision to innovate. Ultimately, profitability hinges on the revenue side, particularly on the output of a PFAL commanding a premium price in the marketplace.

In that context, optimization modeling can achieve better results with a profit maximization strategy based on improved product attributes that lead to different price points, such as the markets for premium leafy greens and microgreens.

Appendix 14.1

TABLE 14.5 Complete table of scenario analysis of changes in Japanese plant factory profitability.

Scenarios based on the Japanese smart plant factory model	Base case (2016 capital cost)	Base case (2020 capital cost)	Scenario 1: half scale	Scenario 2: 2X scale	Scenario 3: reduced growth cycle (4 days)	Scenario 4: increase plant density
Stocks planted	5000	5000	2500	10,000	5600	6000
Stocks harvested	4500	4500	2250	9000	5040	5400
Packs sold (90 g each)	3600	3600	1800	7200	4032	4320
Price/kg ($)	12.00	12.00	12.00	12.00	12.00	12.00
Total revenue ($)	1,399,680	1,399,680	699,840	2,799,360	1,567,642	1,679,616
Total variable cost ($)	610,529	610,530	316,281	1,199,027	682,031	715,010
Total fixed cost ($)	749,943	616,443	386,468	1,076,394	644,721	658,398
EBIT ($)	39,208	172,707	(2908)	523,938	240,890	306,207
EBITDA (EBIT + deprec.) ($)	333,608	333,607	77,542	845,739	403,025	465,692
Equipment cost ($)	2,207,998	1,206,752	603,376	2,413,504	1,216,013	1,196,133
Crop space (m^2)	2168	2168	1084	4336	2184	2149
Facility size (m^2)	997	997	587	1817	1003	990
ROI (based on EBIT)	1.78%	14.31%	−0.48%	21.71%	19.81%	25.60%
Number of employees	9	9	5	17	10	10
Growth cycle (days)	38	38	38	38	34	38
Salable biomass (kg m^{-2} y^{-1})	53.8	53.8	53.8	53.8	59.8	65.1

TABLE 14.6 Complete table of scenario analysis of changes in Japanese plant factory profitability (cont.).

Scenarios based on the Japanese smart plant factory model (cont.)	Base case (2020 capital cost)	Scenario 5: +10% biomass/ plant	Scenario 6: +20% lighting efficiency	Scenario 7: U.S. labor costs	Scenario 8: Crop with 1 25-day cycle	Scenario 9: Increase market price 5%
Stocks planted	5000	5000	5000	5000	7200	5000
Stocks harvested	4500	4500	4500	4500	6480	4500
Packs sold (90 g each)	3600	3600	3600	3600	5184	3600
Price/kg ($)	12.00	12.00	12.00	12.00	12.00	12.60
Total revenue ($)	1,399,680	1,539,648	1,399,680	1,399,680	2,015,539	1,469,664
Total variable cost ($)	610,530	610,530	582,172	762,161	835,979	610,529
Total fixed cost ($)	616,443	616,443	616,443	616,443	715,501	616,443
EBIT ($)	172,707	312,675	201,065	21,076	464,059	242,692
EBITDA (EBIT + deprec.) ($)	333,607	473,575	361,966	181,976	626,381	403,592
Equipment cost ($)	1,206,752	1,206,752	1,206,752	1,206,752	1,217,411	1,206,752
Crop space (m^2)	2168	2168	2168	2168	2187	2168
Facility size (m^2)	997	997	997	997	1004	997
ROI (based on EBIT)	14.31%	25.91%	16.66%	1.75%	38.12%	20.11%
Number of employees	9	9	9	9	11	9
Growth cycle (days)	38	38	38	38	25	38
Salable biomass (kg m^{-2} y^{-1})	53.8	59.2	53.8	53.8	76.8	53.8

References

Agrilyst, 2016. State of Indoor Farming. www.agrilyst.com.

Agrilyst, 2017. State of Indoor Farming. www.agrilyst.com.

Agritecture Consulting, Autogrow, 2019. Global CEA Census 2019. https://www.agritecture.com/census.

Bian, Z.H., Yang, Q.C., Liu, W.K., 2015. Effects of light quality on the accumulation of phytochemicals in vegetables produced in controlled environments: a review. J. Sci. Food Agric. 95 (5), 869–877. https://doi.org/10.1002/jsfa.6789.

Blin, J.-M., 1980. Comments on the economics of markets: a simple model of the market-making process. J. Bus. 53 (3), S193–S197. http://www.jstor.org/stable/2352223.

Clark, S., 2020. Financial viability of an on-farm processing and retail enterprise: a case study of value-added agriculture in rural Kentucky (USA). Sustainability 12 (2). https://doi.org/10.3390/su12020708.

Contain, 2019. Automation, AI & the Next Generation of Indoor Agriculture. https://indoor.ag/white-papers/.

Dalrymple, D.G., 1973. Controlled Environment Agriculture: A Global Review of Greenhouse Food Production. Foreign Agricultural Economic Report No. 89. U.S. Dept. of Agriculture, Economic Research Service, Washington, D.C. http://catalog.hathitrust.org/Record/011392828.

Goodman, W., Minner, J., 2019. Will the urban agricultural revolution be vertical and soilless? A case study of controlled environment agriculture in New York city. Land Use Pol. 83 (June 2018), 160–173. https://doi.org/10.1016/j.landusepol.2018.12.038.

Hatfield, J.L., Prueger, J.H., 2015. Temperature extremes: effect on plant growth and development. Weather Climate Extremes 10, 4–10. https://doi.org/10.1016/j.wace.2015.08.001.

Heuvelink, E., Challa, H., 1989. Dynamic optimization of artificial lighting in greenhouses. Acta Hortic. 260, 401–412. https://doi.org/10.17660/ActaHortic.1989.260.26.

van Iersel, M.W., 2017. Optimizing LED lighting in controlled environment agriculture. In: Gupta, S.D. (Ed.), Light Emitting Diodes for Agriculture: Smart Lighting. Springer, Singapore, pp. 59–80. https://doi.org/10.1007/978-981-10-5807-3_4.

Integrative Economics, 2016. The Vertical Farming Market in North America. Industry Scan and Market Potential Assessment: Food Crops, Bio-Pharmaceuticals, and Medical Cannabis. https://www.verticalfarming.com/.

Kelly, N., Choe, D., Meng, Q., Runkle, E.S., 2020. Promotion of lettuce growth under an increasing daily light integral depends on the combination of the photosynthetic photon flux density and photoperiod. Sci. Hortic. 272, 109565. https://doi.org/10.1016/j.scienta.2020.109565.

Kozai, T., 2018. Smart Plant Factory − The Next Generation Indoor Vertical Farms. Springer, Singapore. https://doi.org/10.1007/978-981-13-1065-2.

Kozai, T., Niu, G., Takagaki, M., 2015. Plant Factory: An Indoor Vertical Farming System for Efficient Quality Food Production. Elsevier.

Kubota, C., Kroggel, M., Both, A.J., Burr, J.F., Whalen, M., 2016. Does supplemental lighting make sense for my crop?-empirical evaluations. Acta Hortic. 1134, 403–411. https://doi.org/10.17660/ActaHortic.2016.1134.52.

Lensing, C., 2018. Controlled Environment Agriculture: Farming for the Future? https://www.cobank.com/knowledge-exchange/specialty-crops/controlled-environment-agriculture.

Mejía, J.W., 2020. Feasibility Study of a Vertical Farm as a Social Enterprise in Pomona. California State Polytechnic University, California.

Newbean Capital, 2017. Indoor Crop Production, Feeding the Future. https://indoor.ag/white-papers/.

Newbean Capital, Local Roots, Proteus Environmental Technologies, 2015. Indoor Crop Production: Feeding the Future. Indoor Ag-Con. http://indoor.ag/white-paper.

Nicole, C.C.S., Charalambous, F., Martinakos, S., Van De Voort, S., Li, Z., Verhoog, M., Krijn, M., 2016. Lettuce growth and quality optimization in a plant factory. Acta Hortic. 1134, 231–238. https://doi.org/10.17660/ActaHortic.2016.1134.31.

de Nijs, B., 2017. Does Vertical Farming Make Sense? Hortidaily.Com. http://www.hortidaily.com/article/35974/Does- vertical-farming-make-sense.

Porter, M.E., 1979. How competitive forces shape strategy. Harv. Bus. Rev.

Ramríez-Arias, A., Rodríguez, F., Guzmán, J.L., Berenguel, M., 2012. Multiobjective hierarchical control architecture for greenhouse crop growth. Automatica 48 (3), 490–498. https://doi.org/10.1016/j.automatica.2012.01.002.

Rodríguez, F., Berenguel, M., Arahal, M.R., 2003. A Hierarchical Control System for Maximizing Profit in Greenhouse Crop Production. European Control Conference, ECC 2003, pp. 2753–2758. https://doi.org/10.23919/ECC.2003.7086458.

Shamshiri, R.R., Kalantari, F., Ting, K.C., Thorp, K.R., Hameed, I.A., Weltzien, C., Ahmad, D., Shad, Z., 2018. Advances in greenhouse automation and controlled environment agriculture: a transition to plant factories and urban agriculture. Int. J. Agric. Biol. Eng. 11 (1), 1–22. https://doi.org/10.25165/j.ijabe.20181101.3210.

USDA Agricultural Marketing Services, 2020. Plant Variety Database. https://www.ams.usda.gov/resources/data.

USDA Economic Research Service, 2019. Price Spreads from Farm to Consumer/Fresh Lettuce, Iceberg. https://www.ers.usda.gov/data-products/price-spreads-from-farm-to-consumer/.

USDA National Agricultural Statistics Service, 2015. USDA Census of Horticultural Specialties (2014), vol. 3.

Zeidler, C., Schubert, D., Vrakking, V., 2017. Vertical Farm 2.0: Designing an Economically Feasible Vertical Farm − A Combined European Endeavor for Sustainable Urban Agriculture. https://www.researchgate.net/publication/321427717.

Business model and cost performance of mini-plant factory in downtown

Na Lu[1], Masao Kikuchi[1], Volkmar Keuter[2] and Michiko Takagaki[1]

[1]Center for Environment, Health and Field Sciences, Chiba University, Kashiwa, Chiba, Japan;
[2]Department Photonics and Environment, Fraunhofer Institute for Environmental, Safety, and Energy Technologies UMSICHT, Oberhausen, Germany

15.1 Introduction

With the rapid development of the plant factory technology, the business model and the application range of plant factories have become wider and more diverse. It is a distinct direction in this trend that many companies have begun to develop mini-type plant factories with artificial lighting (mini-PFALs) with various innovative applications. In a previous article (Takagaki et al., 2020), the definition of mini-PFALs is indoor plant—growing systems, used not only for the commercial plant production and sales but also for various other purposes. The main function of the mini-PFALs is not limited to plant production efficiency, but, in many cases, is shifting to other functional levels. For example, the mini-PFALs can be used as a hobby kit, as a decoration, new concept of art, or as a therapy to improve quality of life, etc. It can also be applied for extreme regions, such as Antarctic and desert regions, to help solving the scarcity in fresh foods. The new applications of mini-PFALs could help achieving many of the sustainable development goals set by the United Nations (2015), such as "good health and well-being," "Clean water and sanitation," "Sustainable cities and communities," and "responsible consumption and production".

In this chapter, we describe the basic composition of mini-PFALs, their functions, applications, and development status in various countries and regions. As the most important core of achieving the sustainability of development, their business models are introduced, and their cost performances are analyzed.

15.2 Concepts and basic elements of mini-PFALs

15.2.1 Concepts

In this chapter, we classify the mini-PFAL as nonwalk-in type and walk-in type.

15.2.1.1 Nonwalk-in type
(1) Closed type, for example, cabinet style, with a certain level of environmental control equipment such as fans, pumps, but no air conditioner and CO_2 supply.
(2) Open shelf type or on-desk type, no environmental control system expects for the light.

15.2.1.2 Walk-in type

The walk-in type mini-PFALs usually have environmental control systems and can control air temperature, light, CO_2, water and fertilizer, etc.

15.2.2 Basic elements

Regardless of the type of mini-PFAL, the following basic elements are indispensable:

(1) support frame or shelf;
(2) tray or pot (fill with nutrient solution or medium);
(3) lighting system;
(4) water;
(5) fertilizer; and
(6) plants.

The material and design of a mini-PFAL are to be selected according to different uses. For example, the PFALs that are placed in kitchens, restaurants, and leisure clubs would use relaxing light color, rather than the light spectrum that is most effective for photosynthesis. Irrigation systems should also be designed in consideration of factors such as convenience, mobility, and noise levels. Sometimes, it is necessary to use substrate or soil too. There are also some very distinctive features with table lamps, as well as special designs for sound therapy and horticulture therapy. As for plants, most of them are vegetables and flowers. The ornamental plants are occasionally used for hobby kits and therapy purposes.

15.3 Business models

15.3.1 Business models in Asia

The nonwalk-in type mini-PFALs are mostly applied in restaurants, homes, offices, and indoor public spaces.

15.3.1.1 Restaurants

Various types of leafy vegetables can be grown together to meet the needs of the menu with different recipes in restaurants. These types of mini-PFALs are usually very exquisitely and beautifully designed, take up little space, and have multiple functions, such as integration into a wall or used as a screen to separate seats (Figs. 15.1 and 15.2).

FIGURE 15.1 The mini-PFALs placed in a restaurant. *(Reprinted from the Figure 6.2A in the book of "Plant Factory", second edition).*

FIGURE 15.2 The mini-PFAL used in a restaurant fixed to the wall and as screens (Planet Co. Ltd., Toyohashi, Japan).

15.3.1.2 Homes

For home use mini-PFALs, the types of plants are usually more extensive, ranging from leafy vegetables, aromatic herbs to small fruity vegetables, flowers, and even to ornamental plants. Most of the mini-PFALs are placed in the kitchen area so that the plants can easily be picked and used when cooking (Figs. 15.3 and 15.4). Some companies developed small shelf type and on-desk type mini-PFALs that can be used in living rooms and bedrooms. The on-desk type becomes popular, attracting kids in particular, due to its cute design and multiple functions (Fig. 15.3).

15.3.1.3 Offices and indoor public spaces

Products for offices and public spaces have also been developed recently. As well known, many people face significant stress in the workplace, which often deteriorates their performance and health. It has been reported that indoor plants reduce the discomfort symptoms of workers in offices (Fjeld et al., 1998; Wood et al., 2002; Bringslimark et al., 2007; Toyoda et al., 2020). The newly developed mini-PFAL products for office use are expected to relieve the stress of workers, allow them to relax moderately, and increase their vitality and productivity (Fig. 15.5). Products placed in public spaces are mainly used to create a green

FIGURE 15.3 The mini-PFALs used in living room, bedroom, and kitchen as hobby kits in homes (SINOIN-NOVO TECH, China).

FIGURE 15.4 The mini-PFALs used in kitchen at homes (J&C Smart-I shape, China).

atmosphere, improve the area of green space, increase the chance of contact between people and nature, and make people feel comfortable. Meanwhile, the produced plants also can be used for cooking or for sale (Fig. 15.6).

15.3.1.4 Horticultural therapy

In modern society, one of the common issues is the increase in medical and welfare budgets due to aging and stressed state. The mini-PFALs are also used for horticultural therapy in Japan. Horticultural therapy focuses on the effects of horticulture on the human mind and body and improves health and life quality of people. The therapy can be used as a tool to

FIGURE 15.5 The mini-PFALs used in offices (SINOINNOVO TECH, China).

FIGURE 15.6 The mini-PFALs used in hotel lobby and shopping malls (SINOINNOVO TECH, China).

relieve stress for people in stress (the so-called preillness stage) and who need assistance due to aging or disabilities. The mini-PFAL can be easily and flexibly adapted in horticultural therapy (Fig. 15.7).

Likewise, mini-PFALs can also be used as a healing tool for rehabilitation of the patients after returning home. For example, under the recent novel coronavirus epidemic, many people, who have suffered from it, survived and left hospitals still face many troubles in the recovery process, which may take months, years, or even the rest of their lives. Some of them are also confronting cognitive and emotional issues (Belluck, 2020). It has been proved that interaction with plants could bring relaxing effects and reduce the risk of disease and medical cost (Thomsen et al., 2011). The mini-PFAL is, therefore, considered as a tool bringing relaxing effects, improving life quality, and reducing medical expenses in such cases.

FIGURE 15.7   The mini-PFALs used for horticulture therapy and horticulture classes (Planet Co. Ltd., Japan).

15.3.1.5 Soundscape design

In the design of many mini-PFALs, water pumps are needed to pump out and circulate nutrient solution in the cultivation system to provide nutrients for plant growth. In most cases, it is unavoidable that the sound of the pump and the flow of the nutrient solution create noise. A research considered how to transform noise into comfortable and beautiful sound (Shono et al., 2018). This soundscape design is integrated into the design of mini-PFAL system which is a fusion of sound and green. Both the sound and plants are expected to bring healing effects to people. It helps to open awareness and connect with the surrounding environment through sound (Fig. 15.8). This type of mini-PFAL can be applied into many public spaces, such as lobby, hotel room, office, hospital, museum, and fitness and recreation centers.

In this system, a specifically designed "sound instrument" is installed into the mini-PFAL in such a way that the nutrient solution is dropping continuously on the surface of the instrument to create restful sound (Fig. 15.8). The examples of the frequency characteristics of the sound pressure level are shown in Fig. 15.8. The sound pressure level can be controlled by adjusting the amount of the nutrient solution flow and the number of instruments for various sound environments.

15.3.1.6 Walk-in type mini-PFAL in supermarket

Some mini-PFALs are also applied in supermarkets. Some supermarkets in China sell fresh vegetables grown in mini-PFALs, placed in their premises, such as in the backyard, directly to their customers. In Japan, the walk-in type product from PLANET company is placed in the vegetable area of a supermarket (Fig. 15.9), where various potted herbs and vegetables (such as pepper mint, basil, and lettuces) grown in the mini-PFAL are sold to consumers for consumption or cultivation.

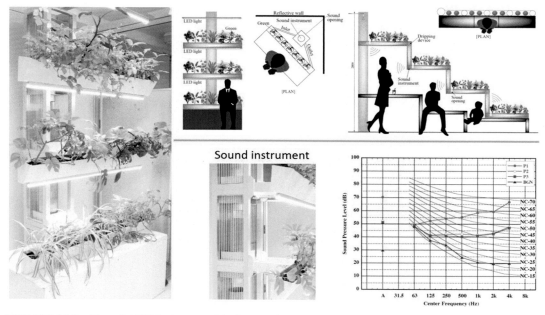

FIGURE 15.8 The mini-PFALs integrated with special soundscape designs inside the water circulation system. *The system was developed by Planet Co., Ltd, TECH-TAIYO KOGYO Co., Ltd, Soundscape Labo and Chiba University. Photos and figures are provided by Taiko Shono and Na Lu.*

FIGURE 15.9 Walk-in type mini-PFAL used in supermarkets. *(Picture Credit: by Planet Co. Ltd., Japan).*

15.3.1.7 Walk-in type in park or urban community

More than one million shipping containers are discarded every year in the world. After careful design, the abandoned shipping containers can be recycled and well integrated with the plant factory technologies. The company SANANBIO has designed a series of integral container–type mini-PFALs that can be placed in urban public and leisure spaces. In Fig. 15.10, a mini-PFAL is displayed in a park for science education and sightseeing purposes. The produced vegetables are supplied to nearby city residents. This allows them to experience freshness in life with clean and pollution-free vegetable products.

FIGURE 15.10   The walk-in type mini-PFAL in a park. *(Picture Credit: Fujian Sanan Sino-Science Photobiotech, Co. Ltd., China)*

15.3.1.8 Supply chain model

Many companies only provide a single type of product, but there are companies like SINOINNOVO TECH, which also provide a full range of product supply chain business models. Fig. 15.11 shows the concept of supply chain business model of SINOINNOVO TECH company. The "FARMBOX" is a medium-scale plant factory (customized) that serves as a hub or transfer station, providing plant seedlings, end consumer vegetables, and other transit products. They are generally placed in a community or business district in town, directly supplying end consumer vegetables (hydroponic vegetables, organic vegetables, and other high-quality vegetables, through designated suppliers) and semiended vegetables (seedlings and live vegetables) that can be cultivated by customers at home. A dedicated operation team maintains and regularly loads the FARMBOXs scattered around the city and also provides customers with know-hows on plant cultivation. The main product is a series of so called "ECOGREEN" mini-PFALs, as shown in Figs. 15.3 and 15.5, which are used on the client side. Customers can buy seedlings or live vegetables from the FARMBOX and continue growing them at home or offices. The mini-PFAL

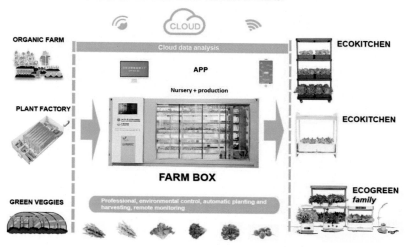

Supply Chain Business Model

FIGURE 15.11 The business model combined with nonwalk-in type and walk-in type mini-PFALs in a city (SINOINNOVO TECH, China).

products are equipped with sensors that can collect growth data and send them to their central cloud data analysis system. As a back-end support, this system is equivalent to a remote enterprise resource planning management system for vegetable production, which is convenient for order management and distribution service based on big data and LBS (location-based services).

15.3.1.9 Future applications: extreme climate area, space exploration, and ships

Although most mini-PFALs are used in urban areas, their flexibility also allows them to be applied to remote isolated regions in the future. Some prototype mini-PFALs are already developed for extreme climate areas, space exploration (Gaind, 2017), and ships for scientific research. Different from ordinary plant factories, the design of the container-type mini-PFAL requires the integration of the operating system and the modularization of components to facilitate transportation and installation navigation (Figs. 15.12 and 15.13). It is usually installed firmly inside, and the outside is relatively flat and tidy without external accessories to suit for long-distance transportation and installation under various environmental conditions. In extreme climate area, such as South Pole station and remote mountainous areas with no access to grid, a full solar power supply system is needed. In design, continuous circulating solar panels can be used for power supply to reduce energy consumption. The DC low voltage network system would be better for safe and reliable reason. The mini-PFALs used in ships are treated with anticorrosion coating to resist salt spray corrosion during navigation (Fig. 15.14).

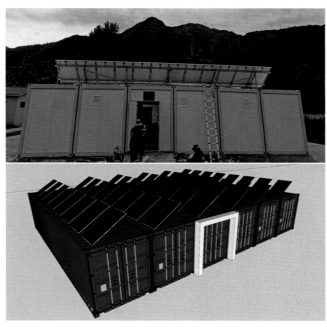

FIGURE 15.12 Container type mini-PFALs applied to remote mountain area. *(Picture Credit: Xiamen Lumigro Technology Co. Ltd., China).*

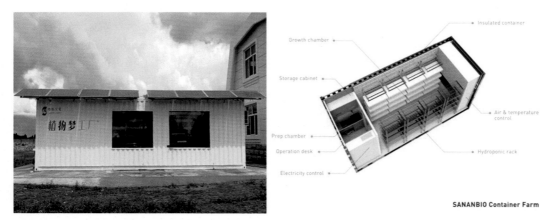

FIGURE 15.13 Container type mini-PFALs applied extreme cold area. (In north area of China) *(Picture Credit: Fujian Sanan Sino-Science Photobiotech, Co. Ltd., China). Some other applications: https://www.youtube.com/watch?v=_jQsxE2BmQE&app=desktop*

FIGURE 15.14   Container-type mini-PFALs applied to ships (Xiamen Lumigro Technology Co. Ltd., China).

15.3.2 Business models in Europe

Achieving a secure supply of fresh and high-quality food while at the same time reducing the environmental impact of food production will place enormous pressure on food production systems in Europe. Several studies currently indicate that the challenges faced by developed countries will not be related to the maximization of yields of agricultural crops, but rather maintaining current yields with reduced environmental impact. In that sense, food security will only be achieved through sustainability.

In Europe, especially in Germany, different projects and concepts are summarized under the term Controlled Environment Agriculture (CEA). The interpretation of CEA ranges from greenhouse structures, as more often seen in the Netherlands and different parts of Germany, over building integrated structures as the "Altmarktgarten" in Germany's Ruhr area (Fig. 15.15, Weidner et al., 2020), where the production area is complemented by a research area, with a total area of more than 1000 m^2. It will be possible to research future urban agriculture systems in the "Altmarktgarten" and develop technical systems within a metropolitan context. Further examples of CEA are the BIGH Farm in Brussels, the "Indoor Farms" of Jones Food Company, or smaller units. A large number of these projects are situated in metropolitan areas and are therefore part of an "urban green revolution." Nevertheless, in contrast to some other countries, the expansion of Indoor Farms is still quite slow. Reasons might be the well-established greenhouse industry in different parts of Europe and the high price sensitivity of the cultivated products.

The market of "Indoor Farming" systems in Europe is emerging, but still young and quite fragmented. It is likely that technology developers and start-ups will fill the market niches until large companies step in.

FIGURE 15.15   The Building integrated CEA *Altmarktgarten*, Oberhausen, Germany. *Picture Credit: Fraunhofer UMSICHT.*

15.3.2.1 *Examples of mini-PFALs in Europe*

Currently, there are a couple of concepts and some products as described above on the worldwide market for mini-PFALs, also known as appliance farms in Europe. Putting mini-PFALs in consumers' homes is not yet a mainstream concept, but indoor gardening is increasingly moving toward a larger consumer groups that go beyond early adoption tech enthusiasts and is becoming a business field for large companies.

To date, mini-PFALs products or concepts have been predominantly offered by Asian and US home appliance manufacturers such as LG, Samsung, and GE. However, European companies are seeing a new field of engagement and gradually coming up with their own product offerings. For instance, in 2019, the German company BOSCH launched Smart Grow a countertop mini-PFAL system designed to simplify the cultivation of herbs and greens that is part of Bosch's "Smart Home" product line. The system is available in two sizes that accommodate either three or six prepackaged seedling capsules of a variety of plants that go from herbs such as basil, parsley, sage, etc., to edible flowers and microfoods. The plants are illuminated and watered by a patented lighting and irrigation system.

Other types of shelf type mini-PFALs are the US AeroGarden or the Estonian Click & Grow. Kitchen-integrated systems as the product concepts by LG, GE, and Samsung are more sophisticated. While the Samsung's "Chef Garden," the LG's "Harvester," or the "Plantbox" by AIPLUS are still product concepts, German company Miele is already on the market with the "Plantcube" (Fig. 15.16). Miele acquired former startup Agrilution in late 2019 and integrated it as a new business unit under the Miele brand. The founders of Agrilution started their business development of personal farming appliances in 2013, and at present, they have 30 employees. Today, they have a product on the European market (Germany, Austria, and Benelux), which is still quite cost-intensive, but, according to one of the founders, within 5 years (from 2020) the price is expected to drop within the range of washing machines (mid-price segment). As many new home appliances, the "Plantcube"

FIGURE 15.16 The mini-PFALs used in homes in Germany. *(Picture Credit: Agrilution (Miele), Germany)*

comes along with an own APP to control the mini-PFAL and is appropriate for the cultivation of a wide variety of herbs, microgreens, and salads. The advantage of the "Plantcube" is the simplicity with which the average user can harvest fresh greens daily.

Besides the price reduction goals, Agrilution's plans for the future are related to expanding in further countries in Europe and worldwide as well as integrating new smart technologies into the "Plantcube."

In contrast to the appliance products described above, the Dutch "Kweeker" is a stand-alone greenhouse style product for indoor as well as outdoor applications. Nevertheless, it comes along with the same features, such as LED lighting panels and an own APP for controlling the cultivation process. The "Kweeker" does apply soil cultivation boxes instead of nutrient solutions and is sold via distributors in the Netherlands.

One of the most prominent representatives of this category in Europe is the Berlin-based company "Infarm". The "Farming as a Service" business also pays off economically for the company: Infarm charges a fee for setting up the mini-PFALs in supermarkets or restaurants. The hardware remains to Infarm company and the crops belongs to the supermarkets. The company is already responsible for more than 200,000 plants per month in Denmark, France, Germany, Luxembourg, the United States of America, and Switzerland (Jendrischik, 2020). Founded in 2013, Infarm meanwhile has underwent a quite impressive expansion. As of September 2020, the company raised a series C round of funding with new VC investors (Heuberger, 2020). Raising more than US$ 150 million in the first two rounds, having about 250 employees (Kapalschinski and Matthes, 2019) and more than 900 installations in seven

FIGURE 15.17   The mini-PFAL of Infarm at a Marks & Spencer supermarket in London, UK.

countries of Europe and Northern America (Kapalschinski, 2020). In addition, Infarm is supplying their decentralized farms via 10 larger and centralized locations. As of September 2020, there are existing contracts with food retailers, e.g., in Germany (e.g., Metro, Aldi, Kaufland, and Edeka), in Denmark (e.g., Irma), or the United Kingdom (e.g., Marks and Spencer, Fig. 15.17) (Infarm, 2020).

15.3.2.2 Future perspectives

Current agricultural systems contribute significantly to exceeding the Planetary Boundaries, particularly due to the excessive and inefficient use of fertilizers (Bönisch et al., 2016; Rockström et al., 2009). In Europe, currently only a small share of the nutrients present in industrially produced fertilizers applied in agriculture for the cultivation of food crops are effectively consumed as food. In addition to food waste and unbalanced diets, this is one of the main causes for the low nutrient use efficiency in the European food system (Schulze, 2016). Achieving a secure supply of fresh and high-quality food while at the same time reducing the environmental impact of food production will place enormous pressure on food production systems.

Current empirical evidence suggests that urban and peri-urban agriculture will play a key role in providing a stable food supply for the cities of the future (Dubbeling et al., 2017; Eigenbrod and Gruda, 2015; Jennings et al., 2015). Existing CRFS (City Region Food Systems) initiatives are already integrating agricultural production into the urban infrastructure by means of vertical and building-integrated approaches and are playing an important role in the Urban Food Revolution (Kraas et al., 2016; Ladner, 2011). Moreover, agriculture consumes around 104 million metric tons of nitrogen, 46 million metric tons of phosphate (P_2O_5), and 33 million metric tons of potassium (K_2O) worldwide every year (Jennings et al., 2015). Due to the rapidly growing world population's need for fresh food and other agricultural products, an average annual increase of 2%–3% per component is currently expected. Europe is not self-sufficient in its supply of phosphorus and nitrogen, but relies on imports from Morocco, China, the United States of America, and other countries, which are in turn influenced by fluctuations in the global fertilizer and energy market (Eigenbrod and Gruda, 2015).

Looking at the national and European regulation and legislation level (AbfKlärV (German Federal Ministry for Environment, Nature Conservation and Nuclear Safety, 2017), DüMV (German Federal Ministry of Food and Agriculture, 2019)), foundations are being laid for the use of greater proportions of phosphorus from secondary sources of raw materials such as wastewater in the agricultural sector. The spreading of wastewater sludge on agricultural land will, however, be phased out. The overarching goal of a German national funded research project (SUSKULT) is to introduce urban agricultural production as a component of the circular urban economy of the future. SUSKULT is specifically developing a hydroponic PFAL-based sustainable and local food production system that will recover the essential resources of water, nitrogen, phosphorus, potassium, CO_2, and heat from the operation of a "wastewater treatment plant of the future," so-called NEWtrient Center (Fig. 15.18). In the future, these centers could treat resource flows that include all nutrients in cities, e.g., also biodegradable waste. However, liquid fertilizers produced within the NEWtrient Centers can play an important role also for the different mini-PFAL types described in this chapter.

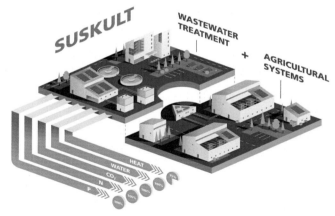

FIGURE 15.18 Concept of the SUSKULT urban circular agricultural production. *Picture Credit: Fraunhofer UMSICHT.*

15.4 Cost performance of Mini-PFALs in downtown

As presented thus far in this chapter, numerous types of mini-PFAL technology have been developed, and many business models using these technologies have been burgeoning in many urban areas in the world; business models to open up new markets for various types newly developed mini-PFALs as well as to apply these mini-PFALs in their businesses, such as plant production, supermarket, restaurant, horticulture therapy, environment control, etc. The fact that many firms for, and with, mini-PFALs are commercially operated as going concerns in Asia as well as in Europe is a prima facie evidence that the businesses related to mini-PFALs could be economically viable and sustainable. In this section, we try to demonstrate how profitable these businesses are and what factors, if any, are constraining the businesses to be so. We take up three cases, all of which are business models that apply mini-PFALs in their businesses, and all the three are of commercially operating firms using the mini-PFALs for plant production. This selection of the cases is due mainly to the data availability; it is easier for these business models to obtain data for examining the cost performance, the benefit in particular, than for such business models as horticulture therapy and environmental control, many of which are not commercial businesses but the projects of public or nonprofit organizations. As the first trial to assess the cost performance, we begin with the commercial business models.

15.4.1 Data

Data were obtained from three commercial firms, one in China and two in Japan.

15.4.1.1 Supermarket in metropolitan area, China (Case I)

A large supermarket operating with a walk-in type mini-PFAL, the same size as shown in Fig. 15.10, made from 20 ft shipping container of $14.88 \, m^2$; placed in the supermarket premises outside of the supermarket shopping zone; with eight LED sets of 280W each; producing every day 45 plants (4.5 kg) of lettuce with 14 h lighting; selling the output in the supermarket directly to final consumers at the fix price of 6 yuan per 100 g.

15.4.1.2 Supermarket in Tokyo, Japan (Case II)

A large supermarket operating with a walk-in, display-window type mini-PFAL, shown in Fig. 15.9, of the size of $3 \, m^2$; with 18 LED lumps of 18 W each; producing every month 872 pots of various herbs and vegetables; with 16 h lighting per day; selling the output in the supermarket directly to final consumers, for consumption or cultivation at home, at the fixed price of 220 yen per pot.

15.4.1.3 Restaurant in Tokyo metropolitan area, Japan (Case III)

A restaurant operating with a mini-PFAL, consisting of a table-type display cabinet for seedlings and two shelf-type display cabinets for plant growth, shown in Fig. 15.1; with two LED lumps of 50 W each for the seedling cabinet and an LED lump of 200 W each for the plant growth cabinet, with 12 h lighting per day; producing 3.2 kg of lettuce per month; serving the output to customers as salad dishes, 50 g of lettuce per dish, at the fixed price of 700 yen per dish, of which 20% is attributed to the costs of materials and services other than lettuce, such as salad dressing, utensils, and serving services.

15.4.2 Cost performance of mini-PFAL: definition

In this study, the economic viability, or the cost performance, of mini-PFALs is measured by means of the benefit—cost ratio. The benefit of a small plant factory is defined as the total revenue of a mini-PFAL:

$$B = P_y \tag{15.1}$$

where B is the benefit of the mini-PFAL, Y is the quantity (kg/year) of the output produced and sold by the mini-PFAL operator, and P_y is the unit price (US $/kg) at which the output is sold. The production cost of the plant production in a mini-PFAL operation is comprised of five groups of costs: current input costs, labor costs, depreciation costs, maintenance costs, and capital interest costs.

Current inputs are the production inputs the values of which are transferred to the output during a production period, included in the case of plant production in mini-PFALs are seeds, media, nutrient solution, water, and electricity for LED lighting, environment control, and pumping for nutrient solution circulation.

Labor inputs are required for seeding, harvesting, preparing nutrient solution, watering, cleaning, and selling/marketing/serving.

The depreciation costs are for the mini-PFAL system installed, obtained by dividing the initial investment in the system by the lifespan of the system. The maintenance costs are the costs for the maintenance and fine-tuning of the mini-PFAL and its equipment, facilities, and devices.

The capital interests are the opportunity cost of funds invested in these capital goods. The working capital used for acquiring current production inputs and hiring labor could bear interests as well, if used elsewhere. However, such interests are not considered in this study, since the working capital for the mini-PFAL operation revolves rapidly, monthly, weekly, or even daily.

The total cost of producing a plant in a mini-PFAL is expressed in the following equation:

$$C = \left[\sum_i P_i \, \text{Input}_i \right] + \left[w \sum_j \text{Labor}_j \right] + [I / LS] + [\alpha I] + [rI] \tag{15.2}$$

where C = total cost (US $/year/factory), Input_i = the quantity of ith current input used in the production per year, P_i = the unit price of ith current input, Labor_j = labor inputs (person-hours/year) used in the production for jth labor activity, w = the wage rate (US $/person-hour), I = the acquisition cost of the mini-PFAL (US $), LS = the lifespan of mini-PFAL (years, assumed 10 years), α = maintenance costs of the small plant factory (in % share of the investment, assumed 2%), and r = the interest rate (%/year, assumed 5% in China and 2.5% in Japan). Eqs. (15.1) and (15.2) give the benefit—cost ratio for the plant production of mini-PFALs as follows:

$$B/C = P_i Y_i / \left\{ \left[\sum_i P_i \, \text{Input}_i \right] + \left[w \sum_j \text{Labor}_j \right] + [I / LS] + [\alpha I] + [rI] \right\} \tag{15.3}$$

The greater the B/C, the better the economic performance of the small plant factory, and if $[(B/C) - 1] \rangle r$, the plant production operation of the mini-PFALs is economically viable.

15.4.3 Cost performance of mini-PFAL: results

The necessary data to estimate the benefit-cost ratio and the results of estimation are shown in Tables 15.1 and 15.2, respectively. All the value terms are expressed in terms of US $, converting from the local currencies by using the exchange rates of US $ 1.0 = Yuan 6.0 = Yen 100.0.

TABLE 15.1 The unit price of output and inputs and the quantity produced and used in the plant production by mini-PFALs operated by commercial firms in China and Japan.[a]

	Unit	Case I (China) walk-in mini-PFAL in the premises of supermarket[b]		Case II (Japan) walk-in, display mini-PFAL in supermarket[c]		Case III (Japan) display mini-PFAL in restaurant[d]	
		Unit price (US $ (per unit)	Quantity (per year)	Unit price (US $ (per unit)	Quantity (per year)	Unit price (US $ (per unit)	Quantity (per year)
Plant output	kg	10.0	1,643	73.3	314	112.0	38
Production input							
Current input							
Seeds	1000 seeds	20	18.0	10	104.6	4.69	8.0
Media	Piece, pot, piece [e] 1		180	0.11	10,464	0.06	3,840
Nutrient	kg litter, litter[e]	4	50	0.01	24,000	20	14
Water	ton	0.5	35	0.5	11.68		
CO_2	kg	1	220				
Electricity	KWh	0.08	12,000	0.25	5,046	0.25	2,019
Labor	hour	3.0	456	10.0	312	12	123
Mini-plant factory							
Acquisition cost		60,000 $		50,000 $		5,000 $	
Lifespan		10 years		10 years		10 years	
Maintenance cost[f]		2.0%/year		2.0%/year		2.0%/year	
Interest rate		5.0%/year		2.5%/year		25%/year	

[a] *The value is expressed in US dollars. The exchange rates between US dollar, yuan, and yen used are US $ 1 = Yuan 6 = Yen 100.*
[b] *A walk-in type mini-plant factory of 15 m² (Fig. 15.10), producing leaf lettuce.*
[c] *A walk-in, display-window—type mini-plant factory of 3 m² (Fig. 15.9), used for producing potted herbs (pepper mint and basil) and vegetables (various kinds of lettuce).*
[d] *A display-window—type mini-plant factory, consisting of two ornamental cabinets for seedlings and plant growth (Fig. 15.1), used for producing leaf lettuce.*
[e] *In the order of Case I, Case II, and Case III.*
[f] *As the percentage of the acquisition cost.*

TABLE 15.2 Benefits and costs of plant production per year by mini-PFALs operated by commercial firms in China and Japan.[a]

	Case I (China) walk-in mini-PFAL in the premises of supermarket		Case II (Japan) walk-in, display mini-PFAL in supermarket		Case III (Japan) display mini-PFAL in restaurant	
	Value ($/year)	% Share	Value ($/year)	% Share	Value ($/year)	% Share
Benefit (B)						
Plant output	16,425		23,021		4301	
Cost						
Production cost						
Current input						
Seeds	360	2.7	1,046	7.3	38	1.2
Media	180	13	1,151	8.0	230	7.1
Nutrient	200	1.5	240	1.7	288	8.8
Water	18	0.1	6	0.0		
CO_2	220	1.6				
Electricity	960	7.1	1,262	8.8	505	15.5
Others	6	0.0	240	1.7		
Total	1,944	14.4	3,945	27.6	1,061	32.5
Labor	1,369	10.1	3,120	21.8	1,476	45.3
Total production cost	3,312	24.5	7,065	49.4	2,537	77.8
Investment-related cost						
Depreciation[b]	6,000	44.4	5,000	34.9	500	15.3
Maintenance cost	1,200	8.9	1,000	7.0	100	3.1
Interest rate	3,000	22.2	1,250	8.7	125	3.8
Total inv.-related cost	10,200	75.5	7,250	50.6	725	22.2
Total cost (C)	13,512	100.0	14,315	100.0	3,262	100.0
B/C ratio[c]	1.22		1.61		1.32	
Rate of return (%/year)	22		61		32	

[a] Computed from the data in Table 15.1. The value is expressed in US dollars. The exchange rates between U S dollar, yuan, and yen used are US $ 1 = Yuan 6 = Yen 100.
[b] Acquisition cost/lifespan.
[c] See Eq. (15.3).

15.4.3.1 Cost performance

For all cases, the mini-PFAL operation is economically viable. Even for Case I for which the rate of return is lowest, it is estimated to be 22% (Table 15.2), higher than the market interest rate of 5% (Table 15.1).

It is remarkable that the cost structure differs greatly among the three business models (Table 15.2). In Case I, the "mini-PFAL in the premises of supermarket" model, the investment-related cost dominates over the production cost, whereas in Case III, the latter dominates over the former. Case II stands just in-between, with nearly equal shares for these two cost groups. This difference could partly be explained by the difference in the business model. As a business model, Case I is most "traditional," in the sense that the mini-PFAL is just to produce vegetables, as the ordinary farming is to produce food crops. In contrast, in Cases II and III, the mini-PFALs, being placed in supermarket or restaurant for display to customers, produce some added values more than mere food crops, and the degree of customers' contact with the "mini-PFAL for display" is higher in the restaurant (Case III) than in the supermarket (Case II). As a result, the value productivity of mini-PFAL (output value/total investment-related cost) increases significantly from Case I to Case II, and to Case III.

It should be noted, however, that the differences in the cost structure are also resulted from the differences in the price structure between China and Japan. In the production cost, labor and electricity are the largest and the second largest cost items, respectively, in the three cases alike (Table 15.2). The unit price of these production inputs, however, is sharply different between the two countries: the price of electricity is US $ 0.08/KWh in China, whereas US $ 0.25/KWh in Japan and the wage rate is US $ 3/h in China, whereas US $ 10–12/h in Japan (Table 15.1). If these prices in China were the same as in Japan, the percentage shares of production cost and investment-related costs in Case I would be nearly the same as for Case II, and the B/C ratio would decline to less than unity.

15.4.3.2 Sensitivity test

The point just explained indicates that the cost performance of certain business model depends on the price structure of the country the model is adopted, as already pointed out in Section 15.3.2 for Europe. In order to check which factors are influential to the cost performance of the mini-PFAL plant production, sensitivity tests are conducted by estimating how the benefit–cost ratio changes when a certain parameter in Eq. (15.3), such as the output price, is allowed to change within a certain rage, for three given levels of the acquisition cost of the mini-PFALs for each case. The assumed levels of acquisition cost are $ 30,000, $ 60,000, and $ 90,000 for Case I, $ 25,000, $ 50,000, and $ 75,000 for Case II, and $ 2,500, $ 5,000, and $ 10,000 for Case III. The center level is the acquisition cost actually paid by the firm in each case. Fig. 15.19 depicts the results of the test for output price, lifespan, wage rate, and electricity price.

15.4.3.2.1 Output price

The output price is the most decisive factor of the cost performance of the mini-PFAL business models. For Case I, with the actual acquisition cost of $ 60,000, a decline in the output price from $ 10/kg to $ 8/kg reduces the B/C ratio to a level less than unity, i.e., below the breakeven level of cost performance. If the acquisition cost were low, $ 30,000, the breakeven

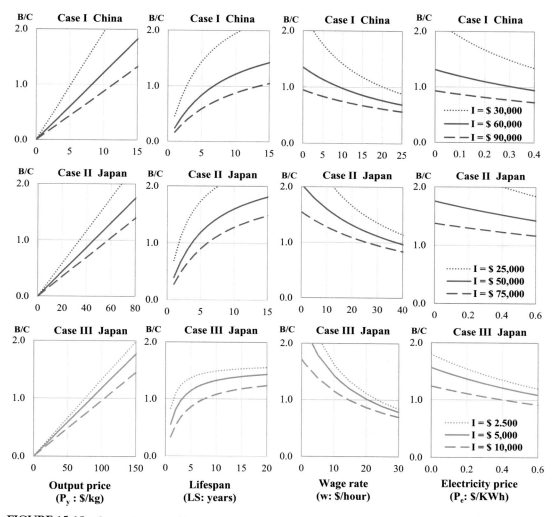

FIGURE 15.19 Sensitivity tests of factors that affect the cost performance of commercial operation of mini-PFALs: three business models in China and Japan.

P_y could be as low as $ 5/kg, whereas if it were high, $90,000, the breakeven P_y would be $ 12/kg. In the case of Case II with a display type mini-PFAL in a supermarket, the breakeven P_y is $ 46/kg for the actual acquisition cost of $ 50,000, and at the actual output price of $ 73.3/kg, the B/C ratio could still be larger than unity, even if the acquisition cost were $ 75,000. For Case III of display type mini-PFAL in a restaurant, the actual output price is $112/kg, which is about five times as high as the lettuce price in ordinary supermarkets in Japan. The breakeven price for the actual acquisition cost of $ 5000 is $ 85/kg, and even if I = $ 2,500, the breakeven price would be as high as $ 75/kg.

15.4.3.2.2 Lifespan of small plant factory

The useable lifespan of a mini-PFAL affects the cost performance through determining the depreciation of the mini-PFAL system. In this study, the lifespan of 10 years is assumed. The sensitivity analysis shows that the breakeven lifespan is less than 10 years for all the cases and all the levels of the acquisition cost of mini-PFALs, except for I = $ 90,000 for Case I and that the breakeven lifespan of the actual acquisition cost of I = $ 60,000 for Case I is 7 years. With these two exceptions, in no cases the breakeven lifespan exceeds 7 years. These results suggest that as long as the lifespan of five to 6 years is assured, it would not be a critical constraint for the introduction of mini-PFALs.

15.4.3.2.3 Wage rate

Labor is the cost item that takes the largest share in the production cost and its unit price, the wage rate, could be a factor that interferes the introduction of the mini-PFALs. For Case I in China, the wage rate at present is $ 3/h. Since the breakeven wage rate is $ 10/h for the actual acquisition cost of the mini-PFAL, some room is still left for the increase in the wage rate. Considering the rapid rate of increase in the wage rate in China in recent years, however, it would soon become a critical constraint to this business model. In fact, if I = $ 90,000, there is no chance for this model to be economically viable, even if w = 0. For the two business models in Japan, the situation in this respect seems not so serious. As long as labor is available at $w \leq$ $ 26/h, Case II could be economically viable and so is Case III for $w \leq$ $15/h.

15.4.3.2.4 Electricity price

The results of the sensitivity analysis indicate that the electricity price in not a constrain at all for two business models in Japan where the electricity price is world' highest (Saengtharatip et al., 2018); for the price region of $ 0/KWh < P_e < $ 0.6/KWh, a breakeven electricity price is seen only for I = $ 10,000 of Case III at the price as high as $ 0.4/KWh. For Case I in China, too, the electricity price seems not to be so serious constraint, unless the mini-PFAL is too expensive. For I = $ 60,000, the breakeven price is $ 0.32/KWh and for I = $30,000, it is $ 0.75/KWh (not seen in Fig. 15.19 because of out of range), not binding at all.

15.4.3.2.5 Interest rate

The sensitivity test is also conducted for the interest rate, though not shown in Fig. 15.19. Reflecting the low-interest regime in the world in the 2010s, the interest rates in the capital market in Japan and China are low. The results of the sensitivity analysis indicate that the interest rate is binding only for I = $90,000 of Case I, with the breakeven interest rate of 2.5% per year. For cases all the rest, the interest rates are not constraining the cost performance of the business models.

References

Belluck, P., 2020. Here's What Recovery from Covid-19 Looks like for Many Survivors. The New York Times. https://www.nytimes.com/2020/07/01/health/coronavirus-recovery-survivors.html. (Accessed 28 September 2020).

Bönisch, A., Engler, S., Leggewie, C., 2016. Fünf Minuten nach Zwölf? In: Steven, E., Oliver, S., Wilfried, B. (Eds.), Regional, innovativ und gesund. Vandenhoeck & Ruprecht, Göttingen, pp. 25–40.

Bringslimark, T., Hartig, T., Patil, G.G., 2007. Psychological benefits of indoor plants in workplaces: putting experimental results into context. Hortscience 42 (3), 581–587.

Dubbeling, M., Santini, G., Renting, H., Taguchi, M., Lançon, L., Zuluaga, J., De Paoli, L., Rodriguez, A., Andino, V., 2017. Assessing and planning sustainable city region food systems: insights from two latin american cities. Sustainability 9 (8), 1455. https://doi.org/10.3390/su9081455.

Eigenbrod, C., Gruda, N., 2015. Urban vegetable for food security in cities. A review. Agronomy for Sustainable Development 35 (2), 483–498.

Fjeld, T., Veiersted, B., Sandvik, L., Riise, G., Levy, F., 1998. The effect of indoor foliage plants on health and discomfort symptoms among office workers. Indoor Built Environ. 7 (4), 204–209.

Gaind, N., December 18, 2017. In pictures: the best science images of the year. Nature 552 (7685), 308–313. https://doi.org/10.1038/d41586-017-08492-y.

German Federal Ministry for Environment, 2017. Nature Conservation and Nuclear Safety. Sewage Sludge Ordinance (in German). https://www.bmu.de/en/law/sewage-sludge-ordinance/. (Accessed 22 September 2020).

German Federal Ministry of Food and Agriculture, 2019. Fertilizer Regulations (2019) (in German). https://www.bgbl.de/xaver/bgbl/start.xav?startbk=Bundesanzeiger_BGBl&jumpTo=bgbl119s1414.pdf. (Accessed 23 September 2020).

Heuberger, S., 2020. Liechtensteiner prinzenhaus beteiligt sich an Infarm (in German). https://www.gruenderszene.de/food/infarm-finanzierung-medienbericht. (Accessed 22 September 2020).

Infarm, 2020. https://www.infarm.de. (Accessed 22 September 2020).

Jennings, S., Cottee, J., Curtis, T., Miller, S., 2015. Food in an urbanised world: the role of city region food systems. Urban Agriculture Magazine (29), 5–7.

Jendrischik, M., 2020. Vertical Farming: wie startups landwirtschaft und supermärkte verändern (in German). https://www.cleanthinking.de/vertical-farming-infarm-startups-veraendern-landwirtschaft-und-supermaerkte/. (Accessed 29 September).

Kapalschinski, C., 2020. Kräuter-start-up infarm gewinnt nach aldi auch kaufland als kunden (in German). https://www.handelsblatt.com/unternehmen/handel-konsumgueter/vertical-farming-kraeuter-start-up-infarm-gewinnt-nach-aldi-auch-kaufland-als-kunden/v_detail_tab_comments/26041852.html. (Accessed 22 September 2020).

Kapalschinski, C., Matthes, S., 2019. 100 millionen dollar für gemüse: berliner start-up überzeugt investoren (in German). https://www.handelsblatt.com/technik/vernetzt/gewaechshaeuser-im-supermarkt-100-millionen-dollar-fuer-gemuese-berliner-start-up-ueberzeugt-investoren/24442702.html. (Accessed 22 September 2020).

Kraas, F., Leggewie, C., Lemke, P., Matthies, E., Messner, D., Nakicenovic, N., Schellnhuber, H.J., Schlacke, S., Schneidewind, U., Brandi, C., Butsch, C., 2016. Der Umzug der Menschheit: Die transformative Kraft der Städte. WBGU-German Advisory Council on Global Change.

Ladner, P., 2011. The urban food revolution: Changing the way we feed cities. New Society Publishers.

Rockström, J., Steffen, W., Noone, K., et al., 2009. A safe operating space for humanity. Nature 461 (7263), 472–475. https://doi.org/10.1038/461472a.

Saengtharatip, S., Lu, N., Takagaki, M., Kikuchi, M., 2018. Productivity and cost performance of lettuce production in a plant factory using various light-emitting-diodes of different spectra. J. ISSAAS 24 (1), 1–9.

Schulze, G., 2016. Growth Within: A Circular Economy Vision for a Competitive Europe. Ellen MacArthur Foundation and the McKinsey Center for Business and Environment, pp. 1–22.

Shono, T., Shimoda, H., Lu, N., Obayashi, S., Hu, J., 2018. Study of soundscape design incorporating sound instrument into Mini-Plant Factory. In: INTER-NOISE and NOISE-CON Congress and Conference Proceedings. Institute of Noise Control Engineering, Chicago, IL, USA, pp. 5495–5504.

Takagaki, M., Hara, H., Kozai, T., 2020. Micro-and mini-PFALs for improving the quality of life in urban areas. In: Plant Factory. Academic Press, pp. 117–128.

Thomsen, J.D., Sønderstrup-Andersen, H.K., Müller, R., 2011. People–plant relationships in an office workplace: perceived benefits for the workplace and employees. Hortscience 46 (5), 744–752.

Toyoda, M., Yokota, Y., Barnes, M., Kaneko, M., 2020. Potential of a small indoor plant on the desk for reducing office workers' stress. Horttechnology 30 (1), 55–63.

United Nations, 2015. Sustainable Development Goals. https://sustainabledevelopment.un.org/sdgs>. (Accessed 28 September 2020).

Weidner, E., Deerberg, G., Keuter, V., 2020. Urban agriculture. In: The Future of Agriculture - Local, High-Quality and Value-Adding. Springer Vieweg, Berlin, Heidelberg.

Wood, R.A., Orwell, R.L., Tarran, J., Torpy, F., Burchett, M., 2002. Potted-plant/growth media interactions and capacities for removal of volatiles from indoor air. J. Hortic. Sci. Biotechnol. 77 (1), 120–129.

Indoor production of tomatoes

Mike Zelkind, Tisha Livingston and Victor Verlage

80 Acres Farms, Hamilton, OH, United States

16.1 Introduction

Vertical farming has come of age, and to accelerate the impact of its contribution on the industry, will have to evolve beyond growing just lettuces, leafy greens, and herbs into fruiting crops: a case for vertically farmed tomatoes. This chapter captures the perspective and insights learned from the founders of 80 Acres Farms (80acresfarms.com) and Infinite Acres (infinite-acres.com). We strive to present the prospect for vertical farming to evolve into massive sustainable growth by establishing fruiting crops as one viable path for scaled up production. After demonstrating success in growing lettuces, leafy greens, and herbs in vertical farms, 80 Acres Farms has now accomplished the first milestones in growing and marketing "commercially viable tomatoes" in a true indoor farming system. This breakthrough frames the opportunity for vertical farming to scale and reach critical mass not only in leafy vegetable crops but also fruiting crops.

A little background story on why we chose indoor farming over greenhouse production. We knew that our current food supply chain was broken. So much of the food we produce gets wasted in the field or throughout the supply chain. What makes it to the store does not meet the consumer's demands for clean, fresh, and consistently available produce. We knew that we wanted to impact the food supply with local food production to satisfy consumer needs. We did not know exactly how to accomplish that task. To find the best solution, we researched all obvious solutions. We traveled to Europe to learn from the best high-tech greenhouses and to Japan to visit the most advanced vertical farms. Imagine the following situation: it is the summer solstice, at high noon, and you are in a greenhouse and all the high pressure sodium (HPS) lights are on. Why? Because after whitewashing the greenhouse to control the heat, additional light is needed. With no insulation, heating and cooling are expensive in addition to the capital costs of purchasing lights and operating costs of lighting the greenhouse. Also, crop quality is difficult to control and requires very sophisticated growers. At that the moment, we knew that we needed to build and grow our crops in a vertical farm.

It requires a very experienced grower to manage a long-term crop such as tomato because of the constant need to adjust, manage, and modify growing strategies to react to the environmental variability, which is due to nature's inconsistency and seasonality. With climate change causing extreme highs and lows, adaptive growing strategies are becoming more challenging. Examples of challenging growing condition include: how do you vent the greenhouse when it rains? How do you manage and balance a greater leaf area index? How do you manage excessive heat inside a greenhouse? How do you manage disease and pest pressure? How do you maintain fruit quality at consistently high levels? Once growers learn to handle such issues, they need to begin thinking about scaling up the business because production volume is key to profitability and relevance in the industry. Every greenhouse is a unique engineering project that is designed for its local conditions. At the end of the day, you can't really control the sweetness or shelf life of the tomato crop and you are always at the mercy of mother nature. Why should that be the case in the 21st century? Technology should enable us to adopt a completely novel and different approach. We need an approach that enables stressing the plants on command, not randomly when the weather decides, and an approach that allows scaling and making an impact on the food supply that the consumers desire.

16.2 Controlled environment agriculture initial assessment

Vertical farming has gained traction over the past few years and continues to build momentum as more start-up projects demonstrate the ability to deliver consumer relevant products at affordable prices. Most of the production capacity in true indoor vertical farms consists of mixed salad, premium baby leaves, and herbs. The time has come to advance vertical farming to the next stage of evolution to enable new capabilities that true indoor farming can offer through the production of fruiting crops. There are many fruiting crops like strawberries that can be considered, but tomato is the perfect candidate to lead the evolution of this young controlled environment agriculture industry from lettuce into a massive scaling opportunity of the broader variety of crops.

16.2.1 Vegetative crops a logical initial step

There exist well understood and tested reasons why growing a crop where the edible portion of the plants is the vegetative stage. This has made lettuce and other leafy greens the perfect first choice for vertical farming as a commercially viable business model. The short architecture of the plant and the ability to plan for many short production cycles in densely populated, tight spaces offer a practical solution for space utilization inside these "three dimensional" vertical farms. Also, the simpler physiology of a lettuce crop and the ease of growing a vegetative crop versus the complexity of nourishing a fruiting crop provided the incentive for vertical farming to get started with leafy greens.

16.2.2 Commoditization of lettuces and leafy greens

Although lettuce offers many advantages for high biomass production with good quality and yield, lettuce is a product that has been heavily "commoditized" making it difficult to achieve profitability and return on investment in acceptable timelines at a significant scale. Beyond a specific group of niche products, the lettuce category needs to be de-commoditized by building functionality into this category. Many efforts are underway to produce crops that are demonstrably high in nutrition and concentrating "health driven" compounds that consumers recognize and are willing to purchase at the appropriate price.

16.2.3 Managing assortment to achieve scale while maintaining proximity to market and freshness value proposition

Although there exists a defined sizable market for lettuce and leafy greens in specific densely populated regions of the world, the aggregated demand for lettuce alone in most markets as a "niche" product is not enough to absorb a scalable vertical farming operation while remaining a "local product." These new "food factories" will need to increase the assortment to maintain the "hyper-fresh" value proposition and to avoid reverting to the longer shipping distances that are currently afflicting the established supply chain with thousands of transportation miles or several days to reach the consumer. This has a direct impact on the freshness, quality, nutrient content, and cost of mixed salads and leafy greens in the market today.

16.3 Justification for fruiting crops

16.3.1 Niche versus relevance

There is a segment of the market with disposable income that, although limited in volume, can afford niche products at a premium price. Increasing the assortment and variety of relevant products that already enjoy a large volume demand at higher prices will help vertical farming consolidate this business model at scale.

16.3.2 Fruiting crops are the next logical step for increasing capacity utilization and achieving scale

To evolve into a mature and more relevant food production industry, vertical farming will have to increase the assortment of produced crops to include fruiting plants. Initially, these additional crops will have a better opportunity to access a market with an already established large volume demand for high value products. The evolution into fruiting crops will require first to select cultivars that are adapted to confined spaces while demonstrating high productivity and marketable yield. The architecture of the crop will determine which engineering designs are most viable to consider. One option is to select fruiting crop types with naturally occurring short plant architecture, like strawberries, that can easily adapt to a "multilayer" farm form. A second alternative, for multilayer farms, is to redesign the plant architecture to a "dwarf" growing habit. In the case of tomato, this can be accomplished by either selecting from old heirloom determinate varieties or by breeding dwarf, determinate growth patterns into existing commercial greenhouse varieties with attractive quality attributes.

16.3.3 Challenge to design vertical farms with versatile environment zones to accommodate for different crop needs in a sensible way at scale

One important reason to expand the crop assortment inside a vertical farm is to offer a more complete selection for local markets. New designs of vertical farms must enable the capability to manage different growing climates, in close proximity, to provide optimal conditions for maximum productivity of very different plant types. The new farms should offer versatility in climate, lighting, and fertigation within the same space.

16.3.4 Need to map the evolution for vertical farming into fruiting crops

Adapting fruiting crops to vertical farming has already started. There are different concepts for vertical farms being tested in the market to optimize space utilization for maximum productivity, quality, and differentiated consumer value. Some vertical farmers have designed growing systems utilizing vertical columns while others, like Infinite Acres and 80 Acres, decided to pursue a "stacked planks" multilayer approach. In all of these and other forms, both the growing system and the architecture of the plant must be adapted in a synergistic way to a farm design that enables commercial viability. The growing model needs to deliver sufficient productivity and efficient use of resources in a practical process to make the final product affordable to consumers while protecting the environment and advancing sustainability.

Over the next few years, as the world's population increases to nine billion, demand for food is projected to grow beyond current capabilities. Resources to produce agricultural crops, namely land and water, in conventional ways will continue to experience pressure from the growing population. Climate change will also play a role affecting the ability to maintain a reliable flow of fruits and vegetables to densely populated urban areas. All these factors will increase costs over traditional agricultural systems and will allow for indoor farming to compete in the marketplace.

16.4 Tomatoes are a very viable initial option to establish fruiting crops into vertical farming

Growing tomatoes in an indoor facility requires a more sophisticated design than the average lettuce vertical farm. Fruiting crops, like tomatoes, will require more steps for a more complex crop physiology as compared to leafy greens where the edible product is the vegetative part of the plant. The long-term nature of a fruiting crop requires added care and design challenges of airflow, lighting, nutrition, and irrigation. The entire crop management system needs to be reengineered to fit the vertical farming form. For illustrative purposes of the requirement for a tomato facility, we have utilized excerpts from a proposal by Infinite Acres for an indoor tomato farm.

This proposal requires a new indoor facility to be built with a footprint off roughly $10,000 \, m^2$. The total growing area will be $6,000 \, m^2$. More specifically, the total growing area is divided in six different zones of around $1,000 \, m^2$ each (Fig. 16.1). For each zone, we will set up the necessary technical areas to create the desired climate related to the crop.

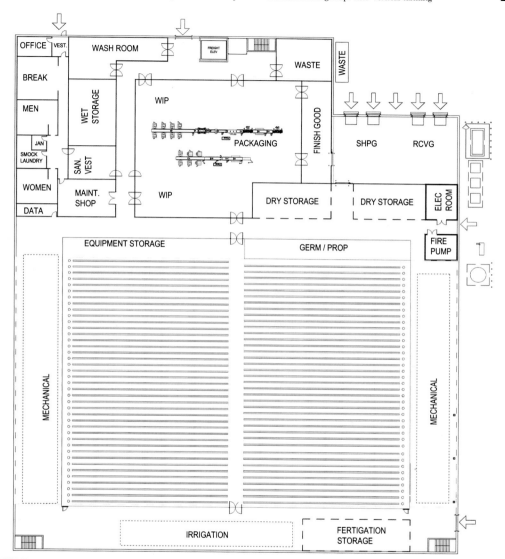

FIGURE 16.1 Top view of a vertical farm for tomato production. Bottom: irrigation technical zone. Center: cultivation rooms with independent climate control capability. Top: packaging lines 1 and 2, storage and employee welfare areas.

16.4.1 Growing zone process

Upon entering the high-tech indoor farm, the module consist of a growing area and a separate technical room. The growing area has a continuous evenly spread airflow controlling temperature, humidity, and CO_2. The strict climate controls in the growing module, combined with a state-of-the-art light-emitting diode (LED) lighting system and irrigation system, provide the optimal environment for any plant. This farm design is modular and can be fitted to any specific design, within parameters, to ensure the most consistent climate and dehumidification.

Compared to other indoor farm systems, we have a proactive control which is needed when working with equipment that can adjust the climate in a short period of time. This is different from a conventional heating system where the response time is more gradual.

We measure the climate parameters before and inside the mixing chamber, before the heating and/or cooling block, and after the air treatment unit. This enables us to direct and adjust the equipment to achieve the optimal climate. The software is fully integrated with all other modules and can monitor when the lights are activated (more heat added), and when irrigation has started (temporarily more humidity), etc.

80 Acres' proprietary computer control system will manage all growing conditions remotely through artificial intelligence to adjust the growing conditions and provide alerts to the growers for final confirmation or human override of the "optimized" system settings. A user interface is provided on a remote computer that allows a user to monitor and control various operations. Examples of such operations include: removing from or inserting into an indoor farming module trays containing crops, starting or stopping irrigation of crops within an indoor farming module, adjusting the temperature, humidity, and/or lighting within an indoor farming module, etc.

Not only does this system offer recipe management and remote setting adjustments but also manages tracking and tracing from seed to distribution, quality checks, unit economics compared to target, crop monitoring through sensor and vision data as well as shop floor data collection.

16.4.2 Infinite Acres indoor growing module

The indoor growing module is a high-tech growing facility. The module consists of a growing area and a separate small technical room. The growing area has a continuous evenly spread airflow control of temperature, humidity, and CO_2. Combined with a state-of-the art LED system and irrigation system, this provides an optimal environment for any plant. The module is modular and to be fitted to the specific design, within parameters for optimal airflow.

16.4.3 Unique technology in a closed loop

- Continuous circulation of air. No direct exchange with outside air.
- Equally divided air distribution by push and pull concept, "laminar air flow."
- 97% less water usage compared to open field production. Most of the transpiration water from the crop will be captured and recycled back to the system as irrigation water.

- Free of chemicals: no herbicides or pesticides required.
- Highest quality and consistency combined with reliable year-round production.
- High flexibility in hardware and controls, climate, irrigation, and light to support a wide diversity of crops growing side by side in the same chamber is possible.
- Proven track-record in food production facilities.

16.4.4 Climate control

Total control of the climate inside the indoor facility gives the grower influence over both the optimal growth of the crop as well as the prevention of diseases and crop damage. Infinite Acres supplies a range of systems and devices for monitoring and controlling the indoor facility, from traditional measuring boxes to highly sensitive sensors, and from air treatment to advanced environmental controls. All these assist the grower in creating the ideal internal climate.

To create a climate for growth, it is crucial that all the installations around the crop work seamlessly together. 'Integrated control logic' is the solution to optimize temperature, humidity, and airflow in all available installations. This integrated control logic is the base of every Infinite Acres climate system. However, this integration does not stop with climate. Infinite Acres' optimal growing environment also includes strict controls over irrigation, plant nutrition, light energy (photoperiod, intensity, and balanced spectrum), and CO_2. All variables are controlled and managed as a cohesive integrated control system. This complete and complex process can be optimized with one single holistic Infinite Acres technology solution.

16.4.5 Climate system

The climate control system creates a stable climate inside the module. The complete air volume is refreshed over 80 times per hour. The optimal air distribution creates an even, gentle flow and prevents wind stress. The temperature difference over the growing racks, between the air inlet and air outlet is 2°C, and the moisture difference is 1.3 g kg^{-1}.

16.4.6 Specifications of single layer system

Table 16.1 shows specifications and examples of environmental parameters of single layer system.

16.4.7 Lighting system

For any crop, light and CO_2 are crucial for photosynthesis and thereby for growth. Optimizing the light and CO_2 conditions around your crop is therefore an important step for a better result. We are using our controls together with multiple products that can help you monitor the conditions and analyze these conditions. This is the first step toward optimization. We also offer the option to optimize light. The strategy to provide the crop optimal photosynthetically active radiation energy includes the control of artificial lighting at the balanced photoperiod, intensity (or photosynthetic photon flux density), and light

TABLE 16.1 Specifications and examples of environmental parameters of a single layer system.

	Values
Growing area	1,000 m^2 per zone
Overall dimensions	6 grow rooms, each 20 × 50 m (interior)
Air speed	0.0–0.2 m s^{-1}
Air refresh rate	40–80 times per hour
CO_2 concentration	0–1,500 ppm
Temperature	18–27 °C
Relative humidity at 24°C	70–90% RH
Light intensity (photosynthetic photon flux density)	0–700 μmol m^{-2} s^{-1}
% Light spectrum customizable	Blue: 7% Green: 4% Red: 85% Far-red: 4%

Parameters can be adjusted to meet customer needs.

wavelength. Connecting the installations into one single integrated control strategy makes controlling easier and more precise. Optimizing the use of lights saves costs and optimizes the photosynthesis process.

16.4.8 Priva water systems

16.4.8.1 Water unit

To maintain a stable electrical conductivity and pH, the fertigation system provides accurate fertilizer and acid/lye injection. The Indoor Grow module is equipped with all commonly known irrigation systems (e.g., ebb/flow, drippers, etc.) With this system you will be able to create different recipes for the specific grow phases of the crop. The distribution will take place automatically and at the desired time or frequency of the client.

For this proposal, we used the Nutrifit which is our most used water unit of our portfolio (Fig. 16.2). For remote areas with extreme climates and where water is scarce, indoor growing is a great opportunity because the water usage is reduced up to 90% in comparison with traditional greenhouse growing.

16.4.8.2 Water disinfection unit Vialux system

"Affordable water disinfection with ultraviolet (UV) light for any horticulture company", that is what the Priva Vialux M-Line stands for (Fig. 16.3). With UV disinfection, significant improvements can be realized regarding disease control and efficient water use. In addition, the Vialux M-Line meets the most stringent demands for food hygiene. The Priva Vialux M-Line UV disinfection system destroys all existing organisms such as fungi, bacteria, and viruses within the irrigation water. The system can be applied for both new entry water or

FIGURE 16.2   Impression of an nutrifit indoor.

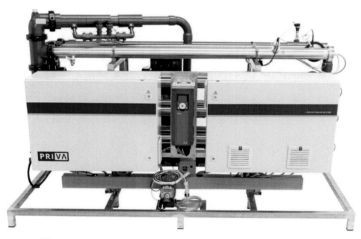

FIGURE 16.3   Vialux M-line ultraviolet disinfection unit.

recycled drain water. UV disinfection allows continuous recirculation of the irrigation water in a safe manner. As a result, expensive fertilizers are not flushed down the drain and do not contaminate the environment.

16.5 Priva Vialux M-Line advantages

- Protect the crop against germs like viruses, bacteria, and fungi by disinfecting surface and drain water.
- Realize savings by reusing drain water and fertilizers at the lowest cost per m^3 of disinfected water compared to other UV disinfection systems.
- Meet regulations and consumer demands.
- Ensure the water quality is safe for healthy produce.
- Prevent water discharge.
- A flexible design.
- A scalable and expandable system for any scale of operation.
- Wide range of operation in terms of level and volume of UV disinfection.
- Easy to replace the lamp, quartz tube, and UV sensor.

16.5.1 Energy consumption of facility

For this indoor facility, the amount of energy that is needed on a 24-hr basis for the growing system is specified in Table 16.2. From our experience, the grow rooms are operational for almost 52 weeks per year. Each growing day has a light period of 16 h. This automatically results in the highest energy consumption for that period. Also, during nighttime, the ventilation strategy is slightly adjusted.

TABLE 16.2 Specifications of electric energy consumption based on our 6,000 m^2 growing area.

User	Max			Nominal			Night		
Air handling units	134	[kW]	8%	134	[kW]	9%	69	[kW]	20%
Heat pumps	643	[kW]	36%	548	[kW]	35%	193	[kW]	56%
Lights	749	[kW]	42%	749	[kW]	47%	0	[kW]	0%
Circulation pumps	43	[kW]	2%	43	[kW]	3%	22	[kW]	6%
Dry coolers	39	[kW]	2%	33	[kW]	2%	20	[kW]	6%
Transport pumps	82	[kW]	5%	70	[kW]	4%	42	[kW]	12%
Irrigation incl. disinfection	72	[kW]	4%	0	[kW]	0%	0	[kW]	0%
Total	**1,762**	**[kW]**		**1,577**	**[kW]**		**346**	**[kW]**	
	100%	[−]		89%	[−]		20%	[−]	
Duration	6	[h]		12	[h]		6	[h]	

Daytime conditions: 23°C and 70% RH.
Nighttime conditions: 22°C and 85% RH.
Installed light intensity 280 μmol m^{-2} s^{-1}.
Light efficiency 3.0 μmol J^{-1}.

Infinite Acres and 80 Acres continue to develop technologies and solutions to establish commercially viable fruiting crops in vertical "indoor" farming. Our Fireworks cherry tomatoes are already in the market and are well underway to lead this evolution with more fruiting crops being developed in the near future.

At Infinite Acres and 80 Acres, we look forward to continuing to support the vertical farming entrepreneurs to succeeding in the market to grow locally in a sustainable and profitable way.

PART IV

Advanced research in PFALs and indoor farms

Toward an optimal spectrum for photosynthesis and plant morphology in LED-based crop cultivation

Shuyang Zhen[1], Paul Kusuma[2] and Bruce Bugbee[2]

[1]Texas A&M University, College Station, TX, United States; [2]Utah State University, Logan, UT, United States

17.1 Introduction

Photobiologically active radiation refers to photons between 280 and 800 nm (Dörr et al., 2020), as these photons can drive photosynthesis (a process that turns water and CO_2 into oxygen and carbohydrates) and/or excite a suite of photoreceptors that modulate plant development. This biologically active region can be broadly split into six regions: UV-B (280–320 nm), UV-A (320–400 nm), blue (400–500 nm), green (500–600 nm), red (600–700 nm), and far-red (700–800 nm).

Sunlight and conventional electric light sources (e.g., high-pressure sodium and metal halide lamps) provide minimal opportunity to manipulate the light spectrum. Recent developments in high efficiency narrow spectrum light-emitting diodes (LEDs) enable precise control of plant growth and development by creating 'light recipes.' In this chapter we discuss general principles of plant spectral responses, focusing on photosynthetic efficiency, plant shape, and secondary metabolite production.

17.2 Photosynthesis overview

Plants evolved under broad spectrum sunlight—the ultimate energy source for all life on the Earth. About half of the photons in solar radiation that reaches the Earth's surface falls into the biologically active region, with peak output in the range that is visible to the human

Plant Factory Basics, Applications and Advances
https://doi.org/10.1016/B978-0-323-85152-7.00018-5

eye (approximately 400—700 nm). For photosynthesis, photons must be of sufficient energy for photochemical reactions but must not contain too much energy to cause damage to the photosynthetic apparatus. Coincidently (but perhaps not surprisingly), photons in the visible region of about 400—700 nm fit this requirement and are known as photosynthetically active radiation (PAR).

Photosynthesis is dependent on the amount of photons plants receive, which is commonly quantified in two ways: (1) the instantaneous photosynthetic photon flux density (PPFD; $\mu mol\ m^{-2}\ s^{-1}$), and (2) the cumulative number of photons received over a day—called the daily light integral (DLI; $mol\ m^{-2}\ d^{-1}$). Crop yield shows a strong linear correlation with the total amount of light intercepted by the plant canopy both in the field under sunlight (Gifford et al., 1984; Monteith, 1977) and in indoor agriculture under LEDs (Zhen and Bugbee, 2020a). However, photons of different wavelengths do not drive photosynthesis with equal efficiency (expressed as moles of CO_2 fixed per mole of photons absorbed). This was known from pioneering studies on photosynthesis under monochromatic light (Emerson and Lewis, 1943; Hoover, 1937). Classic studies by McCree (1971) and Inada (1976) found that red photons are most efficient for photosynthesis, followed by green and blue photons, while photosynthetic efficiency decreases significantly on the edges of the PAR region (Fig. 17.1C). The results from these classic studies should be interpreted with caution because the studies were conducted on single leaves under low light due to technical limitations at the time. More recent studies examine the photosynthetic responses at canopy level under higher light, which we discuss in detail below.

17.3 Morphology overview

Plants are sessile organisms that must acclimate to their local growing environment. Among the environmental factors, light plays a central role in their development. Some characteristics of the light environment are more constant such as modifications caused by an overhead canopy, while others fluctuate substantially on a timescale of seconds (e.g., cloud cover or leaf movement in the wind). Additionally, there are seasonal changes in the PPFD and photoperiod (or daylength). Plants possess a suite of photoreceptors (proteins coupled with a photon absorbing chromophore) that are sensitive to specific photons. Plants sense the light signals (i.e., changes in PPFD, spectrum, and photoperiod) via their photoreceptors and modify their development accordingly. These modifications can include a number of key changes over a plant's life cycle, including seed germination, plant shape, phytochemical composition, biomass allocation, flowering, and seed-setting (Kendrick and Kronenberg, 1994).

Under natural environments, changes in light spectral quality primarily occur between full sun versus vegetative shade. Table 17.1 provides some typical values of the spectral composition of sunlight and shade light. Compared to full sunlight, shade light is notably enriched in far-red, which is caused by filtration of most of the photons in the PAR region by the top layers of a plant canopy for photosynthesis. These shifts in spectral ratios are thus accompanied by a significant overall decrease in total photon flux density that can reach 95% or greater (Casal, 2012). Shade conditions can simply be the presence of neighboring plants that reflect a small amount of additional far-red (Ballare et al., 1987). Species tend to respond differently to shade or changes in spectral quality, such as rapid elongation of stems in order

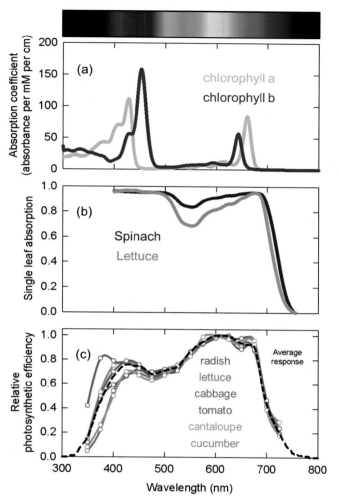

FIGURE 17.1 (A) Absorption coefficients of chlorophylls extracted with diethyl ether, (B) light absorption in leaves, and (C) spectral effect of photosynthesis. Relative photosynthetic efficiency curves of diverse species are redrawn *from data by McCree, K.J., 1971. The action spectrum, absorptance and quantum yield of photosynthesis in crop plants. Agric. Meteorol. 9, 191–216.*

to outgrow neighbors or expansion of leaves (often making the leaves thinner at the same time) to increase light capture (Gommers et al., 2013). The response of a specific species is dependent on the type of environment under which it evolved, for example, forest understories or open fields.

For morphological development, these broad categories can be grouped into photons that make plants more compact (UV-B, UV-A, blue, and red) or less compact (green and far-red). These six categories are admittedly crude, as development is actually dependent on photoreceptor absorption (Fig. 17.2). For example, a UV-A photon at 330 nm is unlikely

TABLE 17.1   Typical spectral composition of the photobiologically active radiation under full sun and canopy shade environments. In addition to the shifts in spectral ratios, both the photosynthetic photon flux density (PPFD) and biological photon flux density (BPFD) decrease substantially under shade. Different photon categories are expressed as a percentage of the total biologically active radiation.

	Full sun	Vegetation shade
UV-B (280–320 nm)	0.2%	0.1%
UV-A (320–400 nm)	5%	3%
Blue (400–500 nm)	19%	6%
Green (500–600 nm)	25%	10%
Red (600–700 nm)	27%	7%
Far-red (700–800 nm)	24%	74%
PPFD (400–700 nm)	2100 (μmol m^{-2} s^{-1})	25 (μmol m^{-2} s^{-1})
BPFD (280–800 nm)	3000 (μmol m^{-2} s^{-1}; 100%)	100 (μmol m^{-2} s^{-1}; 100%)

to have the same effect on plant growth and development as a UV-A photon at 395 nm. Likewise, a far-red photon at 730 nm is expected to have a significantly different effect on development than a far-red photon at 780 nm. Despite this, these coarse categories are still useful as the peak absorbance of different photoreceptors roughly fall into the specified photon categories.

There are three well-studied classes of photoreceptors that modulate development. These are (1) the phytochromes, which have peak absorbance in the red and far-red regions, although they can absorb from 300 to 800 nm; (2) the cryptochromes, which primarily absorb in the UV-A, blue, and green regions; and (3) the phototropins, which primarily absorb in the UV-A and blue regions (Fig. 17.2). Plants also contain other photoreceptors including UV RESISTANCE LOCUS8 (UVR8), which responds to UV-B photons; and zeitlupes, which respond to blue/UV-A photons, but these are less well studied (Folta and Carvalho, 2015; Galvão and Fankhauser, 2015).

17.4 General effects of spectrum

For the remainder of this chapter, we discuss the general effects of the six categories of photons (UV-B, UV-A, blue, green, red, and far-red) on the photosynthesis and development of crops. Note that photons below 280 nm (UV-C; 100–280 nm) can be generated with electric lamps, conventionally with high-pressure mercury lamps and now UV-C LEDs have become available. These high-energy ionizing photons have germicidal effects and may be applied in short flashes to stimulate plant resistance to pathogens (Aarrouf and Urban, 2020) and to control branching and height of ornamental plants (Bridgen, 2016), but they tend to be highly damaging and can reduce yields (Lee et al., 2014). Their applications in crop cultivation need to be further studied.

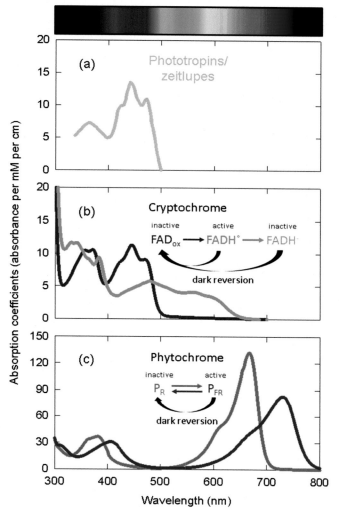

FIGURE 17.2 Absorption coefficients of three major families of photoreceptors that modulate plant development. (A) Phototropins and zeitlupes. (B) Absorption coefficients for the cytochrome chromophore (FAD) in the oxidized (FAD$_{ox}$) and neutral radical (FADH) states. (C) Absorption coefficients for the P$_R$ and P$_{FR}$ forms of phytochrome. Straight colored *arrows* indicate photoreceptor activation or deactivation upon photon absorption and curved black arrows indicate dark (meaning light-independent) reversion. Dark reversion rate is dependent on the concentration of oxygen in cryptochrome (Müller and Ahmad, 2011) and on temperature in phytochrome (Klose et al., 2020). *(A) Data from Ahmad, M., Grancher, N., Heil, M., Black, R.C., Giovani, B., Galland, P., Lardemer, D., 2002. Action spectrum for cryptochrome-dependent hypocotyl growth inhibition in Arabidopsis. Plant Physiol. 129 (2), 774—785; Salomon, M., Christie, J.M., Knieb, E., Lempert, U., Briggs, W.R., 2000. Photochemical and mutational analysis of the plant blue light receptor, phototropin. Biochemistry 39 (31), 9401—9410; (B) data were kindly provided by Müller, P., Bouly, J.P., Hitomi, K., Balland, V., Getzoff, E. D., Ritz, T., Brettel, K. (2014). ATP binding turns plant cryptochrome into an efficient natural photoswitch. Sci. Rep. 4 (1), 1—11 for more detail; (C) curves redrawn from data by Kelly, J.M., Lagarias, J.C., 1985. Photochemistry of 124-kilodalton Avena phytochrome under constant illumination in vitro. Biochemistry 24 (21), 6003—6010; Lagarias, J.C., Kelly, J.M., Cyr, K.L., Smith Jr, W.O., 1987. Comparative photochemical analysis of highly purified 124 kilo-dalton oat and rye phytochromes in vitro. Photochem. Photobiol. 46 (1), 5—13.*

17.4.1 UV-B (280–320 nm)

The classification of UV-B photons begins at 280 nm, as this is the shortest wavelength of photons from the Sun that can penetrate through the Earth's atmosphere. The long wavelength cutoffs for UV-B are either based on studies on sunburn (315 nm) or skin cancer (320 nm) in humans (Kusuma et al., 2020). We use 320 nm since this covers a broader range of photons that are potentially damaging to plants, but perhaps better metrics are needed for plant-based responses. It is important to use UV-B with caution as it creates hazardous conditions to both plants and workers.

17.4.1.1 Photosynthesis

UV-B is generally detrimental to photosynthesis and plant growth as it causes damage to photosynthetic apparatus, primarily photosystem II (Jansen et al., 1998; Tyystjärvi, 2008). Damages to DNA, proteins (including the key photosynthetic enzyme Rubisco) and cell membranes often occur under moderate to high intensity UV-B radiation. Exposure to UV-B can also induce thickening of leaves and cuticle layers, a decrease in chlorophyll content and an increase in UV-B screening pigments, all of which may further affect photosynthesis by altering the leaf optical properties and distribution of PAR within the leaves (Bornman and Vogelmann, 1991).

17.4.1.2 Development

Although UV-B induces damage at high intensities, lower intensity UV-B can act as a beneficial stress (Neugart and Schreiner, 2018), increasing the production of beneficial secondary metabolites including anthocyanins, phenolics, and flavonoids. These compounds act as 'sunscreen,' absorbing the UV radiation primarily in the epidermal layers before it can damage the photosystems. Additionally, these compounds have antioxidant properties, scavenging reactive oxygen species (Jansen et al., 1998). Many of these molecules change the overall pigmentation of the plant and can promote human health, potentially increasing the value of the crop (Schreiner et al., 2012).

Even though UV-B may improve crop quality, its application tends to decrease stem and leaf expansion (Wargent et al., 2009a,b; Yao et al., 2006). Together with the decreases in photosynthesis, these decreases in leaf area (reduced photon capture) can lead to yield reductions. The effects of UV-B on both morphology and secondary metabolite synthesis depend on the background growth light intensities (DLIs), with larger impacts at lower DLIs (Dou et al., 2019; Warner and Caldwell, 1983).

Many of these responses to UV-B are mediated through the UVR8 photoreceptor, which regulates the expression of hundreds of genes (Favory et al., 2009), but some UV-B responses are also mediated by independent mechanisms (Coffey et al., 2017; Wargent et al., 2009a). Measurement of a transcript downstream of UVR8 activation showed that the action spectrum of this photoreceptor drops by about 310 nm (Brown et al., 2009).

Intumescence is a physiological disorder that occurs in certain cultivars of crops (e.g., Maxifort tomato) cultivated in greenhouses and plant factories. It manifests as small blisters on the surface of leaves, and at the microscopic level, these protrusions are caused by cell hypertrophy (Williams et al., 2014). Although the cause for this disorder is not well understood, UV-B has been shown to decrease its severity (Kubota et al., 2017).

Increased cuticle thickness and changes in phytochemical composition in plants grown with UV-B may also increase their resistance to fungal pathogens and diseases (Raviv and Antignus, 2004).

17.4.2 UV-A (320–400 nm)

A commonly used violet LED with a peak at about 405 nm and with 10%–50% of its output below 400 nm (depending on operating conditions) has sometimes been marketed as a UV LED (Samuolienė et al., 2020). LEDs with peaks between 365 and 400 nm can be manufactured with higher efficiency than UV-B LEDs, but the efficiency decreases with decreasing peak wavelength. Distinctions between longer and shorter wavelength UV-A photons are important as the effects of UV-A vary considerably with wavelength.

17.4.2.1 Photosynthesis

UV-A photons are of lower energy than UV-B photons and are thus less damaging (Verdaguer et al., 2017). Additionally, the longer wavelength UV-A photons can be absorbed by photosynthetic pigments (chlorophyll and carotenoids) and drive photosynthesis. McCree (1971) found that photosynthetic efficiency of UV-A photons decreased rapidly with decreasing wavelength from 400 to 350 nm (Fig. 17.1C). Although, some species (e.g., radish and sugar beet) were shown to use UV-A photons with relatively high efficiency, and plants grown in growth chambers without prior UV-A exposure generally used UV-A more efficiently than field grown plants (McCree, 1971). Interestingly, Mantha et al. (2001) showed that shorter wavelength UV-A photons (peak around 340 nm) can enhance leaf photosynthesis through fluorescence emission of violet, blue, and green photons, which are absorbed by photosynthetic pigments. This effect is more pronounced under low light situations when photosynthesis is not light saturated.

Plants accumulate UV-absorbing compounds (notably in the upper epidermis) upon UV exposure as a photoprotective mechanism. Therefore, both the photosynthetic efficiency and the potential photoinhibitory effects of UV-A decrease as crops acclimate to UV-A. It is important to note that there is large variation in species sensitivity to UV radiation (Fig. 17.1C). From the standpoint of optimizing photosynthetic efficiency, more studies are needed to better understand whether it is beneficial to include UV-A photons in greenhouses and indoor agriculture.

17.4.2.2 Development

Plant responses to UV-A photons have been less well studied compared to UV-B, and the studies thus far have shown both beneficial and detrimental effects. With the continued development of LED technology, studies in controlled environments are beginning to fill in the gaps in knowledge regarding the responses to UV-A and the underlying mechanisms (Verdaguer et al., 2017).

Under sole-source LED light, leaf area was increased by supplementation with UV-A photons between 365 and 400 nm in both lettuce (Chen et al., 2019) and tomato (Kang et al., 2018; Khoshimkhujaev et al., 2014). By contrast, other studies have shown no effect of UV-A on leaf area in cucumber, soybean, and lettuce (Samuolienė et al., 2020; Yao et al., 2006).

Supplemental UV-A has also been shown to increase stem length in tomato (Kang et al., 2018), decrease stem length in soybean (Yao et al., 2006), and have no effect in cucumber (Jeong et al., 2020; Yao et al., 2006). The effects were similarly inconsistent on fresh and dry mass, either showing increases (Kang et al., 2018; Lee et al., 2014) or no changes (Jeong et al., 2020; Li and Kubota, 2009; Samuolienė et al., 2020). These inconsistencies are not simply explained by species differences because studies on the same species (e.g., lettuce) have provided conflicting results (Chen et al., 2019; Li and Kubota, 2009; Samuolienė et al., 2020). The specific intensity of both PAR and UV-A likely contributed to the differences.

In addition to altering morphology, UV-A has been shown to contribute to secondary metabolite production. Mariz-Ponte et al. (2019) found that UV-A, not UV-B, increased flavonoids and phenolic concentrations in tomato fruits. UV-A also induces secondary metabolite production in lettuce (Lee et al., 2014; Li and Kubota, 2009), with higher doses being more effective (Chen et al., 2019).

Photons from the violet LED (peak at about 405 nm) are at the cusp of photosynthetic activity (Fig. 17.1) and photoinhibition (Takahashi et al., 2010). In one study these photons were shown to have no effect on leaf area and yield when applied supplementally (Samuolienė et al., 2020).

The UVR8 photoreceptor does not appear to induce responses beyond UV-B, instead the cryptochromes are most likely involved in responses to UV-A photons (Wade et al., 2001). The absorbance spectrum for cryptochrome activation (FAD_{ox} in Fig. 17.2B, described in further detail below) shows one peak around 350 nm and another peak at about 450 nm, with a valley between 375 and 425 nm, indicating that longer wavelength UV-A photons (~375−400 nm) are likely not as effective in regulating plant development compared to shorter wavelength UV-A and longer wavelength blue photons. The studies discussed in this section generally use supplementation of UV-A photons between 365 and 400 nm. It is possible that adding blue photons (around 450 nm) would have had the same or even greater effect.

17.4.3 Blue (400−500 nm)

17.4.3.1 Photosynthesis

Blue photons are most efficiently absorbed by leaves (Fig. 17.1B) and are often thought to be needed for efficient photosynthesis due to their role in inducing stomatal opening. Interestingly, studies on the spectral response of photosynthesis found that blue photons tend to have the lowest photosynthetic efficiency among photons in the PAR region. For example, McCree (1971) reported that the average photosynthetic efficiency of blue photons was 17%−28% lower than red and green photons in diverse species. Similarly, Hogewoning et al. (2012) found that the photosynthetic efficiency of blue photons in cucumber was about 30% lower than red photons. In addition to chlorophylls, blue photons are absorbed by non-photosynthetic pigments (e.g., anthocyanins and flavonoids) as well as photosynthetic carotenoids, which transfer excitation energy to chlorophylls with reduced efficiency (Frank and Cogdell, 1996; Siefermann-Harms, 1985). This largely accounts for the lower photosynthetic efficiency of blue photons.

Compared to red and green photons, absorption of blue photons primarily occurs in the upper cell layers of the leaves at the top of a plant canopy (Vogelmann and Evans, 2002). This nonuniform light distribution within a plant canopy leads to reduced photosynthetic efficiency (Melis, 2009). In addition, high-intensity blue light can induce chloroplast movement to the side walls parallel to the light direction (inducing a 'stacked' orientation), causing a decrease in light absorption, and thus a decrease in photosynthetic rate (Haupt and Scheuerlein, 1990; Kagawa et al., 2001).

It is well known that blue light, often applied as a short pulse, induces fast stomatal opening (Assmann et al., 1985). A recent study found that when varying the fractions of red and blue photons, blue light—induced stomatal opening minimally enhanced steady-state photosynthesis and consistently decreased water use efficiency under medium and high light intensities (Zhen and Bugbee, 2020b). Additional studies are needed to elucidate the role of blue light in photosynthesis, stomatal regulation, and water use of crops grown under electric lights, especially if dynamic lighting strategies (e.g., fluctuating light levels and spectra) are implemented.

17.4.3.2 Development

Blue photons act on three families of photoreceptors to modulate plant development: cryptochromes, phototropins, and zeitlupes. Cryptochromes and zeitlupes control plant development through modulation of gene expression, while phototropins act through association with cell membranes (Galvão and Fankhauser, 2015; Lin, 2000). Responses mediated by phototropins include phototropism, stomatal opening, chloroplast reorientation, and leaf movement (Christie, 2007). Zeitlupes play a role in flowering (Galvão and Fankhauser, 2015). The effects of blue photons on photomorphogenesis and secondary metabolite production (the focus of this section) are primarily controlled by cryptochromes.

Many studies conducted in controlled environments have shown that increasing the fraction of blue photons, especially between 10% and 50%, often causes leaf area and stem length to decrease, thus leading to decreases in yield (Hernández and Kubota, 2016; Kang et al., 2016; Meng et al., 2019, 2020; Snowden et al., 2016; Son and Oh, 2013, 2015; Wang et al., 2016). Less commonly, studies have found no effect on yield from increasing the fraction of blue photons (Li and Kubota, 2009; Snowden et al., 2016).

Because leaf area and yield generally decrease with increasing blue photon fraction, lower fractions of blue photons may be preferred. However, studies show that growing plants in the absence of blue photons often lead to low chlorophyll concentrations and excessive stem elongation (Son and Oh, 2013; Snowden et al., 2016; Yorio et al., 2001). Leaf area and crop yield in the absence of blue photons have been reported to increase in some species (Meng et al., 2020; Son and Oh, 2013; Wang et al., 2016) but decrease in others (Hernández and Kubota, 2016; Snowden et al., 2016; Yorio et al., 2001). Therefore, the decision of whether or not to completely remove blue photons may be species dependent.

One of the key regulators in flavonoid and anthocyanin synthesis is the enzyme chalcone synthase. The expression of this enzyme is partially controlled by cryptochromes (Wade et al., 2001). Thus, blue photons have been shown to increase the production of secondary metabolites including phenolics and flavonoids (Li and Kubota, 2009; Son and Oh, 2013, 2015).

17.4.4 Green (500–600 nm)

17.4.4.1 Photosynthesis

Green photons are often perceived as less efficient for photosynthesis than red and blue photons, largely due to the minimal absorption of green photons by chlorophylls. This lower absorption/higher reflectance is what gives leaves their green color. However, chlorophyll absorption spectra are quantified with extracted pigments in a solvent. Light absorption of a leaf differs from that of extracted chlorophyll solution because leaf chlorophylls are concentrated in chloroplasts (see Fig. 17.1A compared to Fig. 17.1B). This uneven distribution of chlorophylls within leaf cells flattens the absorption of strongly absorbed red and blue photons while only marginally reducing absorption of weakly absorbed green photons (also known as 'sieve effect'; see Terashima et al. (2009) for more details). On the other hand, the diffusive nature of plant tissues (light is reflected/scattered at the interfaces of cell walls and intercellular air spaces) increases the light path length inside the leaf and thus increasing the overall absorptance of photons, especially green (Vogelmann, 1993). Zhen and Bugbee (2020c) reported that leaf absorptance of green photons ranged from 77% to 88% for a number of crop species, including lettuce, spinach, and tomato. In comparison, leaf absorptance of those species was around 95% in the blue region and 88%–93% in the red region. Thus, it is a misconception that 'green leaves do not (efficiently) absorb green photons.'

On the basis of absorbed photons, McCree (1971) found that photosynthetic efficiency of green photons was 20% higher than blue photons and equivalent to red photons. Note that the values were determined from the average response of 22 diverse species grown in growth chambers and the measurements were made at 25 nm intervals under low light conditions. Under high light, strongly absorbed red and blue photons tend to oversaturate upper cell layers of the leaf while 'starving' the bottom leaf cells, resulting in reduced leaf photosynthetic efficiency. Because green photons are not as strongly absorbed by chlorophylls, they can penetrate deeper into the leaf (and canopy) (Brodersen and Vogelmann, 2010), driving photosynthesis deep within the leaf (Sun et al., 1998). As a result, green photons may be more efficient for leaf photosynthesis than red and blue photons when added to high light (Liu and van Iersel, 2021; Terashima et al., 2009).

At canopy level, the difference in light absorption between green and red (and blue) photons is expected to be smaller than the differences within a leaf (Paradiso et al., 2011). Furthermore, the more uniform light distribution within the canopy under green photons likely lead to higher canopy photosynthetic efficiency. Measurements of photosynthetic efficiency at canopy level will help to further elucidate the value of green photons for photosynthesis and crop productivity.

17.4.4.2 Development

The past two decades of photobiological studies have revealed a potential role of green photons in morphogenesis. The flavin adenine dinucleotide (FAD) chromophore embedded in cryptochrome has three states: FAD_{ox}, the oxidized state; FADH, the semireduced neutral radical state; and $FADH^-$, the fully reduced state. Of these three states, FADH is the active form, while FAD_{ox} and $FADH^-$ are both inactive (Ahmad, 2016). FAD_{ox} absorbs most prominently in the blue region, converting it into active FADH (Fig. 17.2B). As the active form, FADH inhibits stem expansion. The absorption spectrum of FADH shows a high absorbance of green photons, which lead to the inactivation of cryptochrome (to FADH⁻). Thus, green photons have been suggested to induce shade avoidance responses by reversing blue photon—induced decreases in plant size. A decrease in B:G ratio tends to occur in canopy shade (Smith et al., 2017). Shade avoidance in response to green has been supported mainly by studies in *Arabidopsis thaliana* (Bouly et al., 2007; Zhang et al., 2011).

Because green photons may reverse the effects of blue photons, it may be expected that increasing the fractions of green photons (while the blue fraction remains constant) will increase leaf area. This effect was observed in an early study that investigated responses of lettuce to green photons (Kim et al., 2004), but subsequent studies in horticultural species have shown minimal responses to green photons (Hernández and Kubota, 2016; Kang et al., 2016; Snowden et al., 2016; Son and Oh, 2015), and in some cases, plant diameter/leaf area actually decrease (Meng et al., 2020; Snowden et al., 2016). Overall, increasing the fraction of green photons has been observed to minimally affect leaf area and stem/petiole elongation in horticultural species.

Green photons have been observed to reverse blue photon—induced anthocyanin accumulation (Meng et al., 2019; Zhang and Folta, 2012), but this effect is not consistent (Meng et al., 2020).

17.4.5 Red (600—700 nm)

17.4.5.1 *Photosynthesis*

Red photons are efficiently absorbed by chlorophylls and are among the most efficient for photosynthesis, especially under low to medium light intensities. Additionally, red LEDs typically have higher photon efficacy (moles of photon output per joule of input energy, see Chapter 7) compared to blue and green LEDs, which contributes to their prevalence in plant factories. However, red photons become less efficient for leaf (and likely canopy) photosynthesis than green photons under high light intensities (Terashima et al., 2009; also see discussion in Section 17.4.4.1 on green photons). Unlike green photons, red photons are primarily absorbed by the top layer of leaves in a canopy, thus exposing them to often excessive light levels. Photosynthetic efficiency decreases with increasing PPFD as plants dissipate an increasing fraction of the absorbed light as heat—a photoprotective mechanism against potential oxidative damages under high light (Ruban, 2015).

Furthermore, although including a high fraction of red photons in grow lights can result in energy savings, the cost of red LEDs (i.e., initial capital investment) is relatively high. White LEDs are fairly efficient and less expensive as they are widely used for human lighting. Kusuma et al. (2020) suggested that a white LED enriched with red photons may be best suited for plant factories considering the photon efficacy and luminaire cost. Broad spectrum white light is gaining popularity over a combination of blue + red light for horticulture lighting as it is easier to diagnose pests, diseases, nutritional disorders under white light; it also creates a more pleasant light environment for workers. The green photons emitted by white LEDs may also improve photosynthetic efficiency at canopy level.

17.4.5.2 *Development*

Red photons act on the photoreceptor phytochrome to modulate plant development. The effect of red photons is often discussed in tandem with far-red, either through the R:FR ratio,

FIGURE 17.3 Photobleaching of medical cannabis inflorescence under a high fraction of red photons (approximately 75% or greater). *Photo courtesy of Mitchell Westmoreland.*

or phytochrome photoequilibrium, which is an estimated ratio of active phytochrome (P_{FR}) to total phytochrome ($P_R + P_{FR}$). Issues with these common metrics were recently reviewed in Kusuma and Bugbee (2021).

Absorbance of red photons by P_R (the inactive form of phytochrome) converts it into the active P_{FR} form. This active form then goes on to modulate plant development via regulation of gene expression. P_{FR} inhibits the expression of genes related to cell wall expansion and the hormones auxin, gibberellin, and brassinosteroids (de Lucas and Prat, 2014). Therefore, red photons tend to decrease cell expansion in leaves and stems.

It has been observed that photobleaching may occur under a high fraction of red photons (Fig. 17.3). The exact cause of this phenomenon is unknown, as longer wavelength photons do not typically induce photobleaching (Takahashi et al., 2010). Due to the widespread use of 660 nm red LEDs in horticulture, research should focus on the mechanism causing this effect in order to identify potential remedies.

17.4.6 Far-red (700—800 nm)

17.4.6.1 *Photosynthesis*

The spectral response of photosynthesis is traditionally quantified under monochromatic light. Far-red photons have long been considered inactive for photosynthesis due to their low photosynthetic efficiency when applied alone (Emerson and Lewis, 1943; McCree, 1971). However, spectral responses determined under monochromatic lights do not account for synergistic action among photons of different wavelengths. Synergism among wavelengths on photosynthesis was first discovered by Emerson et al. (1957), who found that the photosynthetic rate under simultaneous illumination of photons above 680 nm and shorter wavelength light was greater than the sum of the rates from applying each light separately—a phenomenon now known as the Emerson Enhancement Effect. This finding contributed to the identification of two photosystems in photosynthesis, both of which are required to be stimulated in order for photosynthesis to operate (Duysens and Amesz, 1962; Hill and Bendall, 1960; Myers, 1971). In fact, to achieve optimal photosynthetic efficiency, the two photosystems need to operate at matching rates. The significance of wavelength synergy in photosynthetic efficiency has received little attention until recently with the current definition of PAR (400—700 nm) excluding far-red photons.

Several recent studies using narrow spectra LEDs have shown that far-red photons added to shorter wavelength photons (e.g., red and blue) synergistically increase leaf photochemical efficiency (Zhen and van Iersel, 2017) and photosynthetic rate (Hogewoning et al., 2012; Murakami et al., 2018). This is because shorter wavelength photons from 400 to 680 nm tend to overexcite one of the photosystems (PSII), while longer wavelength far-red photons preferentially excite the other photosystem (PSI) (Evans, 1987; Laisk et al., 2014; Zhen et al., 2019). Combining far-red with shorter wavelength photons helps restore the balance of excitation between the two photosystems, leading to synergistic enhancement of photosynthetic efficiency (Zhen and van Iersel, 2017). Furthermore, both short- and long-term studies of canopy photosynthesis indicated that far-red photons (700—750 nm) are equally efficient for photosynthesis at canopy level when up to ~40% of those photons were applied with 400—700 nm photons (Zhen and Bugbee, 2020a, 2020c). These recent findings warrant reconsideration of the photosynthetic value of far-red photons and argue for a new definition of PAR that extends to 750 nm (Zhen et al., 2021).

Because far-red photons are of lower energy, they can achieve a high photon efficacy (Kusuma et al., 2020, Capter 7). As far as photosynthetic efficiency is concerned, it is cost-effective to include far-red photons in the grow light in plant factories.

17.4.6.2 Development

Plants are highly sensitive to far-red photons. As discussed in Section 17.4.5.2 on red photons, far-red interacts with plant development through the photoreceptor phytochrome. While phytochrome activation (P_R to P_{FR}) is most sensitive to red photons, phytochrome inactivation (P_{FR} to P_R) is most sensitive to far-red (Fig. 17.2C). When phytochrome is inactivated, the inhibition of gene expression related to cell wall expansion and hormone synthesis is lifted (de Lucas and Prat, 2014; Legris et al., 2019). This leads to more leaf expansion and/or stem elongation depending on the species.

Supplementation with far-red in sole-source lighting has been shown to increase fresh and dry mass of lettuce in tandem with an increase in leaf area (Lee et al., 2016; Meng and Runkle, 2019), and these responses tend to be more pronounced at higher fractions of far-red. However, due to the increase in photosynthesis from far-red, it is difficult to separate the effect on leaf expansion (thus more photon capture) from that on photosynthesis. Other studies that substitute far-red rather than supplement far-red still show an increase in leaf expansion and dry mass (Fig. 17.4; Zhen and Bugbee, 2020a). In contrast, ornamental species geranium and snapdragon were shown to increase leaf area with far-red substitution but without an increase in dry mass (Park and Runkle, 2017). Although lettuce has been shown to increase leaf area in response to far-red, a decrease in leaf area is often reported in shade-avoiding species (Casal, 2012). In species that are adapted to

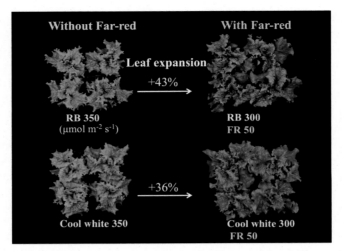

FIGURE 17.4 Lettuce 'Waldmann's Dark Green' grown under blue + red (top left) or cool white LEDs (bottom left). When replacing 15% of the red/blue or white photons with far-red photons (total photon flux remained constant), leaf expansion of lettuce significantly increased without any reductions in canopy photosynthetic efficiency. The increased radiation capture with far-red led to higher yield. *For more details see Zhen, S., Bugbee, B., 2020a. Substituting far-red for traditionally defined photosynthetic photons results in equal canopy quantum yield for CO2 fixation and increased photon capture during long-term studies: implications for re-defining PAR. Front. Plant Sci. 11, 1433.*

high light like tomato and cucumber, far-red tends to increase stem and/or petiole elongation (Kalaitzoglou et al., 2019; Meng et al., 2019; Park and Runkle, 2017). This may not be desirable in plant factories.

Plant responses to far-red also interact with the background PPFD and temperature. The reversion of active P_{FR} back to inactive form of P_R can occur independent of light in a temperature-dependent manner, with faster reversion rate at higher temperature. This is known as thermal reversion of phytochromes, and the effect is more pronounced under lower PPFDs (Sellaro et al., 2019). This effect means that plant responses to far-red (elongation) are expected to be more pronounced at both higher temperatures and lower PPFD.

In addition to these effects on morphology, one recent study found that supplemental far-red increased fruit yield of tomatoes, possibly through increased fruit sink strength and dry mass partitioning to the fruits (Ji et al., 2020).

Far-red can be applied near the end-of-day to mimic the relative increase in far-red under natural conditions (Kasperbauer, 1971). These end-of-day far-red treatments are still used to this day as an energy-saving method to alter development in greenhouse and indoor crop production, but it tends to be less effective than far-red applied over the entire photoperiod (Kalaitzoglou et al., 2019; Morgan and Smith, 1978).

Far-red supplementation was reported to decrease anthocyanins and/or carotenoids (with no effect on phenolics) in lettuce (Li and Kubota, 2009; Zou et al., 2019) and also caused carotenoid concentration to decrease in tomato (Kalaitzoglou et al., 2019). By contrast, Lee et al. (2016) saw an increase in the concentration of phenolics with increasing far-red.

17.5 Concluding remarks

LED-based crop production enables precise control of the light environment (intensity, spectrum, and photoperiod) to optimize growth, modulate plant morphology and beneficial secondary metabolites synthesis. With an increasing number of high efficiency LEDs of distinct spectral peaks becoming available, there is a need for more research-based information on creating light 'recipes' for various crops that have different desirable traits and are harvested for roots, leaves, flowers, fruits, and/or secondary metabolites. In terms of optimal spectrum for photosynthesis, most of the data thus far are collected using monochromatic lights on single leaves under low light. More recent studies show that the responses differ at canopy level and under different light intensities and spectral combinations (synergistic responses). Importantly, photosynthetic responses also interact with morphological and physiological changes mediated by an array of photoreceptors during long-term crop cultivation. Spectral effects on plant development vary significantly among species, spectral peaks, intensity, duration, and timing of application. Additionally, the spectral responses interact with other environmental factors such as temperature, water, and nutrient availability. Lastly, the photon efficacy of the LED luminaire should be considered as it plays a large role in determining the cost of lighting. Continued research efforts in both academia and private sectors are necessary toward the development of optimal spectra for crop production in plant factories.

References

Aarrouf, J., Urban, L., 2020. Flashes of UV-C light: an innovative method for stimulating plant defences. PloS One 15 (7), e0235918.

Ahmad, M., 2016. Photocycle and signaling mechanisms of plant cryptochromes. Curr. Opin. Plant Biol. 33, 108−115.

Ahmad, M., Grancher, N., Heil, M., Black, R.C., Giovani, B., Galland, P., Lardemer, D., 2002. Action spectrum for cryptochrome-dependent hypocotyl growth inhibition in Arabidopsis. Plant Physiol. 129 (2), 774−785.

Assmann, S.M., Simoncini, L., Schroeder, J.I., 1985. Blue light activates electrogenic ion pumping in guard cell protoplasts of Vicia faba. Nature 318 (6043), 285−287.

Ballare, C.L., Sanchez, R.A., Scopel, A.L., Casal, J.J., Ghersa, C.M., 1987. Early detection of neighbour plants by phytochrome perception of spectral changes in reflected sunlight. Plant Cell Environ. 10, 551−557.

Bornman, J.F., Vogelmann, T.C., 1991. Effect of UV-B radiation on leaf optical properties measured with fibre optics. J. Exp. Bot. 42 (4), 547−554.

Bouly, J.-P., Schleicher, E., Dionisio-Sese, M., Vandenbussche, F., Van Der Straeten, D., Bakrim, N., Meier, S., Batschauer, A., Galland, P., Bittl, R., Ahmad, M., 2007. Cryptochrome blue light photoreceptors are activated through interconversion of flavin redox states. J. Biol. Chem. 282 (13), 9383−9391.

Bridgen, M.P., 2016. Using ultraviolet-C (UV-C) irradiation on greenhouse ornamental plants for growth regulation. VIII Int. Symp. Light Horticult. 1134, 49−56.

Brodersen, C.R., Vogelmann, T.C., 2010. Do changes in light direction affect absorption profiles in leaves? Funct. Plant Biol. 37 (5), 403−412.

Brown, B.A., Headland, L.R., Jenkins, G.I., 2009. UV-B action spectrum for UVR8-mediated HY5 transcript accumulation in Arabidopsis. Photochem. Photobiol. 85 (5), 1147−1155.

Casal, J.J., 2012. Shade avoidance. Arabidop. Book/Am. Soci. Plant Biol. 10.

Chen, Y., Li, T., Yang, Q., Zhang, Y., Zou, J., Bian, Z., Wen, X., 2019. UVA radiation is beneficial for yield and quality of indoor cultivated lettuce. Front. Plant Sci. 10, 1563.

Christie, J.M., 2007. Phototropin blue-light receptors. Annu. Rev. Plant Biol. 58, 21−45.

Coffey, A., Prinsen, E., Jansen, M.A.K., Conway, J., 2017. The UVB photoreceptor UVR8 mediates accumulation of UV-absorbing pigments, but not changes in plant morphology, under outdoor conditions. Plant Cell Environ. 40 (10), 2250−2260.

de Lucas, M., Prat, S., 2014. PIF s get BR right: phytochrome interacting factor s as integrators of light and hormonal signals. New Phytol. 202 (4), 1126−1141.

Dörr, O.S., Brezina, S., Rauhut, D., Mibus, H., 2020. Plant architecture and phytochemical composition of basil (*Ocimum basilicum* L.) under the influence of light from microwave plasma and high-pressure sodium lamps. J. Photochem. Photobiol. B Biol. 202, 111678.

Dou, H., Niu, G., Gu, M., 2019. Pre-harvest UV-B radiation and photosynthetic photon flux density interactively affect plant photosynthesis, growth, and secondary metabolites accumulation in basil (*Ocimum basilicum*) plants. Agronomy 9 (8), 434.

Duysens, L.N.M., Amesz, J., 1962. Function and identification of two photochemical systems in photosynthesis. Biochim. Biophys. Acta 64 (2), 243−260.

Emerson, R., Chalmers, R., Cederstrand, C., 1957. Some factors influencing the long- wave limit of photosynthesis. Proc. Natl. Acad. Sci. U. S. A 43 (1), 133.

Emerson, R., Lewis, C.M., 1943. The dependence of the quantum yield of Chlorella photosynthesis on wave length of light. Am. J. Bot. 30 (3), 165−178.

Evans, J.R., 1987. The dependence of quantum yield on wavelength and growth irradiance. Funct. Plant Biol. 14 (1), 69−79.

Favory, J.J., Stec, A., Gruber, H., Rizzini, L., Oravecz, A., Funk, M., Albert, A., Cloix, C., Jenkins, G.I., Oakeley, E.J., Seidlitz, H.K., 2009. Interaction of COP1 and UVR8 regulates UV-B-induced photomorphogenesis and stress acclimation in Arabidopsis. EMBO J. 28 (5), 591−601.

Folta, K.M., Carvalho, S.D., 2015. Photoreceptors and control of horticultural plant traits. Hortscience 50 (9), 1274−1280.

Frank, H.A., Cogdell, R.J., 1996. Carotenoids in photosynthesis. Photochem. Photobiol. 63 (3), 257−264.

Galvão, V.C., Fankhauser, C., 2015. Sensing the light environment in plants: photoreceptors and early signaling steps. Curr. Opin. Neurobiol. 34, 46−53.

Gifford, R.M., Thorne, J.H., Hitz, W.D., Giaquinta, R.T., 1984. Crop productivity and photoassimilate partitioning. Science 225 (4664), 801−808.

Gommers, C.M., Visser, E.J., St Onge, K.R., Voesenek, L.A., Pierik, R., 2013. Shade tolerance: when growing tall is not an option. Trends Plant Sci. 18 (2), 65−71.

Haupt, W., Scheuerlein, R., 1990. Chloroplast movement. Plant. Cell Environ. 13 (7), 595−614.

Hernández, R., Kubota, C., 2016. Physiological responses of cucumber seedlings under different blue and red photon flux ratios using LEDs. Environ. Exp. Bot. 121, 66−74.

Hill, R., Bendall, F., 1960. Function of the two cytochrome components in chloroplasts: a working hypothesis. Nature 186 (4719), 136−137.

Hogewoning, S.W., Wientjes, E., Douwstra, P., Trouwborst, G., Van Ieperen, W., Croce, R., Harbinson, J., 2012. Photosynthetic quantum yield dynamics: from photosystems to leaves. Plant Cell 24 (5), 1921−1935.

Hoover, W.H., 1937. The dependence of carbon dioxide assimilation in a higher plant on wavelength of radiation. Smithson. Inst. Miscell. Collect. 95, 1−13.

Inada, K., 1976. Action spectra for photosynthesis in higher plants. Plant Cell Physiol. 17 (2), 355−365.

Jansen, M.A., Gaba, V., Greenberg, B.M., 1998. Higher plants and UV-B radiation: balancing damage, repair and acclimation. Trends Plant Sci. 3 (4), 131−135.

Jeong, H.W., Lee, H.R., Kim, H.M., Kim, H.M., Hwang, H.S., Hwang, S.J., 2020. Using light quality for growth control of cucumber seedlings in closed-type plant production system. Plants 9 (5), 639.

Ji, Y., Nuñez Ocaña, D., Choe, D., Larsen, D.H., Marcelis, L.F., Heuvelink, E., 2020. Far-red radiation stimulates dry mass partitioning to fruits by increasing fruit sink strength in tomato. New Phytol. 228 (6), 1914−1925.

Kagawa, T., Sakai, T., Suetsugu, N., Oikawa, K., Ishiguro, S., Kato, T., Tabata, S., Okada, K., Wada, M., 2001. Arabidopsis NPL1: a phototropin homolog controlling the chloroplast high-light avoidance response. Science 291 (5511), 2138−2141.

Kalaitzoglou, P., Van Ieperen, W., Harbinson, J., van der Meer, M., Martinakos, S., Weerheim, K., Nicole, C., Marcelis, L.F., 2019. Effects of continuous or end-of-day far-red light on tomato plant growth, morphology, light absorption, and fruit production. Front. Plant Sci. 10, 322.

Kang, W.H., Park, J.S., Park, K.S., Son, J.E., 2016. Leaf photosynthetic rate, growth, and morphology of lettuce under different fractions of red, blue, and green light from light-emitting diodes (LEDs). Horticult., Environ., Biotechnol. 57 (6), 573−579.

Kang, S., Zhang, Y., Zhang, Y., Zou, J., Yang, Q., Li, T., 2018. Ultraviolet-A radiation stimulates growth of indoor cultivated tomato (*Solanum lycopersicum*) seedlings. Hortscience 53 (10), 1429−1433.

Kasperbauer, M.J., 1971. Spectral distribution of light in a tobacco canopy and effects of end-of-day light quality on growth and development. Plant Physiol. 47 (6), 775−778.

Kelly, J.M., Lagarias, J.C., 1985. Photochemistry of 124-kilodalton Avena phytochrome under constant illumination in vitro. Biochemistry 24 (21), 6003−6010.

Kendrick, R.E., Kronenberg, G.H. (Eds.), 1994. Photomorphogenesis in Plants, second ed. Kluwer Academic Publishers, Dordrecht, The Netherlands.

Khoshimkhujaev, B., Kwon, J.K., Park, K.S., Choi, H.G., Lee, S.Y., 2014. Effect of monochromatic UV-A LED irradiation on the growth of tomato seedlings. Horticult., Environ., Biotechnol. 55 (4), 287−292.

Kim, H.H., Goins, G.D., Wheeler, R.M., Sager, J.C., 2004. Green-light supplementation for enhanced lettuce growth under red-and blue-light-emitting diodes. Hortscience 39 (7), 1617−1622.

Klose, C., Nagy, F., Schäfer, E., 2020. Thermal reversion of plant phytochromes. Mol. Plant 13 (3), 386−397.

Kubota, C., Eguchi, T., Kroggel, M., 2017. UV-B radiation dose requirement for suppressing intumescence injury on tomato plants. Sci. Hortic. 226, 366−371.

Kusuma, P., Bugbee, B., 2021. Far-red fraction: an improved metric for characterizing phytochrome effects on morphology. J. Am. Soc. Hortic. Sci. 146 (1), 3−13.

Kusuma, P., Pattison, P.M., Bugbee, B., 2020. From physics to fixtures to food: current and potential LED efficacy. Horticult. Res. 7 (1), 1−9.

Lagarias, J.C., Kelly, J.M., Cyr, K.L., Smith Jr., W.O., 1987. Comparative photochemical analysis of highly purified 124 kilodalton oat and rye phytochromes in vitro. Photochem. Photobiol. 46 (1), 5−13.

Laisk, A., Oja, V., Eichelmann, H., Dall'Osto, L., 2014. Action spectra of photosystems II and I and quantum yield of photosynthesis in leaves in State 1. Biochim. Biophys. Acta Bioenerg. 1837 (2), 315−325.

Lee, M.J., Son, J.E., Oh, M.M., 2014. Growth and phenolic compounds of *Lactuca sativa* L. grown in a closed-type plant production system with UV-A, -B, or -C lamp. J. Sci. Food Agric. 94 (2), 197−204.

Lee, M.J., Son, K.H., Oh, M.M., 2016. Increase in biomass and bioactive compounds in lettuce under various ratios of red to far-red LED light supplemented with blue LED light. Horticult., Environ., Biotechnol. 57 (2), 139−147.

Legris, M., Ince, Y.Ç., Fankhauser, C., 2019. Molecular mechanisms underlying phytochrome-controlled morphogenesis in plants. Nat. Commun. 10 (1), 1−15.

Li, Q., Kubota, C., 2009. Effects of supplemental light quality on growth and phytochemicals of baby leaf lettuce. Environ. Exp. Bot. 67 (1), 59−64.

Lin, C., 2000. Plant blue-light receptors. Trends Plant Sci. 5 (8), 337−342.

Liu, J., van Iersel, M.W., 2021. Photosynthetic physiology of Blue, Green, and Red light: light intensity effects and underlying mechanisms. Front. Plant Sci. 12, 328.

Mantha, S.V., Johnson, G.A., Day, T.A., 2001. Evidence from action and fluorescence spectra that UV-induced violet−blue−green fluorescence enhances leaf photosynthesis. Photochem. Photobiol. 73 (3), 249−256.

Mariz-Ponte, N., Martins, S., Gonçalves, A., Correia, C.M., Ribeiro, C., Dias, M.C., Santos, C., 2019. The potential use of the UV-A and UV-B to improve tomato quality and preference for consumers. Sci. Hortic. 246, 777−784.

McCree, K.J., 1971. The action spectrum, absorptance and quantum yield of photosynthesis in crop plants. Agric. Meteorol. 9, 191−216.

Melis, A., 2009. Solar energy conversion efficiencies in photosynthesis: minimizing the chlorophyll antennae to maximize efficiency. Plant Sci. 177 (4), 272−280.

Meng, Q., Boldt, J., Runkle, E.S., 2020. Blue radiation interacts with green radiation to influence growth and predominantly controls quality attributes of lettuce. J. Am. Soc. Hortic. Sci. 145 (2), 75−87.

Meng, Q., Kelly, N., Runkle, E.S., 2019. Substituting green or far-red radiation for blue radiation induces shade avoidance and promotes growth in lettuce and kale. Environ. Exp. Bot. 162, 383−391.

Meng, Q., Runkle, E.S., 2019. Far-red radiation interacts with relative and absolute blue and red photon flux densities to regulate growth, morphology, and pigmentation of lettuce and basil seedlings. Sci. Hortic. 255, 269−280.

Monteith, J.L., 1977. Climate and the efficiency of crop production in britain. Philosophical transactions of the royal society of London. B. Biol. Sci. 281 (980), 277−294.

Morgan, D.C., Smith, H., 1978. The relationship between phytochrome-photoequilibrium and Development in light grown *Chenopodium album* L. Planta 142 (2), 187—193.

Müller, P., Bouly, J.P., Hitomi, K., Balland, V., Getzoff, E.D., Ritz, T., Brettel, K., 2014. ATP binding turns plant cryptochrome into an efficient natural photoswitch. Sci. Rep. 4 (1), 1—11.

Müller, P., Ahmad, M., 2011. Light-activated cryptochrome reacts with molecular oxygen to form a flavin—superoxide radical pair consistent with magnetoreception. J. Biol. Chem. 286 (24), 21033—21040.

Murakami, K., Matsuda, R., Fujiwara, K., 2018. A mathematical model of photosynthetic electron transport in response to the light spectrum based on excitation energy distributed to photosystems. Plant Cell Physiol. 59 (8), 1643—1651.

Myers, J., 1971. Enhancement studies in photosynthesis. Annu. Rev. Plant Physiol. 22 (1), 289—312.

Neugart, S., Schreiner, M., 2018. UVB and UVA as eustressors in horticultural and agricultural crops. Sci. Hortic. 234, 370—381.

Paradiso, R., Meinen, E., Snel, J.F., De Visser, P., Van Ieperen, W., Hogewoning, S.W., Marcelis, L.F., 2011. Spectral dependence of photosynthesis and light absorptance in single leaves and canopy in rose. Sci. Hortic. 127 (4), 548—554.

Park, Y., Runkle, E.S., 2017. Far-red radiation promotes growth of seedlings by increasing leaf expansion and whole-plant net assimilation. Environ. Exp. Bot. 136, 41—49.

Raviv, M., Antignus, Y., 2004. UV radiation effects on pathogens and insect pests of greenhouse-grown crops. Photochem. Photobiol. 79 (3), 219—226.

Ruban, A.V., 2015. Evolution under the sun: optimizing light harvesting in photosynthesis. J. Exp. Bot. 66 (1), 7—23.

Salomon, M., Christie, J.M., Knieb, E., Lempert, U., Briggs, W.R., 2000. Photochemical and mutational analysis of the FMN-binding domains of the plant blue light receptor, phototropin. Biochemistry 39 (31), 9401—9410.

Samuolienė, G., Viršilė, A., Miliauskienė, J., Haimi, P., Laužikė, K., Jankauskienė, J., Novičkovas, A., Kupčinskienė, A., Brazaitytė, A., 2020. The photosynthetic performance of red leaf lettuce under UV-A irradiation. Agronomy 10 (6), 761.

Schreiner, M., Mewis, I., Huyskens-Keil, S., Jansen, M.A.K., Zrenner, R., Winkler, J.B., O'brien, N., Krumbein, A., 2012. UV-B-induced secondary plant metabolites-potential benefits for plant and human health. Crit. Rev. Plant Sci. 31 (3), 229—240.

Sellaro, R., Smith, R.W., Legris, M., Fleck, C., Casal, J.J., 2019. Phytochrome B dynamics departs from photoequilibrium in the field. Plant Cell Environ. 42 (2), 606—617.

Siefermann-Harms, D., 1985. Carotenoids in photosynthesis. I. Location in photosynthetic membranes and light-harvesting function. Biochim. Biophys. Acta Rev. Bioenerg. 811 (4), 325—355.

Smith, H.L., McAusland, L., Murchie, E.H., 2017. Don't ignore the green light: exploring diverse roles in plant processes. J. Exp. Bot. 68 (9), 2099—2110.

Snowden, M.C., Cope, K.R., Bugbee, B., 2016. Sensitivity of seven diverse species to blue and green light: interactions with photon flux. PloS One 11 (10), e0163121.

Son, K.H., Oh, M.M., 2013. Leaf shape, growth, and antioxidant phenolic compounds of two lettuce cultivars grown under various combinations of blue and red light-emitting diodes. Hortscience 48 (8), 988—995.

Son, K.H., Oh, M.M., 2015. Growth, photosynthetic and antioxidant parameters of two lettuce cultivars as affected by red, green, and blue light-emitting diodes. Horticult., Environ., Biotechnol. 56 (5), 639—653.

Sun, J., Nishio, J.N., Vogelmann, T.C., 1998. Green light drives CO2 fixation deep within leaves. Plant Cell Physiol. 39 (10), 1020—1026.

Takahashi, S., Milward, S.E., Yamori, W., Evans, J.R., Hillier, W., Badger, M.R., 2010. The solar action spectrum of photosystem II damage. Plant Physiol. 153 (3), 988—993.

Terashima, I., Fujita, T., Inoue, T., Chow, W.S., Oguchi, R., 2009. Green light drives leaf photosynthesis more efficiently than red light in strong white light: revisiting the enigmatic question of why leaves are green. Plant Cell Physiol. 50 (4), 684—697.

Tyystjärvi, E., 2008. Photoinhibition of photosystem II and photodamage of the oxygen evolving manganese cluster. Coord. Chem. Rev. 252 (3—4), 361—376.

Verdaguer, D., Jansen, M.A., Llorens, L., Morales, L.O., Neugart, S., 2017. UV-A radiation effects on higher plants: exploring the known unknown. Plant Sci. 255, 72—81.

Vogelmann, T.C., 1993. Plant tissue optics. Annu. Rev. Plant Biol. 44 (1), 231—251.

Vogelmann, T.C., Evans, J.R., 2002. Profiles of light absorption and chlorophyll within spinach leaves from chlorophyll fluorescence. Plant. Cell Environ. 25 (10), 1313–1323.

Wade, H.K., Bibikova, T.N., Valentine, W.J., Jenkins, G.I., 2001. Interactions within a network of phytochrome, cryptochrome and UV-B phototransduction pathways regulate chalcone synthase gene expression in Arabidopsis leaf tissue. Plant J. 25 (6), 675–685.

Wang, J., Lu, W., Tong, Y., Yang, Q., 2016. Leaf morphology, photosynthetic performance, chlorophyll fluorescence, stomatal development of lettuce (*Lactuca sativa* L.) exposed to different ratios of red light to blue light. Front. Plant Sci. 7, 250.

Wargent, J.J., Gegas, V.C., Jenkins, G.I., Doonan, J.H., Paul, N.D., 2009a. UVR8 in *Arabidopsis thaliana* regulates multiple aspects of cellular differentiation during leaf development in response to ultraviolet B radiation. New Phytol. 183 (2), 315–326.

Wargent, J.J., Moore, J.P., Roland Ennos, A., Paul, N.D., 2009b. Ultraviolet radiation as a limiting factor in leaf expansion and development. Photochem. Photobiol. 85 (1), 279–286.

Warner, C.W., Caldwell, M.M., 1983. Influence of photon flux density in the 400–700 nm waveband on inhibition of photosynthesis by UV-B (280–320 nm) irradiation in soybean leaves: separation of indirect and immediate effects. Photochem. Photobiol. 38 (3), 341–346.

Williams, K.A., Craver, J.K., Miller, C.T., Rud, N., Kirkham, M.B., 2014. Differences between the physiological disorders of intumescences and edemata. In XXIX international horticultural congress on horticulture: sustaining lives. Livelihoods Landscapes (IHC2014) 1104, 401–406.

Yao, Y., Yang, Y., Ren, J., Li, C., 2006. UV-spectra dependence of seedling injury and photosynthetic pigment change in Cucumis sativus and Glycine max. Environ. Exp. Bot. 57 (1–2), 160–167.

Yorio, N.C., Goins, G.D., Kagie, H.R., Wheeler, R.M., Sager, J.C., 2001. Improving spinach, radish, and lettuce growth under red light-emitting diodes (LEDs) with blue light supplementation. Hortscience 36 (2), 380–383.

Zhang, T., Folta, K.M., 2012. Green light signaling and adaptive response. Plant Signaling & Behavior 7 (1), 75–78.

Zhang, T., Maruhnich, S.A., Folta, K.M., 2011. Green light induces shade avoidance symptoms. Plant Physiol. 157 (3), 1528–1536.

Zhen, S., Bugbee, B., 2020a. Substituting far-red for traditionally defined photosynthetic photons results in equal canopy quantum yield for CO_2 fixation and increased photon capture during long-term studies: implications for redefining PAR. Front. Plant Sci. 11, 1433.

Zhen, S., Bugbee, B., 2020b. Steady-state stomatal responses of C3 and C4 species to blue light fraction: interactions with CO_2 concentration. Plant. Cell Environ. 43 (12), 3020–3032.

Zhen, S., Bugbee, B., 2020c. Far-red photons have equivalent efficiency to traditional photosynthetic photons: implications for redefining photosynthetically active radiation. Plant. Cell Environ. 43 (5), 1259–1272.

Zhen, S., Haidekker, M., van Iersel, M.W., 2019. Far-red light enhances photochemical efficiency in a wavelength-dependent manner. Physiol. Plant. 167 (1), 21–33.

Zhen, S., van Iersel, M.W., 2017. Far-red light is needed for efficient photochemistry and photosynthesis. J. Plant Physiol. 209, 115–122.

Zhen, S., van Iersel, M., Bugbee, B., 2021. Why far-red photons should be included in the definition of photosynthetic photons and the measurement of horticultural fixture efficacy. Front. Plant Sci. 12, 693445.

Zou, J., Zhang, Y., Zhang, Y., Bian, Z., Fanourakis, D., Yang, Q., Li, T., 2019. Morphological and physiological properties of indoor cultivated lettuce in response to additional far-red light. Sci. Hortic. 257, 108725.

Indoor lighting effects on plant nutritional compounds

Nathan Kelly[1], Viktorija Vaštakaitė-Kairienė[2] and Erik S. Runkle[1]

[1]Department of Horticulture, Michigan State University, East Lansing, MI, United States;
[2]Institute of Horticulture, Lithuanian Research Centre for Agriculture and Forestry, Babtai, Lithuania

18.1 Introduction

In controlled-environment agriculture, specialty crops can be cultivated to have increased concentrations of biologically active substances and can be used as functional foods for medicinal purposes. The changing eating habits and the increasing popularity of raw food, as well as numerous studies confirming the benefits of plant metabolites in the human diet, have increased the demand for nutrient-rich vegetables year round. In many published studies where phytochemical concentrations under various environmental conditions were evaluated, light was one of the main factors affecting metabolism and nutrient uptake in plants. Greater nutritional value can be achieved by activating plant defense systems that respond to environmental stresses, such as light, or by directly stimulating metabolite synthesis through plant photosynthetic and photomorphogenic receptors. Plants accumulate higher contents of compounds with antioxidant properties such as vitamin C, polyphenols, or other antiradical scavengers to reduce reactive oxygen species that are elevated during oxidative stress. The same antioxidants strengthen the human immune system and protect against noninfectious chronic disorders such as cancer and neurodegenerative, cardiovascular, and lymphatic systems diseases (Dasgupta and Klein, 2014). Moreover, plant metabolites create distinct taste profiles that are dependent on the contents of sugars and bitter phenolic compounds. Furthermore, metabolites such as yellow, orange, and red carotenoids, and purple anthocyanins increase leaf pigmentation and can affect the marketability of specialty crops.

The quantity and quality of light delivered to plants can profoundly impact the nutritional content of food crops grown indoors. Light quantity refers to the number of photons within the photosynthetically active radiation (PAR) waveband, which is usually considered to be from 400 to 700 nm. It is expressed per square meter on either an instantaneous ($\mu mol\ m^{-2}\ s^{-1}$) or daily ($mol\ m^{-2}\ d^{-1}$) basis. Light quality can be defined as the spectral distribution of a light source. In other words, the photon spectrum is the specific combination of wavelengths that a light source emits. The photon spectrum can be divided into three main groups of color: blue (B; 400−499 nm), green (G; 500−599 nm), and red (R; 600−699 nm). Ultraviolet-B (UV-B; 280−315 nm), UV-A (315−399 nm), and far-red (FR; 700−800 nm) light are outside of the PAR waveband, but also influence plant growth and quality attributes. For instance, UV-A light can increase anthocyanin accumulation and leaf redness (Li and Kubota, 2009), while FR light increases leaf expansion (Park and Runkle, 2017). In this chapter, we summarize research-based information about how the photon flux density (PFD) and spectrum regulate the concentration of different nutritional compounds and mineral elements in leafy greens and microgreens grown indoors. Complementary studies performed in greenhouses are also mentioned when warranted.

18.2 Regulation by light quantity

18.2.1 Phenolic compounds

18.2.1.1 Total phenolic content

Phenolic compounds are the most ubiquitous secondary metabolites found in various parts of the plant. Phenolics play an important role in plant growth, protect against pathogens and unfavorable environmental conditions such as UV light, and contribute toward pigment accumulation and human sensory traits of horticultural and agronomic plants. It is recommended to supplement daily diets with vegetables that are natural sources of phenolic compounds because of their antiallergenic, antiatherogenic, anti-inflammatory, antimicrobial, antithrombotic, cardioprotective, and vasodilatory properties (Balasudram et al., 2006).

The metabolism of phenolic compounds in horticultural plants is regulated by the light environment. Samuolienė et al. (2013a) compared the total phenolic content in kohlrabi (*Brassica oleracea* var. *gongylodes* 'Delicacy Purple'), mustard (*Brassica juncea* 'Red Lion'), red pak choi (*Brassica rapa* var. *chinensis* 'Rubi F$_1$'), and tatsoi (*Brassica rapa* var. *rosularis*) microgreens grown under a photosynthetic photon flux density (PPFD) of 110, 220 (control), 330, 440, and 545 $\mu mol\ m^{-2}\ s^{-1}$ from B (peak = 455 nm), R (peaks = 638 and 665 nm), and FR (peak = 731 nm) light-emitting diodes (LEDs) indoors. Microgreens grown under the lowest PPFD of 110 $\mu mol\ m^{-2}\ s^{-1}$ had a lower total phenolic content than those grown under the control. Total phenols were greatest under PPFDs of 440 and 545 $\mu mol\ m^{-2}\ s^{-1}$ for tatsoi and kohlrabi, and 330 and 440 $\mu mol\ m^{-2}\ s^{-1}$ for red pak choi. In mustard, the content of phenolic compounds under the highest investigated PPFD of 545 $\mu mol\ m^{-2}\ s^{-1}$ was similar to that of plants grown under the control. In a later study, Zheng et al. (2018) investigated the influence of an increased PPFD provided by supplemental B (peak = 460 nm) light for 12 h d^{-1}, for 10 days before harvest, on total phenolic content in green- and red-leaf pak choi (*Brassica campestris* ssp. *chinensis* var. *communis*) grown in a greenhouse. In green pak choi, the total phenolic content significantly increased with an increasing B PFD from 50 to 150 µmol m^{-2} s^{-1}. The greatest content of phenols in red pak choi was achieved under 100 µmol m^{-2} s^{-1} of supplemental B light.

18.2.1.2 Anthocyanins

Anthocyanins are a class of phenolic compounds with a range of colors varying from orange and red to blue (Tanaka and Ohmiya, 2008). They are water-soluble pigments that are primarily located in the vacuoles of plant cells and are mostly limited to peripheral tissues and structures like the upper mesophyll, which are exposed to a high PFD (Trojak and Skowron, 2017). Foliar anthocyanins function as screening pigments to shade and protect the photosynthetic apparatus from excessive light (Hoch et al., 2003). For human health, anthocyanins are involved in chronic disease prevention (Pojer et al., 2013; Liu et al., 2018).

Anthocyanin accumulation in plants is determined by developmental stage, species and cultivar, and environmental factors. Zheng et al. (2018) reported that the total anthocyanin content in red pak choi was higher than in green-leaf plants; however, the content increased in both cultivars under supplemental B (peak = 460 nm) LED lighting at 50, 100, and 150 µmol m^{-2} s^{-1} compared to plants grown without supplemental B light. Craver et al. (2017) reported that anthocyanin concentration in purple kohlrabi microgreens depended on the spectral composition emitted by LED arrays. For example, total anthocyanin concentration of kohlrabi grown under B + R light was 17% and 18% higher under a PPFD of 210 or 315 µmol m^{-2} s^{-1}, respectively, compared with those grown under 105 µmol m^{-2} s^{-1}. Also, the total anthocyanin concentration in kohlrabi grown under B + R + FR light was 31% and 24% higher under PPFDs of 210 or 315 µmol m^{-2} s^{-1}, respectively, compared with those grown under 105 µmol m^{-2} s^{-1}. Anthocyanin concentration of kohlrabi grown under R + B light was 14% higher when grown under a PPFD of 210 µmol m^{-2} s^{-1} compared with those grown under 105 µmol m^{-2} s^{-1}, however, no further increase occurred in those grown under a PPFD of 315 µmol m^{-2} s^{-1} (Craver et al., 2017). Finally, anthocyanin concentration of kohlrabi microgreens grown under different spectral qualities was similar at light intensities of 105 or 210 µmol m^{-2} s^{-1} (Craver et al., 2017). These results partially agree with those by Samuolienė et al. (2013a), who reported that total anthocyanin content in *Brassica* microgreens increased with PPFD when delivered by B + R + FR LEDs. Total anthocyanin content in red pak choi 'Rubi F$_1$' and tatsoi under a PPFD of 330 µmol m^{-2} s^{-1}, and in kohlrabi 'Delicacy Purple' under 440 µmol m^{-2} s^{-1}, was higher than when the microgreens were grown under a PPFD of 110 or 220 µmol m^{-2} s^{-1}.

18.2.2 Vitamins

18.2.2.1 Vitamin C

Vitamin C (ascorbic acid or ascorbate) is an antioxidant that protects plants from oxidative damage by scavenging free oxygen radicals, acts as a signaling modulator of cells in many physiological processes, and maintains signaling pathways mediated by phytohormones. Increasing vitamin C content at the cellular level can increase plant stress tolerance to adverse environmental conditions (Aziz et al., 2018). In humans, dietary vitamin C has an important role in numerous metabolic functions including tissue growth and maintenance, amelioration of oxidative stress, and immune regulation (Carr and Maggini, 2017).

In green- and red-leaf pak choi, the concentration of vitamin C was influenced by supplemental B (peak = 460 nm) light when applied 10 days before harvest for 12 h d^{-1} in a greenhouse (Zheng et al. (2018). Vitamin C content in both cultivars of pak choi was about six- and sevenfold higher under 100 μmol m^{-2} s^{-1} of supplemental B light compared to those grown without supplemental light. However, there was no significant increase in vitamin C content in green pak choi under 50 and 150 μmol m^{-2} s^{-1}, and red pak choi under 50 μmol m^{-2} s^{-1} of supplemental B light. In a separate study, increasing the PPFD from 50 to 200 μmol m^{-2} s^{-1} of B + R light with a ratio of 4.0, for 48 h before harvest, increased vitamin C content in butterhead lettuce (*Lactuca sativa* var. *capitata*) (Zhou et al., 2012). The effect was most pronounced when the PPFD increased from 50 to 100 μmol m^{-2} s^{-1}; the rate of increase slowed as the PPFD increased beyond 100 μmol m^{-2} s^{-1}. According to Fu et al. (2017), the vitamin C content in lettuce var. *yomaicai* leaves was dependent on the interaction of PPFD from R + B LEDs (ratio 4.0) and nitrogen concentration in the nutrient solution. The vitamin C content in lettuce grown under 140 μmol m^{-2} s^{-1} was higher than plants grown under a PPFD of 60 or 220 μmol m^{-2} s^{-1}. However, increasing the nitrogen concentration in the nutrient solution decreased vitamin C content regardless of PPFD. In another study, Samuolienė et al. (2013a) reported that vitamin C content in *Brassica* microgreens was generally similar under PPFDs ranging from 110 to 545 μmol m^{-2} s^{-1} when delivered by B + R + FR LEDs, except for red pak choi 'Rubi F_1' and tatsoi, where it was greatest under a PPFD of 110 μmol m^{-2} s^{-1}.

18.2.2.2 Total carotenoid content

Plant carotenoids are a large group of lipophilic compounds whose main role is to contribute to light capture in photosynthesis and photoprotection under high-light conditions (Wurtzel, 2019). Carotenoids are both primary metabolites in green tissues, and specialized metabolites in vivid orange, yellow, and red tissues found in many flowers, fruits, and roots (Bartley and Scolnik, 1995; Sun and Li, 2020; Sathasivam et al., 2020). Carotenoids are especially important to human nutrition and health because of their high-antioxidant activity, which help to reduce the risk of chronic diseases like cancer, cardiovascular disease, and age-related eye disease (Hannoufa and Hossain, 2012; Sun and Li, 2020). Carotenoids, such as α-carotene and β-carotene, are precursors for vitamin A. Because of its valuable health properties and industrial importance, interest in carotenoids, particularly β-carotene and lutein, has increased (Sathasivam et al., 2020).

Brazaitytė et al. (2015a) investigated the influence of PPFD at 110, 220 (control), 330, 440, and 545 μmol m^{-2} s^{-1} from R + B + FR LEDs on carotenoid concentration in *Brassica* microgreens. In general, total carotenoid concentration in red pak choi 'Rubi F_1', tatsoi, and mustard 'Red Lion' microgreens was greater when plants were grown under a PPFD of 330 and 440 μmol m^{-2} s^{-1} compared to the control. The highest investigated PPFD led to a decrease in carotenoid concentration in mustard and red pak choi but increased carotenoid concentration in tatsoi microgreens. Total carotenoid accumulation in microgreens grown under a PPFD of 110 μmol m^{-2} s^{-1} was lower than the control in red pak choi and tatsoi. With some exceptions, there were similar tendencies on individual carotenoid content in *Brassica* microgreens. β-carotene content was greater under a PPFD of ≥ 330 µmol m^{-2} s^{-1} than under the control PPFD. These results indicate that the total carotenoid content in plants does not always correspond to the contents of individual carotenoids, and that they do not necessarily increase with PPFD.

18.2.2.3 α-Tocopherols

Vitamin E is a group of major lipid-soluble monophenols in the cell antioxidant defense system. Tocopherols and tocotrienols, generally referred to as "tocols," are synthesized by photosynthetic organisms into four homologs (α, β, δ, and γ) (Brigelius-Flohé and Traber, 1999; Guzman et al., 2012; Shahidi and De Camargo, 2016). In cooperation with the carotenoid cycle, vitamin E fulfills at least two different functions in chloroplasts at two major sites of singlet oxygen production: preserving photosystem II from photoinactivation and protecting membrane lipids from photo-oxidation (Havaux et al., 2005). In the leaves of higher plants, the predominant form is α-tocopherol, whereas γ-tocopherol is often the major isoform in seeds (Savidge et al., 2002; Lushchak and Semchuk, 2012). α-tocopherol reverses vitamin E deficiency symptoms in humans, maintaining proper skeletal muscle homeostasis and promoting plasma membrane repair (Guzman et al., 2012). Because of their high-antioxidant activity, tocols participate in anti-inflammatory and immune-enhancing processes in the human body and help to prevent the human body against atherosclerosis, diabetes, obesity, cataracts, Alzheimer's disease, cancer, and human immunodeficiency virus (Tucker and Townsend, 2005; Shahidi and De Camargo, 2016). Moreover, high vitamin E intake is correlated with reduced risk of cardiovascular disease, whereas intake of other dietary antioxidants (e.g., vitamin C or β-carotene) is not, suggesting that tocols act beyond that of their antioxidant function (Brigelius-Flohé and Traber, 1999).

Due to their cellular location, tocopherols have a protective role for photosynthesis (Trebst, 2003). Photosystem II is inactivated by tocopherol deficiency in green algae under high light (Trebst et al., 2002); however, in plants, the effects of PPFD remain unclear. According to Samuolienė et al. (2013a), the accumulation of α-tocopherol in *Brassica* microgreens under B + R + FR LEDs was PPFD and plant species-dependent. For example, in mustard 'Red Lion' microgreens, α-tocopherol content was greatest under the lowest (110 µmol m^{-2} s^{-1}) and highest (545 µmol m^{-2} s^{-1}) PPFDs. In contrast, it was highest under the lowest PPFD(s) in tatsoi and red pak choi 'Rubi F$_1$'. In another study, Samuolienė et al. (2017) investigated the effects of B (peak = 445 nm) light at 0, 25, 50, 75, and 100 µmol m^{-2} s^{-1}, mostly in combination with R (peaks = 638 and 665 nm) LEDs (PPFD of 300 µmol m^{-2} s^{-1}) on total and individual tocopherol content in mustard 'Red Lion', beet (*Beta vulgaris* 'Bulls Blood'), and parsley (*Petroselinum crispum* 'Plain Leaved' or 'French') microgreens. Among species, there was no consistent or clear trend between the B PFD and individual or total tocopherol content. For example, total tocopherol was greatest under 0 or 50 µmol m^{-2} s^{-1} of blue light for mustard and parsley and 50 or 100 µmol m^{-2} s^{-1} for beet. However, the extreme PFDs of B light (0 and 100 µmol m^{-2} s^{-1}) had the greatest increases of α-, β-, and δ-tocopherol in mustard and beet.

18.2.2.4 Folate

Tetrahydrofolate and its derivates—a group under the name of folates or vitamin B—are vital cofactors for enzymes in plants that mediate one-carbon transfer diseases such as cancer, cardiovascular disease, and neuroreactions (Hanson and Gregory, 2011). Humans cannot synthesize folates and, therefore, need them in the diet, and plants are usually the main dietary source (Bekaert et al., 2008). Folates may be stabilized by antioxidant compounds (such as vitamin C and glutathione) and by binding to proteins (Scott et al., 2000). Okazaki and Yamashita (2019) indicated that the PPFD from R + B + G LEDs regulated folate biosynthesis in lettuce plants, since folate content was greater under 200 μmol m^{-2} s^{-1} compared to 150 μmol m^{-2} s^{-1}. In contrast, PPFD appeared to have only a slight influence on folate content in mustard (Lester et al., 2006). However, information is lacking on whether PPFD can be manipulated to increase folate concentration across a range of plant species, and further research is needed.

18.2.3 Nitrates and nitrites

Green vegetables are an important component of the human diet, but unfortunately, constitute a group of foods that contribute maximally to nitrate (NO_3) consumption by humans. Nutritional, environmental, and physiological factors can be responsible for NO_3 accumulation in plants (Anjana and Iqbal, 2007). Although NO_3 is stable, dietary NO_3 is converted to nitrite (NO_2) through a nonenzymatic process and nitric oxide (NO) by symbiotic bacteria in the oral cavity and stomach. NO plays an important role in protecting the cardiovascular system and gastric mucosa, and in metabolic diseases. Dietary NO_3 serves as an effective donor of NO. Above maximum residue levels, NO_3 is thought to be harmful because of the potential production of carcinogenic nitrosamines under certain conditions such as acidic stomach and methemoglobinemia (blue baby syndrome). The World Health Organization recommended a maximum daily NO_3 and NO_2 uptake of 3.7 and 0.06$-$0.07 mg kg^{-1} body weight, respectively (Ma et al., 2018).

In agriculture production systems, nitrogen fertilization and PPFD influence the NO_3 content in vegetables (Cantliffe, 1973). Fu et al. (2017) demonstrated that the NO_3 content in lettuce var. *yomaicai* leaves was influenced by PPFD (ranging from 60 to 220 μmol m^{-2} s^{-1}) and the nitrogen concentration (7 to 23 mmol L^{-1}) in the nutrient solution. The NO_3 content of leaf tissue increased linearly in response to increased nitrogen in nutrient solutions. The highest NO_3 content in lettuce leaves occurred under a PPFD of 60 μmol m^{-2} s^{-1} and the highest nitrogen concentration in the nutrient solution. Viršilė et al. (2019) demonstrated that NO_3 content in green-leaf 'Lobjoits Green Cos' and red-leaf 'Red Cos' lettuce was highest under a PPFD of 100 μmol m^{-2} s^{-1} from primarily B (peak = 455 nm) and R (peaks = 627 and 660 nm) LEDs, was remarkably lower under 200 μmol m^{-2} s^{-1}, and further decreased as PPFD increased above 300 μmol m^{-2} s^{-1}. Additionally, unsafe concentrations of NO_2 were measured in lettuce grown under a PPFD of 100 μmol m^{-2} s^{-1}, and concentrations decreased remarkably when plants were grown under a PPFD of 200 μmol m^{-2} s^{-1}. Finally, in green- and red-leaf lettuce grown under 300 μmol m^{-2} s^{-1}, the NO_2 content was 12- and sevenfold lower, respectively, compared to plants grown under a PPFD of 100 μmol m^{-2} s^{-1}.

18.2.4 Sensory attributes

18.2.4.1 Coloration

Plant color, as well as shape or size, is a quality trait that can influence buyer decisions. A common way to quantify leaf color is to use the L*, a*, b* scale, which is based on the theory that receptors in the human eye perceive color as three pairs of opposites. First, in the L* scale, low numbers indicate dark shades and high numbers indicate light shades. In the a* scale, red and green are compared, and a positive number indicates redness while a negative number indicate greenness. The b* scale compares yellow versus blue, and a positive number indicates yellowness and a negative number indicates blueness. Several studies have evaluated color values of leafy greens grown in different ways, such as under different PPFDs. When averaged among 10 species of leafy greens, the L* value under a PPFD of 200–400 μmol m^{-2} s^{-1} was 13% greater than those under 800–1200 μmol m^{-2} s^{-1} at harvest (Colonna et al., 2016). Leaves became less green and more yellow when leafy vegetables were harvested from under a high PPFD. Kelly et al. (2020) reported that the effect of PPFD on the redness of red-leaf lettuce 'Rouxai' depended on the daily light integral (DLI), or its two factors, PPFD and photoperiod. For example, at a PPFD of 150 μmol m^{-2} s^{-1}, increasing the photoperiod from 16 to 24 h d^{-1} (and the DLI from 8.6 to 13.0 mol m^{-2} d^{-1}) increased redness by 46%. Under the same photoperiod, increasing PPFD increased the redness of lettuce. For example, under a photoperiod of 20 h d^{-1}, increasing the PPFD from 120 to 150 μmol m^{-2} s^{-1} increased leaf redness by 39%.

18.2.4.2 Sugars and taste

Plant soluble sugars are mostly comprised of glucose, fructose, and sucrose (Magwaza and Opara, 2015). Sucrose consists of the monosaccharides glucose and fructose and is a widespread disaccharide in nature. As the main products of photosynthesis, sugars can have profound effects on plant growth, particularly cell division and expansion, storage, signaling, and stress acclimation (Eveland and Jackson, 2012; Keunen et al., 2013; Salerno and Curatti, 2003). Consumer assessment of initial vegetable quality, which influences decision to purchase, is often based on external attributes like color, shape, and size; however, the decision for subsequent purchases is dependent upon consumer satisfaction based on flavor and internal quality (Opara and Pathare, 2014; Magwaza and Opara, 2015). Sweetness in many vegetables is a desirable attribute (Magwaza and Opara, 2015), and the desired taste characteristics can be improved by light manipulation. Samuolienė et al. (2013a) reported that microgreens grown under a low PPFD, from R + B + FR LEDs, had a low-sucrose content. However, the PPFD that elicited the highest sucrose production varied among microgreen species: 545 μmol m^{-2} s^{-1} in kohlrabi, 440 μmol m^{-2} s^{-1} in red pak choi, and 330 μmol m^{-2} s^{-1} in tatsoi. Zhou et al. (2012) showed that increasing the PPFD from 50 to 200 μmol m^{-2} s^{-1} from B (peak = 460 nm) and R (peak = 630 nm) LEDs for 48 h before harvest increased the contents of soluble sugars in butterhead lettuce. This increase in soluble sugar content increased with increasing PPFD at a relatively constant rate and showed little evidence of reaching a maximum. It may therefore be reasonable to predict that soluble sugar content would continue to increase as PPFD increases beyond 200 μmol m^{-2} s^{-1}. The effect of PPFD on soluble sugar content in leafy greens can be influenced by other environmental factors as well as nutrition. For example, soluble sugar content in lettuce 'Italy' increased as PPFD increased from 150 to 350 μmol m^{-2} s^{-1}, then decreased under 450 μmol m^{-2} s^{-1}, while it decreased with increasing nutrient solution strength (Song et al., 2020).

18.3 Regulation by photon spectrum

The ability to control the photon spectrum of a light source allows growers to regulate plant quality attributes such as leaf coloration and concentrations of secondary metabolites and vitamins, some of which affect the nutritional and taste profiles of the plant. Since an "ideal" photon spectrum is somewhat subjective and situational, spectral manipulations that enhance one attribute may diminish another, such as plant yield or taste. For example, greater ratios of B to R light increased total phenolic and flavonoid concentrations but decreased yields of lettuce 'Sunmang' and 'Grand Rapids TBR' (Son and Oh, 2013). The increased concentrations of phenolic and flavonoid compounds make the lettuce more nutritious, but they are often associated with a less desirable, bitter taste. Therefore, choosing the right spectrum is important to acquire the traits desired while also avoiding undesirable growth attributes. The concentrations of many compounds can be altered by changes in the photon spectrum, but regulation of some compounds is better understood than others. Here, we focus on the effects of individual wavebands, and their interactions, on phenolic compound concentrations, including flavonoids such as anthocyanins and their antioxidant properties; vitamins such as vitamin C, β-carotene, lutein, α-tocopherols, and folate; and nitrates. Finally, we discuss how some of these chemicals affect leaf coloration and taste of leafy greens.

18.3.1 Phenolic compounds

18.3.1.1 *Total phenolic concentration*

18.3.1.1.1 Ultraviolet light

UV-B and UV-A light can both increase the synthesis of secondary metabolites such as flavonoids (a subset of phenolics), which strongly absorb UV light to prevent cellular damage (Chappell and Hahlbrock, 1984). Although UV light increases the concentration of some phytochemicals, it may increase or decrease growth rates, depending on the wavelength. For example, red-leaf lettuce 'Hongyeom' exposed to 11 μmol m^{-2} s^{-1} of UV-A (peak = 352 nm) light had a 30% greater concentration of total phenolic compounds than the white-light control at a PPFD of 185 μmol m^{-2} s^{-1} and a photoperiod of 16 h d^{-1} (Lee et al., 2013). In the same study, the UV-A light increased shoot fresh weight by about 22% after 7 days of application. Finally, UV-B (peak = 306 nm) at a PFD of 11 μmol m^{-2} s^{-1} greatly increased total phenolic concentration after two days of application but inhibited further biomass accumulation (Lee et al., 2013). In another study, increasing the PFD of UV-A (peak = 373 nm) from 5 to 21 μmol m^{-2} s^{-1} did not have any effect on the total phenolic concentration of lettuce 'Red Cross' when exposed for the final 12 days of production (Li and Kubota, 2009). This indicates that different wavelengths of UV light can have different effects on phenolics as a group, responses can depend on cultivar, or both. Additionally, many studies have investigated the effects of UV light transmitted through fully to partly UV-transparent films. For example, total phenolic concentration of lettuce 'Lollo Rosso' increased as shorter wavelengths of UV light were transmitted (Tsormpatsidis et al., 2008).

18.3.1.1.2 Blue and green light

Blue LEDs are commonly used in plant factories. They are typically paired with red LEDs and make up a smaller percentage (often 10%–20%) of the total PPFD since high proportions of B light can suppress leaf size and lead to lower yields. However, B light increases the concentration of many phytochemicals, especially phenolics. For instance, as the percentage of B (peak = 456 nm) light in an R + B LED spectrum increased from 0% to 59%, total phenolic concentration in lettuce 'Sunmang' and 'Grand Rapids TBR' increased by up to 200% (Son and Oh, 2013). Additionally, antioxidant capacity, which is closely related to phenolic concentration, increased with phenolic content (Son and Oh, 2013). In another study, 100 μmol m^{-2} s^{-1} of B (peak = 470 nm) or B + R (peak = 660 nm) LED lighting applied to lettuce 'Banchu Red Fire' during the seedling stage for seven days increased total phenolic concentration by 57%–242% and antioxidant capacity by 64%–286%, compared to only R LEDs or fluorescent light (Johkan et al., 2010). However, when only fluorescent light was applied from the end of the seedling stage until harvest, there were no differences in phenolic content or antioxidant capacity between B + R LED treatments (Johkan et al., 2010), indicating that the effect of B light on phenolic concentration is toward the end of production. Finally, Chinese kale (*Brassica oleracea* var. *alboglabra*) 'Bailey' sprouts grown under B (peak = 470 nm) light at a PFD of 30 μmol m^{-2} s^{-1} had a higher total phenolic concentration and antioxidant capacity (by up to 75% and 50%, respectively) compared to those grown in darkness, under white light, or R LEDs (peak = 660 nm) (Qian et al., 2016). These results indicate that, similar to UV light, B light can increase phenolic concentration and antioxidant capacity.

Narrow-band green LEDs are rarely included in LED fixtures for plant applications because of the paradigm that G light has a lower quantum yield than B or R light. Furthermore, G LEDs are not energy efficient, making them less practical for commercial operations. Therefore, there is relatively little research on its effects on plant quality. However, while G LEDs are not used often, white LEDs are commonly used and emit substantial proportions of G light.

Supplemental G (peak = 510 nm) light at a PFD of 15 μmol m^{-2} s^{-1} slightly increased the antioxidant capacity of lentil (*Lens esculenta*) and wheat (*Triticum aestivum*) sprouts as well as total phenolic concentrations (Samuolienė et al., 2011). In another study, a PFD of 30 μmol m^{-2} s^{-1} of G (peak = 520 nm) light added to an R + B + FR LED background did not affect total phenolic concentration, but G (peaks = 505 or 530 nm) light added to high-pressure sodium (HPS) lighting decreased total phenolic concentration in lettuce 'Thumper' by 13% and 29%, respectively (Samuolienė et al., 2013b). Similarly, supplemental G (peak = 526 nm) light in a white-light background did not affect total phenolic concentration of lettuce 'Red Cross' (Li and Kubota, 2009). More research is needed to better understand the effects of G light on additional plant species and cultivars, but research to date indicates that it has little to no effect on concentrations of phenolic compounds.

18.3.1.1.3 Red and far-red light

Typical LED fixtures, whether consisting of broad- and/or narrow-band LEDs, emit large proportions of R light because it efficiently drives photosynthesis and, thus, plant growth.

R light can also increase phenolic compounds in various plants. For instance, in lettuce 'Red Cross', increasing the PFD from R (peak = 658 nm) LEDs from 75 to 177 µmol m^{-2} s^{-1} in a white-light background increased (by up to 8%) phenolic concentration compared to white light alone or other supplemental wavelengths (Li and Kubota, 2009). Supplemental R (peak = 622 nm) light also increased phenolic concentration in baby leaf lettuce 'Thumper' by 83% compared to the basal spectrum (Samuolienė et al., 2013b), but monochromatic R (peak = 660 nm) light did not affect the phenolic concentration of Chinese kale 'Bailey', compared to white LEDs (Qian et al., 2016). In another study, 210 µmol m^{-2} s^{-1} of R (peak = 638 nm) light applied to multiple microgreen species three days before harvest increased total phenolic concentration by 9% to 40% and antioxidant capacity of mustard, basil (*Ocimum basilicum*), kale (*Brassica oleracea* var. *sabellica*), and parsley by 7% to 55% (Samuolienė et al., 2012). Therefore, R light can increase total phenolics in some species, while also promote growth.

The PFD of FR light and its ratio with R light (R:FR) can affect plant growth and quality attributes. In basil 'Lettuce Leaf' and 'Red Rubin', the total phenolic concentrations decreased by about 52% and 25%, respectively, when the R:FR decreased from 8.2 to 2.5 in a broad photon spectrum (Bantis et al., 2016). Furthermore, total phenolic concentration of lettuce 'Red Cross' slightly decreased when FR (peak = 734 nm) was supplemented in a white-light background instead of other wavelengths (Li and Kubota, 2009). This response is likely, in part, due to a dilution effect, since FR light increases extension growth of leafy greens. Interestingly, when multiple PFDs of FR were added to a narrow-band R + B spectrum (therefore changing the R:FR from 8.6 to 0.7), total phenolic concentration and antioxidant capacity increased in lettuce 'Sunmang' by up to threefold and 3.2-fold, respectively (Lee et al., 2016). Little research has been published on FR light effects on plant phenolics, but the effects of FR (or the R:FR) could depend on plant species as well as the background photon spectrum.

18.3.1.2 Anthocyanins

Anthocyanins are a major and important group of phenolic compounds in leafy greens because of their antioxidant properties as well as their role in pigment accumulation in plant leaves. A photon spectrum containing higher proportions of B light can promote anthocyanin biosynthesis. Supplementing UV or B light to a broad photon spectrum can also be effective. Anthocyanin concentration can be altered by supplemental wavebands, as described below, but effects are likely dependent on light wavelength, plant species, and cultivar.

18.3.1.2.1 Ultraviolet light

Like phenolics, anthocyanins absorb UV light and accumulate during exposure. For instance, UV-A (peak = 373 nm) added to a broad photon spectrum increased anthocyanin concentration of lettuce 'Red Cross' by 11% (Li and Kubota, 2009). Furthermore, the timing and the peak wavelength of UV light determined the magnitude of anthocyanin biosynthesis responses. UV light with peaks of 310, 325, or 340 nm applied at a PFD of about 13 µmol m^{-2} s^{-1}, for one to three days before harvest, increased anthocyanin concentrations in red-leaf lettuce 'Red Fire' by about 20% to 160%, although 310 nm was the most effective (Goto et al., 2016). Additionally, UV-A (peak = 352 nm) light increased anthocyanin concentration in lettuce 'Hongyeom' by over 300% when applied three days before harvest

(Lee et al., 2013). In that study, compared to the control, UV light of any peak wavelength studied increased anthocyanin concentration to a greater extent than total phenolic concentration, indicating that while anthocyanins are a phenolic compound, they could be more responsive to UV light than phenolics as a group.

18.3.1.2.2 Blue and green light

Blue light can also increase anthocyanin concentration in plants. Specifically, the R to B ratio (R:B) of a photon spectrum can considerably affect lettuce anthocyanin concentration. For example, in baby leaf lettuce 'Hongha', increasing the percentage of B light in an R + B spectrum to 43% B increased anthocyanin content by 6.9-fold compared to plants grown under fluorescent light (Lee et al., 2010). Similarly, in basil 'Sweet Genovese' and 'Red Rubin' microgreens, R + B lighting with various R:B ratios (2:1, 1:1, and 1:2) had 70%, 52%, and 55% higher concentrations of anthocyanins compared to a white-light spectrum at the same PPFD of 120 µmol m^{-2} s^{-1} (Lobiuc et al., 2017). Additionally, in lettuce 'Red Cross', increasing the percentage of B light in a white-light spectrum from 23% to 55% increased anthocyanin concentration by 30% (Li and Kubota, 2009). Finally, in other crops, such as Chinese kale sprouts, anthocyanin concentration was twofold greater when plants were grown under B light compared to white light, and threefold greater compared to R light (Qian et al., 2016).

There is considerably less research on the effects of G light on plant anthocyanin concentrations, and therefore it is unclear whether G light has any effect on anthocyanin accumulation. In lettuce 'Red Cross', supplemental G light led to lower anthocyanin levels than B light, but they were similar to the white-light control (Li and Kubota, 2009). Conversely, 30 µmol m^{-2} s^{-1} of G (peak = 530 nm) light added to HPS lamps or 12 µmol m^{-2} s^{-1} of G (peak = 520 nm) added to R + B LEDs significantly increased anthocyanin concentration in lettuce 'Thumper', by 39% and 22%, respectively (Samuolienė et al., 2013b). G light can also interact with other wavebands, such as FR and B light, to affect anthocyanin concentration. For example, 1 to 50 µmol m^{-2} s^{-1} of G light added to a very low PFD (1 µmol m^{-2} s^{-1}) of FR light increased anthocyanin concentration of kale and broccoli (*Brassica oleracea*) microgreens by about threefold, while the addition of G light to a low PFD (10 µmol m^{-2} s^{-1}) of B light did not affect B light-induced anthocyanin production in either species (Carvalho and Folta, 2016). Interestingly, when the PFD of FR and B increased to 10 and 50 µmol m^{-2} s^{-1}, respectively, supplemental G light did not increase anthocyanin concentration in either species.

18.3.1.2.3 Red and far-red light

Generally, R and FR light have less of an effect on plant anthocyanin concentrations compared to shorter photon wavelengths. As stated before, as the percentage of R light in an R + B spectrum increased, anthocyanin concentration decreased in lettuce 'Hongha' (Lee et al., 2010). Furthermore, supplemental R light had no effect, while FR light decreased anthocyanin concentration by 40% in lettuce 'Red Cross' (Li and Kubota, 2009). In another experiment, 28 µmol m^{-2} s^{-1} of additional R light decreased anthocyanin concentration by 39% in lettuce 'Thumper' when added to an R + B spectrum (Samuolienė et al., 2013b). Additionally, lettuce 'Outredgeous' grown under 300 µmol m^{-2} s^{-1} of R or R + FR light had lower anthocyanin concentrations (by 46% and 72%, respectively) compared to those grown under

R + B light (Stutte et al., 2009). Finally, in Chinese kale, monochromatic R light increased anthocyanin concentration compared to plants grown in darkness by about fourfold, but white light and B light were more effective (Qian et al., 2016). These results indicate that while R light can induce anthocyanin biosynthesis, it is not as effective as B or UV light. One possible explanation is that plants grown under a high PFD of R and/or FR light typically are larger, and thus there could be a dilution effect taking place.

18.3.2 Vitamins

18.3.2.1 Vitamin C

There is not a specific waveband that increases vitamin C concentration in all species and cultivars. A broad photon spectrum with a moderate PFD of G light to penetrate the leaf canopy may be effective at increasing vitamin C concentration, although more research is needed.

18.3.2.1.1 Ultraviolet light

Unlike phenolics, the effects of supplemental UV light on plant vitamin C concentrations are less predictable and depend on plant species. For instance, $4\,\mu mol\ m^{-2}\ s^{-1}$ of UV-A (peak = 380 nm) light added to $175\,\mu mol\ m^{-2}\ s^{-1}$ of B + R + FR LEDs decreased vitamin C concentration of green-leaf lettuce 'Thumper' by 14% (Samuolienė et al., 2013b). However, at a PPFD of 300, $6\,\mu mol\ m^{-2}\ s^{-1}$ of UV-A light (peak = 366 nm) increased concentration in beet and pak choi microgreens by 53% and 19%, respectively, but decreased concentration in beet by about 25% when UV-A was delivered at $12\,\mu mol\ m^{-2}\ s^{-1}$ (Brazaitytė et al., 2015b). Additionally, in the previous study, UV-A delivered at either PFD decreased vitamin C concentration of basil by 26% to 35%, but longer wavelengths (peaks = 390 and 402 nm) did not influence its concentration (Brazaitytė et al., 2015b). Finally, increasing the PFD of UV-A light (peak = 373 nm) from 5 to $21\,\mu mol\ m^{-2}\ s^{-1}$ did not influence vitamin C concentration in lettuce 'Red Cross' (Li and Kubota, 2009). Therefore, UV light appears to have little to no influence on vitamin C concentration in leafy greens, although more research is needed on additional species.

18.3.2.1.2 Blue and green light

The effects of B light on plant vitamin C concentrations seem to be dependent on plant species and cultivar. In general though, B light is effective at promoting vitamin C accumulation. For instance, lettuce 'Redfire', spinach (*Spinacia oleracea*) 'Okame', or komatsuna (*Brassica campestris*) "Komatsuna" grown under $300\,\mu mol\ m^{-2}\ s^{-1}$ of R + B light or B alone had higher vitamin C concentrations than those grown under white light or R light, by up to threefold in lettuce but with much smaller increases in spinach and komatsuna (Ohashi-Kaneko et al., 2007). Additionally, lettuce 'Oak Leaf' grown under $300\,\mu mol\ m^{-2}\ s^{-1}$ of R + B light had higher vitamin C content than plants grown under R or fluorescent light (by about 65% and 10%, respectively) (Shin et al., 2014). The peak wavelength of B light might differentially affect vitamin C concentration in baby leaf lettuce. B light with a peak at 470 nm supplemented to HPS lamps increased vitamin C concentration of lettuce 'Thumper' by 53%, but at a peak of 455 nm, concentration decreased by 19% (Samuolienė et al., 2013b). However, supplemental B (peak = 476 nm) light added to white light did not affect vitamin C concentration of lettuce 'Red Cross' (Li and Kubota, 2009).

There is very little published research about the effects of G light on plant vitamin C content. When G (peaks = 520, 505, or 530 nm) light was added to either an R + B LED spectrum or HPS lamps, it did not affect vitamin C concentration of lettuce 'Thumper' (Samuolienė et al., 2013b). Similar results were found in 'Red Cross' lettuce; supplemental G (peak = 526 nm) light added to white light did not have any effect (Li and Kubota, 2009). Although supplemental G light might not affect vitamin C concentration in leaves directly exposed to light, some research indicates that it can enhance accumulation in inner leaves because G light penetrates the leaf canopy better than B or R light. For instance, while none of the wavebands studied influenced vitamin C concentration in the outer four leaves of butterhead lettuce (cultivar not reported), 300 μmol m^{-2} s^{-1} of G (peak = 520 nm) light slightly increased vitamin C concentration of inner leaves leaf (Saengtharatip et al., 2020).

18.3.2.1.3 Red and far-red light

R light generally has little effect on vitamin C concentration. In 'Oak Leaf' lettuce, plants grown under 300 μmol m^{-2} s^{-1} of R light had the lowest vitamin C concentration, by at least 33%, compared to B, R + B, or fluorescent light (Shin et al., 2014). Similarly, lettuce 'Redfire', spinach 'Okame', and komatsuna 'Komatsuna' grown under R light had similar or slightly lower vitamin C concentrations compared to plants grown under white light, R + B, or B light (Ohashi-Kaneko et al., 2007). Furthermore, supplemental R light added to white light or R + B light did not affect vitamin C concentration in lettuce 'Thumper' or 'Red Cross' (Li and Kubota, 2009; Samuolienė et al., 2013b). There is very little research on the effect of FR light on vitamin C accumulation, but the limited research published suggests that it has little to no effect. For example, FR (peak = 734 nm) light supplemented to white light did not increase concentration in lettuce 'Red Cross', but surprisingly, LEDs with a peak at 850 nm did in a green-leaf cultivar, 'Green Oak Leaf', by about 45% (Chen et al., 2016; Li and Kubota, 2009). Interestingly, in the red-leaf cultivar, no supplemental wavebands affected vitamin C concentration, but in the green-leaf cultivar, all supplemental wavebands except yellow (peak = 596 nm) did (Chen et al., 2016; Li and Kubota, 2009). It is possible that vitamin C synthesis may be more regulated by specific wavelengths of light in green-leaf cultivars compared to red-leaf ones.

18.3.2.2 *β-carotene and lutein*

There are few consistent trends on the effects of the photon spectrum on plant carotenoid content, due to in part to species and cultivar differences. However, based on studies published to date, at least a small percentage of UV and/or B light in the spectrum generally promotes carotenoid accumulation in some plant species.

18.3.2.2.1 Ultraviolet light

Light plays an essential role in promoting the biosynthesis of β-carotene and lutein in plants (Pizarro and Stange, 2009). While the effect of the photon spectrum on β-carotene and lutein synthesis is less clear, there is some evidence that suggests alterations in the photon spectrum can modify their concentrations. For instance, in mustard 'Red Lion' microgreens, 12 μmol m^{-2} s^{-1} of UV-A (peak = 366 nm) light added to 300 μmol m^{-2} s^{-1} of B + R + FR LED lighting for 10 h d^{-1} increased β-carotene (by 70%) and lutein (by 56%) concentrations, while the same UV-A light delivered for 16 h d^{-1} decreased β-carotene (by 45%)

and increased lutein (by 35%) concentrations (Brazaitytė et al., 2019). Additionally, UV-A (peak = 390 nm) added for 10 h d^{-1} slightly decreased β-carotene concentration, but it increased by nearly twofold when added for 16 h d^{-1}. Lutein concentration increased with both exposure durations by about 40% and 70% (Brazaitytė et al., 2019). In contrast, in lettuce 'Thumper' and 'Red Cross', UV-A (peaks = 380 and 373 nm) added to either an R + B or a white-light background did not affect β-carotene concentration (Li and Kubota, 2009; Samuolienė et al., 2013b).

18.3.2.2.2 Blue and green light

Blue light can also increase carotenoid content in some plants. B light applied to a white-light background increased β-carotene concentration in lettuce 'Red Cross' by 8% compared to white light alone, or all other wavebands tested (Li and Kubota, 2009). Additionally, increasing the percentage of B light in an R + B spectrum (PPFD of 200 μmol m^{-2} s^{-1}) from 0% to 17% increased total carotenoid content in lettuce, spinach, kale, basil, and pepper (*Capsicum basilicum*) by 34% to 85% (Naznin et al., 2019). Furthermore, 100 μmol m^{-2} s^{-1} of monochromatic B light applied during the seedling stage increased total carotenoid concentration of red-leaf lettuce by 24% compared to R light, but the difference diminished after fluorescent light was applied for the rest of the growing cycle (Johkan et al., 2010). In some species, B light decreased carotenoid contents. For instance, B (peaks = 455 and 470 nm) light supplemented to light from HPS fixtures decreased β-carotene content in lettuce 'Thumper' by 28% and 68% (Samuolienė et al., 2013b). In another study, monochromatic B light decreased total carotenoid content in lettuce 'Redfire' by about 30%, increased it in spinach 'Okame' by about 40%, but had no effect in komatsuna 'Komatsuna' (Ohashi-Kaneko et al., 2007).

There is little published research on the effects of G light on plant carotenoid content, but most research suggests it has little to no effect. In lettuce 'Red Cross', G light added to white light did not affect β-carotene content (Li and Kubota, 2009). Furthermore, G light added to an R + B background did not affect β-carotene content in lettuce 'Thumper' (Samuolienė et al., 2013b). In another study, G (peak = 520) light added to an R + B background increased β-carotene content in mizuna (*Brassica rapa*) microgreens by 121% but decreased it in broccoli and kohlrabi microgreens by 38% and 32%, respectively (Samuolienė et al., 2019). Finally, G light added to an R + B + FR spectrum increased β-carotene and lutein contents of mustard 'Red Lion' and beet 'Bulls Blood' microgreens (Brazaityte et al., 2016). More research is needed to better understand the effects of G light on carotenoid contents of additional plant species and production stages.

18.3.2.2.3 Red and far-red light

Depending on the crop, R light can have differential effects on plant carotenoid content. However, effects attributed to R light could instead be caused by simultaneous changes in the photon spectrum, such as reduced B. For instance, total carotenoid concentrations of lettuce, spinach, kale, basil, and pepper decreased by 24% to 46% as the percentage of R light in an R + B spectrum increased from 83% to 100% (or as B decreased from 17% to 0%) (Naznin et al., 2019). Similarly, lettuce seedlings provided with R light for seven days had 12% to 24% lower carotenoid content than those provided with fluorescent light, B, or R + B light (Johkan et al., 2010). In lettuce 'Red Cross', R light added to a white-light background did not affect β-carotene concentration, while supplemental FR light decreased it by 16% (Li and Kubota, 2009). In another study, β-carotene concentration in mizuna microgreens was lower under an R + B spectrum compared to those grown under R + B with supplemental G (peak = 520 nm), yellow (peak = 595 nm), or orange (peak = 622 nm) LEDs (by up to 40%), while concentrations were substantially greater under the R + B spectrum in kohlrabi and broccoli microgreens (Samuolienė et al., 2019). Finally, R light supplemented to a fluorescent background PPFD of 275 μmol m^{-2} s^{-1} increased lutein concentration by up to 62% in kale 'Winterbor' but did not affect β-carotene (Lefsrud et al., 2008). FR light slightly decreased concentrations of both lutein and β-carotene compared to other supplemental wavelengths.

18.3.2.3 α-Tocopherols

To date, there are not many publications on the effects of the photon spectrum on α-tocopherol content, but some evidence suggests that it can be affected by the photon spectrum on some crops. Whether the photoperiod was 10 or 16 h d^{-1} and the peak of supplemental UV-A was 366 or 390 nm, there was little (~5%) to no effect on α-tocopherol content of mustard microgreens 'Red Lion' (Brazaitytė et al., 2019). However, at a slightly longer wavelength (402 nm) of light, α-tocopherol content of mustard microgreens increased by about 10% to 40% (Brazaitytė et al., 2019). In microgreens of basil 'Sweet Genovese', beet 'Bulls Blood', and pak choi 'Rubi', α-tocopherol content was differentially affected by wavelength and intensity of UV-A light when added to an R + B + FR spectrum (Brazaitytė et al., 2015b). For example, at an intensity of 6 μmol m^{-2} s^{-1}, UV-A light with peaks of 366, 390, and 402 nm decreased α-tocopherol contents of basil by 42% to 64% and beet by 16% to 27%, but 390 and 402 nm UV-A increased concentration in pak choi by 40% and 61% (Brazaitytė et al., 2015b). In another study with lettuce 'Thumper', all supplemental wavelengths (peaks = 380, 520, 595, and 622 nm) added to R + B + FR LEDs decreased α-tocopherol contents by 64%, 34%, 57%, and 53%, respectively (Samuolienė et al., 2013b). In the same study, supplemental B light (peak = 455 nm) added to an HPS background increased α-tocopherol content by 41%, while supplemental B or G light (peaks = 470, 505, or 530 nm) had no effect or slightly decreased (~7%) α-tocopherol content.

18.3.2.4 Folate

Leafy greens such as lettuce and spinach are among the best sources of dietary folate. Particularly, butterhead, romaine, and red-leaf lettuce varieties have some of the highest folate concentrations (Kim et al., 2016). Despite leafy greens having high-folate concentrations and their importance in human nutrition, little research has investigated how changing environmental factors, such as the photon spectrum, can affect them. One study suggests that the photon spectrum interacts with air temperature to affect folate concentration (Okazaki and Yamashita, 2019). In lettuce 'Fancy Green', an R + G + B spectrum increased folate concentration by about 47% compared to an R + B spectrum at the same PPFD of 150 μmol m^{-2} s^{-1}, but only at an air temperature of 25°C (not at 20 or 28°C) (Okazaki and Yamashita, 2019). In a greenhouse study with lamb's lettuce (*Valerianella locusta*) 'Nordhollandse', supplemental LED lighting delivered during different seasons influenced plant folate concentration (Długosz-Grochowska et al., 2016). In autumn and winter, supplemental R + B (90% R+10% B and 70% R+30% B) light increased folate concentration by 19% to 25% more than 100% R LED light when the PPFD was about 200 μmol m^{-2} s^{-1}

(Długosz-Grochowska et al., 2016). In another greenhouse study, approximately 135 μmol m^{-2} s^{-1} of supplemental R + B LED light increased total folate concentrations of lettuce 'Frillice' by up to 21% compared to HPS light, R + cool-white LEDs, warm-white LEDs, and warm-white + B LEDs (Hytönen et al., 2018).

18.3.3 Nitrates

Altering the photon spectrum is an effective method to decrease NO$_3$ concentration in leafy greens. In green-leaf lettuce 'Lobjoits Green Cos', 30 μmol m^{-2} s^{-1} of G (peak = 510 nm) light added to a PPFD of 270 μmol m^{-2} s^{-1} from B + R + FR LEDs decreased NO$_3$ concentration by 45%, while UV-A (peak = 380 nm), yellow (peak = 595 nm), and R (peak = 622 nm) additions had no effect (Viršilė et al., 2020). In the same experiment, NO$_3$ concentration in red-leaf lettuce 'Red Cos' increased by 58% when UV-A light was added to the spectrum but decreased by up to 58% when G or R light was added. In another study, spinach 'Okame' plants grown under a PPFD of 270 μmol m^{-2} s^{-1} from R + B LEDs had about a 27% lower concentration of nitrates than those grown under white fluorescent lights (Ohashi-Kaneko et al., 2007). Similarly, lettuce 'Red Fire' plants grown under R or B LEDs had an approximately 40% lower NO$_3$ concentration than plants grown under white fluorescent lights (Ohashi-Kaneko et al., 2007). In another study with lettuce 'Green Oak Leaf', increasing the PFD of infrared (peak = 850 nm) from 0.5 to 30 μmol m^{-2} s^{-1} and yellow (peak = 596 nm) light from 53 to 71 μmol m^{-2} s^{-1} in a white LED background (PPFD = 135 μmol m^{-2} s^{-1}) increased NO$_3$ concentrations by about 12% and 34%, while R (peak = 660 nm), G (peak = 522 nm), and B (peak = 450 nm) light decreased concentration by up to 33% (Chen et al., 2016). Finally, G (peak = 530 nm) light added to an R + B spectrum totaling 200 μmol m^{-2} s^{-1} decreased NO$_3$ concentration in hydroponically grown lettuce 'Butterhead' by up to 45% when plants were exposed to the photon spectrum for 24 or 48 h (Bian et al., 2018). Based on these results, shorter wavelength G light can decrease NO$_3$ concentration in some leafy greens but may depend on plant species or cultivar. In addition, plants grown under LEDs, such as an R + B spectrum, might have lower NO$_3$ concentrations than plants grown under fluorescent lights.

18.3.4 Sensory attributes

18.3.4.1 Coloration

Plant coloration is an important quality attribute for some leafy greens. For instance, red-leaf lettuce needs to have red coloration to be marketable. Plants accumulation pigments, such as anthocyanins and β-carotene, in response to genetic and environmental factors. β-carotene is a red-orange pigment that plays an important role in the coloration of many yellow, orange, and red fruits and vegetables, while anthocyanins are red-purple pigments that are found in darker red, purple, and blue fruits and vegetables.

In indoor production, the photon spectrum influences plant coloration. UV and B light can be incorporated into the photon spectrum to increase plant coloration, while FR and G light have little to no positive effect, and may actually decrease pigmentation. For instance, the red coloration of red-leaf lettuce 'Rouxai' increased by up to 63% under higher proportions of B light (20, 40, or 60 µmol m^{-2} s^{-1}) in an R + B spectrum, but decreased as G or FR light was substituted for B light (Meng et al., 2019). In the same study, incorporating more FR light into the spectrum, and thus decreasing the R:FR, increased yield but decreased leaf redness, likely due to a dilution effect. Similarly in a greenhouse study, 100 µmol m^{-2} s^{-1} of end-of-production supplemental lighting with R and/or B LEDs increased leaf coloration of four lettuce cultivars ('Cherokee', 'Magenta', 'Ruby Sky', and 'Vulcan') compared to no end-of-production lighting (Owen and Lopez, 2015). Additionally, supplemental end-of-day lighting (white + R + FR LEDs) or end-of-production (R + B LED) lighting increased the redness of lettuce 'Salanova Red Sweet Crisp' and 'Salanova Red Incised' compared to treatments without either form of supplemental lighting (Zhang et al., 2019).

18.3.4.2 *Sugars, phenolics, and taste*

While few studies have investigated the effects of the photon spectrum on plant metabolites and taste profiles, some studies indicate it can modify plant chemical contents and influence taste. Sugar and phenolic compounds play large roles in affecting taste, with sugars leading to increased sweetness and phenolic compounds typically increasing bitterness (Soares et al., 2013). Sucrose concentrations increased in both red- and green-leaf lettuce cultivars 'Red Cos' and 'Lobjoits Green Cos' by up to 60% when 30 µmol m^{-2} s^{-1} of G (peak = 510 nm) or R (peak = 622 nm) light was added to a PFD of 270 µmol m^{-2} s^{-1} from R + B + FR LEDs (Viršilė et al., 2020). Furthermore, 270 µmol m^{-2} s^{-1} from R + B + white LEDs or fluorescent light were roughly twofold more effective at increasing soluble sugar concentration in lettuce 'Capitata' than a narrow-band R + B spectrum (Lin et al., 2013). Additionally, sensory analysis indicated that plants with higher soluble sugar concentrations also had the highest taste ratings (5 and 5.5 out of 6) (Lin et al., 2013). However, increasing the concentrations of some nutritious compounds, such as phenolics, can increase bitterness. Therefore, including UV, B, and/or R light in a spectrum could increase plant phenolic compounds and possibly perceived bitterness. More research is needed to determine specific phenolic compounds that are responsible for the bitter taste, and whether or not they are affected by changing the photon spectrum. Furthermore, more research is needed to identify a photon spectrum that optimizes both nutritional quality and sensory attributes like coloration and taste, without diminishing yield, if in fact one exists.

18.4 Conclusion

Nutritional value, coloration, and taste are all important quality attributes of leafy greens and can be manipulated by altering the photon spectrum and flux density. Increasing the PPFD can increase the concentration of nutritious phenolic compounds that also increase leaf coloration in many plant species. Furthermore, high PPFDs can decrease potentially harmful nitrates found in high concentrations in leafy greens. While increasing the PPFD can increase biomass and thus harvestable yield, concentrations of other nutritious chemicals and vitamins, such as β-carotene and α-tocopherols, can decrease, depending on the plant species and cultivar.

Altering the photon spectrum can also increase the concentrations of beneficial nutrients. Incorporating UV and/or B light in the spectrum sometimes increases the concentrations of phenolic compounds, β-carotene, and vitamin C. While R and FR light have less of an effect on those chemicals, they can increase sugar content, and thus plant taste. Therefore, both the PPFD and photon spectrum are critical environmental factors to consider when creating an indoor growing environment. There is no such thing as a "perfect" or "ideal" light environment that elicits all the traits a grower and consumer seek and thus, one must consider such tradeoffs. Depending on the production goals and desired attributes, whether those are yield or quality-related, the light environment should be designed to maximize them.

References

Anjana, S.U., Iqbal, M., 2007. Nitrate accumulation in plants, factors affecting the process, and human health implications. A review. Agron. Sustain. Dev. 27, 45–57.

Aziz, A., Akram, N.A., Ashraf, M., 2018. Influence of natural and synthetic vitamin C (ascorbic acid) on primary and secondary metabolites and associated metabolism in quinoa (Chenopodium quinoa Willd.) plants under water deficit regimes. Plant Physiol. Biochem. 123, 192–203.

Balasundram, N., Sundram, K., Samman, S., 2006. Phenolic compounds in plants and agri-industrial by-products: antioxidant activity, occurrence, and potential uses. Food Chem. 99, 191–203.

Bantis, F., Ouzounis, T., Radoglou, K., 2016. Artificial LED lighting enhances growth characteristics and total phenolic content of Ocimum basilicum, but variably affects transplant success. Sci. Hortic. 198, 277–283.

Bartley, G.E., Scolnik, P.A., 1995. Plant carotenoids: pigments for photoprotection, visual attraction, and human health. Plant Cell 7 (7), 1027–1038.

Bekaert, S., Storozhenko, S., Mehrshahi, P., Bennett, M.J., Lambert, W., Gregory III, J.F., Schubert, K., Hugenholtz, J., Van Der Straeten, D., Hanson, A.D., 2008. Folate biofortification in food plants. Trends Plant Sci. 13 (1), 28–35.

Bian, Z., Cheng, R., Wang, Y., Yang, Q., Lu, C., 2018. Effect of green light on nitrate reduction and edible quality of hydroponically grown lettuce (Lactuca sativa L.) under short-term continuous light from red and blue light-emitting diodes. Environ. Exp. Bot. 153, 63–71.

Brazaitytė, A., Sakalauskienė, S., Samuolienė, G., Jankauskienė, J., Viršilė, A., Novičkovas, A., Sirtautas, R., Miliauskienė, J., Vaštakaitė, V., Dabašinskas, L., 2015a. The effects of LED illumination spectra and intensity on carotenoid content in Brassicaceae microgreens. Food Chem. 173, 600–606.

Brazaitytė, A., Viršilė, A., Jankauskienė, J., Sakalauskienė, S., Samuolienė, G., Sirtautas, R., Novičkovas, A., Dabašinskas, L., Miliauskienė, J., Vaštakaitė, V., Bagdonavičienė, A., Duchovskis, P., 2015b. Effect of supplemental UV-A irradiation in solid-state lighting on the growth and phytochemical content of microgreens. Int. Agrophys. 29, 13–22.

Brazaitytė, A., Viršilė, A., Samuolienė, G., Jankauskienė, J., Sakalauskienė, S., Sirtautas, R., Novičkovas, A., Dabašinskas, L., Vaštakaitė, V., Miliauskienė, J., Duchovskis, P., 2016. Light quality: growth and nutritional value of microgreens under indoor and greenhouse conditions. Acta Hortic. 1134, 277–284.

Brazaitytė, A., Viršilė, A., Samuolienė, G., Vaštakaitė-Kairienė, V., Jankauskienė, J., Miliauskienė, J., Novičkovas, A., Duchovskis, P., 2019. Response of mustard microgreens to different wavelengths and durations of UV-A LEDs. Front. Plant Sci. 10, 1153.

Brigelius-Flohé, R., Traber, M.G., 1999. Vitamin E: function and metabolism. FASEB J 13 (10), 1145–1155.

Cantliffe, D.J., 1973. Nitrate accumulation in table beets and spinach as affected by nitrogen, phosphorus, and potassium nutrition and light intensity 1. Agron. J. 65 (4), 563–565.

Carr, A.C., Maggini, S., 2017. Vitamin C and immune function. Nutrients 9 (11), 1211.

Carvalho, S.D., Folta, K.M., 2016. Green light control of anthocyanin production in microgreens. Acta Hortic. 1134, 13–18.

Chappell, J., Hahlbrock, K., 1984. Transcription of plant defence genes in response to UV light or fungal elicitor. Nature 311 (5981), 76–78.

Chen, X.-L., Xue, X.-Z., Guo, W.-Z., Wang, L.-C., Qiao, X.-J., 2016. Growth and nutritional properties of lettuce affected by mixed irradiation of white and supplemental light provided by light-emitting diode. Sci. Hortic. 200, 111—118.

Colonna, E., Rouphael, Y., Barbieri, G., De Pascale, S., 2016. Nutritional quality of ten leafy vegetables harvested at two light intensities. Food Chem. 199, 702—710.

Craver, J.K., Gerovac, J.R., Lopez, R.G., Kopsell, D.A., 2017. Light intensity and light quality from sole-source light-emitting diodes impact phytochemical concentrations within Brassica microgreens. J. Am. Soc. Hortic. Sci. 142 (1), 3—12.

Dasgupta, A., Klein, K., 2014. Antioxidants in Food, Vitamins and Supplements: Prevention and Treatment of Disease. Academic Press, Cambridge, MA.

Długosz-Grochowska, O., Kołton, A., Wojciechowska, R., 2016. Modifying folate and polyphenol concentrations in Lamb's lettuce by the use of LED supplemental lighting during cultivation in greenhouses. J. Funct. Foods 26, 228—237.

Eveland, A.L., Jackson, D.P., 2012. Sugars, signalling, and plant development. J. Exp. Bot. 63 (9), 3367—3377.

Fu, Y., Li, H., Yu, J., Liu, H., Cao, Z.Y., Manukovsky, N.S., Liu, H., 2017. Interaction effects of light intensity and nitrogen concentration on growth, photosynthetic characteristics and quality of lettuce (*Lactuca sativa* L. Var. youmaicai). Sci. Hortic. 214, 51—57.

Goto, E., Hayashi, K., Furuyama, S., Hikosaka, S., Ishigami, Y., 2016. Effect of UV light on phytochemical accumulation and expression of anthocyanin biosynthesis genes in red leaf lettuce. Acta Hortic. 1134, 179—185.

Guzman, I., Yousef, G.G., Brown, A.F., 2012. Simultaneous extraction and quantitation of carotenoids, chlorophylls, and tocopherols in Brassica vegetables. J. Agric. Food Chem. 60 (29), 7238—7244.

Hannoufa, A., Hossain, Z., 2012. Regulation of carotenoid accumulation in plants. Biocatal. Agric. Biotechnol. 1 (3), 198—202.

Hanson, A.D., Gregory III, J.F., 2011. Folate biosynthesis, turnover, and transport in plants. Annu. Rev. Plant Biol. 62, 105—125.

Havaux, M., Eymery, F., Porfirova, S., Rey, P., Dörmann, P., 2005. Vitamin E protects against photoinhibition and photooxidative stress in *Arabidopsis thaliana*. Plant Cell 17 (12), 3451—3469.

Hoch, W.A., Singsaas, E.L., McCown, B.H., 2003. Resorption protection. Anthocyanins facilitate nutrient recovery in autumn by shielding leaves from potentially damaging light levels. Plant Physiol. 133 (3), 1296—1305.

Hytönen, T., Pinho, P., Rantanen, M., Kariluoto, S., Lampi, A., Edelmann, M., Joensuu, K., Kauste, K., Mouhu, K., Piironen, V., Halonen, L., Elomaa, P., 2018. Effects of LED light spectra on lettuce growth and nutritional composition. Light. Res. Technol. 50 (6), 880—893.

Johkan, M., Shoji, K., Goto, F., Hashida, S.-N., Yoshihara, T., 2010. Blue light-emitting diode light irradiation of seedlings improves seedling quality and growth after transplanting in red leaf lettuce. HortScience 45 (12), 1809—1814.

Kelly, N., Choe, D., Meng, Q., Runkle, E.S., 2020. Promotion of lettuce growth under an increasing daily light integral depends on the combination of the photosynthetic photon flux density and photoperiod. Sci. Hortic. 272, 109565.

Keunen, E., Peshev, D., Vangronsveld, J., Van Den Ende, W., Cuypers, A., 2013. Plant sugars are crucial players in the oxidative challenge during abiotic stress: extending the traditional concept. Plant Cell Environ. 36 (7), 1242—1255.

Kim, M.J., Moon, Y., Tou, J.C., Mou, B., Waterland, N.L., 2016. Nutritional value, bioactive compounds and health benefits of lettuce (*Lactuca sativa* L.). J. Food Compos. Anal. 49, 19—34.

Lee, J.-G., Oh, S.-S., Cha, S.-H., Jang, Y.-A., Kim, S.-Y., Um, Y.-C., Cheong, S.-R., 2010. Effects of red/blue light ratio and short-term light quality conversion on growth and anthocyanin contents of baby leaf lettuce. J. Bio-Environ. Control 19 (4), 351—359.

Lee, M.-J., Son, J.E., Oh, M.-M., 2013. Growth and phenolic compounds of *Lactuca sativa* L. grown in a closed-type plant production system with UV-A, -B, or -C lamp. J. Sci. Food Agric. 94 (2), 197—204.

Lee, M.-J., Son, K.-H., Oh, M.-M., 2016. Increase in biomass and bioactive compounds in lettuce under various ratios of red to far-red LED light supplemented with blue LED light introduction. Hortic. Environ. Biotechnol. 57 (2), 139—147.

Lefsrud, M.G., Kopsell, D.A., Sams, C.E., 2008. Irradiance from distinct wavelength light-emitting diodes affect secondary metabolites in kale. HortScience 43 (7), 2243—2244.

Lester, G.E., 2006. Environmental regulation of human health nutrients (ascorbic acid, β-carotene, and folic acid) in fruits and vegetables. HortScience 41 (1), 59—64.

Li, Q., Kubota, C., 2009. Effects of supplemental light quality on growth and phytochemicals of baby leaf lettuce. Environ. Exp. Bot. 67 (2), 59–64.

Lin, K.-H., Huang, M.-Y., Huang, W.-D., Hsu, M.-H., Yang, Z.-W., Yang, C.-M., 2013. The effects of red, blue, and white light-emitting diodes on the growth, development, and edible quality of hydroponically grown lettuce (*Lactuca sativa* L. var. capitata). Sci. Hortic. 150, 86–91.

Liu, Y., Tikunov, Y., Schouten, R.E., Marcelis, L.F.M., Visser, R.G.F., Bovy, A., 2018. Anthocyanin biosynthesis and degradation mechanisms in Solanaceous vegetables: a review. Front. Chem. 6, 52.

Lobiuc, A., Vasilache, V., Oroian, M., Stoleru, T., Burducea, M., Pintilie, O., Zamfirache, M.-M., 2017. Blue and red led illumination improves growth and bioactive compounds contents in acyanic and cyanic *Ocimum basilicum* L. microgreens. Molecules 22 (12), 2111.

Lushchak, V.I., Semchuk, N.M., 2012. Tocopherol biosynthesis: chemistry, regulation and effects of environmental factors. Acta Physiol. Plant. 34, 1607–1628.

Ma, L., Hu, L., Feng, X., Wang, S., 2018. Nitrate and nitrite in health and disease. Aging Dis. 9, 938.

Magwaza, L.S., Opara, U.L., 2015. Analytical methods for determination of sugars and sweetness of horticultural products-A review. Sci. Hortic. 184, 179–192.

Meng, Q., Kelly, N., Runkle, E.S., 2019. Substituting green or far-red radiation for blue radiation induces shade avoidance and promotes growth in lettuce and kale. Environ. Exp. Bot. 162, 383–391.

Naznin, M., Lefsrud, M., Gravel, V., Azad, M., 2019. Blue light added with red LEDs enhance growth characteristics, pigments content, and antioxidant capacity in lettuce, spinach, kale, basil, and sweet pepper in a controlled environment. Plants 8 (4), 93.

Ohashi-Kaneko, K., Tarase, M., Noya, K.O.N., Fujiwara, K., Kurata, K., 2007. Effect of light quality on growth and vegetable quality in leaf lettuce, spinach and komatsuna. Environ. Control Biol. 45 (3), 189–198.

Okazaki, S., Yamashita, T., 2019. A manipulation of air temperature and light quality and intensity can maximize growth and folate biosynthesis in leaf lettuce. Environ. Control Biol. 57 (2), 39–44.

Opara, U.L., Pathare, P.B., 2014. Bruise damage measurement and analysis of fresh horticultural produce—a review. Postharvest Biol. Technol. 91, 9–24.

Owen, G.W., Lopez, R.G., 2015. End-of-production supplemental lighting with red and blue light-emitting diodes (LEDs) influences red pigmentation of four lettuce varieties. HortScience 50 (5), 676–684.

Park, Y., Runkle, E.S., 2017. Far-red radiation promotes growth of seedlings by increasing leaf expansion and whole-plant net assimilation. Environ. Exp. Bot. 136, 41–49.

Pizarro, L., Stange, C., 2009. Light-dependent regulation of carotenoid biosynthesis in plants. Cienc. Investig. Agrar. 36 (2), 143–162.

Pojer, E., Mattivi, F., Johnson, D., Stockley, C.S., 2013. The case for anthocyanin consumption to promote human health: a review. Compr. Rev. Food Sci. Food Saf. 12 (5), 483–508.

Qian, H., Liu, T., Deng, M., Miao, H., Cai, C., Shen, W., Wang, Q., 2016. Effects of light quality on main health-promoting compounds and antioxidant capacity of Chinese kale sprouts. Food Chem. 196, 1232–1238.

Saengtharatip, S., Goto, N., Kozai, T., Yamori, W., 2020. Green light penetrates inside crisp head lettuce leading to chlorophyll and ascorbic acid content enhancement. Acta Hortic. 1273, 261–269.

Salerno, G.L., Curatti, L., 2003. Origin of sucrose metabolism in higher plants: when, how and why? Trends Plant Sci. 8 (2), 63–69.

Samuolienė, G., Urbonavičiute, A., Brazaitytė, A., Šabajeviene, G., Sakalauskaitė, J., Duchovskis, P., 2011. The impact of LED illumination on antioxidant properties of sprouted seeds. Cent. Eur. J. Biol. 6 (1), 68–74.

Samuolienė, G., Brazaitytė, A., Sirtautas, R., Sakalauskienė, S., Jankauskienė, J., Duchovskis, P., Novičkovas, A., 2012. The impact of supplementary short-term red LED lighting on the antioxidant properties of microgreens. Acta Hortic. 956, 649–655.

Samuolienė, G., Brazaitytė, A., Jankauskienė, J., Viršilė, A., Sirtautas, R., Novičkovas, A., Sakalauskienė, S., Sakalauskaitė, J., Duchovskis, P., 2013a. LED irradiance level affects growth and nutritional quality of Brassica microgreens. Open Life Sci. 8 (12), 1241–1249.

Samuolienė, G., Brazaitytė, A., Sirtautas, R., Viršilė, A., Sakalauskaitė, J., Sakalauskienė, S., Duchovskis, P., 2013b. LED illumination affects bioactive compounds in romaine baby leaf lettuce. J. Sci. Food Agric. 93 (13), 3286–3291.

Samuolienė, G., Viršilė, A., Brazaitytė, A., Jankauskienė, J., Sakalauskienė, S., Vaštakaitė, V., Novičkovas, A., Viškelienė, A., Sasnauskas, A., Duchovskis, P., 2017. Blue light dosage affects carotenoids and tocopherols in microgreens. Food Chem. 228, 50–56.

Samuolienė, G., Brazaitytė, A., Viršilė, A., Miliauskienė, J., Vaštakaitė-Kairienė, V., Duchovskis, P., 2019. Nutrient levels in Brassicaceae microgreens increase under tailored light-emitting diode spectra. Front. Plant Sci. 10, 1475.

Sathasivam, R., Radhakrishnan, R., Kim, J.K., Park, S.U., 2020. An update on biosynthesis and regulation of carotenoids in plants. South Afr. J. Bot. https://doi.org/10.1016/j.sajb.2020.05.015.

Savidge, B., Weiss, J.D., Wong, Y.-H.H., Lassner, M.W., Mitsky, T.A., Shewmaker, C.K., Post-Beittenmiller, D., Valentin, H.E., 2002. Isolation and characterization of homogentisate phytyltransferase genes from *Synechocystis* sp. PCC 6803 and arabidopsis. Am. Soc. Plant Physiol. 129, 321–332.

Scott, J., Rébeillé, F., Fletcher, J., 2000. Folic acid and folates: the feasibility for nutritional enhancement in plant foods. J. Sci. Food Agric. 80 (7), 795–824.

Shahidi, F., De Camargo, A.C., 2016. Tocopherols and tocotrienols in common and emerging dietary sources: occurrence, applications, and health benefits. Int. J. Mol. Sci. 17 (10), 1745.

Shin, Y.-S., Lee, M.-J., Lee, E.-S., Ahn, J.-H., Kim, M.-K., Lee, J.-E., Do, H.-W., Cheung, J.-D., Park, J.-U., Um, Y.-G., Park, S.-D., Chae, J.-H., 2014. Effect of light emitting diodes treatment on growth and quality of lettuce (*Lactuca sativa* L. 'Oak Leaf'). J. Life Sci. 24 (2), 148–153.

Soares, S., Kohl, S., Thalmann, S., Mateus, N., Meyerhof, W., De Freitas, V., 2013. Different phenolic compounds activate distinct human bitter taste receptors. J. Agric. Food Chem. 61 (7), 1525–1533.

Son, K.-H., Oh, M.-M., 2013. Leaf shape, growth, and antioxidant phenolic compounds of two lettuce cultivars grown under various combinations of blue and red light-emitting diodes. HortScience 48 (8), 988–995.

Song, J., Huang, H., Hao, Y., Song, S., Zhang, Y., Su, W., Liu, H., 2020. Nutritional quality, mineral and antioxidant content in lettuce affected by interaction of light intensity and nutrient solution concentration. Sci. Rep. 10, 1–9.

Stutte, G., Edney, S., Skerritt, T., 2009. Photoregulation of bioprotectant content of red leaf lettuce with light-emitting diodes. HortScience 44 (1), 79–82.

Sun, T., Li, L., 2020. Toward the 'golden' era: the status in uncovering the regulatory control of carotenoid accumulation in plants. Plant Sci. 290, 110331.

Tanaka, Y., Ohmiya, A., 2008. Seeing is believing: engineering anthocyanin and carotenoid biosynthetic pathways. Curr. Opin. Biotechnol. 19 (2), 190–197.

Trebst, A., 2003. Function of β-carotene and tocopherol in photosystem II. Z. Naturforsch., C: Biosci. 58 (9–10), 609–620.

Trebst, A., Depka, B., Holländer-Czytko, H., 2002. A specific role for tocopherol and of chemical singlet oxygen quenchers in the maintenance of photosystem II structure and function in *Chlamydomonas reinhardtii*. FEBS Lett. 516 (1–3), 156–160.

Trojak, M., Skowron, E., 2017. Role of anthocyanins in high-light stress response. World Sci. News 81 (2), 150–168.

Tsormpatsidis, E., Henbest, R.G.C., Davis, F.J., Battey, N.H., Hadley, P., Wagstaffe, A., 2008. UV irradiance as a major influence on growth, development and secondary products of commercial importance in Lollo Rosso lettuce 'Revolution' grown under polyethylene films. Environ. Exp. Bot. 63 (1–3), 232–239.

Tucker, J.M., Townsend, D.M., 2005. Alpha-tocopherol: roles in prevention and therapy of human disease. Biomed. Pharmacother. 59, 380–387.

Viršilė, A., Brazaitytė, A., Vaštakaitė-Kairienė, V., Miliauskienė, J., Jankauskienė, J., Novičkovas, A., Samuolienė, G., 2019. Lighting intensity and photoperiod serves tailoring nitrate assimilation indices in red and green baby leaf lettuce. J. Sci. Food Agric. 99 (14), 6608–6619.

Viršilė, A., Brazaitytė, A., Vaštakaitė-Kairienė, V., Miliauskienė, J., Jankauskienė, J., Novičkovas, A., Laužikė, K., Samuolienė, G., 2020. The distinct impact of multi-color LED light on nitrate, amino acid, soluble sugar and organic acid contents in red and green leaf lettuce cultivated in controlled environment. Food Chem. 310, 125799.

Wurtzel, E.T., 2019. Changing form and function through carotenoids and synthetic biology. Plant Physiol. 179 (3), 830–843.

Zhang, M., Whitman, C.M., Runkle, E.S., 2019. Manipulating growth, color, and taste attributes of fresh cut lettuce by greenhouse supplemental lighting. Sci. Hortic. 252, 274–282.

Zheng, Y.-J., Zhang, Y.-T., Liu, H.-C., Li, Y.-M., Liu, Y.-L., Hao, Y.-W., Lei, B.-F., 2018. Supplemental blue light increases growth and quality of greenhouse pak choi depending on cultivar and supplemental light intensity. J. Integr. Agric. 17 (10), 2245–2256.

Zhou, W.L., Liu, W.K., Yang, Q.C., 2012. Quality changes in hydroponic lettuce grown under pre-harvest short-duration continuous light of different intensities. J. Hortic. Sci. Biotechnol. 87 (5), 429–434.

Indoor production of ornamental seedlings, vegetable transplants, and microgreens

Yujin Park[1], Celina Gómez[2] and Erik S. Runkle[3]

[1]College of Integrative Sciences and Arts, Arizona State University, Mesa, AZ, United States;
[2]Environmental Horticulture Department, University of Florida, Gainesville, FL, United States;
[3]Department of Horticulture, Michigan State University, East Lansing, MI, United States

Abbreviations

B, blue
CRI, color rendering index
DLI, daily light integral
EOD end of day
FR, far red
G, green
HPS, high-pressure sodium
LEDs, light emitting diodes
PFD, photon flux density
PPE, phytochrome photoequilibrium
PPFD, photosynthetic photon flux density
R, red
SL, supplemental lighting
SSL, sole-source lighting
W, white

19.1 Introduction

Seedling and transplant production are an essential part of vegetable and floriculture crop production. Quality young plants are a high-value product that can improve early establishment of crops; increase finish crop quality, uniformity, and yield; and decrease production time. Growing high-quality young plants requires careful management of the environment,

Plant Factory Basics, Applications and Advances
https://doi.org/10.1016/B978-0-323-85152-7.00020-3

particularly light and temperature, as well as cultural management such as water and fertilizer. Indoor plant production systems, where all environmental conditions can be controlled, have been used for at least a few decades to produce high-quality seedlings and transplants (Kozai et al., 2004). Until recently, though, production has only utilized conventional lighting fixtures with a fixed and broad spectrum (e.g., fluorescent lamps).

Indoor plant production systems consist of six principal components, including a thermally insulated and airtight structure, a multilayer cultivation system with lighting devices, air conditioners and fans, a CO_2 supply unit, a nutrient solution supply unit, and an environmental control unit (Kozai, 2020). Each component has been designed to maximize crop quality and productivity while minimizing the use of resources, such as electric energy, water, CO_2, and fertilizer (Kozai, 2013). In particular, lighting typically consumes 75%—80% of the total electrical cost for indoor plant production systems (Fisher et al., 2019), and thus, lighting optimization in terms of energy use and plant growth has been an important commercial consideration.

Light emitting diodes (LEDs) have become the primary light source for indoor plant production systems because they have several technical advantages compared to conventional lamps [e.g., high-pressure sodium (HPS) and fluorescent], including high energy efficacy, low heat emission, and long operating lifetime (Goto, 2012; Fujiwara, 2020). In addition, the possibility of controlling the photon spectrum with LEDs offers a way to regulate plant growth and development processes and, thus, produce more consistent crops with more desirable characteristics.

For plant applications, light quantity is best described by the photosynthetic photon flux density (PPFD) (in μmol m^{-2} s^{-1}) for instantaneous measurements or daily light integral (DLI, in mol m^{-2} d^{-1}) as an integrated value. It primarily regulates plant photosynthesis, biomass, quality, and overall yield. In addition, plants acclimate to light quantity by adjusting their architecture, leaf morphology, and photosynthetic capacity. In general, plants grown under a low light quantity have larger leaves to increase light interception, while those grown under higher light have thicker leaves to increase photosynthetic capacity (Walters, 2005; Poorter et al., 2019). In this chapter, we discuss recent research findings on the use of LED sole-source lighting (SSL) for the production of floriculture and vegetable transplants and microgreens in indoor plant production systems.

19.2 Floriculture plugs

Floriculture plugs are seedlings grown in small individual cells of a propagation tray. The desired quality attributes for trays of plugs generally include a high propagation success rate (e.g., high germination percentage), compact growth (i.e., minimal internode elongation), dark green leaves, a well-developed root system, and a high shoot dry mass. The quality of floriculture plugs greatly influences subsequent growth and development and thus, quality of finished crops and the overall crop time. Here, we review how blue (B, 400—500 nm), far red (FR, 700—800 nm), and a broad spectrum (white; W) of SSL regulate the quality attributes of ornamental seedlings and their subsequent growth after transplant. In addition, we highlight recent studies that compared the quality of young plants grown indoors under SSL with those grown conventionally in a greenhouse.

19.2.1 Effects of blue light

B and red (R, 600–700 nm) photons are especially effective at driving photosynthesis and are the most electrically efficient colors of LEDs based on photon efficacy (μmol J^{-1}) (McCree, 1972; Kusuma et al., 2020). Thus, B + R LEDs were commonly used in early generations of commercial horticultural LED fixtures (Mitchell et al., 2012). To elucidate appropriate B:R ratios in SSL for growing ornamental young plants, Wollaeger and Runkle (2015) investigated how increasing the portion of light from B LEDs (from 0% to 100%; peak of 446 nm) when combined with R (peaks of 634 and 664 nm) LEDs influenced seedling growth of impatiens (*Impatiens walleriana*), salvia (*Salvia splendens*), and petunia (*Petunia × hybrida*) under a PPFD of 160 μmol m^{-2} s^{-1}. For comparison, seedlings were also grown under cool-W fluorescent lamps. Seedlings grown under 100% R had a similar seedling height, leaf area, and shoot fresh mass with those grown under fluorescent lamps. However, the seedlings of impatiens and salvia grown with $\geq$6% B light were 37%–48% or 29%–50% shorter than those under 100% R, respectively. In general, seedling height was similar when the portion of B ranged between 6% and 100%, indicating the inhibitory effect of B light was saturated by about 10 μmol m^{-2} s^{-1} or 6% B. In addition, seedlings of impatiens grown under $\geq$50% B and salvia grown under $\geq$25% B light had 45%–51% smaller leaves and 36%–49% less fresh shoot mass than those grown under 100% R. Shoot dry mass showed a similar trend as fresh shoot mass, but in general, dry shoot mass under fluorescent lamps was less than that under 100% R and similar to that with $\geq$25% B light. The results of this study suggested that LED SSL that included a B photon flux density (PFD) of 10 μmol m^{-2} s^{-1} (or 6% B) with R light can produce compact ornamental seedlings without an excessive inhibition in biomass accumulation.

In another study, Craver et al. (2020) grew petunia 'Dreams Midnight' for 28 days under 50% B+50% R (peaks of 451 and 660 nm) or 10% B+90% R LEDs at a PPFD of 150 or 300 μmol m^{-2} s^{-1} and under two CO_2 concentrations of 450 or 900 μmol mol^{-1}. Regardless of PPFD and CO_2 concentration, seedling shoot and root dry mass, leaf area index, and internode length were greater under 10% B+90% R compared to 50% B+50% R. In addition, when light-use efficiency was calculated as the ratio of leaf gross photosynthesis to incident PPFD, it was higher under 10% B compared to under 50% B when the incident PPFD was lower than 500 μmol m^{-2} s^{-1}, regardless of CO_2 concentration. The results of this study suggested that reduced seedling growth under a higher portion of B light could be at least partly attributed to a decrease in light interception from a lower leaf area index, shorter internode length, and decrease in light-use efficiency.

19.2.2 Effects of far-red light

In plants, FR light is perceived by phytochrome photoreceptors and regulates a wide range of plant processes, including seed germination, deetiolation, shade-avoidance responses (e.g., stem and leaf expansion), and photoperiodic flowering (Franklin, 2008; Kami et al., 2010). Upon absorption of FR light, phytochromes are converted from a biologically active Pfr (FR-absorbing) form into the inactive Pr (R-absorbing) form, while R light converts Pr back to Pfr. Thus, the PFD of FR relative to R (or the R:FR) regulates the concentration of the active Pfr form of phytochrome, which determines the extent of phytochrome-mediated plant responses (Sager et al., 1988). The phytochrome photoequilibrium (PPE) can be estimated based on the photon spectrum and absorption by Pr and Pfr (Sager et al., 1988). In a wide range of species, stems were progressively shorter as the estimated PPE increased (Smith, 1982). Also, flowering of some long-day plants was accelerated when lighting during the night had an intermediate PPE (Craig and Runkle, 2016). Thus, FR in the photon spectrum influences the PPE and regulates plant morphology and flowering time of at least some long-day plants.

Sunlight and conventional light sources used for plant growth emit at least a small portion of FR light. For example, the portion of FR PFD in the 400–800 nm waveband is 25% for sunlight, 7% for fluorescent lamps, and 6% for warm-W LEDs (Meng and Runkle, 2016; Lopez et al., 2017; Kelly et al., 2020). However, B + R LED arrays for plant application emit essentially no FR light. Park and Runkle (2017, 2018a, 2019) investigated the inclusion of FR to B + R SSL in a series of experiments with several annual bedding plants.

Park and Runkle (2017) compared seedling growth and subsequent flowering of geranium (*Pelargonium × hortorum*), petunia, snapdragon (*Antirrhinum majus*), and impatiens under 32 μmol m^{-2} s^{-1} of B + 128 μmol m^{-2} s^{-1} of R (peaks of 447 and 660 nm) with those grown with the addition of 16, 32, or 64 μmol m^{-2} s^{-1} of FR (peak of 731 nm). Two additional treatments substituted 32 or 64 μmol m^{-2} s^{-1} of R with FR. Adding FR light or partially substituting R with FR decreased the R:FR (PPE) from 1:0 (0.88) to 1:1 (0.69). At the transplant stage, as the estimated PPE decreased from 0.88 to 0.69 with the inclusion of FR, seedling height of all species linearly increased by 30%–70% and leaf expansion of geranium and snapdragon linearly increased by 30%–40%. In geranium and snapdragon, the greater leaf area with FR was accompanied by greater shoot dry mass (by 28%–50%) with the addition of FR to the same PPFD. In addition, shoot dry mass was similar when R was partially substituted with FR, despite the decreases in PPFD by 40%. When the carry-over effects of SSL treatments were investigated after transplant in a greenhouse environment, the seedlings of long-day snapdragon grown with ≥16 μmol m^{-2} s^{-1} of FR flowered 10–12 days earlier than those previously grown without FR (Fig. 19.1). In contrast, there was little to no SSL effect on flowering in petunia, impatiens, and geranium.

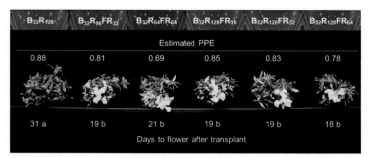

FIGURE 19.1   **Subsequent flowering of snapdragon after transplant.** Seedlings of snapdragon were grown for 26 days with an 18-hour photoperiod under six sole-source lighting treatments delivered from blue (B), red (R), and far-red (FR) light emitting diodes. The values after each waveband indicate their photon flux densities in μmol m^{-2} s^{-1}. PPE is the phytochrome photoequilibrium estimated under each lighting treatment (Sager et al., 1988). Seedlings were then transplanted and subsequently grown in a common greenhouse environment with a 16-hour photoperiod until they flowered. Data represent the means of days to flower after transplant of two replications with 10 plants per replication. Means with different letters are significantly different by Tukey's honestly significant difference test (*P* < .05). *For experimental details, see Park, Y., Runkle, E.S., 2017. Far-red radiation promotes growth of seedlings by increasing leaf expansion and whole-plant net assimilation. Environ. Exp. Bot. 136, 41–49.*

Park and Runkle (2018a, 2019) performed two additional experiments to investigate how the effects of FR light on seedling growth and flowering responses were influenced by PPFD and B light in three popular bedding plants, petunia, geranium, and coleus (*Solenostemon scutellarioides*). In one experiment, they grew seedlings under three R:FR (1:0, 2:1, and 1:1) and two PPFDs (96 or 288 µmol m^{-2} s^{-1}) (Park and Runkle, 2018a). The two PPFDs delivered 32 µmol m^{-2} s^{-1} of B and 64 or 256 µmol m^{-2} s^{-1} of R. In this study, the B PFD was kept constant to avoid confounding effects from changes in B light. In general, the addition of FR increased seedling height and shoot dry mass similarly, regardless of PPFD. For example, as R:FR (PPE) decreased from 1:0 (0.88) to 1:1 (0.70), seedling height and shoot dry mass of coleus linearly increased by 121% and 31%, respectively, under the low PPFD and by 138% and 33%, respectively, under the high PPFD (Fig. 19.2). However, an R:FR $\leq$ 2 (PPE $\leq$ 0.78) during the seedling stage accelerated subsequent flowering in long-day petunia by 11 days under the lower PPFD but by 7 days under the higher PPFD. This indicates that the additional FR had greater effects on accelerating subsequent flowering under the low PPFD. This study and others performed in greenhouses (Kohyama et al., 2014; Owen et al., 2018) indicate that the effect of FR on promoting flowering of long-day plants decreases as the PPFD (or DLI) increases.

In another experiment, seedlings were grown under an R PFD of 160 µmol m^{-2} s^{-1} or 80 µmol m^{-2} s^{-1} of R + 80 µmol m^{-2} s^{-1} of B without or with FR light that delivered an R:FR = 1:0, 8:1, or 1:1 (Park and Runkle, 2019). Similar to the previous two experiments, the addition of FR light generally increased seedling height and shoot dry mass in all species at transplant and accelerated the subsequent flowering of long-day petunia after transplant. However, while B light attenuated the effects of FR on stem elongation, it did not affect dry mass accumulation or subsequent flowering. For example, decreasing the R:FR (PPE) from 1:0 (0.89) to 1:1 (0.69) increased stem length of petunia by 674% without B light but by only 129% with B light. This indicates that FR effects on extension growth were suppressed by a B PFD of 80 µmol m^{-2} s^{-1} (Fig. 19.3). In contrast, shoot dry mass of petunia seedlings increased with

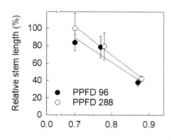
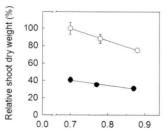

Estimated PPE

FIGURE 19.2 **Influence of the estimated phytochrome photoequilibrium (PPE) of six sole-source lighting treatments on relative stem length and shoot dry mass of coleus seedlings grown under a photosynthetic photon flux density (PPFD) of 96 or 288 µmol m^{-2} s^{-1}.** The PPE was estimated under each lighting treatment according to Sager et al. (1988). Data points represent the relative value of the means and standard errors of two replications with 10 plants per replication. *Modified from Park, Y., Runkle, E.S., 2018a. Far-red radiation and photosynthetic photon flux density independently regulate seedling growth but interactively regulate flowering. Environ. Exp. Bot. 155, 206–216.*

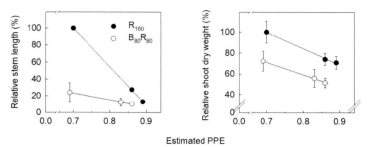

FIGURE 19.3   Influence of the estimated phytochrome photoequilibrium (PPE) of six sole-source lighting treatments on relative stem length and shoot dry mass of petunia seedlings grown under a red photon flux density of 160 μmol m^{-2} s^{-1} (R$_{160}$) or a blue photon flux density of 80 μmol m^{-2} s^{-1} plus a red photon flux density of 80 μmol m^{-2} s^{-1} (B$_{80}$R$_{80}$). The PPE was estimated under each lighting treatment according to Sager et al. (1988). Data points represent the relative value of the means and standard errors of three replications with 10 plants per replication. *Modified from Park, Y., Runkle, E.S., 2019. Blue radiation attenuates the effects of the red to far-red ratio on extension growth but not on flowering. Environ. Exp. Bot. 168, 103871.*

decreasing R:FR from 1:0 to 1:1 regardless of B light, increasing by 41% without B light and by 40% with B light (Fig. 19.3). In addition, after transplant, petunia treated with FR light at an R:FR (PPE) of 1:1 (0.69 or 0.70) flowered 7—10 days earlier, regardless of B light.

In an effort to determine saturation levels of FR, Elkins and van Iersel (2020) evaluated 18 PFDs of supplemental FR light on the growth and morphology of foxglove (*Digitalis purpurea*) seedlings grown for 55 days under a PPFD of 186 μmol m^{-2} s^{-1} and a 16-hour photoperiod provided by W LEDs. As the FR PFD increased from 4 to 69 μmol m^{-2} s^{-1}, shoot and root dry mass, plant height, and number of leaves increased by 38%, 20%, 38%, and 34%, respectively, while root mass fraction decreased by 16%. Surprisingly, the addition of FR had little to no effect on specific leaf area and compactness. Given that a 37% increase in total PFD (PPFD + FR) correlated well with a 34% increase in total plant dry mass, the authors attributed the increased growth to increased photosynthesis rather than a shade acclimation response. Growth responses were linear across the range of FR light intensities used, and thus, they were unable to identify a saturating FR PFD.

19.2.3 Comparison of white light with blue + red light

The photon spectrum of LED SSL influences plant growth attributes as well as the efficacy of the fixtures and human vision. B + R LEDs have been considered desirable for horticultural application when considering normal plant growth and high electrical efficiency, but their purplish spectrum can make it difficult to identify the colors of plants or detect pathogens such as botrytis. Adding green (G, 500—600 nm) to B + R light and using W LEDs are ways to create broad-band spectra and increase the low color rendering ability of B + R LEDs (Kim et al., 2004; Runkle, 2018).

Park and Runkle (2018b) investigated the merits of using LED arrays that contain mint W LEDs or G LEDs for growing ornamental seedlings, compared to using B + R LEDs. They grew seedlings of begonia (*Begonia × semperflorens*), geranium, petunia, and snapdragon under a PPFD of 160 μmol m^{-2} s^{-1} delivered from 100% W (peak of 558 nm), 75%

W+25% R (peak of 660 nm), 45% W+55% R, 25% W+75% R, 20% B (peak of 447 nm)+40% G (peak of 531 nm)+40% R, and 15% B+85% R LEDs (Fig. 19.4). In all species, at the transplant stage, seedlings grown with W or G LEDs generally had a similar seedling height, total leaf area, and fresh and dry mass compared to those grown under B + R LEDs (Fig. 19.5).

They also evaluated the electrical efficiency and visual appearance of the spectra using three metrics, including photosynthetic photon efficacy, dry mass gain per electric energy consumption, and color rendering index (CRI), which are suggested for horticultural LED lighting (Nelson and Bugbee, 2014; Hernández and Kubota, 2016; Both et al., 2017). B + R LEDs had a higher photosynthetic photon efficacy (2.25 μmol J^{-1}) than W or W + R LEDs (1.52–2.13 μmol J^{-1}) or B + G + R LEDs (1.51 μmol J^{-1}). This confirmed that LED arrays that contain W or G LEDs are less efficient at converting electricity into photosynthetic photons than B + R LEDs (Table 19.1). This is consistent with recent information on more efficacious LEDs (Kusuma et al., 2020). However, when the dry mass gain per unit of electric energy consumption (g kWh^{-1}) was calculated under each photon spectrum for each crop species, the values were similar under B + R and W + R LEDs in all species studied (Table 19.1). Thus, W + R LEDs produced plant biomass as efficiently as B + R LEDs. For visual appearance, W LEDs had a higher CRI value (64) than B + G + R (56) or B + R (−175)

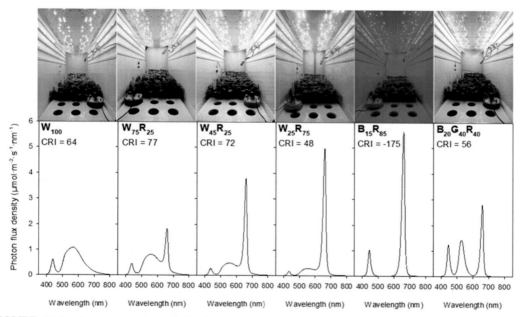

FIGURE 19.4 **Appearance, spectral distribution, and color rendering index (CRI) of six sole-source lighting treatments delivered from mint white (W), red (R), blue (B), and green (G) light emitting diodes (LEDs) at a total photosynthetic photon flux density of 160 μmol m^{-2} s^{-1}.** The subscript values after each LED type indicate the percentages of the total PPFD delivered from each LED type. *Modified from Park, Y., Runkle, E.S., 2018b. Spectral effects of light-emitting diodes on plant growth, visual color quality, and photosynthetic photon efficacy: white versus blue plus red radiation. PloS One 13 (8), e0202386.*

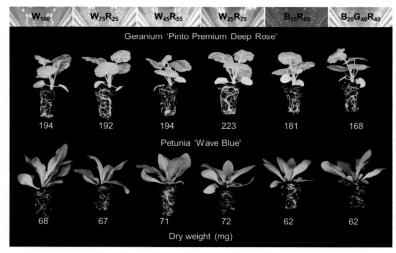

FIGURE 19.5 Seedlings of geranium and petunia at the transplant stage. Geranium and petunia were grown for 25 and 19 days, respectively, under six sole-source lighting treatments delivered from mint white (W), red (R), blue (B), and green (G) light emitting diodes (LEDs) at a total photosynthetic photon flux density (PPFD) of 160 μmol m^{-2} s^{-1}. The values after each LED type represent their percentages of the total PPFD. Data represent the means of dry mass of two replications with 10 plants per replication per species, and all means within a species are statistically similar ($P \geq .05$). *For experimental details, see Park, Y., Runkle, E.S., 2018b. Spectral effects of light-emitting diodes on plant growth, visual color quality, and photosynthetic photon efficacy: white versus blue plus red radiation. PloS One 13 (8), e0202386.*

TABLE 19.1 Photosynthetic photon efficacy and dry mass gain per unit electric energy consumption for sole-source lighting treatments. Begonia, geranium, snapdragon, and petunia seedlings were grown for 34, 25, 24, and 19 days, respectively, under six sole-source lighting treatments delivered from mint white (W), red (R), blue (B), and green (G) light emitting diodes at a total photosynthetic photon flux density (PPFD) of 160 μmol m^{-2} s^{-1}. The values after each LED type represent their percentages of the total PPFD.

Lighting treatments	Photosynthetic photon efficacy (μmol J^{-1})	Dry mass gain per unit electric energy consumption (g kWh^{-1})			
		Begonia	**Geranium**	**Snapdragon**	**Petunia**
W$_{100}$	1.52	0.78 b[a]	2.18 cd	1.37	1.00 cd
W$_{75}$R$_{25}$	1.72	0.92 ab	2.44 bcd	1.53	1.12 bcd
W$_{45}$R$_{55}$	1.88	1.17 a	2.70 bc	1.58	1.30 abc
W$_{25}$R$_{75}$	2.13	1.11 a	3.52 a	1.78	1.50 a
B$_{15}$R$_{85}$	2.25	1.14 a	3.02 ab	1.76	1.36 ab
B$_{20}$G$_{40}$R$_{40}$	1.51	0.79 b	1.88 d	1.32	0.91 d
Significance		**[b]	**	NS	**

[a]*Means with different letters are significantly different by Tukey's honestly significant difference test ($P < .05$) and lack of mean separation indicates nonsignificance.*
[b]*NS or ** Nonsignificant or significant at $P < .01$, respectively.*
Modified from Park, Y., Runkle, E.S., 2018b. Spectral effects of light-emitting diodes on plant growth, visual color quality, and photosynthetic photon efficacy: white versus blue plus red radiation. PloS One 13 (8), e0202386.

LEDs, and fixtures that partially contained W LEDs with 25% or 55% R LEDs further improved the CRI to 77 or 72, respectively (Fig. 19.4). Therefore, the results suggested that LED arrays that partially contain W LEDs can be used to produce floriculture transplants using similar electric energy with higher visual color quality compared to B + R LEDs.

19.2.4 Comparison of seedlings grown indoors with sole-source lighting and in a greenhouse with supplemental lighting

The largest market period for ornamental annual and herbaceous perennial crops occurs during the spring. To meet the market demand, most young ornamental plants are grown during late winter and early spring. During this major production time, the photosynthetic DLI inside the greenhouses in northern latitudes is often lower than the recommended minimum DLI of 10 mol m^{-2} d^{-1} for high-quality ornamental young plants (Runkle, 2007; Lopez and Runkle, 2008). Commercial greenhouse growers use supplemental lighting (SL), which is usually delivered by HPS lamps and more recently by LED fixtures, to increase the DLI and plant growth. For LED SSL to be used to grow floriculture transplants, seedlings grown under SSL and their finished plants must be of comparable or greater quality compared to those grown conventionally in a greenhouse with SL. A few studies have compared the effects of SSL and SL on the growth of ornamental seedlings at a similar total DLI, photoperiod, and average daily temperature.

Randall and Lopez (2015) compared seedlings of vinca (*Catharanthus roseus*), impatiens, geranium, petunia, and French marigold (*Tagetes patula*) grown under greenhouse ambient light with SL from HPS lamps or 13% B+87% R (peaks of 466 and 660 nm) LED arrays delivering a PPFD of 70 μmol m^{-2} s^{-1} and a 16-hour photoperiod (average DLI of 10.6 mol m^{-2} d^{-1}). The seedlings were also grown in a growth chamber under SSL at a PPFD of 185 μmol m^{-2} s^{-1} with 16-hour photoperiod (DLI of 10.6 mol m^{-2} d^{-1}) from 13% B+87% R (peaks of 466 and 660 nm) or 30% B+70% R LED arrays. With a few exceptions, seedlings grown under SSL were of similar or shorter in height and had similar or greater root and shoot dry mass and relative chlorophyll content compared to those grown in the greenhouse with SL. With a higher portion of B under LED SL (13%) than that under HPS (5%), as well as less FR, the seedlings grown under LED SL were more compact compared to those grown under HPS SL. Similarly, seedlings grown under SSL generally were shorter and had a higher relative chlorophyll content compared to those grown under HPS SL. After transplant to a common environment, flowering time of vinca, geranium, petunia, and French marigold was generally similar among SL and SSL treatments.

In another study, Zhang et al. (2020) compared seedling growth of eight species of annual bedding plants grown under SSL from B + R + FR or W LEDs with those grown in a greenhouse under SL from HPS lamps. They grew seedlings indoors with an 18-hour photoperiod at a PPFD of 180 μmol m^{-2} s^{-1} (DLI of 12 mol m^{-2} d^{-1}) delivered by two combinations of B + R (peaks of 449 and 664 nm) LEDs without or with 10, 20, or 40 μmol m^{-2} s^{-1} of FR LEDs (peak of 733 nm). Additional treatments were warm-W LEDs (which emitted

14 $\mu mol\ m^{-2}\ s^{-1}$ of B, 54 $\mu mol\ m^{-2}\ s^{-1}$ of G, 95 $\mu mol\ m^{-2}\ s^{-1}$ of R, and 17 $\mu mol\ m^{-2}\ s^{-1}$ of FR) delivered indoors or ambient sunlight in a greenhouse with SL from HPS lamps at a PPFD of 100 $\mu mol\ m^{-2}\ s^{-1}$ to establish a similar DLI (12 $mol\ m^{-2}\ d^{-1}$) and photoperiod. At the transplant stage, SSL and SL treatments had little to no effect on shoot dry mass and average leaf area in all eight species, but they affected the stem length and relative chlorophyll concentration in a few species. For example, stem length of snapdragon was similar under SSL with an FR PFD of $\leq 20\ \mu mol\ m^{-2}\ s^{-1}$ and in the greenhouse, but stem length under SSL with 40 $\mu mol\ m^{-2}\ s^{-1}$ of FR or W SSL was 86%−140% taller than those grown in the greenhouse. In petunia, seedlings under W SSL or SSL without or with 10 $\mu mol\ m^{-2}\ s^{-1}$ of FR had higher relative chlorophyll concentration than those in the greenhouse. However, increasing FR decreased the relative chlorophyll concentration, and seedlings grown under SSL with an FR PFD $\geq 20\ \mu mol\ m^{-2}\ s^{-1}$ had a similar leaf relative chlorophyll concentration as those in the greenhouse. After transplant and finishing in a common greenhouse environment, long-day plant snapdragon grown under W SSL or SSL with FR at $\geq 20\ \mu mol\ m^{-2}\ s^{-1}$ flowered 7−11 days earlier than those grown under SSL without FR or in the greenhouse with SL. In the other species, SL and SSL treatments during the young plant stage did not affect the subsequent flowering. Thus, seedlings grown indoors with SSL from B + R + FR or W LEDs, or in a greenhouse with HPS SL, generally had similar growth attributes and subsequent flowering. However, in the most sensitive long-day plants (e.g., snapdragon), the inclusion of FR in an SSL spectrum accelerated subsequent flowering compared to seedlings grown in the greenhouse or under SSL without FR light.

19.3 Floriculture liners

Liners are grown from cuttings stuck in small cells of a propagation tray and are often challenging to propagate because they are prone to dehydration before roots fully develop. During callusing, unrooted cuttings are typically propagated under light quantities lower than those used for seedling propagation, with PPFD and DLI values typically ranging from 120 to 200 $\mu mol\ m^{-2}\ s^{-1}$ and 3−8 $mol\ m^{-2}\ d^{-1}$, respectively (Faust et al., 2017). Here, we summarize studies that investigated the effect of light quantity during indoor production of liners. In addition, we review results from photon spectrum studies that have demonstrated significant potential from using B, R, or FR light to induce faster rooting of cuttings. We also highlight studies that compared carry-over effects when cuttings were propagated indoors under SSL versus those conventionally propagated in a greenhouse with SL.

19.3.1 Effects of light quantity

Early studies evaluating the effect of light quantity during indoor poinsettia (*Euphorbia pulcherrima*) propagation from cuttings indicated that increasing the PPFD from 180 to 250 $\mu mol\ m^{-2}\ s^{-1}$ using a 16-hour photoperiod provided by cool-W fluorescent lamps and incandescent light bulbs was not beneficial until after root formation (Svenson and Davies, 1990). In an industry report, van Dalfsen et al. (2011) described results from a study comparing growth and rooting of several hard-to-propagate tree species using PPFDs from 10 to 36 $\mu mol\ m^{-2}\ s^{-1}$ and a 16-hour photoperiod provided by a combination B + R LEDs without or with FR light (no peak wavelengths reported). They concluded that blueblossom ceanothus (*Ceanothus thyrsiflorus*) and boxwood (*Buxus sempervirens*) cuttings needed a PPFD of $\geq 25\,\mu mol\,m^{-2}\,s^{-1}$ to induce root formation, whereas PPFD had no effect on growth or rooting of Lawson cypress (*Chamaecyparis lawsoniana*) cuttings. They noted that a higher PPFD (e.g., $50\,\mu mol\,m^{-2}\,s^{-1}$) could improve rooting success in multilayer propagation systems.

Park et al. (2011) compared growth of cuttings from four rose (*Rosa hybrida*) cultivars grown indoors under PPFDs of 90, 180, or $270\,\mu mol\,m^{-2}\,s^{-1}$ provided by a combination of HPS and fluorescent lamps using a 13.5-hour photoperiod. Rooting responses were measured 45 days after sticking cuttings. In three of the four cultivars, rooting percentage, number of roots per cutting, and the length of the longest root significantly increased as PPFD increased from 90 to $270\,\mu mol\,m^{-2}\,s^{-1}$. In another study, *Calibrachoa* (*Calibrachoa* 'MiniFamous Neo Royal Blue') cuttings were propagated under a PPFD of 40 or $80\,\mu mol\,m^{-2}\,s^{-1}$ with a 16-hour photoperiod provided by B (peak of 440 nm), R (peak of 660 nm), cool-W (no peak wavelength reported), a mixture of B, R, and cool-W LEDs, or HPS lamps (Olschowski et al., 2016). Cuttings had the highest shoot dry mass and total root length under a PPFD of $80\,\mu mol\,m^{-2}\,s^{-1}$ provided by HPS lamps, which was partly attributed to a warmer substrate temperature. There were few effects of the photon spectrum among cuttings grown under a PPFD of $40\,\mu mol\,m^{-2}\,s^{-1}$, with the exception of those grown under only B light, which had shoots that were nearly twice as long as those in the other treatments. In general, shoot and root growth was greater under the higher PPFD than under the lower PPFD (Olschowski et al., 2016).

Gil et al. (2020) compared rooting of chrysanthemum (*Chrysanthemum* × *morifolium*) cuttings grown indoors under B (peak of 450 nm) or R (peak of 625 nm) LEDs or fluorescent lamps providing PPFDs of 5, 35, or $65\,\mu mol\,m^{-2}\,s^{-1}$ with a 16-hour photoperiod. Regardless of the photon spectrum, a PPFD of $65\,\mu mol\,m^{-2}\,s^{-1}$ increased the number of adventitious roots and root dry mass compared to lower PPFDs. However, the root dry mass of cuttings treated with a monochromatic B PFD of $5\,\mu mol\,m^{-2}\,s^{-1}$ was similar to that of those treated with a PFD of $65\,\mu mol\,m^{-2}\,s^{-1}$ from R LEDs or fluorescent lamps. It appears that a desirable PPFD for propagating liners from cuttings depends (at least in part) on light quality and its potential to quickly induce adventitious rooting and growth. Furthermore, other environmental factors that can minimize cutting dehydration (e.g., high humidity) might play a key role in determining desirable PPFDs during indoor propagation.

19.3.2 Effects of blue + red light

Most studies evaluating the use of LEDs for SSL on indoor production of liners have focused on using different combinations of B and R light wavebands, which have shown potential to stimulate adventitious root formation and subsequent growth of unrooted cuttings. When evaluating the effect of the photon spectrum on root initiation of woody species propagated indoors, van Dalfsen and Slingerland (2012) reported that monochromatic B or R LEDs (no peak wavelengths reported) providing a PPFD of $40\,\mu mol\,m^{-2}\,s^{-1}$ with a 16-hour photoperiod induced earlier root formation of boxwood cuttings compared to 25% B+75% R LEDs without or with an FR PFD of $15\,\mu mol\,m^{-2}\,s^{-1}$ or monochromatic R supplemented with an FR PFD of $15\,\mu mol\,m^{-2}\,s^{-1}$. In addition, monochromatic B light improved rooting uniformity of leucothoe (*Leucothoe* sp.) cuttings compared to greenhouse propagation.

Similarly, Christiaens et al. (2015) compared rooting quality of chrysanthemum, lavender (*Lavandula angustifolia*), and azalea (*Rhododendron simsii*) propagated indoors under 100% B (peak of 450 nm), 10% B+90% R (peak of 660 nm), 50% B+50% R, 90% B+10% R, or 100% R LEDs. The tested PPFD and photoperiod varied among species and included a PPFD of 60 μmol m^{-2} s^{-1} with 19- or 14-hour photoperiods for chrysanthemum or lavender, respectively, and a PPFD of 30 μmol m^{-2} s^{-1} with an 18-hour photoperiod for azalea. 3 weeks after sticking chrysanthemum cuttings, the percentage of root formation was similar across treatments, but the number of roots per cutting was 49%–81% higher under monochromatic B light compared to all other SSL treatments. In addition, root dry mass was the greatest under monochromatic B or R light. Similarly, root dry mass of lavender measured 3 weeks after sticking cuttings was higher under monochromatic B compared to 10% B+90% R and 50% B+50% R LEDs, but no other treatment differences were reported. Christiaens et al. (2015) also evaluated the effect of two auxin transport inhibitors on adventitious rooting of chrysanthemum. Results suggest that monochromatic B or R light stimulated auxin biosynthesis and/or transport to a greater extent than the rooting inhibition effect of the auxin transport inhibitors. In contrast, the photon spectrum had no effect on rooting quality or rooting reduction by auxin transport inhibitors for azalea when measured 8 weeks after sticking cuttings.

In a series of follow-up experiments, Christiaens et al. (2019) confirmed that monochromatic B or R SSL can stimulate adventitious root formation of chrysanthemum cuttings compared to when they are combined. This is in agreement with other studies that have reported inhibitory effects of B + R light on rooting of tissue culture microcuttings (Kurilčik et al., 2008; Iacona and Muleo, 2010; Daud et al., 2013). However, monochromatic B or R light is not always the most effective at inducing early root formation. For example, Cho et al. (2019) compared the effect of monochromatic B (peak of 446 nm), R (peak of 663 nm), G (peak of 530 nm), yellow (peak of 602 nm), and W (peaks of 437 and 560 nm) LEDs on root growth and overall liner quality of seven coleus cultivars propagated indoors for 28 days under a PPFD of 100 μmol m^{-2} s^{-1} and a 16-hour photoperiod. The number of adventitious roots from cuttings grown under monochromatic R light was highest in three cultivars compared to those of cuttings grown in the other treatments and more than doubled in five cultivars when compared to those grown under monochromatic G. In contrast, the effect of monochromatic B light on root formation was inconsistent across cultivars. Monochromatic R light also increased cutting fresh mass by 56% compared to G light, but generally led to similar shoot and root growth compared to monochromatic B light. Another study compared various photon spectrum treatments for indoor production of six difficult-to-propagate succulent (*Echeveria* spp.) cultivars (Kim et al., 2018). Leaf cuttings were grown for 60 days under a PPFD of 200 μmol m^{-2} s^{-1} with a 16-hour photoperiod provided by LEDs delivering monochromatic R (peak of 660 nm) without or with different combinations of B (peak of 450 nm), G (peak of 530 nm), and/or FR (peak of 730 nm) light. In general, higher proportions of R light promoted root formation and inhibited shoot growth, whereas the opposite trend occurred when cuttings were grown under higher B light.

In contrast to the studies by Christiaens et al. (2015, 2019), Gil et al. (2020) showed that survival rate, root length, and root dry mass of chrysanthemum cuttings grown under a

PPFD of 35 μmol m^{-2} s^{-1} with a 16-hour photoperiod were lower under monochromatic R (peak of 625 nm) compared to monochromatic B (peak of 450 nm) light from LEDs or fluorescent lamps. In addition, monochromatic B light induced the most uniform rooting from terminal, middle, and bottom leaf-bud cuttings. They also evaluated the relative expression level of seven genes involved in the auxin signaling pathway or in the formation of lateral roots in chrysanthemum. One day after sticking, the relative expression of CmLBD1 (chrysanthemum class I LATERAL ORGAN BOUNDARIES DOMAIN), which is involved in the induction of adventitious root formation, was greatest under monochromatic B light. This trend was maintained when cuttings were measured 5 days after sticking, leading to earlier root formation compared to the two other treatments, which only showed a significant upregulation of this gene on day 14 after sticking.

The photon spectrum of LED SSL can also affect stomatal conductance and transpiration, with higher proportions of B light typically leading to more water loss in some species (Clavijo-Herrera et al., 2018; Pennisi et al., 2019; Matthews et al., 2020). Davis (2016) propagated silverberry (*Elaeagnus* sp.), photinia (*Photinia* sp.), and rhododendron (*Rhododendron* sp.) cuttings under a PPFD of 100 μmol m^{-2} s^{-1} delivered from 100% B, 64% B+36% R, 35% B+65% R, 11% B+89% R, or 100% R LEDs (peak wavelengths not reported). Survival was higher in treatments containing a B PFD between 11% and 35%, and decreased as B light increased, which was attributed to cutting dehydration in response to SSL. Similar findings were reported by Navidad (2015) when comparing survival of subalpine fir (*Abies lasiocarpa*) and Norway spruce (*Picea abies*) cuttings propagated under a PPFD of 300 μmol m^{-2} s^{-1} and a 16-hour photoperiod provided by HPS lamps supplemented with incandescent light bulbs without or with a B PFD of ~75 μmol m^{-2} s^{-1} from LEDs (peak of 460 nm). 4 weeks after sticking, 70% of cuttings from both species had survived under HPS lamps, whereas only 48% of subalpine fir and 59% of Norway spruce cuttings survived when grown under HPS + B LEDs. In addition, water loss of Norway spruce propagated by seed almost tripled when propagated under higher B light.

19.3.3 Effects of far-red light

FR light can affect the rooting capacity of several vegetatively propagated species and is thought to influence auxin homeostasis and signaling that regulate adventitious root formation of cuttings (Ruedell et al., 2015; Christiaens et al., 2016). van Dalfsen and Slingerland (2012) reported that adventitious rooting of rhododendron cuttings was superior under a PFD of 40 μmol m^{-2} s^{-1} from R LEDs with an FR PFD of 15 μmol m^{-2} s^{-1} provided with a 16-hour photoperiod compared to greenhouse propagation without or with 20 μmol m^{-2} s^{-1} of SL from 20% B+80% R LEDs (no peak wavelengths reported). For example, based on an estimated rooting success rate that considered several parameters relevant to root initiation, rhododendron cuttings grown under R + FR LEDs had a 96% success rate, whereas greenhouse-propagated cuttings had an 88% success rate. In that same study, Chinese thuja (*Platycladus orientalis*) and leucothoe cuttings also showed positive rooting responses under R LEDs supplemented with FR light compared to greenhouse propagation, whereas boxwood cuttings seemed to have been negatively affected by FR light.

Christiaens et al. (2019) evaluated the effect of FR light on adventitious rooting of chrysanthemum cuttings grown under LEDs (no peak wavelengths reported) providing a PPFD of

60 μmol m^{-2} s^{-1} and a 19-hour photoperiod. In general, decreasing the PPE by lowering the R:FR without or with B light promoted rooting and decreased the effect of auxin transport inhibitors. More specifically, the addition of an FR PFD of 30 or 60 μmol m^{-2} s^{-1} to R light accelerated rooting and increased root dry mass and number of primary roots. Moreover, root dry mass significantly increased when an FR PFD of 60 μmol m^{-2} s^{-1} was added to a PPFD of 60 μmol m^{-2} s^{-1} of B + R light compared to monochromatic R or B + R + FR light with a total PFD (PPFD + FR) of 60 μmol m^{-2} s^{-1}. The authors suggested that the beneficial rooting effect of FR light is associated with a shade-avoidance response that increases auxin biosynthesis, which, in addition to increasing stem elongation and leaf area expansion, can improve rooting success of cuttings (Ahkami et al., 2013). Furthermore, the PPE affects the available pool of active auxin within plants and, thus, is likely a significant driver in the positive rooting responses under higher FR light (Tanaka et al., 2002). However, although rooting responses may be improved by adding FR light in some species, the acclimation capacity to a greenhouse or field environment may be lower when grown under a low R:FR compared to a higher ratio because their typical leaf structure (e.g., thinner, lower chlorophyll concentration, and decreased antioxidants) could limit photosynthetic activity (Oguchi et al., 2003).

19.3.4 Effects of liner treatments after transplanting

Although the photon spectrum from SSL can influence early root formation and growth, few studies have evaluated carry-over treatment effects and subsequent acclimation of liners. Zheng et al. (2020) aimed to understand how liners propagated indoors under low-light conditions adapt to a dynamic light environment in the greenhouse. Cuttings were grown indoors for 4 weeks under a PPFD of 100 μmol m^{-2} s^{-1} and a 16-hour photoperiod delivered from 100% B (peak of 450 nm), 40% B+60% R (peak of 660 nm), or 100% R LEDs or broadband W light delivered from plasma lamps. Their study included two plant species with contrasting light saturation levels: chrysanthemum (which light-saturates at a PPFD of 500—600 μmol m^{-2} s^{-1}) and peace lily (*Spathiphyllum wallisii*) (which light-saturates at 200—300 μmol m^{-2} s^{-1}). Plants grown under monochromatic B or R light had a lower leaf mass per area when compared to B + R light for chrysanthemum and B + R and W light for peace lily. After being propagated indoors, liners were transferred to the greenhouse where the DLI from sunlight ranged from 16.4 to 19.7 mol m^{-2} d^{-1}. On the first day of transfer, photosynthesis was significantly lower in liners propagated under monochromatic compared to broadband W light. However, photosynthesis of chrysanthemum liners propagated under monochromatic B light increased after 1 week in the greenhouse, approaching levels similar to liners grown under B + R light, but was generally lower than liners grown under broadband W light. In addition, the electron transport rate of peace lily leaves was lower under monochromatic R compared to B light. After 1 month in the greenhouse, chrysanthemum liners propagated under broadband W light had the highest shoot dry mass, followed by those propagated under B + R or monochromatic B light. There were no treatment effects on the shoot dry mass of peace lily. It appears that although monochromatic B and R light can induce early root formation, subsequent acclimation in the greenhouse can take longer. Dynamic lighting treatments that induce early root formation and stimulate development of the photosynthetic apparatus in leaves might improve liners propagated indoors.

19.3.5 Comparison of liners grown indoors with sole-source lighting and in a greenhouse with supplemental lighting

Owen and Lopez (2019) compared the growth of cuttings of herbaceous perennial sage (*Salvia nemorosa* 'Lyrical Blues') and wand flower (*Gaura lindheimeri* 'Siskiyou Pink') in a greenhouse with day-extension SL provided by HPS lamps delivering 40 μmol m^{-2} s^{-1} to provide a 16-hour photoperiod (average DLI of 3.3 mol m^{-2} d^{-1}) or in a walk-in growth chamber with SSL at 60 μmol m^{-2} s^{-1} and a 16-hour photoperiod (DLI of 3.4 mol m^{-2} d^{-1}) from 100% B (peak of 460 nm), 50% B+50% R (peak of 660 nm), 25% B+75% R, or 100% R LEDs. 10 days after sticking cuttings, callus diameter and rooting percentage of perennial sage and wand flower cuttings were similar under SL and SSL treatments. In general, as the portion of B light of SSL increased from 0% to 50%, stem length decreased and relative leaf chlorophyll content increased in both species. Stem length and relative leaf chlorophyll content under SSL with ≤25% B light were similar with cuttings under SL, but relative leaf chlorophyll content was greater and stem length was shorter under 50% B+50% R SSL than those under SL. Therefore, the results suggest that B + R SSL indoors and SL from HPS lamps in the greenhouse have generally similar effects on the growth of ornamental liners when the DLI is similar, while increasing the portion of B light of SSL can produce more compact transplants and greener leaves compared with SL with HPS lamps in a greenhouse.

19.4 Vegetable transplants

Using transplants is an increasingly popular way to establish vegetable crops in the field, greenhouses, and indoor plant production systems. Some vegetable crops commonly grown from transplants include tomato (*Solanum lycopersicum*), pepper (*Capsicum annuum*), lettuce (*Lactuca sativa*), cucumber (*Cucumis sativus*), and watermelon (*Citrullus lanatus*) (Welbaum, 2015). Characteristics of high-quality vegetable transplants are similar to those for ornamental plugs and liners. High-quality vegetable transplants usually have compact and thick stems, healthy green leaves, and actively growing white roots (Prunty, 2015). We summarize recent studies that investigated the effects of light quantity and B and FR light from SSL on growth attributes of vegetable transplants and the carry-over effects of SSL after transplant.

19.4.1 Effects of light quantity

Fan et al. (2013) investigated how increasing the PPFD of SSL from 50 to 550 μmol m^{-2} s^{-1} from 50% B+50% R LEDs (peaks of 460 and 658 nm) with a 12-hour photoperiod influenced leaf morphology, photosynthesis, and plant growth in young cherry tomato plants. Increasing the PPFD from 50 to 300 μmol m^{-2} s^{-1} increased stem diameter (by 16%), fresh mass (by 96%), and dry mass (by 195%) and suppressed plant height (by 47%) and specific leaf area (by 55%). However, the responses were generally saturated at a PPFD of 300 μmol m^{-2} s^{-1}. Leaves grown under a PPFD of 300 or 450 μmol m^{-2} s^{-1} had the thickest leaves, longest palisade and parenchyma cells, and greatest stomatal frequency and stomatal area per unit leaf area, which were indicative of a high photosynthetic capacity. Net leaf photosynthesis also increased with increasing PPFD up to 300 μmol m^{-2} s^{-1} but then decreased at 450 or 550 μmol m^{-2} s^{-1}. Thus, a PPFD of 300 μmol m^{-2} s^{-1} generally elicited desirable plant growth characteristics in young tomato plants. In addition, this study showed that dry mass gain per unit electric power consumption (g kWh^{-1}) was 13%–56% higher under a PPFD of 300 μmol m^{-2} s^{-1} compared to under lower or higher PPFDs.

19.4.2 Effects of blue light

Plant growth and biomass accumulation depend on light interception by its canopy and photosynthesis (Bugbee, 2016). Increasing B light has somewhat contrasting effects on plant growth. B light perception and signaling mediated by cryptochromes, phototropins, and phytochrome photoreceptors inhibit the elongation of hypocotyls, stems, petioles, and leaves (Huché-Thélier et al., 2016). Increasing B light can increase leaf photosynthesis through changes in leaf characteristics, such as increasing stomatal conductance and leaf chlorophyll concentration (Hogewoning et al., 2010). Hernández and Kubota (2016) investigated the effects of increasing B light on the photosynthesis, morphology, and overall plant growth of cucumber seedlings. They grew them indoors for 17 days under different percentages of B (from 0% to 100%) using B + R LEDs (peaks of 455 and 661 nm) at a PPFD of 100 μmol m^{-2} s^{-1} and a 16-hour photoperiod. Increasing B light generally increased the net photosynthetic rate, stomatal conductance, and chlorophyll content per unit leaf area. However, increasing the B light up to 75% produced more compact cucumber seedlings, with smaller leaves and lower biomass. For example, shoot height and hypocotyl length decreased linearly, by 62% and 67%, respectively, as B increased from 0% to 75%. Increasing B between 10% and 75% also reduced leaf area and shoot fresh and dry mass of cucumber seedlings by 41%, 34%, and 22%, respectively. The results indicated that the suppressive effects of B light on plant architecture had a greater effect on overall plant growth than the promotive effects of B light on photosynthesis.

In addition, this study made an interesting observation that B light without R light had little to no inhibitory effect on morphological and growth response (Hernández and Kubota, 2016). Under 100% B (without R) light, cucumber seedlings had greater plant height, hypocotyl length, epicotyl length, and shoot fresh mass compared to those grown under B + R light. Similarly, leaf area and shoot dry mass under 100% B were similar to or lower than those under 10% $\leq$ B $\leq$ 75%. Jeong et al. (2020) showed similar results for plant height and hypocotyl length of cucumber 'Joeunbaekdadagi' seedlings grown for 22 days under B + R LEDs (no peak wavelengths reported) at a PPFD of 200 μmol m^{-2} s^{-1} and a 16-hour photoperiod.

19.4.3 Effects of far-red light

Because of the potentially important roles of FR light in photoperiodic flowering, the first popular application of FR LEDs was to control flowering time of floriculture crops produced in greenhouses (Craig and Runkle, 2013, 2016). FR LEDs have also been suggested for indoor and greenhouse use to regulate plant morphology, growth, and pigmentation in vegetable crops and their transplants (Kubota et al., 2012). Meng and Runkle (2019) performed two experiments to investigate the merits of adding FR to B + R SSL for growing lettuce and basil (*Ocimum basilicum*) seedlings under different B:R ratios and PPFD conditions. In the first experiment, they grew seedlings of lettuce 'Rex' and 'Cherokee' and basil 'Genovese' under four combinations of B and/or R (peaks of 447 and 661 nm) SSL [$B_{30}R_{150}$ (low B:R), $B_{90}R_{90}$ (high B:R), R_{180}, or B_{180}, where the subscripts after each LED color indicate their PFD in μmol m^{-2} s^{-1}] with or without 30 μmol m^{-2} s^{-1} of FR (peak of 732 nm). In their second experiment, they compared the effects of adding 30 or 75 μmol m^{-2} s^{-1} of FR on seedling growth in lettuce 'Rex' and 'Rouxai' under $B_{90}R_{90}$ (low PPFD) or $B_{180}R_{180}$ (high PPFD).

In general, adding FR light to B + R light increased leaf length and shoot fresh and dry mass, while it decreased relative specific chlorophyll content and red coloration (Meng and Runkle, 2019). However, the effects of FR light were more pronounced under the high B:R and the lower PPFD than under the low B:R and the higher PPFD. For example, the addition of 30 μmol m^{-2} s^{-1} of FR increased shoot fresh and dry mass of lettuce 'Cherokee' by 48% and 44%, respectively, under $B_{90}R_{90}$, but only by 17% and 17%, respectively, under $B_{30}R_{150}$. In addition, FR light increased shoot dry mass of lettuce 'Rex' and 'Rouxai' by up to 35% and 57%, respectively, under the lower PPFD, but only up to 20% and 24%, respectively, under the higher PPFD.

Another application of FR light during the production of vegetable transplants involves the reduction of intumescence injury. This physiological disorder can occur in some cultivars of some plants in the Solanaceae plant family (such as tomato), is thought to be caused by a lack of UV light, and can be aggravated by high relative humidity (Lang and Tibbitts, 1983; Morrow and Tibbitts, 1988). In a series of experiments, Eguchi et al. (2016a,b) showed that end-of-day (EOD) FR light mitigated intumescence development compared to monochromatic B and R light. They first grew tomato 'Beaufort' seedlings for 16–18 days under a PPFD of 69 or 103 μmol m^{-2} s^{-1} and an 18-hour photoperiod using B and R LEDs (peaks of 455 and 661 nm) providing either 10% B+90% R (control) or 75% B+25% R without or with different EOD dosages from FR LEDs (peak of 738 nm) (Euguchi et al., 2016b). Seedlings grown under the control had severe intumescence, but adding 67 mmol m^{-2} d^{-1} of EOD-FR decreased the percentage of leaves exhibiting intumescence from 71% to 39%. In addition, compared to the control seedlings, those grown under 75% B light without EOD-FR had a lower number of leaves with intumescence. Seedlings had the fewest number of leaves exhibiting intumescence (5%) and the least severity of intumescence injury when EOD-FR light was added to the 75% B+25% R treatment.

In a follow-up study, Eguchi et al. (2016a) evaluated the combination of high B + EOD-FR light for mitigating intumesce injury on tomato 'Maxifort' and 'Beaufort', two highly susceptible rootstock cultivars. They grew seedlings under a PPFD of 199 μmol m^{-2} s^{-1} with an 18-hour photoperiod delivered by B + R LEDs (peaks of 455 and 661 nm) providing either 10% B+90% R or 50% B+50% R without or with 1.1 mmol m^{-2} d^{-1} of EOD-FR from LEDs (peak of 742 nm). Increasing the percentage of the PPFD as B light from 10% to 50% reduced the percentage of leaves with intumescence from 68% to 57% for 'Beaufort' and from 55% to 41% for 'Maxifort'. Furthermore, EOD-FR suppressed intumescence injury, but seedlings grown under the 10% B treatment with EOD-FR had elongated stems, which is an undesirable characteristic for tomato transplants. In contrast, the combination of high B light and EOD-FR further suppressed intumescence injury for both cultivars and led to relatively compact growth. The suppressive effect of EOD-FR on intumescence was more pronounced in 'Beaufort' than in 'Maxifort', suggesting that the FR dose required to suppress intumescence can be cultivar specific.

19.4.4 Effects of seedling treatments after transplanting

The quality of transplants can affect their establishment, growth, and yield after transplant. Thus, the lighting conditions during propagation can affect plant performance after transplant. To investigate the effects of SSL treatments during the seedling stage at transplant and at harvest, Yan et al. (2019) grew seedlings of lettuce for 20 days under 12 combinations of 2 PPFDs (200 and 250 μmol m^{-2} s^{-1}), 2 photoperiods (14 or 16 h), and 3 lighting spectra (florescent lamp and W LEDs with an R:B ratio of 1.2 or 2.2). Seedlings were then transplanted and subsequently grown in the same environment at a PPFD of 250 μmol m^{-2} s^{-1} with a 16-hour photoperiod. At transplant, under the same PPFD and photoperiod, lettuce seedlings grown under fluorescent lamps generally had larger leaves and leaf and root dry mass compared to those grown under two types of W LEDs. Under each photon spectrum, increasing the PPFD from 200 to 250 μmol m^{-2} s^{-1} or photoperiod from 14 to 16 hours had little effect or slightly increased leaf and root dry mass of transplants. At harvest, mature plants showed slightly different trends from seedling growth. In general, leaf fresh mass of mature lettuce plants was higher when previously grown under florescent lamps or W LEDs with an R:B of 2.2 than those grown under W LEDs with an R:B of 1.2. In addition, irrespective of the photon spectrum, the seedlings grown at a PPFD of 200 μmol m^{-2} s^{-1} and a 16-hour photoperiod had higher leaf and root fresh mass at harvest than the other PPFD and photoperiod combinations. These results indicated that photon spectrum, photoperiod, and PPFD of lighting treatments during the seedling stage affect seedling growth as well as the yield and quality of mature lettuce at harvest, but not necessarily in the same way.

Photoperiod is one of the environmental factors that can regulate flowering of strawberry (*Fragaria × ananassa*). For long-day strawberry cultivars, flowering is induced or accelerated by long-day conditions (Sønsteby and Heide, 2009). Tsuruyama and Shibuya (2018) investigated how the photoperiod of SSL during the seedling stage regulated growth and subsequent flowering of seed-propagated long-day strawberry 'Elan' and 'Yotsuboshi'. The strawberry plants were grown under B + R (peaks of 470 and 625 nm, B:R ratio of 1:2) LEDs in a growth chamber or under sunlight in a greenhouse at the same DLI of 10 mol m^{-2} d^{-1}. In a growth chamber, the DLI of 10 mol m^{-2} d^{-1} was delivered with 8-, 12-, 16-, and 24-hour photoperiods (the PPFD under each photoperiod was 350, 230, 175, and 115 μmol m^{-2} s^{-1}, respectively). In the greenhouse, the average ambient photoperiod was 13.6 hours during the experimental period. At transplant, the seedling dry mass of both strawberry cultivars generally increased as the SSL photoperiod increased and the PPFD decreased. For example, seedlings grown under SSL with a 16- and 24-hour photoperiod had 49%−142% or 34%−86% larger dry mass in 'Elan' or 'Yotsuboshi', respectively, than those under SSL with 8- and 12-hour photoperiods or under sunlight with an average photoperiod of 13.6 hours. The relative growth rate and net assimilation rate showed similar trends as with seedling dry mass. After transplant, strawberry 'Elan' grown under SSL with 16- and 24-hour photoperiods during the seedling stage had flower buds 14−23 days earlier than those grown under SSL with 8- and 12-hour photoperiods or under sunlight. In contrast, lighting treatments during the seedling stage had little to no effect on flower bud formation in 'Yotsuboshi'.

19.5 Microgreens

Microgreens are very young seedlings of vegetables and herbs. They are densely seeded and harvested from 7 to 21 days after seeding (Treadwell et al., 2016), when the cotyledons have developed either with or without the emergence of the first true leaves, depending on species. Some of the most popular plant species, subspecies, and varieties grown as microgreens include radish (*Raphanus sativus*), broccoli (*Brassica oleracea* var. *italic*), kale (*B. oleracea*), cabbage (*B. oleracea*), mizuna (*Eruca sativa*), bok choy (*Brassica chinensis*), arugula (*Brassica eruca*), and mustard (*Brassica juncea*) in the Brassicaceae family and beet (*Beta vulgaris* subsp. *vulgaris*), chard (*B. vulgaris* subsp. *vulgaris*, cicla group), and amaranth (*Amaranthus tricolor*) in the Amaranthaceae family (Verlinden, 2020; Turner et al., 2020). Important traits for microgreens include fresh yield, hypocotyl length, color, flavor, phytochemical composition, and nutritional value (Xiao et al., 2015; Ying et al., 2020). Recently, several studies have investigated the use of SSL to produce high-quality microgreens indoors.

19.5.1 Comparison of blue + red light versus darkness

During seed germination and seedling development, plants are highly sensitive to light and dark conditions, which can produce different growth forms, otherwise known as phenotypic plasticity (von Arnim and Deng, 1996; Sultan, 2000). For example, seedlings grown in darkness developed elongated hypocotyls with unopened yellow cotyledons, while those grown under light had shorter hypocotyls and expanded green cotyledons (von Arnim and Deng, 1996). Given that B + R SSL is commonly used for vegetable production, Kong and Zheng (2019) examined how providing 15% B+85% R (peaks of 455 and 660 nm) LED lighting at a PPFD of 316 μmol m^{-2} s^{-1} with a 17-hour photoperiod affected plant traits during and after germination in 18 vegetable genotypes compared to those grown in darkness. The 18 vegetable genotypes included green- and red-leaf cultivars of amaranth, basil, lettuce, bok choy, arugula, cabbage, mustard, kale, and radish.

In general, regardless of the genotype, each plant trait showed similar trends in response to lighting, although the response magnitude depended on the genotype (Kong and Zheng, 2019). Compared to darkness, providing B + R light decreased the germination rate, shoot length, and shoot fresh mass and increased root length, root diameter, lateral root number, and coloration of leaves and stems. In particular, when the magnitude of changes of each plant trait in response to B + R light was quantified using a phenotypic plasticity index (which was calculated as the coefficients of variation under B + R light and darkness), shoot color, shoot length, and root branching were the most sensitive traits to lighting. When the sensitivity of genotype to light was quantified using mean phenotypic plasticity (calculated by averaging the indices of plasticity obtained for all of the plant traits in each genotype), the genotypes with lowest seed mass, such as amaranth and basil, generally had higher mean phenotypic plastic values than those with higher seed mass, such as radish and cabbage. This indicates that small-seeded microgreens are more sensitive to light. In addition, within the same species, the red-leaf cultivar showed higher phenotypic plasticity than the green-leaf cultivar.

19.5.2 Effects of blue light

The desired quality attributes for microgreens often include long hypocotyls, high fresh yield, and dark green or light red cotyledons (for green- or red-leafed microgreen species, respectively) (Ying et al., 2020). Considering the involvement of B light in regulating hypocotyl elongation, pigmentation, and biomass accumulation, Ying et al. (2020) quantified how increasing the percentage of the PPFD as B light from 5% to 30% under B + R LEDs (peaks were not indicated) influenced the yield and appearance quality of cabbage, kale, arugula, and mustard microgreens at a PPFD of 300 μmol m^{-2} s^{-1} with a 16-hour photoperiod. Increasing the percentage of the PPFD as B light from 5% to 30% linearly decreased the hypocotyl length of kale (by 7%) and mustard (by 14%) and promoted the coloration of cotyledons in all species tested. However, increasing B light had little to no effect on fresh and dry mass in kale, arugula, and mustard. In cabbage, fresh and dry mass showed quadratic responses, with peaks at 15% of B light. Fresh mass (and dry mass) of cabbage microgreens grown under 15% B light was 19% or 12% (10% and 7%) greater than those grown under 5% or 30% B light.

However, other wavebands in the photon spectrum can influence plant responses to B light. For example, the cryptochrome-mediated inhibition of hypocotyl elongation required phytochrome activity (Ahmad and Cashmore, 1997). Thus, with low phytochrome activation (i.e., without R or with FR light), B light can have little to no suppressive effects on extension growth. Actually, in some species, plants grown under monochromatic B light showed greater elongation growth compared to those grown under B + R light (Hernández and Kubota, 2016; Kong et al., 2018).

Kong et al. (2019a) investigated this phenomenon in microgreens by comparing the effects of monochromatic B light (peak of 450 nm) without or with FR (10% of B light, peak of 730 nm) and UV-A (7.5% of B light, peak of 370 nm) light versus the effects of monochromatic R (peak of 660 nm) light on the morphological traits of arugula, cabbage, mustard, and kale seedlings at PPFDs of 50 and 100 μmol m^{-2} s^{-1} with a 24-hour photoperiod. Consistent with the paradigm, compared to monochromatic R light, monochromatic B light increased hypocotyl length by 39%−53% and petiole length by 31%−54% in arugula, cabbage, and kale under both PPFDs. Compared to monochromatic B light, adding UV-A or FR only slightly inhibited or promoted, respectively, elongation of hypocotyls and petioles, depending on species. Furthermore, in a following study, Kong et al. (2019b) observed hypocotyl elongation was greater under monochromatic B compared to monochromatic R light, especially under a 24-hour photoperiod.

19.5.3 Effects of light quantity

There is a general paradigm that a 1% increase in light quantity increases yield by 1%. Upon closer inspection, in a number of floriculture and vegetable crops, a 1% light quantity increment led to a 0.5%−1% increase in harvestable yield (Marcelis et al., 2006). However, in the very early stage of seedling development, when plants have a small leaf area, increasing light quantity can have little effect on increasing light interception and thus, plant biomass accumulation (Oh et al., 2010). Jones-Baumgardt et al. (2019) investigated how increasing the PPFD of 15% B+85% R (peaks of 445 and 660 nm) LED SSL from 100 to

$600 \ \mu mol \ m^{-2} \ s^{-1}$ with a 16-hour photoperiod influenced growth, yield, and quality of kale, cabbage, arugula, and mustard microgreens. In all four species, fresh and dry mass increased asymptotically with increasing PPFD. The magnitude of increase in fresh and dry mass with increasing PPFD was greater in species with smaller seed mass, such as arugula (76% and 122%) and mustard (82% and 145%), than those with higher seed mass, such as kale (by 36% and 65%) and cabbage (56% and 69%). In addition, increasing the PPFD inhibited hypocotyl length of kale, cabbage, arugula, and mustard and increased the yellow and red pigments in the green-leaf kale, cabbage, and arugula and purple coloration in the purple-leaf mustard, which suggests a corresponding increase in concentrations of carotenoids and anthocyanins.

19.6 Conclusion

Light quality and quantity interact to influence growth and development of plugs, liners, transplants, and microgreens. With indoor plant production systems and LED lighting, growers can consistently produce these crops year round, with the desired morphologies, and in some cases even inhibit or promote flowering. For many crops, a higher portion of B light inhibits extension growth of leaves and stems. While this can decrease light interception and potentially biomass accumulation, it creates more compact plants, which is generally desirable. Inclusion of FR light generally increases extension growth and for some long-day plants, induces early flowering. In sensitive crops (e.g., tomato), FR and B photons can also mitigate the physiological disorder intumescence. However, the morphological effects of light quality often diminish as the quantity of light delivered to plants increases. Nevertheless, growth of roots and shoots increases with PPFD and DLI, and therefore production time may be shorter (or harvestable yield greater) when plants are grown under relatively high light levels. It is becoming possible for growers of these young plants to first consider the desired attributes of their crops and then determine what light quality, intensity, and photoperiod are needed to produce them.

References

Ahkami, A.H., Melzer, M., Ghaffari, M.R., Pollmann, S., Javid, M.G., Shahinnia, F., Hajirezaei, M.R., Druege, U., 2013. Distribution of indole-3-acetic acid in *Petunia hybrida* shoot tip cuttings and relationship between auxin transport, carbohydrate metabolism and adventitious root formation. Planta 238 (3), 499—517.

Ahmad, M., Cashmore, A.R., 1997. The blue-light receptor cryptochrome 1 shows functional dependence on phytochrome A or phytochrome B in *Arabidopsis thaliana*. Plant J. 11 (3), 421—427.

Both, A.J., Bugbee, B., Kubota, C., Lopez, R.G., Mitchell, C., Runkle, E.S., Wallace, C., 2017. Proposed product label for electric lamps used in the plant sciences. HortTechnology 27 (4), 544—549.

Bugbee, B., 2016. Toward an optimal spectral quality for plant growth and development: the importance of radiation capture. Acta Hortic. 1134, 1—12.

Cho, K.H., Laux, V.Y., Wallace-Springer, N., Clark, D.G., Folta, K.M., Colquhoun, T.A., 2019. Effects of light quality on vegetative cutting and in vitro propagation of coleus (*Plectranthus scutellarioides*). HortScience 54 (5), 926—935.

Christiaens, A., Gobin, B., Van Huylenbroeck, J., Van Labeke, M.C., 2019. Adventitious rooting of *Chrysanthemum* is stimulated by a low red:far-red ratio. J. Plant Physiol. 236, 117—123.

Christiaens, A., Gobin, B., Van Labeke, M.C., 2016. Light quality and adventitious rooting: a mini-review. Acta Hortic. 1134, 385—393.

Christiaens, A., Van Labeke, M.C., Gobin, B., Van Huylenbroeck, J., 2015. Rooting of ornamental cuttings affected by spectral light quality. Acta Hortic. 1104, 219–224.

Clavijo-Herrera, J., van Santen, E., Gómez, C., 2018. Growth, water-use efficiency, stomatal conductance, and nitrogen uptake of two lettuce cultivars grown under different percentages of blue and red light. Horticulturae 4 (3), 1–16.

Craig, D.S., Runkle, E.S., 2013. A moderate to high red to far-red light ratio from light-emitting diodes controls flowering of short-day plants. J. Am. Soc. Hortic. Sci. 138 (3), 167–172.

Craig, D.S., Runkle, E.S., 2016. An intermediate phytochrome photoequilibria from night-interruption lighting optimally promotes flowering of several long-day plants. Environ. Exp. Bot. 121, 132–138.

Craver, J.K., Nemali, K.S., Lopez, R.G., 2020. Acclimation of growth and photosynthesis in petunia seedlings exposed to high-intensity blue radiation. J. Am. Soc. Hortic. Sci. 145 (3), 152–161.

Daud, N., Faizal, A., Geelen, D., 2013. Adventitious rooting of *Jatropha curcas* L. is stimulated by phloroglucinol and by red LED light. In Vitro Cell. Dev. Biol. Plant 49 (2), 183–190.

Davis, P.A., 2016. The use of light-emitting diode systems for improving plant propagation and production. Acta Hortic. 1140, 347–350.

Eguchi, T., Hernández, R., Kubota, C., 2016a. End-of-day far-red lighting combined with blue-rich light environment to mitigate intumescence injury of two interspecific tomato rootstocks. Acta Hortic. 1134, 163–170.

Eguchi, T., Hernández, R., Kubota, C., 2016b. Far-red and blue light synergistically mitigate intumescence injury of tomato plants grown under ultraviolet-deficit light environment. HortScience 51 (6), 712–719.

Elkins, C., van Iersel, M., 2020. Supplemental far-red light-emitting diode light increases growth of foxglove seedlings under sole-source lighting. HortTechnology 30 (5), 564–569.

Fan, X., Xu, Z., Liu, X., Tang, C., Wang, L., Han, X., 2013. Effects of light intensity on the growth and leaf development of young tomato plants grown under a combination of red and blue light. Sci. Hortic. 153, 50–55.

Faust, J.E., Dole, J.M., Lopez, R.G., 2017. The floriculture vegetative cutting industry. Hortic. Rev. 44, 121–172.

Fisher, P.R., Gómez, C., Poudel, M., Runkle, E.S., 2019. The economics of lighting young plants indoors. Grower 83 (2), 50–52.

Franklin, K.A., 2008. Shade avoidance. New Phytol. 179 (4), 930–944.

Fujiwara, K., 2020. Light sources. In: Kozai, T., Niu, G., Takagaki, M. (Eds.), Plant Factory: An Indoor Vertical Farming System for Efficient Quality Food Production. Academic Press, London, England, pp. 139–151.

Gil, C.S., Jung, H.Y., Lee, C., Eom, S.H., 2020. Blue light and NAA treatment significantly improve rooting on single leaf-bud cutting of *Chrysanthemum* via upregulated rooting-related genes. Sci. Hortic. (Amst.) 274, 109650.

Goto, E., 2012. Plant production in a closed plant factory with artificial lighting. Acta Hortic. 956, 37–50.

Hernández, R., Kubota, C., 2016. Physiological responses of cucumber seedlings under different blue and red photon flux ratios using LEDs. Environ. Exp. Bot. 121, 66–74.

Hogewoning, S.W., Trouwborst, G., Maljaars, H., Poorter, H., van Ieperen, W., Harbinson, J., 2010. Blue light dose–responses of leaf photosynthesis, morphology, and chemical composition of *Cucumis sativus* grown under different combinations of red and blue light. J. Exp. Bot. 61 (11), 3107–3117.

Huché-Thélier, L., Crespel, L., Gourrierec, J.L., Morel, P., Sakr, S., Leduc, N., 2016. Light signaling and plant responses to blue and UV radiations—perspectives for applications in horticulture. Environ. Exp. Bot. 121, 22–38.

Iacona, C., Muleo, R., 2010. Light quality affects in vitro adventitious rooting and ex vitro performance of cherry rootstock Colt. Sci. Hortic. 125 (4), 630–636.

Jeong, H.W., Lee, H.R., Kim, H.M., Kim, H.M., Hwang, H.S., Hwang, S.J., 2020. Using light quality for growth control of cucumber seedlings in closed-type plant production system. Plants 9 (5), 639.

Jones-Baumgardt, C., Llewellyn, D., Ying, Q., Zheng, Y., 2019. Intensity of sole-source light-emitting diodes affects growth, yield and quality of *Brassicaceae* microgreens. HortScience 54 (7), 1168–1174.

Kami, C., Lorrain, S., Hornitschek, P., Fankhauser, C., 2010. Light regulated plant growth and development. Curr. Top. Dev. Biol. 91, 29–66.

Kelly, N., Choe, D., Meng, Q., Runkle, E.S., 2020. Promotion of lettuce growth under an increasing daily light integral depends on the combination of the photosynthetic photon flux density and photoperiod. Sci. Hortic. 272, 109565.

Kim, H.-H., Goins, G.D., Wheeler, R.M., Sager, J.C., 2004. Green light supplementation for enhanced lettuce growth under red- and blue-light–emitting diodes. HortScience 39 (7), 1617–1622.

Kim, S., Kim, J., Oh, W., 2018. Propagation efficiencies at different LED light qualities for leaf cutting of six *Echeveria* cultivars in a plant factory system. Prot. Hortic. Plant Fact. 27 (4), 363–370.

Kohyama, F., Whitman, C., Runkle, E.S., 2014. Comparing flowering responses of long-day plants under incandescent and two commercial light-emitting diode lamps. HortTechnology 24 (4), 490–495.

Kong, Y., Kamath, D., Zheng, Y., 2019b. Blue versus red light can promote elongation growth independent of photoperiod: a study in four brassica microgreens species. HortScience 54 (11), 1955–1961.

Kong, Y., Schiestel, K., Zheng, Y., 2019a. Pure blue light effects on growth and morphology are slightly changed by adding low-level UVA or far-red light: a comparison with red light in four microgreen species. Environ. Exp. Bot. 157, 58–68.

Kong, Y., Stasiak, M., Dixon, M.A., Zheng, Y., 2018. Blue light associated with low phytochrome activity can promote elongation growth as shade-avoidance response: a comparison with red light in four bedding plant species. Environ. Exp. Bot. 155, 345–359.

Kong, Y., Zheng, Y., 2019. Variation of phenotypic responses to lighting using combination of red and blue light-emitting diodes versus darkness in seedlings of 18 vegetable genotypes. Can. J. Plant Sci. 99 (2), 159–172.

Kozai, T., 2013. Resource use efficiency of closed plant production system with artificial light: concept, estimation and application to plant factory. Proc. Japan Acad. Ser. B 89 (10), 447–467.

Kozai, T., 2020. Main components and their function. In: Kozai, T., Niu, G., Takagaki, M. (Eds.), Plant Factory: An Indoor Vertical Farming System for Efficient Quality Food Production. Academic Press, London, England, pp. 299–304.

Kozai, T., Chun, C., Ohyama, K., 2004. Closed systems with lamps for commercial production of transplants using minimal resources. Acta Hortic. 630, 239–254.

Kubota, C., Chia, P., Yang, Z., Li, Q., 2012. Applications of far-red light emitting diodes in plant production under controlled environments. Acta Hortic. 952, 59–66.

Kurilcik, A., Miklušytė-Čanova, R., Dapkūnienė, S., Zilinskaitė, S., Kurilčik, G., Tamulaitis, G., Duchovskis, P., Žukauskas, A., 2008. In vitro culture of Chrysanthemum plantlets using light emitting diodes. Cent. Eur. J. Biol. 3 (2), 161–167.

Kusuma, P., Pattison, P.M., Bugbee, B., 2020. From physics to fixtures to food: current and potential LED efficacy. Hortic. Res. 7, 56.

Lang, S.P., Tibbitts, T.W., 1983. Factors controlling intumescence development on tomato plants. J. Am. Soc. Hortic. Sci. 108, 93–98.

Lopez, R., Fisher, P., Runkle, E.S., 2017. Introduction to specialty crop lighting. In: Lopez, R., Runkle, E.S. (Eds.), Light Management in Controlled Environments. Meister Media Worldwide, Willoughby, OH, pp. 119–134.

Lopez, R.G., Runkle, E.S., 2008. Photosynthetic daily light integral during propagation influences rooting and growth of cuttings and subsequent development of New Guinea impatiens and petunia. HortScience 43 (7), 2052–2059.

Marcelis, L.F.M., Broekhuijsen, A.G.M., Meinen, E., Nijs, L., Raaphorst, M.G.M., 2006. Quantification of the growth response to light quantity of greenhouse grown crops. Acta Hortic. 711, 97–104.

Matthews, J.S.A., Vialet-Chabrand, S., Lawson, T., 2020. Role of blue and red light in stomatal dynamic behaviour. J. Exp. Bot. 71 (7), 2253–2269.

McCree, K.J., 1972. The action spectrum, absorbance and quantum yield of photosynthesis in crop plants. Agric. Meteorol. 9, 191–216.

Meng, Q., Runkle, E.S., 2016. Control of flowering using night-interruption and day-extension LED lighting. In: Kozai, T., Fujiwara, K., Runkle, E.S. (Eds.), LED Lighting for Urban Agriculture. Springer Singapore, Singapore, pp. 191–201.

Meng, Q., Runkle, E.S., 2019. Far-red radiation interacts with relative and absolute blue and red photon flux densities to regulate growth, morphology, and pigmentation of lettuce and basil seedlings. Sci. Hortic. 255, 269–280.

Mitchell, C., Both, A.J., Bourget, M., Burr, J., Kubota, C., Lopez, R., Morrow, R., Runkle, E., 2012. LEDs: the future of greenhouse lighting! Chron. Hortic. 52 (1), 6–12.

Morrow, R.C., Tibbitts, T.W., 1988. Evidence for involvement of phytochrome in tumor development on plants. Plant Physiol. 88 (4), 1110–1114.

Navidad, H., 2015. Impact of Additional Blue Light in the Production of Small Plants of Abies Laciocarpa and Picea Abies Propagated by Seeds and Stem Cuttings (Master's Thesis, Nor. Univ. Life Sci., Ås, Nor.).

Nelson, J.A., Bugbee, B., 2014. Economic analysis of greenhouse lighting: light emitting diodes vs. high intensity discharge fixtures. PloS One 9 (6), e99010.

Oguchi, R., Hikosaka, K., Hirose, T., 2003. Does the photosynthetic light-acclimation need change in leaf anatomy? Plant Cell Environ. 26 (4), 505–512.

Oh, W., Runkle, E.S., Warner, R.M., 2010. Timing and duration of supplemental lighting during the seedling stage influence quality and flowering in petunia and pansy. HortScience 45 (9), 1332–1337.

Olschowski, S., Geiger, E.M., Herrmann, J.V., Sander, G., Grüneberg, H., 2016. Effects of red, blue, and white LED irradiation on root and shoot development of *Calibrachoa* cuttings in comparison to high pressure sodium lamps. Acta Hortic. 1134, 245–250.

Owen, W.G., Lopez, R.G., 2019. Comparison of sole-source and supplemental lighting on callus formation and initial rhizogenesis of *Gaura* and *Salvia* cuttings. HortScience 54 (4), 684–691.

Owen, W.G., Meng, Q., Lopez, R.G., 2018. Promotion of flowering from far-red radiation depends on the photosynthetic daily light integral. HortScience 53 (4), 465–471.

Park, S.M., Won, E.J., Park, Y.G., Jeong, B.R., 2011. Effects of node position, number of leaflets left, and light intensity during cutting propagation on rooting and subsequent growth of domestic roses. Hortic. Environ. Biotech. 52 (4), 339–343.

Park, Y., Runkle, E.S., 2017. Far-red radiation promotes growth of seedlings by increasing leaf expansion and whole-plant net assimilation. Environ. Exp. Bot. 136, 41–49.

Park, Y., Runkle, E.S., 2018a. Far-red radiation and photosynthetic photon flux density independently regulate seedling growth but interactively regulate flowering. Environ. Exp. Bot. 155, 206–216.

Park, Y., Runkle, E.S., 2018b. Spectral effects of light-emitting diodes on plant growth, visual color quality, and photosynthetic photon efficacy: white versus blue plus red radiation. PloS One 13 (8), e0202386.

Park, Y., Runkle, E.S., 2019. Blue radiation attenuates the effects of the red to far-red ratio on extension growth but not on flowering. Environ. Exp. Bot. 168, 103871.

Pennisi, G., Blasioli, S., Cellini, A., Maia, L., Crepaldi, A., Braschi, I., Spinelli, F., Nicola, S., Fernandez, J.A., Stanghellini, C., Marcelis, L.F.M., Orsini, F., Gianquinto, G., 2019. Unraveling the role of red:blue LED lights on resource use efficiency and nutritional properties of indoor grown sweet basil. Front. Plant Sci. 10, 305.

Poorter, H., Niinemets, Ü., Ntagkas, N., Siebenkäs, A., Mäenpää, M., Matsubara, S., Pons, T., 2019. A meta-analysis of plant responses to light intensity for 70 traits ranging from molecules to whole plant performance. New Phytol. 223 (3), 1073–1105.

Prunty, R.M., 2015. Characteristics of Good Quality Transplants. Virginia Cooperative Extension, 2906–1383. http://hdl.handle.net/10919/75411. accessed 10.08.20.

Randall, W.C., Lopez, R.G., 2015. Comparison of bedding plant seedlings grown under sole-source light-emitting diodes (LEDs) and greenhouse supplemental lighting from LEDs and high-pressure sodium lamps. HortScience 50 (5), 705–713.

Ruedell, C.M., Rodrigues de Almeida, M., Fett-Neto, A.G., 2015. Concerted transcription of auxin and carbohydrate homeostasis-related genes underlies improved adventitious rooting of microcuttings derived from far-red treated *Eucalyptus globulus* Labill mother plants. Plant Physiol. Biochem. 97, 11–19.

Runkle, E.S., 2007. Maximizing supplemental lighting. Greenh. Prod. News 17 (12), 66.

Runkle, E.S., 2018. White LEDs for plant applications. Greenh. Prod. News 28 (11), 42.

Sager, J.C., Smith, W.O., Edwards, J.L., Cyr, K.L., 1988. Photosynthetic efficiency and phytochrome photoequilibria determination using spectral data. Trans. Am. Soc. Agric. Eng. 31 (6), 1882–1889.

Smith, H., 1982. Light quality, photoreception, and plant strategy. Annu. Rev. Plant Physiol. 33 (1), 481–518.

Sønsteby, A., Heide, O.M., 2009. Long-day flowering response of everbearing strawberries. Acta Hortic. 842, 777–780.

Sultan, S.E., 2000. Phenotypic plasticity for plant development, function and life history. Trends Plant Sci. 5 (12), 537–542.

Svenson, S.E., Davies Jr., F.T., 1990. Relation of photosynthesis, growth, and rooting during poinsettia propagation. Proc. Fla. State Hortic. Soc. 103, 174–176.

Tanaka, S.I., Mochizuki, N., Nagatani, A., 2002. Expression of the *AtGH3a* gene, an arabidopsis homologue of the soybean *GH3* gene, is regulated by phytochrome B. Plant Cell Physiol. 43 (3), 281–289.

Treadwell, D.D., Hochmuth, R., Landrum, L., Laughlin, W., 2016. Microgreens: A New Specialty Crop. Univ. Florida IFAS Ext. Bul. HS1164.

Tsuruyama, J., Shibuya, T., 2018. Growth and flowering responses of seed-propagated strawberry seedlings to different photoperiods in controlled environment chambers. HortTechnology 28 (4), 453–458.

Turner, E.R., Luo, Y., Buchanan, R.L., 2020. Microgreen nutrition, food safety, and shelf life: a review. J. Food Sci. 85 (4), 870–882.

van Dalfsen, P., Slingerland, L., Roelofs, P.F.M.M., 2011. Stekken onder LED-belichting: verkenning naar de mogelijkheden van het stekken van boomkwekerijgewassen onder LED in een meerlaagssysteem zonder daglicht. PPO Bloembollen en Bomen 42 (in Dutch).

van Dalfsen, P., Slingerland, L., 2012. Stekken onder LED-Verlichting 2. Praktijkonderzoek Plant en Ongeving BBF 29 (in Dutch).

Verlinden, S., 2020. Microgreens: definitions, product types, and production practices. In: Warrington, I. (Ed.), Horticultural Reviews, vol. 47. John Wiley & Sons, Hoboken, NJ, pp. 85–124.

von Arnim, A., Deng, X.W., 1996. Light control of seedling development. Annu. Rev. Plant Biol. 47 (1), 215–243.

Walters, R.G., 2005. Towards an understanding of photosynthetic acclimation. J. Exp. Bot. 56 (411), 435–447.

Welbaum, G.E., 2015. Vegetable seeds and crop establishment. In: Vegetable Production and Practices. CAB International, Wallingforth, Oxfordshire, UK, pp. 27–46.

Wollaeger, H.M., Runkle, E.S., 2015. Growth and acclimation of impatiens, salvia, petunia, and tomato seedlings to blue and red light. HortScience 50 (4), 522–529.

Xiao, Z., Lester, G.E., Park, E., Saftner, R.A., Luo, Y., Wang, Q., 2015. Evaluation and correlation of sensory attributes and chemical compositions of emerging fresh produce: Microgreens. Postharvest Biol. Technol. 110, 140–148.

Yan, Z., He, D., Niu, G., Zhai, H., 2019. Evaluation of growth and quality of hydroponic lettuce at harvest as affected by the light intensity, photoperiod, and light quality at seedling stage. Sci. Hortic. (Amst.) 248, 138–144.

Ying, Q., Kong, Y., Jones-Baumgardt, C., Zheng, Y., 2020. Responses of yield and appearance quality of four Brassicaceae microgreens to varied blue light proportion in red and blue light-emitting diodes lighting. Sci. Hortic. 259, 108857.

Zhang, M., Park, Y., Runkle, E.S., 2020. Regulation of extension growth and flowering of seedlings by blue radiation and the red to far-red ratio of sole-source lighting. Sci. Hortic. 272, 109478.

Zheng, L., Steppe, K., Van Labeke, M.C., 2020. Spectral quality of monochromatic LED affects photosynthetic acclimation to high-intensity sunlight of chrysanthemum and spathiphyllum. Physiol. Plant. 169 (1), 10–26.

CHAPTER

20

Molecular breeding of miraculin-accumulating tomatoes with suitable traits for cultivation in plant factories with artificial lightings and the optimization of cultivation methods

Kyoko Hiwasa-Tanase[1,2], *Kazuhisa Kato*[3] *and Hiroshi Ezura*[1,2]

[1]Faculty of Life and Environmental Sciences, University of Tsukuba, Tsukuba, Ibaraki, Japan;
[2]Tsukuba-Plant Innovation Research Center, University of Tsukuba, Tsukuba, Ibaraki, Japan;
[3]Graduate School of Agricultural Science, Tohoku University, Sendai, Miyagi, Japan

20.1 Introduction

A plant factory with artificial lighting (PFAL) is a space completely isolated from the outside environment, where light intensity, light source, temperature, humidity, nutrients, and atmospheric composition can be controlled at will. Because the occurrence of pests and diseases can be controlled, PFALs are able to produce visually appealing, high-quality horticultural crops, and pesticide free. In addition, productivity per unit area is increased by using multilayer rack cultivation.

For the successful initial introduction and adoption of PFALs, the major considerations are the initial investment costs and the running costs. PFALs, which control a variety of environmental conditions, require various types of equipment and isolated space. Moreover, these facilities consume large amounts of electricity for cultivation, which results in high energy costs. Therefore, it is difficult to generate sufficient revenue or profit over the operational expenses in crops that require high intensity light, a large space for growing, or a long production period. For this reason, crop species that are grown in PFALs have been limited to those that meet the following conditions: (1) the plant shape and height are compact, and cultivation is possible in a multilayer rack; (2) the time between planting and harvesting is relatively short; and (3) the plant is able to be grown at relatively low light. Because of the high number of crops per unit area and the short life cycle of these crops, high productivity per unit area per year can be expected. Currently, horticultural crops that satisfy these requirements are mainly limited to leafy vegetables such as lettuce and kale and herbs such as basil (Kozai, 2018).

On the other hand, the cost/return problem could also be solved by growing high value-added crops, which could increase revenues (Hiwasa-Tanase and Ezura, 2016). There are two types of crops that are highly profitable: (1) crops that are nutritionally and functionally improved (health benefit crops) and (2) crops that produce substances of high value such as bioreactors (e.g., edible vaccines). Since PFALs allow for artificial control of the growing environment, it is possible to set optimal growing conditions for each crop and for each purpose. This means that crops (and/or their products) can be produced systematically and consistently, with consistent quality, under conditions that allow for higher accumulation of specific nutritional and functional components and high value-added substances. Additionally, if the high value-added crops are transgenic plants, the accidental spread of transgenic plants or pollen can be prevented when cultivated in a PFAL.

In the past, crop breeding has been performed with a goal to improve crop yield and quality in open field cultivation. As a result, many currently available crops are often unsuitable for cultivation in PFALs. Higher profits can be expected if the high value-added crop has traits suitable for PFAL cultivation. Thus, for more efficient cultivation, it is necessary to breed varieties specifically designed for cultivation in PFALs.

In this chapter, based on our prior research, we discuss the development of tomatoes that are high accumulators of miraculin and are also suitable for cultivation in PFALs (Kato et al., 2010). We also discuss the optimization of cultivation conditions in PFALs using these newly developed lines (Kato et al., 2011).

20.2 Molecular breeding to optimized tomato lines in a plant factory with artificial lighting

20.2.1 High value-added transgenic tomato

Miraculin is a taste-modifying protein found in red berries of the miracle fruit (*Synsepalum dulcificum*), which is native to West Africa. Miraculin itself is not sweet, but it makes you perceive as sweet what is actually a sour taste. With its unique properties, it has potential as a natural, safe, and low-calorie alternative to artificial sweeteners for people who are diabetic or on a restricted diet. However, despite its great potential, miracle fruit production is limited because it is a tropical plant that is difficult to cultivate in Japan (Kurihara and Nirasawa, 1997). In addition, the low fruiting rate of the miracle fruit, even in tropical regions, makes it difficult to produce this fruit systematically and consistently. Therefore, a single miracle fruit, which is a small red berry approximately 2.0—2.5 cm long, is worth more than \$2 in Japan.

To expand its market potential, heterologous production of miraculin has been attempted with various hosts (Hiwasa-Tanase et al., 2012). In 2007, Sun et al. successfully developed a transgenic tomato that expresses the miraculin gene driven by a constitutive promoter. Transgenic miraculin accumulation was genetically stable from generation T_1 to T_5 in the developed tomato (Yano et al., 2010). Subsequently, various studies have used genetically modified tomatoes as a platform for miraculin production (Hirai et al., 2011a,b; Hiwasa-Tanase et al., 2011).

Tomatoes are usually cultivated in a greenhouse or field only once or twice per year in Japan, and the fruit yield and quality depend on the environment. The transgenic tomato developed for accumulation of miraculin was the medium-sized tomato cultivar "Money-maker." The cultivar is an old cultivated species with upright and indeterminate types. The transgenic tomato is usually grown in a netted greenhouse. It is common knowledge that fruit yield and quality vary from year to year and season to season under typical production systems. In contrast, the fruit yield and quality of plants cultivated in PFALs are relatively constant and less affected by environmental conditions (Morimoto et al., 1995).

In fact, when the transgenic tomatoes were grown in a PFAL, the miraculin content, on a fresh weight (FW) basis, showed little variation, while plants grown in a netted greenhouse showed a variation in miraculin content (Hirai et al., 2010a). This result shows that PFAL cultivation can be used for the stable production of the target material. However, the transgenic tomato was hosted by the medium-sized tomato cultivar "Moneymaker" and was not equipped with suitable traits for PFAL cultivation. Therefore, a single truss production system (Okano et al., 2005) was used due to the limited cultivation space available. In this system, the tomato plants were pinched, leaving a few leaves above the first truss, and only the first truss was harvested.

To increase the number of plantings, increase productivity, and reduce the amount of effort required for PFAL cultivation, it is desirable to breed new varieties suitable for cultivation in these facilities.

20.2.2 Selection of mating parent

Most cultivated tomato varieties have higher plant heights (over 1 m) when grown in limited space and relatively longer life cycles (90—110 days from seed germination to first fruit maturation) (Emmanuel and Levy, 2002; Meissner et al., 1997). In contrast, "Micro-Tom," which was originally bred for home gardening (Scott and Harbaugh, 1989), exhibits a relatively small overall size (approximately 10—20 cm height), a short life cycle (70—90 days from seed germination to first fruit maturation), and a determinate-type growth habit. In addition to its dwarfism, it is used as a model cultivar of tomato (Meissner et al., 1997; Saito et al., 2011; Sun et al., 2006) because of its ability to grow under fluorescent lights, making it ideal for indoor cultivation.

When crossbreeding for tomatoes that accumulate miraculin, it is important to know how the genetic background of tomato varieties affects miraculin accumulation. Kim et al. (2010b) created F_1 hybrids from transgenic tomato that accumulate miraculin (cv. "Moneymaker") through crosses with six tomato cultivars, including "Micro-Tom." In all F_1 hybrids, the accumulation pattern in the fruit tissue was similar to that of the original transgenic tomato, with the highest accumulation in the exocarp (Kim et al., 2010a). However, the level of miraculin accumulation in the fruit varied with the different genetic backgrounds.

Miraculin is secreted extracellularly and accumulates in the intercellular layer in transgenic tomato fruits and leaves, similar to the miracle fruit (Hirai et al., 2010b). The number of cells per FW in the exocarp is considerably higher than in other tissues due to the small size of the cells. In other words, the ratio of intercellular layers per area in the exocarp is larger than in other tissues, and this is thought to be the reason why miraculin accumulation is highest in the exocarp.

The ratio of each fruit tissue varies among tomato varieties. In fact, the percent of exocarp in "Micro-Tom," on a weight basis, was 8.16% which was at least 5% higher than in other varieties (Kim et al., 2010a). A relatively high accumulation of miraculin per FW was observed in F_1 plants crossed with "Micro-Tom."

Based on this information, "Micro-Tom" was selected as the breeding parent to breed a variety suitable for PFAL cultivation. Below, we discuss results previously reported by our group (Kato et al., 2010).

20.2.3 Selection of crossed lines

"Micro-Tom" not only has a phenotype that is suitable for PFAL cultivation, but also has a property that is potentially suitable for miraculin production. Cross breeding was performed between the dwarf tomato cultivar "Micro-Tom" and a transgenic tomato line 56B ("Moneymaker") that accumulates miraculin. 56B possesses the miraculin gene driven by the CaMV 35S promoter that induces gene expression throughout plant tissues, and the miraculin protein accumulates in all plant tissues. The homozygous line from the T_3 generation of 56B was used for crossing. After 56B was crossed with "Micro-Tom" and allowed to self-pollinate, the F_2 generation was evaluated and selected for small plant size, miraculin accumulation, and determinate traits as candidate lines for further breeding. Two selected F_2 lines were named cross #1 and cross #2 and were bred by self-pollination to the F_7 and F_6 generations, respectively. Fruit yield was used as a selection indicator when generations were promoting. The homozygous lines of the miraculin gene were fixed at the F_4 generation.

20.2.4 Evaluation of crossed lines for suitability in plant factory with artificial lighting cultivation

To evaluate the lines for suitability in PFAL cultivation, one-month-old seedlings of line 56B (T_7), cross #1 (F_7), and cross #2 (F_6) were transferred and cultivated in a two-layer cultivation system that was previously developed for tomato production (Hirai et al., 2010a, Fig. 20.1). Crossed lines were cultivated at two-fold higher planting densities than 56B

FIGURE 20.1 Plant factory with artificial lighting for tomato cultivation. Seedlings were cultivated in ready-made growing equipment (A) for one month, and they were transferred to the two-layer culture system (B) (Hirai et al., 2010a).

because the seedlings of the crossed lines were more compact than those of 56B (Table 20.1). The 56B plants were pruned, leaving three leaves above the first truss for cultivation in limited space, and the axillary buds were removed during cultivation.

20.2.4.1 Labor for plant factory with artificial lighting cultivation

Cross #1 plants showed succulent and bushy growth, and the extra-axillary buds and leaves had to be removed because they reduced light interception. In contrast, cross #2 plants grew normally even under fluorescent light conditions similar to "Micro-Tom" (Sun et al., 2006), and it was not necessary to remove axillary buds and leaves.

20.2.4.2 Phenotypes of crossed lines

The plant height and fruit size of cross #2 were smaller than those of cross #1 (Fig. 20.2). However, the fruit number in cross #2 was higher than that in cross #1. In the PFAL, some fruits with blossom-end rot were observed in line 56B. However, no fruits with blossom-end rot were found in cross #1 and #2. In addition, there were some dehiscent fruits in line 56B and cross #1, but no dehiscent fruits were observed in cross #2.

20.2.4.3 Fruit yields

After transplanting to the PFAL, the highest fruit yield per plant was observed in cross #1, among the three tested lines, at 76 days. The fruit yield per plant of 56B was higher than that of cross #2 at 76 days but was approximately the same at 90 days. Cross #1 and 56B had almost no yield after day 76 due to removal of axillary buds and pruning, while cross #2, which did not require removal of axillary buds, continued to flower and bear fruit.

TABLE 20.1 Comparison of fruit yield and recombinant miraculin production of cross-bred tomatoes under different cultivation conditions in a plant factory with artificial lighting.

Transgenic tomato	PPFD (μmol m^{-2} s^{-1})[a]	CO_2 concentration (ppm)	Planting density (plants m^{-2} layer^{-1})	Light period (h day^{-1})	Miraculin concentration (μg gFW^{-1})	Number of layers (at same height)[b]	Fruit yield (kgFW m^{-2} year^{-1})[c]	Weight ratio of pericarp (%)[d]	Miraculin production (g m^{-2} year^{-1})[e]	References
56B	450	600	13.3	16	140	2	26.2	60.4	2.2	Kato et al. (2010)
Cross #1	450	600	26.7	16	367	2	73.6	61.5	16.6	
Cross #2	450	600	26.7	16	343	2	45.9	62.0	9.8	
Cross #2	100	1000	44.4	12	472	3	25.5	66.4	8.0	Kato et al. (2011)
	200	1000	44.4	12	362	3	51.0	68.8	12.7	
	300	1000	44.4	12	272	3	83.1	67.2	15.2	
	400	1000	44.4	12	211	3	91.5	65.4	12.6	

[a]PPFD: photosynthetic photon flux density.
[b]The number of layers available for installation based on the height of the closed cultivation system.
[c]The annual fruit yield when grown on two or three layers in the same space.
[d]The ratio of exocarp and mesocarp by weight of the whole fruit.
[e]Calculated from the data of miraculin concentration, fruit yield, and weight ratio of pericarp.

FIGURE 20.2 Seedlings at 34 days after germination (A) and average fruit size (B) of line 56B and the crossed lines. *Revised from Kato, K., Yoshida, R., Kikuzaki, A., Hirai, T., Kuroda, H., Hiwasa-Tanase, K., et al., 2010. Molecular breeding of tomato lines for mass production of miraculin in a pant factory. J. Agric. Food Chem. 58 (17), 9505–9510.*

The maximum fruit yields per area per year of 56B, cross #1, and cross #2 were 26.2, 76.3, and 45.9 kg FW m^{-2} year^{-1}, respectively (Table 20.1). In PFAL cultivation, the crossed lines showed a significant increase in fruit productivity compared to that of 56B because the crossed lines could be grown at two-fold the density of line 56B.

20.2.4.4 Miraculin accumulation and production

The miraculin concentrations in the pericarp of crossed lines were approximately 2.5-fold higher than that of 56B (Table 20.1). Fruit weight appeared to be significantly correlated with cell size in the pericarp (Cheniclet et al., 2005). In fact, the fruit size of the crossed lines was smaller than that of 56B, and the cell size was smaller than that of 56B. It is known that miraculin accumulates in the intercellular layer and at high concentrations especially in the exocarp which is a collection of smaller cells. In general, the crossed lines with smaller fruit had a higher percentage of intercellular layers per FW. This may have led to an increase in miraculin accumulation per unit weight.

20.2.4.5 *Total evaluation*

Cross #1 had a self-pruning growth habit, a suitable plant size for cultivation in PFALs and showed excellent fruit and miraculin production. However, the plants showed bushy growth in the PFAL and required the removal of extra-axillary buds and leaves. There was no blossom-end rot, but dehiscent fruits were observed.

Although cross #2 was inferior to cross #1 in terms of fruit and miraculin production, the plant habit was as stable under fluorescent light as the parent "Micro-Tom" and did not require extra labor. There were also no defective fruits, such as blossom-end rot or dehiscent fruits. Because the plant height of cross #2 was shorter than that of cross #1. It may be possible to grow this cross in a three-layer system in the two-layer cultivation system used in this study. If a three-layer cultivation system can be used in the same space, the fruit yield and miraculin productivity would be 1.5 times higher than with a two-layer cultivation system. In this way, it is possible to increase productivity to a level almost the same as that obtained with cross #1.

Overall, it can be said that cross #2 has traits that are more suitable for cultivation in a PFAL, but cross #1 was superior if yield is more desirable than labor reduction. For the use of these varieties in practice, one will need to make decisions regarding which traits are most important for the given purpose.

20.3 Optimization for production of recombinant miraculin in a plant factory with artificial lighting

20.3.1 Introduction

Plant growth is affected by various environmental factors, such as light, temperature, concentration, and composition of the culture medium, and the type and condition of the medium. Changes in the environment also affect the production of recombinant proteins in transgenic plants (Colgan et al., 2010; Jamal et al., 2009). Since all these factors can be controlled in a PFAL, it is possible to set ideal growing conditions for the production of recombinant proteins. On the other hand, even for plants that have been cultivated in a PFAL, suitable growing and harvesting conditions are likely to differ depending on the target protein. Therefore, it is necessary to first examine the conditions for production of the target protein. The optimization of light conditions is particularly important. Among other environmental factors, light is an essential element for photosynthesis, and photosynthesis is a crucial reaction for the growth and development of plants. Naturally, differences in light conditions affect the productivity of recombinant proteins. For example, in chloroplast transgenic tobacco, light conditions affected the accumulation level of recombinant human cardiotrophin-1 (Farran et al., 2008), human serum albumin (Fernández-San Millán et al., 2003), and immunogenic *Bacillus anthracis* protective antigen (Watson et al., 2004). Arlen et al. (2007) reported that different varieties of transformed tobacco, which accumulate the same protein with the same construct, have different responses to light, resulting in differences in the production of recombinant proteins.

Moreover, although artificial light can be adjusted for quality, quantity, and intensity through the use of a wide variety of light sources, the associated electricity costs account for a large part of the operating costs in PFAL cultivation. For commercial production of recombinant proteins in a large-scale plant factory, reducing the energy cost of light is important. Optimization of light conditions will reduce excess energy and will also lead to lower production costs.

In this chapter, we introduce the effects of various photosynthetic photon flux densities (PPFDs) on plant growth, fruit yield, and miraculin accumulation in PFAL cultivation using cross #2, a miraculin-accumulating tomato bred for PFAL cultivation. Below we discuss results previously reported by our group (Kato et al., 2011).

20.3.2 Effect of various light conditions on the phenotype of cross #2 plants

Seedlings of cross #2 were grown for 34 days after germination at a PPFD of $400\ \mu mol\ m^{-2}\ s^{-1}$ (PPFD 400) and then transferred to PPFDs of 100, 200, 300, and $400\ \mu mol\ m^{-2}\ s^{-1}$. Plant height was significantly higher at PPFD 400 than at PPFD 100 and 200 at four weeks after changing light conditions, but there was no significant difference from PPFD 300. Because of the shade avoidance response, plants expanded more laterally to access more light at lower PPFDs, resulting in a maximum plant diameter at PPFD 100 (Fig. 20.3). Leaves were darker at higher PPFDs, and the photosynthesis rates were positively correlated with PPFD. Fruit numbers were also positively correlated with PPFD until six weeks, but the difference between PPFD 300 and 400 had almost disappeared at 14 weeks. The time to harvest of the first fruit was shorter at higher PPFDs (Table 20.2).

20.3.3 Fruit yields of cross #2 plants grown at various photosynthetic photon flux densities

Fruit size was significantly small at PPFD 100 but almost the same at PPFD 200, 300, and 400 (Table 20.2). Fruit yield per plant was higher at higher PPFDs, but in the last week of harvest, yields were similar at PPFDs 300 and 400. The fruit yields at PPFDs 100, 200, 300, and

PPFD (μmol/m²/s)   100   200   300   400

FIGURE 20.3   Phenotypes of cross #2 grown at various photosynthetic photon flux densities for two weeks after changing light conditions in the plant factory with artificial lighting (Kato et al., 2011). Scale bar indicates 5 cm.

TABLE 20.2   Fruit characteristics of cross #2 grown at various photosynthetic photon flux densities in a plant factory with artificial lighting[c].

PPFD (μmol m^{-2}s^{-1})[a]	Harvesting time of first fruit (days after change of PPFD)	Number of fruits per plant[b]	Fruit weight (g FW)
100	61.3	12.9	4.0
200	51.3	18.4	5.3
300	49.8	28.8	5.5
400	45.4	30.4	5.6

[a]PPFD: photosynthetic photon flux density.
[b]Data from 14 weeks after changing light conditions.
[c]Described data are from Kato et al. (2011).

400 at the end of the cultivation period were 51, 98, 158, and 171 g FW per plant, respectively. The annual yield per unit area tended to be similar to the yield per plant, and the maximum fruit yields per layer at PPFDs 100, 200, 300, and 400 were 8.5, 17.0, 27.7, and 30.5 kg FW m^{-2} year^{-1}, respectively.

When we investigated the suitability of the crossed lines for PFAL cultivation (Section 20.2.4), the yield of cross #2 was 45.9 kg FW m^{-2} year^{-1} in the space of the two-layer cultivation system at a planting density of 26.7 plants m^{-2} under PPFD 450 (Table 20.1). The space of the cultivation system was approximately equal to three layers in this study. In addition, the planting density was 44.4 plants m^{-2} in this study. As a result, the maximum yield per three layers of planting space at PPFD 400 was 91.5 kg FW m^{-2} year^{-1}, which was approximately twice as high as the previous result in the same space (Table 20.1). On the other hand, the fruit yield per plant was lower than previous results because of the high planting density. Efficient use of planting space is important to increase tomato fruit yield in plant factories. Light interception is also affected by planting density, so further yield increases may be expected by optimization of planting density.

20.3.4 Effect of various light conditions and harvesting times on miraculin concentration

Miraculin levels were highest in tomatoes grown at low PPFDs, and the concentrations of miraculin in the pericarp at PPFDs 100, 200, 300, and 400 were 472, 362, 272, and 211 μg g^{-1} FW, respectively (Table 20.1). The concentration of miraculin in each tissue was higher in the exocarp than in the mesocarp, which was similar to the results obtained in a study using the parental line 56B. Miraculin is secreted out of the cell and accumulates in the intercellular layer and is thought to move toward the exocarp during fruit growth (Hirai et al., 2010b). Under different light conditions, miraculin concentrations in the exocarp were almost similar under all conditions, while tomatoes grown at lower PPFDs showed higher levels of miraculin in the mesocarp (Fig. 20.4). The period of fruit development and ripening was the longest at PPFD 100 and the shortest at PPFD 400 (Table 20.2). Fruit growth was dependent on PPFD, and a lower PPFD increased the fruit growth period. In this tomato,

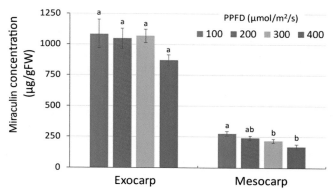

FIGURE 20.4 Miraculin concentration in the exocarp and mesocarp of first fruits of cross #2 grown at various photosynthetic photon flux densities (PPFDs) in a plant factory with artificial lighting (Kato et al., 2011). Letters indicate significant differences from the other PPFDs (Tukey's test, $P < 0.05$). Vertical bars indicate the standard error (n = 8).

the miraculin gene is regulated by a constitutive CaMV 35S promoter, so miraculin is constantly accumulating throughout the fruit growth period. Therefore, it is likely that the fruit of the PPFD 100 group, which required a long fruit growth period, accumulated the highest amount of miraculin. The concentration of miraculin in the exocarp was almost the same regardless of the light conditions, suggesting that the limit of miraculin accumulation in the exocarp was approximately 1000 µg per g FW and that miraculin occurring above this limit was likely to accumulate in the mesocarp (Fig. 20.4). At the time of harvest, the miraculin accumulation in the exocarp was nearly the same under all light conditions, suggesting that the difference in miraculin concentration between light conditions was reflected in the accumulation in the mesocarp.

Under PPFD 400 conditions, extending the harvest time by six weeks from the orange stage increased the concentration of miraculin in the exocarp and mesocarp. In addition, the fruits of cross #2 maintained their shape even after six weeks from the orange stage. Recombinant miraculin in the transgenic tomato is probably stable in the fruit since the miraculin concentration does not decrease during fruit ripening or during the overripe period. The accumulation of the antigen fusion protein F1−V was reduced in the recombinant tomato fruit during fruit development (Matsuda et al., 2010). Several papers reported that the degradation of foreign proteins by proteolysis in plant tissues may lead to a decrease in the yield of foreign proteins (Benchabane et al., 2008; Doran, 2006). It is important to consider the subcellular localization of recombinant proteins for the accumulation of stable recombinant proteins. In general, protein may be less prone to degradation in locations such chloroplasts, the endoplasmic reticulum, and apoplasts than in the cytoplasm (Benchabane et al., 2008; Karg and Kallio, 2009; Meyers et al., 2008; Streatfield et al., 2003). The highest expression levels of LALF32-51-E7, which is a candidate therapeutic vaccine for high-risk human papillomaviruses, were obtained by using chloroplast targeting, which increased its accumulation by 27-fold compared to cytoplasmic localization (Yanez et al., 2017). In the case of miraculin, it is thought that it is secreted extracellularly by its own signal peptide and stabilized. The stability of the recombinant protein in such tissues is an important factor for mass production.

20.3.5 Evaluation of light conditioning for cross #2 cultivation

Fruit productivity was highest at PPFD 400, and the miraculin concentration in fruit was highest at PPFD 100 (Table 20.1). Based on these data, when calculating annual miraculin productivity per unit area, PPFD 300 had the highest miraculin productivity among the tested light conditions. However, when calculating the annual electrical energy cost under each light condition, the highest miraculin productivity per energy was found for PPFD 100, which had the lowest energy cost and the highest miraculin concentration in the fruit. This means that miraculin can be produced at the lowest cost under PPFD 100. Cultivation conditions need to be determined based on the balance between selling price and supply and demand and thus whether productivity or production costs need to be emphasized.

As shown in Section 20.2.4, cross #2 did not develop dehiscent fruit or blossom-end rot, thus extending the harvest time by at least six weeks after the orange fruit stage. This result suggests that it is possible to extend the harvest time to accumulate more miraculin under all light conditions. The result also means that the harvesting does not need to be done regularly at short intervals and can be done in several batches, thus reducing labor costs.

20.4 Conclusions

We bred miraculin-accumulating tomatoes suitable for cultivation in a PFAL by cross-breeding with Micro-Tom, a dwarf tomato. In this study, we demonstrated that crops that were originally unsuitable for PFAL cultivation can be made into varieties with characteristics that are optimized for cultivation in these facilities. The high running costs and initial investment costs of PFAL cultivation have been bottlenecks in the introduction of equipment. However, by breeding cultivars suitable for cultivation in PFALs and expressing the desired recombinant protein or adding high functionality to them, it is possible to create high value-added crops that can be profitably grown in PFALs.

We also examined the effect of light conditions on miraculin accumulation under PFAL cultivation in the cross-bred line. As a result of the investigation, the optimal light conditions for miraculin productivity and production costs were determined. In addition, the timing of fruit harvesting can be optimized. On the other hand, the planting density was approximately 1.7 times higher than that reported previously (Kato et al., 2010; Section 20.2.4), resulting in an increase in productivity per area but a decrease in productivity per plant. Since seedling production has a cost, it is better to reduce the number of seedlings if the same productivity can be achieved. Therefore, it will be important to optimize planting density in the future. There were also differences in CO_2 concentration and light period between the Sections 20.2.4 and 20.3 experiments (Table 20.1). It may also be possible to optimize the CO_2 concentration and photoperiod.

In this chapter, we have introduced the breeding methods and optimized cultivation conditions of cross-bred lines of miraculin-accumulated tomatoes, but even if the lines are crossed in the same way, optimal light conditions, harvest time, temperature, and day length vary depending on the promoter, the type of recombinant protein to be expressed and the localization. Even for varieties that claim to be highly functional and accumulate high levels of certain components, the optimal growing conditions to achieve greater performance vary. Based on the premise that optimal growth conditions do not always coincide with generally accepted growth conditions, it is important to establish optimal growth conditions to maximize profits while taking full advantage of the benefits of PFAL cultivation.

References

Arlen, P.A., Falconer, R., Cherukumilli, S., Cole, A., Cole, A.M., Oishi, K.K., et al., 2007. Field production and functional evaluation of chloroplast-derived interferon-alpha2b. Plant Biotechnol. J. 5, 511–525.

Benchabane, M., Goulet, C., Rivard, D., Faye, L., Gomord, V., Michaud, D., 2008. Preventing unintended proteolysis in plant protein biofactories. Plant Biotechnol. J. 6, 633–648.

Cheniclet, C., Rong, W.Y., Causse, M., Frangne, N., Bolling, L., Carde, J.-P., et al., 2005. Cell expansion and endoreduplication show a large genetic variability in pericarp and contribute strongly to tomato fruit growth. Plant Physiol. 139 (4), 1984–1994.

Colgan, R., Atkinson, C.J., Paul, M., Hassan, S., Drake, P.M.W., Sexton, A.L., et al., 2010. Optimisation of contained Nicotiana tabacum cultivation for the production of recombinant protein pharmaceuticals. Transgenic Res. 19, 241–256.

Doran, P.M., 2006. Foreign protein degradation and instability in plants and plant tissue cultures. Trends Biotechnol. 24, 426–432.

Emmanuel, E., Levy, A.A., 2002. Tomato mutants as tools for functional genomics. Curr. Opin. Plant Biol. 5, 112–117.

Farran, I., Rio-Manterol, F., Iniguez, M., Garate, S., Prieto, J., Mingo-Castel, A.M., 2008. High-density seedling expression system for the production of bioactive human cardiotrophin-1, a potential therapeutic cytokine, in transgenic tobacco chloroplasts. Plant Biotechnol. J. 6, 516–527.

Fernández-San Millán, A., Mingo-Castel, A., Miller, M., Daniell, H., 2003. A chloroplast transgenic approach to hyper-express and purify human serum albumin, a protein highly susceptible to proteolytic degradation. Plant Biotechnol. J. 1, 71–79.

Hirai, T., Fukukawa, G., Kakuta, H., Fukuda, N., Ezura, H., 2010a. Production of recombinant miraculin using transgenic tomato in a closed-cultivation system. J. Agric. Food Chem. 58, 6096–6101.

Hirai, T., Sato, M., Toyooka, K., Sun, H.J., Yano, M., Ezura, H., 2010b. Miraculin, a taste-modifying protein is secreted into intercellular spaces in plant cells. J. Plant Physiol. 167, 209–215.

Hirai, T., Duhita, D., Hiwasa-Tanase, K., Kato, K., Kato, K., Ezura, H., 2011a. The HSP terminator of Arabidopsis thaliana induces extremely high-level accumulation of miraculin protein in transgenic tomato. J. Agric. Food Chem. 59, 9942–9949.

Hirai, T., Kim, Y., Kato, K., Hiwasa-Tanase, K., Ezura, H., 2011b. Uniform accumulation of recombinant miraculin protein in transgenic tomato fruit using a fruit-ripening-specific E8 promoter. Transgenic Res. 20, 1285–1292.

Hiwasa-Tanase, K., Ezura, H., 2016. Molecular breeding to create optimized crops: from genetic manipulation to potential applications in plant factories. Front. Plant Sci. 7, 539.

Hiwasa-Tanase, K., Nyarubona, M., Hirai, T., Kato, K., Ichikawa, T., Ezura, H., 2011. High-level accumulation of recombinant miraculin protein in transgenic tomatoes expressing a synthetic miraculin gene with optimized codon usage terminated by the native miraculin terminator. Plant Cell Rep. 30, 113–124.

Hiwasa-Tanase, K., Hirai, T., Kato, K., Duhita, N., Ezura, H., 2012. From miracle fruit to transgenic tomato: mass production of the taste-modifying protein miraculin in transgenic plants. Plant Cell Rep. 31, 513–525.

Jamal, A., Ko, K., Kim, H.S., Choo, Y.K., Joung, H., Ko, K., 2009. Role of genetic factors and environmental conditions in recombinant protein production for molecular farming. Biotechnol. Adv. 27, 914–923.

Karg, S.R., Kallio, P.T., 2009. The production of biopharmaceuticals in plant systems. Biotechnol. Adv. 27 (6), 879–894.

Kato, K., Yoshida, R., Kikuzaki, A., Hirai, T., Kuroda, H., Hiwasa-Tanase, K., et al., 2010. Molecular breeding of tomato lines for mass production of miraculin in a pant factory. J. Agric. Food Chem. 58 (17), 9505–9510.

Kato, K., Maruyama, S., Hirai, T., Hiwasa-Tanase, K., Mizoguchi, T., Goto, E., et al., 2011. A trial of production of the plant-derived high-value protein in a plant factory. Plant Signal. Behav. 6 (8), 1172–1179.

Kim, Y.W., Kato, K., Hirai, T., Hiwasa-Tanase, K., Ezura, H., 2010a. Spatial and developmental profiling of miraculin accumulation in transgenic tomato fruits expressing the miraculin gene constitutively. J. Agric. Food Chem. 58, 282—286.

Kim, Y.W., Hirai, T., Kato, K., Hiwasa-Tanase, K., Ezura, H., 2010b. Gene dosage and genetic background affect miraculin accumulation in transgenic tomato fruits. Plant Biotechnol. 27, 333—338.

Kozai, T., 2018. Current status of plant factories with artificial lighting (PFALs) and smart PFALs. In: Kozai, T. (Ed.), Smart Plant Factory. The Next Generation Indoor Vertical Farms. Springer Nature Singapore Pte Ltd., pp. 3—14

Kurihara, Y., Nirasawa, S., 1997. Structures and activities of sweetness-inducing substances (miraculin, curculin, strogin) and the heat-stable sweet protein, mabinlin. FFI. J. Jpn. 174, 67—74.

Matsuda, R., Kubota, C., Alvarez, M.L., Cardineau, G.A., 2010. Determining the optimal timing of fruit harvest in transgenic tomato expressing F1-V, a candidate subunit vaccine against plague. Hortscience 45, 347—351.

Meissner, R., Jacobson, Y., Melamed, S., Levyatuv, S., Shalev, G., Ashri, A., 1997. A new model system for tomato genetics. Plant J. 12, 1465—1472.

Meyers, A., Chakauya, E., Shephard, E., Tanzer, F.L., Maclean, J., Lynch, A., 2008. Expression of HIV-1 antigens in plants as potential subunit vaccines. BMC Biotechnol. 8, 53.

Morimoto, T., Torii, T., Hashimoto, Y., 1995. Optimal control of physiological processes of plants in a green plant factory. Contr. Eng. Pract. 3, 505—511.

Okano, K., Nakano, Y., Watanabe, S., 2005. Single-truss tomato system—a labor-saving management system for tomato production JARQ—Jpn. Agric. Resour. Q. 35, 177—184.

Saito, T., Ariizumi, T., Okabe, Y., Asamizu, E., Hiwasa-Tanase, K., Fukuda, N., et al., 2011. TOMATOMA: a novel tomato mutant database distributing micro-tom mutant collections. Plant Cell Physiol. 52 (2), 283—296.

Scott, J.W., Harbaugh, B.K., 1989. Micro-Tom—a miniature dwarf tomato. Fla. Agric. Exp. Stn. Circ. 370, 1—6.

Streatfield, S.J., Lane, J.R., Brooks, C.A., Barker, D.K., Poage, M.L., Mayor, J.M., et al., 2003. Corn as a production system for human and animal vaccines. Vaccine 21, 812—815.

Sun, H.J., Uchii, S., Watanabe, S., Ezura, H., 2006. A highly efficient transformation protocol for Micro-Tom, a model cultivar of tomato functional genomics. Plant Cell Physiol. 47, 426—431.

Sun, H.J., Kataoka, H., Yano, M., Ezura, H., 2007. Genetically stable expression of functional miraculin, a new type of alternative sweetener, in transgenic tomato plants. Plant Biotechnol. J. 5, 768—777.

Watson, J., Koya, V., Leppla, S.H., Daniell, H., 2004. Expression of *Bacillus anthracis* protective antigen in transgenic chloroplasts of tobacco, a non-food/feed crop. Vaccine 22, 4374—4384.

Yanez, R.J.R., Lamprecht, R., Granadillo, M., Weber, B., Torrens, I., Rybicki, E.P., et al., 2017. Expression optimization of a cell membrane-penetrating human papillomavirus type 16 therapeutic vaccine candidate in *Nicotiana benthamiana*. PloS One 12 (8), e0183177.

Yano, M., Hirai, T., Kato, K., Hiwasa-Tanase, K., Fukuda, N., Ezura, H., 2010. Tomato is a suitable material for producing recombinant miraculin protein in genetically stable manner. Plant Sci. 178, 469—473.

C H A P T E R

21

Environmental control of PFALs

Ying Zhang[1] and Murat Kacira[2]

[1]Department of Agricultural and Biological Engineering, University of Florida, Gainesville, FL, United States; [2]Department of Biosystems Engineering, The University of Arizona, Tucson, AZ, United States

21.1 Introduction

In Plant Factories with Artificial Lighting (PFALs), Heating, Ventilation, and Air Conditioning (HVAC) systems provide essential services and processes to maintain desired optimal environmental conditions for plant production. The HVAC systems most often control air temperature and relative humidity, but the design of the airflow distribution system also affects energy and mass exchanges at the canopy of plants and thus influences the photosynthetic rate and transpiration rate of plants. With intensive energy inputs from artificial lighting, pumps, fans, and other equipment, significant cooling loads are needed within PFALs. Cooling is not uncommon even in cold regions. The significant cooling demands make PFALs comparable to data centers. However, the humidity built up inside caused by plant transpiration creates an even challenging environmental control task. With the rapid development of indoor farming technology, PFALs have been applied to production of different types of plants. This demands for control of the indoor environment to meet various design criteria and conditions for quality plant production and desirably with strategies leading to improved resource use efficiency (Kozai, 2013). For example, cooling and dehumidification with adequate air movement is commonly required for leafy greens production in PFALs; however, for indoor plant propagation, cooling and humidification with small amount of air movement is desired. The profitability of indoor farming will require environmental control systems and strategies focusing on minimizing resource inputs while maximizing plant production.

21.2 Airflow and distribution

In PFALs, airflow distribution and management have a great association with the efficiency of an HVAC system. The purpose of the air distribution system is to create a proper combination of temperature, humidity, and air current speed in the occupied zone. The optimal temperature, at which plant development is most rapid, can be around 21°C for cool-season crops, or as high as 32°C for warm-season crops. Air movement alters the mean thickness of the boundary layer and thus affects the rates of diffusion of heat, water, and carbon dioxide (CO_2) at leaf surfaces. With increasing air current speeds, leaf boundary layer resistance decreases, and the heat transfer coefficient of leaves increases (Salisbury, 1979; Drake et al., 2008). Insufficient air circulation decreases the rate of photosynthesis and transpiration by suppressing the gas and water diffusion in the leaf boundary layer and thus limits plant growth and development. Air current speed and airflow direction also affect transpiration and photosynthesis (Kitaya et al., 2000, 2004). Desirable air current speed range for optimum plant growth is between 0.3 and 1 m s^{-1} for general use (Kitaya et al., 2004). However, higher air current speeds may be set to facilitate temperature control. Indoor farming has been developed into different forms, from growth chamber experimental stations to shipping container and warehouse-based commercial farms. The greater the space and volume with more complex structures, the more challenging it becomes to control the environment at the crop canopy level. The designs of the air distribution system in PFALs have been changed from originally utilizing traditional vent configurations with vents on a ceiling or walls (with/without ceiling fans) to a localized climate control with perforated tubes, horizontal airflow fans, or closed shelves with airflow slots. To effectively deliver conditioned air to crop canopies, it is important to understand the internal air flow in an indoor plant factory in detail. Computational fluid dynamic (CFD) is a powerful tool to analyze fluid flows using numerical solution methods. The CFD modeling techniques have been used to evaluate different airflow distribution designs to provide in-depth engineering analysis of complex fluid flows in PFALs. Studies are conducted in growth chambers, shipping container farms, and large-scale vertical farms. Lim and Kim (2014) performed a numerical simulation with CFD to predict airflow pattern in a plant factory with different inlet and outlet locations. It was shown that the airflow pattern in the plant factory with multiple layer production shelves was greatly affected by the locations of the inlet and the outlet. Stagnant zones were observed in the shelves where the airflow was not able to reach. Zhang et al. (2016) designed a perforated air tube that can be placed between light features at each production shelf to provide vertical airflow to the area of plant canopies for the improvement of air movement and tipburn prevention (Fig. 21.1). The implementation of perforated air tubes showed a significant improvement on air movement with a higher average air speed at the canopy zone compared to a system with natural convection. This study showed the possibility of utilizing perforated air tubes to create downwards airflow to the top of the crop canopy and provide an acceptable range of air speed and improved environmental uniformity as long as there is a sufficient head space above the crop canopy to install such air distribution system. The authors then further evaluated a warehouse-based vertical farm to study the effects of vent configurations on airflow distribution at each level of the production shelf in terms of air speed, air temperature, and their uniformity (Zhang and Kacira, 2018). The study compared five air distribution system configurations (Fig. 21.1) and determined that perforated air tubes installed at the side of each shelf sending conditioned air horizontally at just above the surface of the crop canopy can be an effective way to provide desired air current speeds and climate uniformity. The revised design of the perforated air tube providing airflow horizontally overcomes the restriction of the previous design that can be only fitted between light features and therefore enhances the flexibility of the implementation of

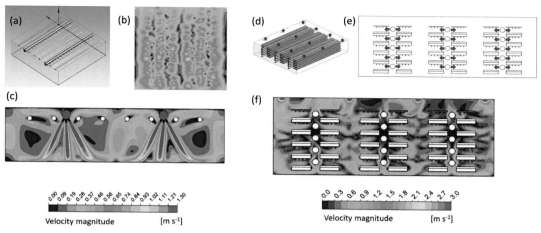

FIGURE 21.1 Two examples of air distribution systems with perforated air tubes. Study 1 (Zhang et al., 2016): (A) an overview of a production shelf with two perforated air tubes installed parallel to light bars; (B) a velocity contour at the height of the surface of plant canopy; and (C) a vertical plane inside the production shelf with a velocity contour. Study 2 (Zhang and Kacira, 2018): (D) an overview of a PFAL with outlet vents placing on the ceiling; (E) a vertical plane of the PFAL showing the locations of perforated air tubes and the directions of airflow coming out of the tubes; and (F) a velocity contour at a vertical plane inside the PFAL.

perforated air tubes in PFALs. A similar design was studied by Fang et al. (2020) considering the airflow resistance of crop canopy and the effects of different types of perforated tubes (pores diameters and numbers) on the airflow pattern in a single shelf production system. With the proper combination of holes, perforated air tubes could deliver desired air speeds on the surface of the crop canopy. Ahmed et al. (2020) evaluated a multifan system for single culture beds, installed on both the front and back sides of culture beds to generate airflow from two opposite horizontal directions, to improve the airflow in a plant factory. Although an engineered design of an air distribution system can greatly improve climate uniformity and air movement at the crop canopy, there is no standard design that can fit various production systems. Further development of airflow distribution systems in PFALs should be focused on smart control applications utilizing image processing techniques, artificial intelligence (AI), and sensors to provide dynamic biofeedback control with variable air speeds.

21.3 Humidity control

Humidity control is important as it can directly affect transpiration, nutrient uptake by the plants, and energy required to manage the humidity control in the growing space. In PFALs, humidity levels are controlled by HAVC equipment based on relative humidity values. For leafy greens, the relative humidity is usually set between 50% and 70%. Lower humidity helps to limit pathogen issues, whereas high humidity favors fungal diseases, such as powdery mildew and botrytis. High humidity with lack of sufficient air movement within the leaf boundary layer can also lead to physiological disorders, such as tipburn. Decreasing vapor pressure deficit (VPD), in other words increasing relative humidity, during the night period was reported to be effective to reduce tipburn incidence of lettuce plants grown under artificial lighting (Collier and Tibbitts, 1984). Therefore, VPD management can be applicable in indoor plant factories as an alternative means to vertical or horizontal airflow; however, there is still research needed to establish standard practices and guidelines for effective use of this approach. On the contrary, for indoor propagation facilities, it is common to maintain the relative humidity at a minimum of 85%. Therefore, depending on the types of plants grown and the objective of the production system, humidity control strategies can vary in PFALs. In the HVAC, humidification and dehumidification systems are used to transfer moisture to or from the air. In a PFAL, if humidity is not under control, only a small amount of moisture generated in the room due to evapotranspiration can be removed. Moisture in the air can be reduced at the cooling coil in the HVAC system if the surface temperature of the cooling coil is below the dew point temperature (the temperature that the air is unable to hold moisture) of the passing air or through air exchanges between indoor and outdoor when outdoor air is at a lower humidity level. If the relative humidity is not under control, the relative humidity value could be significantly higher, e.g., greater than 90%. Therefore, humidity management systems are critical.

Dehumidification control can be achieved by condensing and desiccant dehumidifiers. Condensing dehumidifiers work by decreasing the temperature of the incoming air below the dew point temperature. Reheating after dehumidification is a common practice to raise the temperature of the overcooled air before the air enters the room. This process is typically energy intensive. Harbick and Albright (2016) conducted an energy consumption analysis for greenhouses and plant factories. Large amounts of energy were consumed in PFALs for reheating after dehumidification, which was comparable to the total energy required for lighting. Similar results were reported by Graamans et al. (2020) with an average energy demand of 50% for lighting, 2% for heating, 34% for dehumidification, and 14% for sensible cooling for the PFALs evaluated. Fig. 21.2 shows the effect of different control levels of relative humidity on the total energy consumption of PFALs (Zhang and Kacira, 2020a). With the

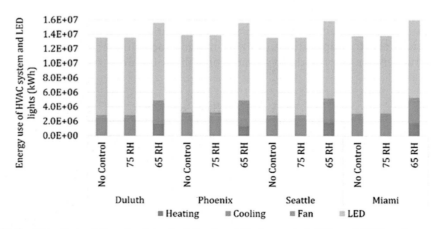

FIGURE 21.2   The effects of the set points of relative humidity (no control, 75%, and 65%) on the total energy use of HVAC system at four locations (with different climates) in the United States.

humidity setpoints from no control to 75% and then to 65%, the total energy consumed by the HVAC system in each PFAL increased. Heating demands were increased for the dehumidification process. Desiccant dehumidifiers use the dehumidification methods that utilize a chemical (desiccant) to absorb moisture in the air and reduce humidity. They are widely studied due to their ability to dehumidify with no need to first cool down the air and the ability to reuse the heat from the regeneration process. Although desiccant systems have been proven as an effective dehumidification method in various greenhouse applications, the energy performance of a dehumidification method depends on technical purposes, system configuration, and operating conditions. Desiccant-based dehumidification system can work well or be preferred when compressor- or refrigerant-based systems are not able to achieve the desired level of dehumidification. The performance of desiccant dehumidifiers in plant factories has not been well studied. Due to the characteristic of high latent heat content in PFALs and the fact that desiccant dehumidification is most effective when the latent load is high (Ge and Wang, 2020), desiccant systems may be a good alternative to reduce electrical requirement compared to the traditional refrigerant systems. The challenge with this system might be the large size and significant weight of the equipment for installation.

When humidification in a PFAL is needed, for instance, in transplant propagation systems or chambers, this can be achieved by a fogging system that atomizes treated water into tiny fog droplets that provide an accurate humidity control without overwetting. Fog nozzles or other fogging devices are commonly spaced in a room to humidify the entire room or send fog directly to the closed production shelves for localized control of the humidity. One of the challenges of humidification is to maintain a consistent temperature and humidity level. The tiny water droplets (fog) can quickly evaporate and become water vapor (gas). At the same time, the heat in the environment is absorbed during the evaporation process and the air is cooled. Using heating mats under the tray/bench or mixing the air in the room are common practices to eliminate a sudden temperature drop and maintain the desired temperature levels and uniformity. Another challenge is condensation. Condensation can form on the ground, chilled production shelves, even on the plants. This can create a breeding ground of diseases, reduce water use efficiency, and increase energy needed to eliminate such condensation. Condensation on the covering of a closed shelf also can affect the light radiation received (a small reduction) and change the thermal properties of the covering and thus change the energy balance of the environment at the plant canopy. Advanced control strategies and algorithms should be developed for improved humidity control considering the plant requirements and the thermodynamics of the indoor environment.

21.4 CO$_2$ control

CO$_2$ is an essential requirement for photosynthesis, and ambient CO$_2$ levels in the air are around 400 μmol mol^{-1}. Most plants grown indoors require a minimum CO$_2$ concentration of 330 μmol mol^{-1}. Furthermore, CO$_2$ enrichment is commonly applied in controlled environment agriculture to boost yield, quality, and growth rates of crops. CO$_2$ control methods are often studied with other environmental parameters to optimize plant growth and resource use efficiencies, such as air circulation (Thongbai et al., 2010), temperature (Ge et al., 2012), light levels (Caplan, 2018), and salinity in the soil (Brito et al., 2020).

Three main ways to supplement CO_2 are generating CO_2 by the combustion of fossil fuels with air heaters, releasing pure bottled CO_2, and using a decomposition process. In PFALs, CO_2 enrichment is necessary for plant production as the CO_2 levels in the photoperiod can be depleted rapidly and become the limiting variable to plant growth. Supplying CO_2 from the combustion of fuel adds heat and moisture. This method is not used and thus discouraged for use in the PFALs. As heat and moisture are undesired products for PFALs, the most common CO_2 source used is tank-type compressed liquid CO_2. The CO_2 levels are usually controlled between 800 and 1200 μmol mol^{-1} with intermittent injection, which is based on the level as determined by a CO_2 monitor/controller. When supplementing CO_2 to an enclosed environment, depending on the location of injection point, there might be differences in levels of CO_2 at high and low areas as CO_2 is heavier than the air. In the PFALs, pure CO_2 is mostly supplied at the outlet of air conditioner and/or air circulation fans are used to mix and diffuse pure CO_2 gas there with room air before air delivery to the plants. A proper air circulation is essential in preventing differences in levels of CO_2. Ejecting CO_2 locally close to the plant canopy can be another option to optimize CO_2 distribution. Zhang et al. (2020) compared two CO_2 enrichment methods in a greenhouse with a CO_2 generator, with one method releasing CO_2 directly into the center of the greenhouse and the other injecting CO_2 around crops with a tube system. The study showed that crop-local CO_2 enrichment could enhance the CO_2 concentration just around the crops and improve the efficiency of CO_2 enrichment. Although the air-tight feature of PFALs has greatly improved the CO_2 use efficiency, studies of CO_2 enrichment methods in PFALs are still needed, which could potentially optimize the CO_2 distribution in the production system and further enhance the efficiency of CO_2 use. CO_2 can also be strategically controlled along with light intensity leading to electrical energy savings in PFALS. For instance, Caplan (2018) evaluated effects of various daily light integral (DLI) (9, 11, 13, 15, 17, and 19 mol m^{-2} d^{-1}) and CO_2 concentrations (400, 550, 700, 850, 1000, and 1300 μmol mol^{-1}) combinations on crop growth and electrical energy use efficiency. This study demonstrated the potential electrical savings from changing the DLI and CO_2 concentration combination from 17/400 to 13/850 with similar plant fresh biomass yields. Since the cost of injecting CO_2 is significantly lower than the cost of electrical energy for lighting, thus elevated CO_2, within a reasonably high level, within safety measures, and with reduced DLIs, can be considered as means of energy savings.

21.5 HVAC optimization

HVAC is another major energy consumer in PFALs in addition to lighting. Energy savings are a major consideration when determining environmental control strategies. The energy consumption of HVAC systems is affected by various factors, including the outdoor climates, building characteristics, operational schedules, thermostat setpoints, and internal heat gains/losses in PFALs. An analysis of each parameter can lead to the identification of specific energy savings strategies. However, the optimal settings vary for different operations. In general, most of the PFALs with stacked production shelves require much more cooling (reduce air temperatures) than heating (increase air temperatures). The heat released from the lights can compensate most if not all the heat losses in cold months. Although the cooling effect of evapotranspiration by plants can convert a significant amount of sensible cooling loads (to reduce the dry bulb temperature of the air) to latent cooling loads (to reduce the moisture level in the air), which can be removed by dehumidification processes, sensible cooling is still needed to lower the air temperature in a PFAL during most of the year (Zhang and Kacira, 2020b). HVAC optimization can be done by optimizing the heat gains and losses in a building envelope to reduce cooling loads. As lighting is the major component accounting for energy consumption in PFALs, choosing lights with higher efficacy is the most effective method to reduce energy consumption for lighting and cooling (Harbick and Albright, 2016; Zhang and Kacira, 2020a) (Fig. 21.3). Graamans et al. (2020) simulated the effects of façade properties on the energy use in PFALs, including U-values, albedo values, opaque/transparent, and wall-to-floor area ratios. This study suggested that the façade design of PFALs should consider local climates. By adjusting façade properties to allow heat, the dissipation of heat across the façade is the most efficient design strategy in terms of energy expenditure in the three locations studied with different climates. Zhang and Kacira (2020a) evaluated other control strategies to enhance energy use efficiency in a PFAL, including the schedule of photoperiod and dark period over the course of day and the set points of daytime and nighttime air temperatures. It was determined that shifting the photoperiod from daytime to nighttime could be a strategy to reduce the cooling loads. The study showed that the setpoints of air temperature affected the total energy consumption of the HVAC system and the production of lettuce crops. Decisions should be made considering the two factors. For HVAC selections, except for choosing an HVAC system with high efficiency, choosing an HVAC system that has an economizer integrated can potentially reduce the energy consumption of PFALs. An economizer is an air conditioning cycle that takes advantage of the free cooling capacity of outdoor air to partially remove the cooling load and results in energy savings. It has been shown that integrating economizers can reduce cooling loads and therefore reduce the total energy consumption of HVAC systems under different climates (Harbick and Albright, 2016; Zhang and Kacira, 2020b) (Fig. 21.4). However, a cold and dry outdoor climate, compared to the hot and humid climates evaluated, has been shown to be ideal for PFALs with high sensible and latent cooling loads. However, the cost of electricity and access to water and its quality in a given climate or region should also be considered when considering PFAL systems and their installations and operations. HVAC optimization should consider these influencing factors to minimize energy consumption.

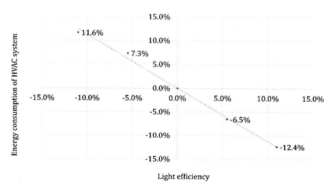

FIGURE 21.3   Sensitivity analysis of the effect of light efficiency on the energy consumption of HVAC system in PFALs (Zhang and Kacira, 2020a).

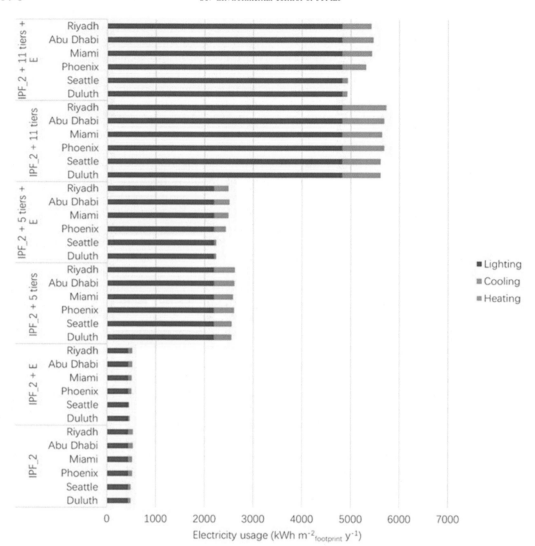

FIGURE 21.4 The comparison of electrical usage in multitier PFALs with different tiers and with/without the integration of economizers at six locations (Zhang and Kacira, 2020b).

21.6 Technology trends

The real-time monitoring and control of the environment in PFALs make this production system a desirable platform for the implementation of advanced technologies, such as AI, computer vision, the internet of things (IoT), and robotics implementation. AI is a technique that can make decisions or detect and derive patterns on data-driven models. AI has much potential and can empower the operators/growers to enhance resource use efficiencies in

PFALs in conjunction with other technologies to optimize control algorithms. Kim and Cho (2020) studied a modular smart plant factory integrating intelligent solar module to maximize the efficiency of power generation, developing an LED module with selective wavelengths to maximize lighting efficiency, IoT module control systems to optimize control algorithms related to the plant growth, and image processing algorithm to extract the information on plant growth. The modular smart plant factory is a good example showing how advanced technologies can improve the efficiency of the system to enhance energy efficiency and accelerate plant growth. With the development and advancements in sensors, devices, machines, and information technology, the future of farming in PFALs will use these sophisticated and smart technology integrations to minimize resource inputs and enable more profitable, resource use efficient, safe, and environmentally friendly production practices. Resource use efficiency of PFALS can be improved by environmental control systems engineered for crop-specific requirements and local climates, with improved climate uniformity, distributed wireless and smart sensors, dynamic- and biofeedback-based environmental control strategies, and AI-integrated environmental controls. Resource use efficiency can be further improved by high-efficiency lighting systems, crop spacing during growth cycles to maximize light use efficiency, and crop varieties bred for the unique environments in PFALs.

References

Ahmed, H.A., Yuxin, T., Yang, Q., 2020. Lettuce plant growth and tipburn occurrence as affected by airflow using a multi-fan system in a plant factory with artificial light. J. Therm. Biol. 88, 102496. https://doi.org/10.1016/j.jtherbio.2019.102496.

Brito, F.A.L., Pimenta, T.M., Henschel, J.M., Martins, S.C.V., Zsögön, A., Ribeiro, D.M., April 2020. Elevated CO_2 improves assimilation rate and growth of tomato plants under progressively higher soil salinity by decreasing abscisic acid and ethylene levels. Environ. Exp. Bot. 176, 104050. https://doi.org/10.1016/j.envexpbot.2020.104050.

Caplan, B.A., 2018. Optimizing Carbon Dioxide Concentration and Daily Light Integral Combination in a Multi-Level Electrically Lighted Lettuce Production System. The University of Arizona. http://hdl.handle.net/10150/630127.

Collier, G.F., Tibbitts, T.W., 1984. Effects of relative humidity and root temperature on calcium concentration and tipburn development in lettuce. J. Am. Soc. Hortic. Sci. 109, 128—131.

Drake, B.G., Raschke, K., Salisbury, F.B., 2008. Temperature and transpiration resistances of xanthium leaves as affected by air temperature, humidity, and wind speed. Plant Physiol. 46, 324—330. https://doi.org/10.1104/pp.46.2.324.

Fang, H., Li, K., Wu, G., Cheng, R., Zhang, Y., Yang, Q., 2020. A CFD analysis on improving lettuce canopy airflow distribution in a plant factory considering the crop resistance and LEDs heat dissipation. Biosyst. Eng. 200, 1—12. https://doi.org/10.1016/j.biosystemseng.2020.08.017.

Ge, F., Wang, C., September 2020. Exergy analysis of dehumidification systems: a comparison between the condensing dehumidification and the desiccant wheel dehumidification. Energy Convers. Manag. 224, 113343. https://doi.org/10.1016/j.enconman.2020.113343.

Ge, Z.M., Zhou, X., Kellomäki, S., Peltola, H., Biasi, C., Shurpali, N., Martikainen, P.J., Wang, K.Y., 2012. Measured and modeled biomass growth in relation to photosynthesis acclimation of a bioenergy crop (Reed canary grass) under elevated temperature, CO_2 enrichment and different water regimes. Biomass Bioenergy 46, 251—262. https://doi.org/10.1016/j.biombioe.2012.08.019.

Graamans, L., Tenpierik, M., Dobbelsteen, A., Stanghellini, C., 2020. Plant factories: reducing energy demand at high internal heat loads through façade design. Appl. Energy 262, 114544. https://doi.org/10.1016/j.apenergy.2020.114544.

Harbick, K., Albright, L.D., 2016. Comparison of energy consumption: greenhouses and plant factories. Acta Hortic. 1134, 285—292. https://doi.org/10.17660/ActaHortic.2016.1134.38.

Kim, B., Cho, J., 2020. A study on modular smart plant factory using morphological image processing. Electronics 9 (10), 1661. https://doi.org/10.3390/electronics9101661.

Kitaya, Y., Tsuruyama, J., Kawai, M., 2000. Effects of air current on transpiration and net photosynthetic rates of plants in a closed plant production system. Transpl. Prod. 21st Century 83−90. https://doi.org/10.1007/978-94-015-9371-7_13.

Kitaya, Y., Shibuya, T., Yoshida, M., Kiyota, M., 2004. Effects of air velocity on photosynthesis of plant canopies under elevated CO_2 levels in a plant culture system. Adv. Space Res. 34 (7 Spec. Iss.), 1466−1469. https://doi.org/10.1016/j.asr.2003.08.031.

Kozai, T., 2013. Resource use efficiency of closed plant production system with artificial light: concept, estimation and application to plant factory. Proc. Jpn. Acad. Ser. B 89, 447−461. https://doi.org/10.2183/pjab.89.447.

Lim, T., Kim, Y.H., 2014. Analysis of airflow pattern in plant factory with different inlet and outlet locations using computational fluid dynamics. J. Biosyst. Eng. 39 (4), 310−317. https://doi.org/10.5307/JBE.2014.39.4.310.

Salisbury, F.B., 1979. Temperature. In: Controlled Environment Guidelines for Plant Research, pp. 75−116.

Thongbai, P., Kozai, T., Ohyama, K., 2010. CO_2 and air circulation effects on photosynthesis and transpiration of tomato seedlings. Sci. Hortic. 126 (3), 338−344. https://doi.org/10.1016/j.scienta.2010.07.018.

Zhang, Y., Kacira, M., 2018. Analysis of environmental uniformity in a plant factory using computational fluid dynamics (CFD) analysis. Acta Hortic. 1227, 607−614. https://doi.org/10.17660/ActaHortic.2018.1227.77.

Zhang, Y., Kacira, M., 2020a. Enhancing resource use efficiency in plant factory. Acta Hortic. 1271, 307−314. https://doi.org/10.17660/ActaHortic.2020.1271.42.

Zhang, Y., Kacira, M., 2020b. Comparison of energy use efficiency of greenhouse and indoor plant factory system. Eur. J. Hortic. Sci. 85, 310−320. https://doi.org/10.17660/eJHS.2020/85.5.2.

Zhang, Y., Kacira, M., An, L., July 2016. A CFD study on improving air flow uniformity in indoor plant factory system. Biosyst. Eng. 147, 193−205. https://doi.org/10.1016/j.biosystemseng.2016.04.012.

Zhang, Y., Yasutake, D., Hidaka, K., Kitano, M., Okayasu, T., May 2020. CFD analysis for evaluating and optimizing spatial distribution of CO_2 concentration in a strawberry greenhouse under different CO_2 enrichment methods. Comput. Electron. Agric. 179, 105811. https://doi.org/10.1016/j.compag.2020.105811.

Human-centered perspective on urban agriculture

Harumi Ikei

Center for Environment, Health and Field Sciences, Chiba University, Chiba, Japan

22.1 Introduction

According to a report released by the United Nations, the world's population is expected to reach 9.7 billion by 2050 (United Nations, 2018), 68% of whom will live in urban areas (United Nations: Department of Economic and Social Affairs, 2017). Such population growth necessitates an increase in food production capacity (Food and Agriculture Organization of the United Nations, 2009). Food production in or near cities will significantly reduce the costs of transport and the fuel consumption and CO_2 emissions associated with transportation.

Advances in plant factory technology can enable more efficient food production than can traditional agriculture. Plant factories can produce fresh, safe, nutrient-dense crops all year round within cities without being affected by climate change. Compared with traditional agriculture, their use of water and land is more efficient, and less labor is required (Kozai and Niu, 2016a). Furthermore, urban dwellers can participate in agricultural production activities that are located on the periphery or within cities. Horticultural activities conducted in cities, such as community gardens, improve the health and quality of life of the participants (Wakefield et al., 2007). Urban agriculture does not replace urban parks; rather, it represents green space with food value (Contesse et al., 2018).

Urban agriculture is currently based primarily on a productivity and management perspective. Plant factories can be established in the future as a sustainable industry by incorporating human-centered ideas such as health and well-being. The Center for Environment, Health and Field Sciences at Chiba University, Chiba, Japan, has been involved in research on sustainable urban agriculture since 2003. In 2020, in an opinion piece in a special issue of *Sustainability* titled *"Sustainable Nature Therapy: Accumulation of Physiological Data on the Wellbeing Effect of Nature,"* Lu et al. described the concept (Lu et al., 2020). The establishment of human-centered plant factories is a promising solution for meeting the Sustainable Development Goals set by the United Nations General Assembly in 2015 (United Nations: Department of Economic and Social Affairs, 2015), specifically, goals 2 "zero hunger," 3 "good health and well-being," 11 "sustainable cities and communities," and 12 "responsible consumption and production."

This chapter proposes the use of nature therapy in the form of sustainable urban agriculture, in which the human-centered approach would be integrated with the conventional agricultural produce—centered approach, whereby the next generation of urban agriculture is represented by the plant factory. First, the background and hypothesis about nature therapy are explained. Second, the benefits of urban green spaces in nature therapy are discussed, focusing on the most recent research findings. Third, the current status, benefits, and problems of urban agriculture are described.

22.2 Nature therapy

22.2.1 Modern society and human stress

Advanced urbanization has improved living conditions (Vlahov et al., 2007) and contributed to an increase in life expectancy worldwide (Dye, 2008; Zuckerman, 2014). However, rapid urbanization has also led to a variety of environmental changes that threaten health and quality of life: large-scale urban development is reducing open spaces, such as the natural environment and agricultural land (Pronczuk and Surdu, 2008), and is causing anthropogenic climate change (Basara et al., 2010). In urban areas, the temperature has risen significantly (Changnon et al., 1996; Kosatsky, 2003), causing discomfort and heat stress to urban dwellers (Kovats and Hajat, 2008; Abdel-Ghany et al., 2013). As a result of urban air pollution, the prevalence of cardiovascular and respiratory diseases has also increased, as have rates of the associated mortality (Brunekreef and Holgate, 2002). The availability of convenient transportation and the sedentary lifestyles typical of office work have reduced physical activity (Lee et al., 2012; Kohl et al., 2012). The rapid increase in information technology has increased levels of stress, labeled "technostress" by the American clinical psychologist Craig Brod in 1984 (Brod, 1984). Mental health is adversely affected by technostress (Salanova et al., 2013; Misra and Stokols, 2012).

The negative physiological effects of stress caused by this modern society—reducing immune function (Herbert and Cohen, 1993), increasing cardiovascular inflammatory activity (Gémes et al., 2008), and adversely affecting the amygdala and related brain areas responsible for stress processing (Lederbogen et al., 2011)—have been reported extensively. Urban environments with little green space have also been reported to induce mental disorders (Engemann et al., 2019; Purtle et al., 2019; Tost et al., 2019); for example, urban areas characterized by little green space have a higher incidence of mental disorders (Tost et al., 2019); children who grow up in urban areas with less green space are at a higher risk for developing mental disorders (Engemann et al., 2019); and in developed countries, urban residence increases the risk of depression in elderly people (Purtle et al., 2019).

22.2.2 Back-to-nature theory

Against the background of such a high-stress society, the relaxing effect of the natural environment on humans has gained much attention. Physiological anthropologist Miyazaki Yoshifumi advocates "Back-to-Nature Theory" (O'Grady, 2015; Miyazaki et al., 2011; Miyazaki 2012) (Fig. 22.1). Humans have been evolving into their present form for 6–7 million years (Brunet et al., 2002). The Industrial Revolution prompted the onset of advanced urbanization and artificial environments; therefore, living in a modern artificial environment represents less than 0.01% of human evolutionary history. The other 99.99% of humans' evolutionary history was spent in a natural environment. The physiological function of humans is most adapted to the natural environment, and the current environment of many people, which is highly urbanized and artificial, is a major stressor.

To improve this stressful situation, "nature therapy" is a health promotion method with medically proven techniques, such as relaxation through exposure to natural stimuli such as forests, urban green spaces, and plants. Nature therapy is defined as "a set of practices aimed at achieving 'preventive medical effects' through exposure to natural stimuli that render a state of physiological relaxation and boost the weakened immune functions to prevent diseases" (Song et al., 2016). Nature therapy is intended to improve immune functions, prevent illnesses, and maintain and promote health through exposure to nature, which results from the attainment of a state of relaxation (Song et al., 2016; Hansen et al., 2017; Miyazaki, 2018) (Fig. 22.2). In other words, according to the back-to-nature theory, exposure to the natural environment returns people's bodies to a more natural state. Song et al. (2015b) found that the natural environment has a physiological adjustment effect: In their study, participants with higher initial blood pressure and pulse rate showed a decrease in these values after walking in a forest environment, whereas participants with lower initial values showed an increase. When participants were walking in an urban environment, however, no physiological adjustment occurred. These effects were clearly brought about by the natural environment of the forest, and these results support the back-to-nature theory.

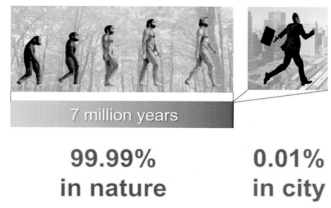

FIGURE 22.1 Evolution-based relationship between humans and nature. *Miyazaki, Y. (2017). Relationship among timber, forest, and comfort: clarifying the effects of physiological relaxation. KIZOKU Networks with permission.*

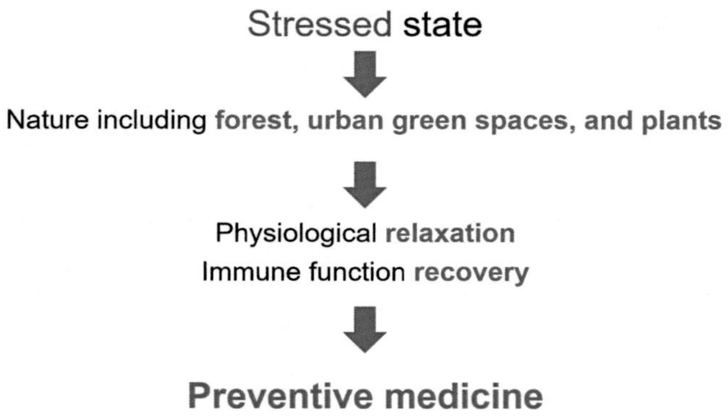

Stressed state

Nature including forest, urban green spaces, and plants

Physiological relaxation
Immune function recovery

Preventive medicine

FIGURE 22.2   Concept of nature therapy. *Revised from Song, C., Ikei, H., Miyazaki, Y., 2016. Physiological effects of nature therapy: a review of the research in Japan. Int. J. Environ. Res. Public Health 13 (8), 781. https://doi.org/10.3390/ijerph13080781 with permission.*

22.3 Benefits of urban green space in nature therapy

In the studies of the assessment of natural therapies, the physiological indices commonly used can be divided into four broad categories: central nervous activity, autonomic nervous activity, endocrine activity, and immune activity (Measurement Research Group of Japan Society of Physiological Anthropology, 1996; Ikei et al., 2017). The traditional measure of central nervous activity commonly used was electroencephalography; more recent studies have been dominated by the use of near-infrared spectroscopy to measure oxygenated hemoglobin concentration in the prefrontal cortex. In initial studies, the indicators of autonomic activity that were used included blood pressure, heart rate, pupil diameter, and pupillary light reflex; now, investigators more commonly use heart rate variability, which is a measure of both sympathetic and parasympathetic nervous activity. As for endocrine indicators, stress hormones, such as cortisol levels in saliva, can be measured. Natural killer cell activity often serves as an indicator of immune activity. An example of a field experiment conducted in an actual natural environment is shown in Fig. 22.3, and an example of a laboratory experiment conducted in an artificial climate chamber with constant temperature, humidity, and illumination is shown in Fig. 22.4.

The use of the forest environment as a place for recreation and health improvement is called "Shinrin-yoku" (forest bathing) in Japan, which means "feeling the atmosphere of the forest through the five senses" (Selhub and Logan, 2012). Several studies have shown that, compared with viewing an urban landscape, sitting in a forest environment for a short period of time and looking at the landscape produces more physiological relaxation; manifestations of such relaxation include calming of prefrontal cortex activity (Park et al., 2007; Song et al., 2020); increased parasympathetic activity, which reflects a relaxed state (Lee et al., 2011; Park et al., 2008, 2009, 2010, 2012; Song et al., 2019b; Tsunetsugu et al., 2013); suppression of

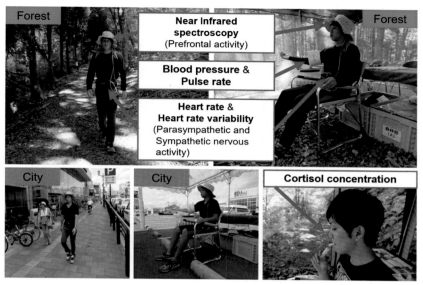

FIGURE 22.3 Physiological measurements in field experimental settings. *Revised from Miyazaki, Y., 2018. Shinrin-Yoku: The Art of Japanese Forest Bathing. Octopus Publishing Group with permission.*

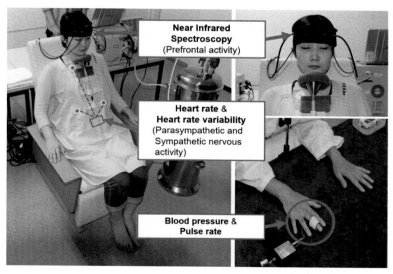

FIGURE 22.4 Physiological measurements in a laboratory experimental setting. *Revised from Miyazaki, Y., 2018. Shinrin-Yoku: The Art of Japanese Forest Bathing. Octopus Publishing Group with permission.*

sympathetic activity, which is heightened in a stressed state (Lee et al., 2011; Park et al., 2009, 2010, 2012; Song et al., 2019b; Tsunetsugu et al., 2013); lower heart and pulse rates (Lee et al., 2011; Park et al., 2008, 2010, 2012; Song et al., 2019b; Tsunetsugu et al., 2007, 2013); lower blood pressure (Park et al., 2010, 2012); and lower levels of cortisol in saliva (Lee et al., 2011; Park et al., 2007, 2008, 2010, 2012; Tsunetsugu et al., 2007). Short walks in a forest environment have similarly been reported to increase parasympathetic activity (Park et al., 2009, 2010, 2012; Song et al., 2019a), suppress sympathetic activity (Park et al., 2009, 2010, 2012; Song et al., 2019a), lower heart or pulse rate (Ochiai et al., 2015b; Park et al., 2010, 2012; Song et al., 2019b), lower blood pressure (Park et al., 2010, 2012), and lower salivary cortisol concentrations (Park et al., 2010, 2012) compared with short walks of the same intensity in urban areas.

Several studies have also been conducted on the effects of participating in forest bathing programs, which include deep breathing and meditation (Ohe et al., 2017; Ochiai et al., 2015a; Ochiai et al., 2015b; Song et al., 2017b). Notably, in a report using immune activity as an indicator, participation in a three-day and two-night forest bathing program improved human natural killer cell activity in both male (Li et al., 2007) and female (Li et al., 2008a) participants; moreover, this effect lasted for approximately 1 month in male participants (Li et al., 2008b). These findings demonstrate that exposure to natural environments, such as forests, enables physiological relaxation and thereby reduces a decline in immune activity.

Today, however, it is difficult for many city dwellers to receive regular exposure to a large-scale natural environment such as a forest because of time and physical constraints. Demographic studies of the potential health promotion effects of urban green space have demonstrated that exposure to urban green space is related to the improved general health of urban residents (Maas et al., 2006; Mitchell and Popham, 2008; Takano et al., 2002). For example, older adults who resided in an area with walkable green space lived longer than those who did not, regardless of age, gender, marital status, baseline functional status, and socioeconomic status (Takano et al., 2002). In an experiment of male university students, walking in an urban park for a short time increased parasympathetic activity, which reflects a relaxed state, and reduced sympathetic activity, which reflects a stressed and arousal state (Song et al., 2013, 2014, 2015a) (Fig. 22.5). Middle-aged females who sat in and gazed at an urban farm garden exhibited increased parasympathetic nervous activity (Igarashi et al., 2015). In elderly patients who required nursing care, observing a hospital rooftop garden also increased parasympathetic activity and decreased sympathetic activity (Matsunaga et al., 2011). Therefore, the physiological effects of urban green spaces on humans are significant and are considered essential for future health promotion.

Besides urban green spaces, nature-derived stimuli that can be used daily can provide stress relief and relaxation. Fresh flowers and foliage plants can provide daily contact with nature in indoor environments such as homes, workplaces, and schools. Previous studies involving male and female high school students (Ikei et al., 2013), female health care workers (Komatsu et al., 2013), and male office workers (Ikei et al., 2014a), have reported the physiological effects of the visual stimulation with odorless pink roses (Fig. 22.6). Compared with the absence of fresh flowers, visual stimulation with fresh roses increased the parasympathetic nervous activity by 15% and decreased the sympathetic nervous activity by 16% (Fig. 22.7). Female university students also exhibited calmer prefrontal brain activity in response to viewing fresh red roses (Song et al., 2017a). Responses to foliage plants

FIGURE 22.5 Walking in an urban park: experimental setting. *The top left photo (spring) is revised from Song, C., Ikei, H., Igarashi, M., Miwa, M., Takagaki, M., Miyazaki, Y., 2014. Physiological and psychological responses of young males during spring-time walks in urban parks. J. Physiol. Anthropol. 33 (1). https://doi.org/10.1186/1880-6805-33-8 with permission. The bottom left photo (fall) is revised from Song, C., Ikei, H., Igarashi, M., Takagaki, M., Miyazaki, Y., 2015a. Physiological and psychological effects of a walk in Urban parks in fall. Int. J. Environ. Res. Publ. Health 12 (11), 14216—14228. https://doi. org/10.3390/ijerph121114216; The bottom right photo (winter) is revised from Song, C., Ikei, H., Igarashi, M., Miwa, M., Takagaki, M., Miyazaki, Y., 2014. Physiological and psychological responses of young males during spring-time walks in urban parks. J. Physiol. Anthropol. 33 (1). https://doi.org/10.1186/1880-6805-33-8 with permission.*

commonly found in homes and offices have also been studied among male and female high school students. A study found that viewing three pots of foliage plants had a physiological relaxing effect of increasing parasympathetic nervous activity and suppressing sympathetic nervous activity, compared to the absence of plants (Ikei et al., 2014b) (Figs. 22.8 and 22.9). High school students' reactions to fresh and artificial yellow pansies in planters have been reported as well; viewing fresh flowers was shown to inhibit sympathetic nervous activity and alleviate physiological stress conditions more than did viewing artificial flowers (Igarashi et al., 2015).

22.4 The status, benefits, and problems of urban agriculture

22.4.1 The state of urban agriculture

Worldwide, the demand for fresh and safe food has grown rapidly, in parallel to the growing interest in improving health (Tsakiridou et al., 2011). Agricultural products with increased levels of vitamins and other nutrients have been developed through improved production processes, cultivation methods, and breeding and have been shown to be clinically relevant for various types of diseases, including anticancer, antiallergenic, antiinflammatory,

FIGURE 22.6   Visual stimulation by fresh pink roses experimental setting. *The top left photo is revised from Ikei, H., Lee, J., Song, C., Komatsu, M., Himoro, E., Miyazaki, Y., 2013. Physiological relaxation of viewing rose flowers in high school students [Article in Japanese, English abstract]. Japanese J. Physiol. Anthropol. 18 (3), 97–103. https://doi.org/10.20718/jjpa.18.3_97 with permission. The bottom left photo is revised from Komatsu, M., Matsunaga, K., Lee, J., Ikei, H., Song, C., Himoro, E., et al., 2013. The physiological and psychological relaxing effects of viewing rose flowers in medical staff [Article in Japanese, English abstract]. Japanese J. Physiol. Anthropol. 18, 1–7 with permission. The bottom right photo is revised from Ikei, H., Komatsu, M., Song, C., Himoro, E., Miyazaki, Y., 2014a. The physiological and psychological relaxing effects of viewing rose flowers in office workers. J. Physiol. Anthropol. 33; with permission.*

and antidepressant effects (Balsano and Alisi, 2009). Sustainable production of high-quality and high-function agricultural products is an urgent challenge (Kozai and Niu, 2016a).

Plant factories meet specific requirements for plant growth and the accumulation of bioactive compounds. Plant factories are used for commercial production of vegetables, herbs, and medicinal plants in Japan, the United States, China, and many other countries (Hayashi, 2016). Environmental factors within all plant factories can be controlled without climate or location restrictions (Kozai and Niu, 2016a, b). By regulating the light from light emitting diodes (LEDs) and the environment in the root zone, the production of bioactive compounds in plants can be greatly enhanced (Lu et al., 2017, 2018; Nguyen et al., 2019).

22.4.2 Benefits of urban agriculture

In a plant factory, weather changes, such as differences in the amount of sunlight and in temperature, do not occur; plant production is therefore simpler and easier than traditional agriculture (Kozai and Niu, 2016a). Plant factories can provide activities for elderly and disabled people. Working with plants can provide physiological relaxation effects for employees, and these effects should be studied.

The comfort-enhancing effects of nature-related activities, including gardening and horticultural activities, are empirically known. To date, various studies have shown that

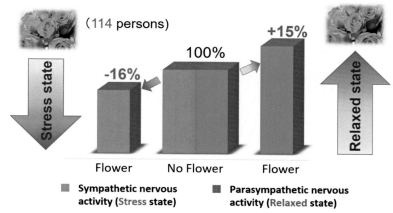

FIGURE 22.7 Physiological relaxation effects of viewing roses. *Revised from Ikei, H., Lee, J., Song, C., Komatsu, M., Himoro, E., Miyazaki, Y., 2013. Physiological relaxation of viewing rose flowers in high school students [Article in Japanese, English abstract]. Japanese J. Physiol. Anthropol. 18 (3), 97—103. https://doi.org/10.20718/jjpa.18.3_97, Komatsu, M., Matsunaga, K., Lee, J., Ikei, H., Song, C., Himoro, E., et al., 2013. The physiological and psychological relaxing effects of viewing rose flowers in medical staff [Article in Japanese, English abstract]. Japanese J. Physiol. Anthropol. 18, 1—7, and Ikei, H., Komatsu, M., Song, C., Himoro, E., Miyazaki, Y., 2014a. The physiological and psychological relaxing effects of viewing rose flowers in office workers. J. Physiol. Anthropol. 33; with permission.*

FIGURE 22.8 Visual stimulation by foliage plants in an experimental setting. *Revised from Ikei, H., Song, C., Igarashi, M., Namekawa, T., Miyazaki, Y., 2014b. Physiological and psychological relaxing effects of visual stimulation with foliage plants in high school students. Adv. Hortic. Sci. 28 (2), 111—116. https://oaj.fupress.net/index.php/ahs/article/view/3009 with permission.*

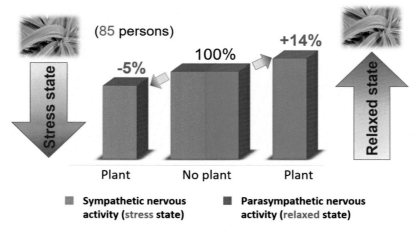

FIGURE 22.9 Physiological relaxation effects of viewing foliage plants. *Revised from Ikei, H., Song, C., Igarashi, M., Namekawa, T., Miyazaki, Y., 2014b. Physiological and psychological relaxing effects of visual stimulation with foliage plants in high school students. Adv. Hortic. Sci. 28 (2), 111—116. https://oaj.fupress.net/index.php/ahs/article/view/3009 with permission.*

FIGURE 22.10 Pot transfer experimental setting. *Revised from Park, S.A., Song, C., Oh, Y.A., Miyazaki, Y., Son, K.C., 2017. Comparison of physiological and psychological relaxation using measurements of heart rate variability, prefrontal cortex activity, and subjective indexes after completing tasks with and without foliage plants. Int. J. Environ. Res. Publ. Health 14 (9). https://doi.org/10.3390/ijerph14091087 with permission.*

gardening and horticultural activities can reduce stress and improve self-esteem (van den Berg and Custers, 2011), social interaction (Cammack et al., 2002), and cognitive health (Cimprich, 1993). The physiological effects of active plant interaction, such as transplanting and potting, have also been investigated, and the calming of prefrontal cortex activity and a reduction in sympathetic nervous activity have been reported (Lee et al., 2013, 2015; Park et al., 2017) (Fig. 22.10).

Various small-scale plant factory systems have been developed for urban dwellers who have otherwise limited access to horticultural activities (Takagaki et al., 2016). These systems allow residents to enjoy indoor farming and can provide restaurants, offices, schools, and hospitals with fresh vegetables and with educational or healing activities. In miniature plant factories, hydroponic systems are commonly used instead of soil; hence, the pleasant sound of water can be incorporated into the design of such factories (Shono et al., 2018).

22.4.3 The problems of plant factories

In the future, the effects of controlled artificial environmental conditions within plant factories on the health of humans, especially the producers, must be elucidated. To increase plant production, the CO_2 concentration of a plant factory is three to five times higher than that of the open natural environment (Lu and Shimamura, 2018). Previous research has shown that CO_2 and volatile organic compounds can negatively affect human cognition and decision-making (Allen et al., 2016). Moreover, although LED lighting is an integral part of plant factories, a mixture of blue (400–500 nm) and red (600–700 nm) LED lights appears darker than white light, and prolonged exposure has been shown to cause headaches and dizziness in factory employees (Wilkins, 2016). Broad-spectrum white LEDs for home and office use have been developed and are now being incorporated into plant factories (Cho et al., 2017). Typically, however, these white LEDs contain blue light, which has been reported to induce photoreceptor-derived cell damage (Kuse et al., 2014), and blue light excitation in the retina can induce cytotoxicity (Ratnayake et al., 2018). Therefore, the effects of high CO_2 concentrations and LED-based light sources on plant factory workers must be investigated in order to minimize risk.

22.5 Conclusion

Research and development related to information and artificial intelligence technologies have been active in the agricultural sector, but these innovations alone cannot develop urban agriculture into a sustainable industry. In 2015, the United Nations General Assembly (United Nations: Department of Economic and Social Affairs, 2015) indicated that human health and well-being must be considered in the development of sustainable urban horticulture. The perspectives of those who should be at the heart of the practical sciences, such as agriculture and horticulture, have not been valued in the past. This problem is currently in a transitional period of a global paradigm shift. By incorporating human-centered thinking into conventional agriculture produce—centered thinking in the agricultural sciences field, the interaction between humans and objects (agriculture produce) will occur. This will enable the further development of the next generation of urban agriculture, represented by plant factories.

Acknowledgment

I am very grateful for the assistance provided by Yoshifumi Miyazaki through insightful comments and suggestions.

References

Abdel-Ghany, A.M., Al-Helal, I.M., Shady, M.R., 2013. Human thermal comfort and heat stress in an outdoor urban arid environment: a case study. Adv. Meteorol. 2013 https://doi.org/10.1155/2013/693541.

Allen, J.G., MacNaughton, P., Satish, U., Santanam, S., Vallarino, J., Spengler, J.D., 2016. Associations of cognitive function scores with carbon dioxide, ventilation, and volatile organic compound exposures in office workers: a controlled exposure study of green and conventional office environments. Environ. Health Perspect. 124 (6), 805–812. https://doi.org/10.1289/ehp.1510037.

Balsano, C., Alisi, A., 2009. Antioxidant effects of natural bioactive compounds. Curr. Pharmaceut. Des. 15 (26), 3063–3073. https://doi.org/10.2174/138161209789058084.

Basara, J.B., Basara, H.G., Illston, B.G., Crawford, K.C., 2010. The impact of the urban heat island during an intense heat wave in Oklahoma city. Adv. Meteorol. 1–10 https://doi.org/10.1155/2010/230365.

Brod, C., 1984. Technostress: The Human Cost of the Computer Revolution.

Brunekreef, B., Holgate, S.T., 2002. Air pollution and health. Lancet 360 (9341), 1233–1242. https://doi.org/10.1016/S0140-6736(02)11274-8.

Brunet, M., Guy, F., Pilbeam, D., Mackaye, H.T., Likius, A., Ahounta, D., Beauvilain, A., Blondel, C., Bocherens, H., Boisserie, J.R., De Bonis, L., Coppens, Y., Dejax, J., Denys, C., Duringer, P., Eisenmann, V., Fanone, G., Fronty, P., Geraads, D., et al., 2002. A new hominid from the upper Miocene of Chad, Central Africa. Nature 418 (6894), 145–151. https://doi.org/10.1038/nature00879.

Cammack, C., Waliczek, T.M., Zajicek, J.M., 2002. The Green Brigade: the psychological effects of a community-based horticultural program on the self-development characteristics of juvenile offenders. HortTechnology 12 (1), 82–86. https://doi.org/10.21273/horttech.12.1.82.

Changnon, S.A., Kunkel, K.E., Reinke, B.C., 1996. Impacts and responses to the 1995 heat wave: a call to action. Bull. Am. Meteorol. Soc. 77 (7), 1497–1506. https://doi.org/10.1175/1520-0477(1996)077<1497:IARTTH>2.0.CO;2.

Cho, J., Park, J.H., Kim, J.K., Schubert, E.F., 2017. White light-emitting diodes: history, progress, and future. Laser Photon. Rev. 11 (2) https://doi.org/10.1002/lpor.201600147.

Cimprich, B., 1993. Development of an intervention to restore attention in cancer patients. Cancer Nurs. 16 (2), 83–92. https://doi.org/10.1097/00002820-199304000-00001.

Contesse, M., van Vliet, B.J.M., Lenhart, J., 2018. Is urban agriculture urban green space? A comparison of policy arrangements for urban green space and urban agriculture in Santiago de Chile. Land Use Pol. 71, 566–577. https://doi.org/10.1016/j.landusepol.2017.11.006.

Dye, C., 2008. Health and urban living. Science 319 (5864), 766–769. https://doi.org/10.1126/science.1150198.

Engemann, K., Pedersen, C.B., Arge, L., Tsirogiannis, C., Mortensen, P.B., Svenning, J.C., 2019. Residential green space in childhood is associated with lower risk of psychiatric disorders from adolescence into adulthood. Proc. Natl. Acad. Sci. U.S.A. 116 (11), 5188–5193. https://doi.org/10.1073/pnas.1807504116.

Food and Agriculture Organization of the United Nations, 2009. Global Agriculture towards 2050. High-Level Expert Forum: How to Feed the World 2050. http://www.fao.org/fileadmin/templates/wsfs/docs/Issues_papers/HLEF2050_Global_Agriculture.pdf.

Gémes, K., Ahnve, S., Janszky, I., 2008. Inflammation a possible link between economical stress and coronary heart disease. Eur. J. Epidemiol. 23 (2), 95–103. https://doi.org/10.1007/s10654-007-9201-7.

Hansen, M.M., Jones, R., Tocchini, K., 2017. Shinrin-yoku (Forest bathing) and nature therapy: a state-of-the-art review. Int. J. Environ. Res. Publ. Health 14 (8). https://doi.org/10.3390/ijerph14080851.

Hayashi, E., 2016. Current status of commercial plant factories with led lighting market in asia, europe, and other regions. In: LED Lighting for Urban Agriculture. Springer Singapore, pp. 295–308. https://doi.org/10.1007/978-981-10-1848-0_22.

Herbert, T.B., Cohen, S., 1993. Stress and immunity in humans: a meta-analytic review. Psychosom. Med. 55 (4), 364–379. https://doi.org/10.1097/00006842-199307000-00004.

Igarashi, M., Aga, M., Ikei, H., Namekawa, T., Miyazaki, Y., 2015. Physiological and psychological effects on high school students of viewing real and artificial pansies. Int. J. Environ. Res. Publ. Health 12 (3), 2521–2531. https://doi.org/10.3390/ijerph120302521.

Ikei, H., Komatsu, M., Song, C., Himoro, E., Miyazaki, Y., 2014a. The physiological and psychological relaxing effects of viewing rose flowers in office workers. J. Physiol. Anthropol. 33, 6. https://doi.org/10.1186/1880-6805-33-6.

Ikei, H., Lee, J., Song, C., Komatsu, M., Himoro, E., Miyazaki, Y., 2013. Physiological relaxation of viewing rose flowers in high school students [Article in Japanese, English abstract]. Japanese J. Physiol. Anthropol. 18 (3), 97–103. https://doi.org/10.20718/jjpa.18.3_97.

Ikei, H., Song, C., Igarashi, M., Namekawa, T., Miyazaki, Y., 2014b. Physiological and psychological relaxing effects of visual stimulation with foliage plants in high school students. Adv. Hortic. Sci. 28 (2), 111–116. https://oaj.fupress.net/index.php/ahs/article/view/3009.

Ikei, H., Song, C., Miyazaki, Y., 2017. Physiological effects of wood on humans: a review. J. Wood Sci. 63 (1), 1–23. https://doi.org/10.1007/s10086-016-1597-9.

Kohl, H.W., Craig, C.L., Lambert, E.V., Inoue, S., Alkandari, J.R., Leetongin, G., Kahlmeier, S., Andersen, L.B., Bauman, A.E., Blair, S.N., Brownson, R.C., Bull, F.C., Ekelund, U., Goenka, S., Guthold, R., Hallal, P.C., Haskell, W.L., Heath, G.W., Katzmarzyk, P.T., et al., 2012. The pandemic of physical inactivity: global action for public health. Lancet 380 (9838), 294–305. https://doi.org/10.1016/S0140-6736(12)60898-8.

Komatsu, M., Matsunaga, K., Lee, J., Ikei, H., Song, C., Himoro, E., et al., 2013. The physiological and psychological relaxing effects of viewing rose flowers in medical staff [Article in Japanese, English abstract]. Japanese J. Physiol. Anthropol. 18, 1–7.

Kosatsky, T., 2003. European heat waves. In: Euro Surveillance : Bulletin Europeen sur les Maladies Transmissibles, vol. 10, pp. 148–149.

Kovats, R.S., Hajat, S., 2008. Heat stress and public health: a critical review. In: Annual Review of Public Health, vol. 29, pp. 41–55. https://doi.org/10.1146/annurev.publhealth.29.020907.090843.

Kozai, T., Niu, G., 2016a. Introduction. In: Kozai, T., Niu, G., Takagaki, M. (Eds.), Plant Factory: An Indoor Vertical Farming System for Efficient Quality Food Production. Academic Press, pp. 3–5. https://www.sciencedirect.com/book/9780128017753/plant-factory.

Kozai, T., Niu, G., 2016b. Role of the plant factory with artificial lighting (PFAL) in urban areas. In: Kozaki, T., Niu, G., Takagaki, M. (Eds.), Plant Factory: An Indoor Vertical Farming System for Efficient Quality Food Production. Academic Press, pp. 7–33. https://www.sciencedirect.com/book/9780128017753/plant-factory.

Kuse, Y., Ogawa, K., Tsuruma, K., Shimazawa, M., Hara, H., 2014. Damage of photoreceptor-derived cells in culture induced by light emitting diode-derived blue light. Sci. Rep. 4 https://doi.org/10.1038/srep05223.

Lederbogen, F., Kirsch, P., Haddad, L., Streit, F., Tost, H., Schuch, P., Wüst, S., Pruessner, J.C., Rietschel, M., Deuschle, M., Meyer-Lindenberg, A., 2011. City living and urban upbringing affect neural social stress processing in humans. Nature 474 (7352), 498–501. https://doi.org/10.1038/nature10190.

Lee, I.M., Shiroma, E.J., Lobelo, F., Puska, P., Blair, S.N., Katzmarzyk, P.T., Alkandari, J.R., Andersen, L.B., Bauman, A.E., Brownson, R.C., Bull, F.C., Craig, C.L., Ekelund, U., Goenka, S., Guthold, R., Hallal, P.C., Haskell, W.L., Heath, G.W., Inoue, S., et al., 2012. Effect of physical inactivity on major non-communicable diseases worldwide: an analysis of burden of disease and life expectancy. Lancet 380 (9838), 219–229. https://doi.org/10.1016/S0140-6736(12)61031-9.

Lee, J., Park, B.J., Tsunetsugu, Y., Ohira, T., Kagawa, T., Miyazaki, Y., 2011. Effect of forest bathing on physiological and psychological responses in young Japanese male subjects. Publ. Health 125 (2), 93–100. https://doi.org/10.1016/j.puhe.2010.09.005.

Lee, M.S., Lee, J., Park, B.J., Miyazaki, Y., 2015. Interaction with indoor plants may reduce psychological and physiological stress by suppressing autonomic nervous system activity in young adults: a randomized crossover study. J. Physiol. Anthropol. 34 (1) https://doi.org/10.1186/s40101-015-0060-8.

Lee, M.S., Park, B.j., Lee, J., Park, K.t., Ku, J.h., Lee, J.w., Oh, K.o., Miyazaki, Y., 2013. Physiological relaxation induced by horticultural activity: transplanting work using flowering plants. J. Physiol. Anthropol. 32 (1) https://doi.org/10.1186/1880-6805-32-15.

Li, Q., Morimoto, K., Kobayashi, M., Inagaki, H., Katsumata, M., Hirata, Y., Hirata, K., Shimizu, T., Li, Y.J., Wakayama, Y., Kawada, T., Ohira, T., Takayama, N., Kagawa, T., Miyazaki, Y., 2008a. A forest bathing trip increases human natural killer activity and expression of anti-cancer proteins in female subjects. J. Biol. Regul. Homeost. Agents 22 (1), 45–55.

Li, Q., Morimoto, K., Kobayashi, M., Inagaki, H., Katsumata, M., Hirata, Y., Hirata, K., Suzuki, H., Li, Y.J., Wakayama, Y., Kawada, T., Park, B.J., Ohira, T., Matsui, N., Kagawa, T., Miyazaki, Y., Krensky, A.M., 2008b. Visiting a forest, but not a city, increases human natural killer activity and expression of anti-cancer proteins. Int. J. Immunopathol. Pharmacol. 21 (1), 117–127. https://doi.org/10.1177/039463200802100113.

Li, Q., Morimoto, K., Nakadai, A., Inagaki, H., Katsumata, M., Shimizu, T., Hirata, Y., Hirata, K., Suzuki, H., Miyazaki, Y., Kagawa, T., Koyama, Y., Ohira, T., Takayama, N., Krensky, A.M., Kawada, T., 2007. Forest bathing enhances human natural killer activity and expression of anti-cancer proteins. Int. J. Immunopathol. Pharmacol. 20 (2), 3–8. https://doi.org/10.1177/03946320070200s202.

Lu, N., Bernardo, E.L., Tippayadarapanich, C., Takagaki, M., Kagawa, N., Yamori, W., 2017. Growth and accumulation of secondary metabolites in perilla as affected by photosynthetic photon flux density and electrical conductivity of the nutrient solution. Front. Plant Sci. 8 https://doi.org/10.3389/fpls.2017.00708.

Lu, N., Shimamura, S., 2018. Protocols, issues and potential improvements of current cultivation systems. In: Kozai, T. (Ed.), Smart Plant Factory: The Next Generation Indoor Vertical Farms. Springer, pp. 31–49.

Lu, N., Song, C., Kuronuma, T., Ikei, H., Miyazaki, Y., Takagaki, M., 2020. The possibility of sustainable urban horticulture based on nature therapy. Sustainability 12 (12), 5058. https://doi.org/10.3390/su12125058.

Lu, N., Takagaki, M., Yamori, W., Kagawa, N., 2018. Flavonoid productivity optimized for green and red forms of perilla frutescens via environmental control technologies in plant factory. J. Food Qual. 2018 https://doi.org/10.1155/2018/4270279.

Maas, J., Verheij, R.A., Groenewegen, P.P., De Vries, S., Spreeuwenberg, P., 2006. Green space, urbanity, and health: how strong is the relation? J. Epidemiol. Community Health 60 (7), 587–592. https://doi.org/10.1136/jech.2005.043125.

Matsunaga, K., Park, B.J., Kobayashi, H., Miyazaki, Y., 2011. Physiologically relaxing effect of a hospital rooftop forest on older women requiring care. J. Am. Geriatr. Soc. 59, 2162–2163. https://doi.org/10.1111/j.1532-5415.2011.03651.x.

Measurement Research Group of Japan Society of Physiological Anthropology, 1996. Handbook for the Scientific Measurement on Human Science. Gihodo Shuppan Co., Ltd [in Japanese].

Misra, S., Stokols, D., 2012. Psychological and health outcomes of perceived information overload. Environ. Behav. 44 (6), 737–759. https://doi.org/10.1177/0013916511404408.

Mitchell, R., Popham, F., 2008. Effect of exposure to natural environment on health inequalities: an observational population study. Lancet 372 (9650), 1655–1660. https://doi.org/10.1016/S0140-6736(08)61689-X.

Miyazaki, Y., 2017. Relationship among timber, forest, and comfort: clarifying the effects of physiological relaxation. KIZOKU Networks [in Japanese].

Miyazaki, Y., 2018. Shinrin-Yoku: The Art of Japanese Forest Bathing. Octopus Publishing Group.

Miyazaki, Y., Park, B., Lee, J., 2011. Nature therapy. In: Osaki, M., Braimoh, A., Nakagami, K. (Eds.), Designing Our Future: Local Perspectives on Bioproduction, Ecosystems and Humanity. United Nations University Press, pp. 407–412.

Nguyen, D.T.P., Lu, N., Kagawa, N., Takagaki, M., 2019. Optimization of photosynthetic photon flux density and root-zone temperature for enhancing secondary metabolite accumulation and production of coriander in plant factory. Agronomy 9 (5). https://doi.org/10.3390/agronomy9050224.

Ochiai, H., Ikei, H., Song, C., Kobayashi, M., Miura, T., Kagawa, T., Li, Q., Kumeda, S., Imai, M., Miyazaki, Y., 2015a. Physiological and psychological effects of a forest therapy program on middle-aged females. Int. J. Environ. Res. Publ. Health 12 (12), 15222–15232. https://doi.org/10.3390/ijerph121214984.

Ochiai, H., Ikei, H., Song, C., Kobayashi, M., Takamatsu, A., Miura, T., Kagawa, T., Li, Q., Kumeda, S., Imai, M., Miyazaki, Y., 2015b. Physiological and psychological effects of forest therapy on middle-aged males with high-normal blood pressure. Int. J. Environ. Res. Publ. Health 12 (3), 2532–2542. https://doi.org/10.3390/ijerph120302532.

Ohe, Y., Ikei, H., Song, C., Miyazaki, Y., 2017. Evaluating the relaxation effects of emerging forest-therapy tourism: a multidisciplinary approach. Tourism Manag. 62, 322–334. https://doi.org/10.1016/j.tourman.2017.04.010.

O'Grady, M.A., 2015. Silence: because what's missing is too absent to ignore. J. Soc. Cult. Res 1, 1–25.

Park, B.J., Tsunetsugu, Y., Ishii, H., Furuhashi, S., Hirano, H., Kagawa, T., Miyazaki, Y., 2008. Physiological effects of Shinrin-yoku (taking in the atmosphere of the forest) in a mixed forest in Shinano Town, Japan. Scand. J. For. Res. 23 (3), 278–283. https://doi.org/10.1080/02827580802055978.

Park, B.J., Tsunetsugu, Y., Kasetani, T., Hirano, H., Kagawa, T., Sato, M., Miyazaki, Y., 2007. Physiological effects of Shinrin-yoku (taking in the atmosphere of the forest) - using salivary cortisol and cerebral activity as indicators-. J. Physiol. Anthropol. 26 (2), 123–128. https://doi.org/10.2114/jpa2.26.123.

Park, B.J., Tsunetsugu, Y., Kasetani, T., Kagawa, T., Miyazaki, Y., 2010. The physiological effects of Shinrin-yoku (taking in the forest atmosphere or forest bathing): evidence from field experiments in 24 forests across Japan. Environ. Health Prev. Med. 15 (1), 18–26. https://doi.org/10.1007/s12199-009-0086-9.

Park, B.J., Tsunetsugu, Y., Kasetani, T., Morikawa, T., Kagawa, T., Miyazaki, Y., 2009. Physiological effects of forest recreation in a young conifer forest in Hinokage Town, Japan. Silva Fenn. 43 (2), 291–301. http://www.metla.fi/silvafennica/full/sf43/sf432291.pdf.

Park, B.J., Tsunetsugu, Y., Lee, J., Kagawa, T., Miyazaki, Y., 2012. In: Li, Q. (Ed.), Effect of the Forest Environment on Physiological Relaxation-The Results of Field Tests at 35 Sites throughout Japan. Nova Science Publishers Inc, pp. 55–65.

FIGURE 23.1 Head vegetables grown in the open field, just before harvest. A certain percentage of photosynthates such as carbohydrates accumulated in outer leaves are translocated to inner leaves (or head) for their growth. About 20 flat and half-curled outer leaves are not edible, so that they are not harvested and left in the open field as organic fertilizer. (A): Green cabbage (Inner leaves are light green-, yellow-, or white-colored), (B): Purple cabbage (All inner leaves are purple-colored), (C): Chinese cabbage. *Photos were taken in January 2020.*

rain, flooding, strong winds, hail, frost, fluctuating temperatures, and insufficient or excess sunshine; (3) soil deterioration due to continuous cropping, excessive application of agrochemicals, insufficient supply of organic substances, soil erosion, and physical destruction of soil structure due to heavy compression, (4) damage from pest insects, birds, and wild animals; and (5) cultivar-dependent sensitivity (genetic traits) to environmental and disease stresses. Production in PFALs can avoid the influence of the above factors.

In addition, significant postharvest yield loss may occur due to damage during harvesting, loading, grading, packaging at the production site, long distance transportation from production to consumption sites, unpacking for display at shops, and preparation at home and restaurants. Local production for local consumption using PFALs can reduce the above postharvest losses considerably.

23.1.4 Factors affecting production cost and working environment in the open field

Head vegetables are edible only when harvested before or at the early stage of bolting which is characterized by the growth of a flower stalk with differentiated flower buds elongated from the shortened stem at the base of head (Fig. 23.2).

Flower bud differentiation of cabbage and Chinese cabbage is induced by low temperatures, while that of head lettuce is induced by high temperatures. Bolting of cabbage, Chinese cabbage, and head lettuce is enhanced by long days and high air temperatures. The critical day length and air temperature for bolting depends mainly on the cultivar (genetic traits).

In general, the cultivation period (from sowing to harvesting) of head vegetables in an open field is four—six months for a spring crop, and seven—eight months for a fall crop. Man-hours required for cultivation vary with the season, weather, cultivar, and cultivation method, so that the man-hours largely depend on seasonal or part-time workers. Since fully automated harvesting of the heads in the open field is difficult to achieve, the heads (fresh weight of about 1.5 kg per head) are mostly harvested by hand using a knife (Cabbage and Chinese cabbage are generally heavier than head lettuce) (Table 23.3).

TABLE 23.3 Fresh weight, size, shape, etc., of some head vegetables purchased in Kashiwa, Japan. One USD = 105 JPY as of 2021.

No.	Variables	Unit	Winter cabbage 1	Winter cabbage 2	Winter cabbage 3	Chinese cabbage	Head lettuce
1	Fresh weight/head	kg	1.157	1.324	1.317	2.041	0.556
2	Height	m	0.12	0.13	0.14	0.31	0.18
3	Circumstance of head (C)	m	0.53	0.63	0.60	0.53	0.51
4	Diameter calculated (= C/3.14)	m	0.17	0.20	0.19	0.17	0.16
5	Diameter of head measured	m	0.22	0.20	0.21	0.19	0.16
6	Price with tax/head	JPY	101	140	184	259	171
7	Price with tax/kg	JPY kg^{-1}	87	106	140	127	307
8	Volume/head	m^3	0.00179	0.00263	0.00279	0.00455	0.00237
9	Density	kg m^{-3}	646	503	472	449	235
10	Photographs						
11	Shop purchased in Kashiwa	–	Maruetsu	Kashiwade	Kashiwade	Wakuwaku Plaza	Tokyu store
12	Date purchased (yy/mm/dd)	–	2021/2/16	2021/2/25	2021/2/25	2021/2/17	2021/2/17

TABLE 23.4 General characteristics of winter and spring cabbages.

Characteristics	Winter cabbage	Spring cabbage
Harvest time	January, February, and March	March, April, and May
Shape	Flat ball (oval)	Ball
Color inside the head	Whitish	Pale yellowish
Specific weight (kg m^{-3})	High (heavy)	Low (light)
Taste	A little sweet	Not sweet
Texture	Crispy	Soft
Used as	Mostly boiled	Often as salad

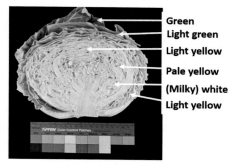

FIGURE 23.2 Longitudinal section of standard-size cabbage purchased at a greengrocery in Kashiwa city, Japan, in February 2021(Note the change in color by position inside the head).

At the same time, it is becoming more difficult for growers to maintain or secure an adequate number of workers to do the manual harvest work in open fields, resulting in higher wages per hour. Thus, employers tend to introduce automatic or semiautomatic harvest machinery to reduce labor costs, which in turn increases production costs in terms of additional capital equipment and operating expenses.

In general, costs of harvesting the head vegetables, application of agrochemicals, and soil improvement account for around 40% of the total production costs. In addition, costs of grading, packing, loading, transportation, and unloading of the heads also account for a significant percentage of the total production costs. Reducing the use of agrochemicals is also another important factor requiring consideration for workers' and consumers' health and environmental conservation.

By producing head vegetables all year round using PFALs, working hours can remain relatively constant throughout the year without the influence of seasonality or weather.

23.1.5 Trends in consumer demands and preferences

Demand for large/heavy head vegetables with a weight of about 1.5 kg per head seems to be gradually declining due to the recent contraction in family size, increase in the population living alone, and changes in dietary life. Consumers in these demographics tend to prefer smaller heads or heads cut in halves or quarters with a fresh weight of around 0.5 kg per head (Figs. 23.3–23.6). A growing number of consumers also tends to prefer clean, precut (sliced, shredded, or chopped) vegetables that can be used or served without washing before cooking or eating fresh (Fig. 23.7). Such demand for clean and small head vegetables seems to be growing due to an increase in health-conscious lifestyle by busy consumers living in large cities.

In addition, there has been a growing demand for prewashed, precut head vegetables packaged in a plastic bag and weighing 1–2 kg for commercial use at restaurants, ready-made meal shops, supermarkets, etc., for use as ingredients for salad bars, gyoza and shumai dumplings, and hot pots. Using such prepared produce helps these businesses cut labor costs and save time.

Consumer demand for vegetables is also becoming more diverse as consumers become more discerning in terms of nutrition (vitamins, polyphenols, etc.), color, taste, texture, size, weight, cleanness, freshness, etc. On the other hand, a contracting agricultural population characterized by aging farmers/growers and decreasing agricultural land area in close proximity to large cities make it difficult to match supply with the above diverse demands. Thus, there seems to be an opportunity to develop a new supply chain of head vegetables to match the consumers' recent demands and to minimize food mileage by local production for local consumption.

Taiki-sya Ltd. developed a PFAL (called Vege-Factory) for producing head lettuce in 2011 and transferred its technology to Noumann, Inc. in 2016. Since then, head lettuce has been produced for sale there. Fig. 23.8 shows head lettuce commercially grown and sold by Noumann, Inc.

FIGURE 23.3　Field-grown baby (Wawasai) Chinese cabbage (about 0.5 kg head⁻¹) and standard-size Chinese cabbage (about 2 kg head⁻¹) for sale at a greengrocery in Chiba, Japan. Baby cabbage and heads cut in halves or quarters of standard-size Chinese cabbage are wrapped with thin plastic films. Full-size standard size heads were unwrapped. *Photos were taken February 2021.*

Price: 88 JPY or 0.92 USD/half-cut   B: 178 JPY or 1.88 USD/head

FIGURE 23.4 Half-cut heads (about 0.5 kg each) of small-size cabbage (left) and full-size heads (about 1.0 kg head^{-1}) of small size cabbage (right) for sale at a greengrocery store in Chiba, Japan. Half-cut one is wrapped with thin-plastic film. Full-size cabbage is not wrapped because one or two most outer leaves are removed by consumer when purchasing (Fig. 23.5) and the rest of the head is washed with tap water before cooking at home. All of them are field-grown. *Photos were taken in February 2021.*

Trash box   **Shopping plastic bag**

FIGURE 23.5 Field-grown small-size cabbages unwrapped. A consumer removes one or two most outer leaves and discards it in the trash box, and then put the head in a shopping plastic bag before going to the cashier. *Photos were taken in February 2021.*

23.1.6 Characteristics of winter, spring, and summer cabbages

In a region with a temperate climate like Japan, cabbages are often classified into three groups based on their harvest time: winter, spring, and summer cabbage. Summer cabbage

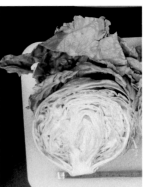

FIGURE 23.6   Medium-size field-grown head lettuce [about 0.5 kg, and 158 JPY (1.5 USD) per head] wrapped with thin plastic film displayed at a greengrocery, Chiba, Japan. Inner leaves are light yellow-colored except for several most outer leaves with light green-colored. *Photos were taken in February 2021.*

A: Thin-shredded cabbage.
198 JPY or 1.88 USD/bag (300g)

B:Thin-shredded cabbage mixed
with some thin-sliced carrot.
100 JPY or 0.95 USD/bag (120 g)

C: Coarsely chopped (square-
shaped) Chinese cabbage.
158 JPY or 1.5 USD/bag (200 g)

FIGURE 23.7   Thin-sliced, shredded, or chopped head vegetables in plastic bags for sale at a greengrocery, in Chiba, Japan. All of them are field-grown but prewashed with a disinfectant and rinsed with clean water, so that they can be served without washing before cooking. *Photos were taken in February 2021.*

is grown in highland areas (altitude 700—800 m) and is harvested in July and August. Characteristics of winter and spring cabbages are cultivar dependent and are presented in Table 23.2. People living in a colder region tend to prefer soft and yellowish-type cabbage for use as fresh salad in February. Then, they must purchase the cabbages from a southern or a warmer region and have it shipped a long distance. One advantage of PFALs is that they can produce any type of cabbage or other vegetables throughout the year to meet the local demand.

23.2 Advantages of plant factories with artificial lighting in head vegetable production

23.2.1 Advantages of plant factories with artificial lighting over open fields

One advantage of PFAL cultivation over field cultivation is the ability to control head formation and bolting through photoperiod and air temperature adjustments. Thus, head vegetables can be grown and harvested in PFALs according to a plan throughout the year.

FIGURE 23.8 Head lettuce (350 g /head) commercially grown for sale in a PFAL operated by Noumann Inc., Fukui Prefecture, Japan. *Photo courtesy of Noumann Inc. (https://www.noumann.com/) and Taiki-sya (https://www.taikisha-group.com/service/vegefactory.html).*

Future research in environmental controls will find means to enhance the growth and quality of heads. Chemical components which affect color, taste, and texture can also be controlled through photon flux and spectral distribution (light quality) of the light source (Chapter 17). Cultivars grown in PFALs do not need to be resistant to disease and environmental stresses, since the PFAL cultivation room is pesticide-, insect-, and pathogen-free, and the environment is always optimized. Adjusting plant spacing (Numbe of plants per m^2) as plants increase in size and using multitier cultivation racks will increase the annual yield per cultivation area by over 100-fold compared with the open field, due to the use of 10–20 tier cultivation racks, shortening the cultivation period through environmental control, increasing planting density, multiple cropping throughout the year, eliminating damage to plants by pest insects and weather, etc. Water consumption for washing before cooking or serving can be considerably reduced in addition to the savings in irrigation water.

23.2.2 Advantages of plant factories with artificial lighting for small-size head vegetable production

A small-size head of cabbage and Chinese cabbage has a fresh weight of 0.5–1.0 kg, and a diameter and height of 0.10–0.15 m, and head lettuce commonly grown in PFALs has a fresh weight of 0.3—0.4 kg and a diameter and height of 0.10—0.15 m. At this size, the vertical distance between tiers inside a PFAL can be reduced to about 0.35 m from a standard distance of 0.5 m or longer. A high planting density (16 plants per m², or 0.25 by 0.25 m per head) at the last stage of growth can also be achieved. The first stage from seeding to the one to three true leaves lasts about two weeks; the second stage (around three weeks) ends at the start of head formation; and the third and final stage is the growth period (around three weeks) until harvest. Through appropriate spacing adjustments, the annual cultivation bed utilization efficiency (ratio of perpendicularly projected leaf area to the cultivation bed area) can probably be doubled compared with the fixed spacing of 0.25 by 0.25 m for the entire cultivation period of eight weeks.

The outer leaves of head vegetables grown in open fields are not edible and are discarded as plant residue. However, the same outer leaves of head vegetables grown in PFALs are edible if they are harvested just before the start of head formation. In many cases, their taste and texture are similar to those of the head. Thus, these leaves can be used or served as fine strips or small pieces in salads. In this way, the cultivation period can be nearly halved and planting density at harvest would be several times higher than that for conventional head vegetable cultivation.

It can be assumed that people living alone or in a small family unit would prefer small-size and precut head vegetables than large ones. Retailers tend to prefer to sell small heads than half- or quarter-cuts of large heads from a labor-saving and loss reduction viewpoints. Consumers tend to prefer the upper leafy parts of the light green Chinese cabbage head than the lower parts of the head with firm, thick white midribs, especially when served in salad (Fig. 23.9). Presliced, shredded, or chopped head vegetables are served as fresh salads,

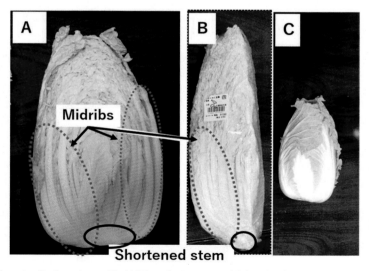

FIGURE 23.9 Longitudinal sections of half-(A) and quarter-cut (B) heads of standard-size Chinese cabbage, and whole head of small-size (baby or wawasai) Chinese cabbage (C). Color of central inner leaves of Chinese cabbage A is light yellow or light yellowish white, and its outer most ones are light green. Increasing number of consumers prefer small-size, $^1/_2$ cut or $^1/_4$ cut head vegetables to full-size standard size heads. Thinner and short midribs in small head are preferable when served as a salad.

pickles, or kimchi (Korean fermented spicy vegetables) and ingredients in hot pots and Chinese foods such as gyoza and shumai dumplings. Preslicing, shredding, and chopping of head vegetables can be almost automated.

23.3 Mini review of head formation, green light effect, and tipburn

23.3.1 Head formation

23.3.1.1 Cabbage

In general, head formation starts when the number of outer (or rosette) leaves reaches 15—20 and the air temperature drops to 15—20°C (Kato and Pak, 1966). Head formation is initiated when the outer leaves shade the shoot tips (Kato and Sooen, 1980). Head growth is enhanced by the translocation of photosynthates (mainly carbohydrates) to the head from the outer leaves and is enhanced by root pruning (Kato and Sooen, 1979). The yield and quality are higher when plants are grown at 20°C with a high nitrogen supply and when the outer leaves have a carbohydrate/nitrogen ratio of about seven (Hara and Sonoda, 1982). Most of these phenomena in head formation are affected by phytohormone balance of the whole plant (Kato and Sooen, 1980). All of these traits are cultivar-dependent, so that these traits need to be confirmed by experiments using recently prevailed cultivars.

23.3.1.2 Chinese cabbage

Head formation is brought about through the incurving development of hyponastic inner leaves shaded by the outer broad green rosette leaves (Ito and Kato, 1957). Nishijima and Fukino (2005a,b, 2006) developed a geometric "double-truncated" model to simulate the hyponastic bend of the basal part of the midrib. They examined the factors involved in head formation, including the role of light and darkness on the bending, both by simulation and experimentation. This kind of work on growth modeling is expected to be advanced considering environmental effects and planting density.

23.3.2 Translocation of photoassimilates from outer to inner leaves

Growth (or biomass increase) rate of inner leaves inside the head largely depends on the translocation rate of photoassimilates (e.g., carbohydrates) from outer to inner leaves. The net photosynthetic rate of most inner leaves should be negative because they contain little or no chlorophyll (Fig. 23.2). In addition, photosynthetic photon flux density (PPFD) at inner leaves decreases exponentially as the leaf number increases and reaches a level lower than light compensation point (about 5 μmol m^{-2} s^{-1}). Thus, promotion of photosynthesis of outer leaves and enhancement of translocation of photoassimilates from outer to inner leaves are essential to promote head growth. For this purpose, it is important to develop a lighting system to promote photosynthesis in the outer leaves (Fig. 23.10). Another role of outer leaves is to shade the lower part of inner leaves to enhance head formation (Kato and Sooen, 1978).

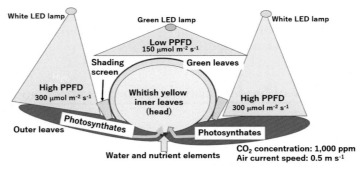

FIGURE 23.10 Rough sketch showing an estimated favorable photosynthetic photon flux density (PPFD) distribution over a head vegetable. High PPFD at outer leaves and relatively low PPFD over the head. Drymass increase in whitish yellow leaves is attributable to the translocation of photosssimilates from outer leaves. Another role of outer leaves is to shade a lower part of head surface to enhance head growth. This idea of lighting the head vegetables is not experimentally proven.

23.3.3 Effect of green light illumination on secondary metabolites and color

Natural leaf etiolation takes place at the central part of cabbage heads due to limited light access, causing inhibition of chlorophyll biosynthesis (Kruk, 2005). Concentrations of ascorbic acid, total polyphenolic substance, and chlorophyll in cabbage leaves were affected by illumination of blue, green, and/or red lights when stored at 4–5°C for 15–18 days (Lee et al., 2014). Green light illumination for two days to postharvest cabbage heads at 20°C changed the color and chemical components of the inner leaves (Amagai et al., unpublished).

Green light penetrates the lettuce head deeper than blue and red light (Fig. 3.15 in Chapter 3). Green light illumination at 400 μmol m^{-2} s to the head lettuce for four days before harvest increased the concentrations of chlorophyll and ascorbic acid of inner leaves of head lettuce grown under artificial light only (Saengtharatip et al., 2020). Senescence of the lower (outer) leaves of romaine lettuce was retarded by supplemental upward lighting from beneath and improved yield (Saengtharatip et al., 2021).

23.3.4 Tipburn symptoms

Lettuce plants grown in PFALs are often injured by tipburn, a brown spot at the leaf margin caused by calcium (Ca) deficiency (Fig. 23.11), which is a crucial defect affecting the appearance and shelf life. Barta and Tibbitts (1991) found that Ca concentrations were significantly lower in the enclosed internal leaves that exhibited tipburn symptoms compared to the exposed leaves that did not exhibit tipburn, and that the reduced levels of Ca in lettuce plants grown in a controlled environment were associated with faster tipburn development rates compared with field grown plants. The tipburn symptoms have been observed frequently also in Chinese cabbage and cabbage plants grown under artificial light as well as in the open field (Parzkill et al., 1976; Aloni, 2015).

Tipburn symptoms of butterhead lettuce plants are often suppressed by blowing air directly around the meristem (Goto and Takakura, 2003; Ahmed et al., 2020). A lack of Ca

FIGURE 23.11 Tipburn symptoms at leaf edges of a loose-headed lettuce (Romaine) plant grown in a plant factory with artificial lighting (PFAL). *Photo by courtesy of Mr. O. Nunomura.*

in the inner leaves resulting from rapid growth under high PPFD often causes frequent tipburn development (e.g., Sago, 2016), although tipburn development is significantly cultivar dependent. Thus, breeding of tipburn-resistant cultivars suited for PLALs is anticipated.

23.4 Research topics on production of head vegetables in plant factories with artificial lighting

To popularize the head vegetable production in commercial PFALs widely, its monetary productivity (ratio of economic yield to production cost) needs to be at least doubled for head lettuce and tripled for cabbage and Chinese cabbage compared with the current monetary productivity of head vegetables in PFALs. Some research topics on the productivity improvement of head vegetables in PFALs, excluding genome editing and biotechnology are:

(1) Development of a lighting system (spectral distribution, lighting cycle, photon flux of LEDs, photon flux density at plant canopy surface, lighting direction, etc.) to enhance photosynthesis, translocation of carbohydrates and other substances from outer to inner leaves, head formation and its growth, and target secondary metabolite production.

(2) More efficient use of outer leaves to enhance head formation and growth (Figs. 23.10 and 23.12) or production of edible outer leaves by harvesting at an early stage of head formation.

(3) Measurement, modeling, and simulation of spatial (three-dimensional) distribution of photon spectrum, spectral photon flux density, concentrations of CO_2, ethylene (C_2H_4), and water vapor gases, and temperature inside the head (Fig. 23.12). CO_2 concentration should be significantly higher inside the head than outside (or room air), meaning that the CO_2 balance of the head is negative during head growth.

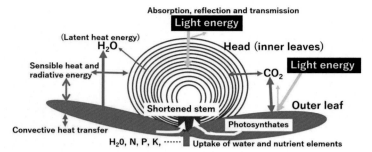

FIGURE 23.12 Scheme showing the balance of gas, liquid, solid materials and energy, translocation of photosynthates from outer to inner leaves (or head) and uptake of water and nutrient elements. Changes with time in spatial distributions of spectral photons, gases [CO_2, H_2O, (C_2H_4), etc.], and other substances (water, ions, carbohydrates, etc.) inside the head are not well known. Those spatial distributions inside the head must be considerably uneven due to the physical structure of the head.

Most carbohydrates needed to grow the head are translocated from the outer leaves.

(4) Changes in optical characteristics of leaves inside the head over time. Note that: (1) innermost leaves are pale yellow, yellow, or white, and the outer leaves are light green or green in color; (2) transmittance and reflectance of green, yellow, and white inner leaves vary considerably with ultraviolet (UV), blue, green, red, and far-red photons (Fig. 23.13). Reflectance of green, yellow, and white leaves (Nos. 1, 4, and 5 in Fig. 23.13) of visible light emitted by white LED lamp (ISL-150X150-series, CCS Inc) were 5.5%, 15%, and 19%, respectively (Amagai et al., unpublished).

(5) Interactions among environmental, physiological, chemical, optical, and economic value characteristics of inner leaves (Fig. 23.14). Note that chlorophyll concentration of white leaves increases soon after light illumination and that the white-colored leaf turns light green soon (Fig. 23.15).

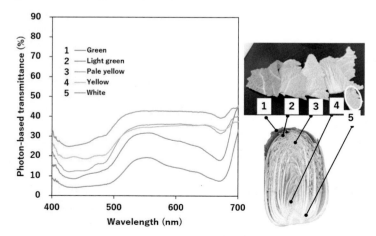

FIGURE 23.13 Spectral transmittances of inner leaves at 5 different positions with various colors of Chinese cabbage. Leaf discs (12 mm in diameter) at 5 different positions were used for measurement under white LED light (ISL-150X150-series, CCS Inc., Burlington, MA, USA). (Amagai et al., unpublished).

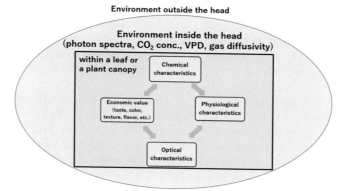

FIGURE 23.14 Scheme showing interactions among chemical, physiological, optical, and economic characteristics within the head, as affected by its environment.

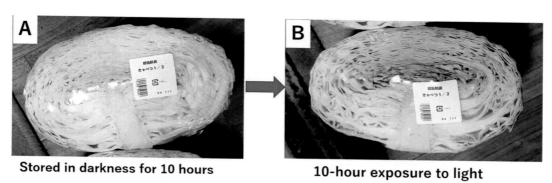

FIGURE 23.15 Longitudinal cross sections of half-cut cabbages purchased at a greengrocery. The cabbages (A) and (B) were the same color (pale green) when purchased. The cabbage A was left in darkness at room temperature for 10 h, while the cabbage B was exposed to weak solar light at window side for 10 h. Then, the color of B turned light green due to chlorophyll synthesis. The color of A remained the same. *Photos were taken in September 2020.*

(6) Gas exchange characteristics of the head in terms of number of air exchanges per hour of the head, gas diffusion coefficient of inner leaves, and dynamics of mass and energy balance of the head and outer leaves.

(7) Effect of UV, blue, green, red, and far-red photon illumination on the chemical components (ascorbic acids, chlorophylls, carotenoids, total/soluble sugars, polyphenols, etc.), color, texture, and taste of inner leaves. Note that green light penetrates deeper inside the head than blue and red photons (Fig. 23.12 and Fig. 3.15 in Chapter 3).

(8) Effect of green photon illumination as a photomorphological stimulant on bolting of the shortened stem.

(9) Effect of root pruning on head formation and growth (Kato and Sooen, 1979).

(10) Methods of maximizing harvest index (fresh weight ratio of marketable head to whole plants) through environmental control, cultivar selection and breeding, improvement of cultivation system, and cultivation management.

(11) Reduction in occurrence of tipburn in lettuce and other head vegetables through environmental control (Franz et al., 2002; Goto and Takakura, 2003; Ahmed et al., 2020).

(12) Breeding of head vegetables suited for production in PFALs: tipburn-resistant cultivars, high growth rate at high CO_2 concentration and low PPFD, etc.

(13) Development of cultivation systems particularly suited for production of head vegetables (e.g., dwarf type) because those for head vegetables and root vegetables are considerably different from those of leafy vegetables and fruit vegetables.

(14) Environmental control and selection of cultivars for optimizing leaf color, taste, texture, or mouth feeling, nutritional value, head shape, specific weight of head ($kg\ m^{-3}$ head volume), size, etc.

(15) Development of software to optimize production scheduling and control of PFALs for head vegetable.

(16) Phytohormonal balance relating to head formation and bolting.

(17) Effect of temperature on hardness/softness or spatial density of the head. Soft head/ inner leaves are said to be suitable for use as salad.

(18) Life Cycle (Inventory) Assessment of field-grown and PFAL-grown head vegetables, considering CO_2 emissions during transport from production to the consumption sites, water consumption for irrigation and washing, chemical fertilizer and agrochemicals, electric energy and fossil fuel, etc (Kikuchi and Kanematsu, 2020).

23.5 Production of root vegetables in plant factories with artificial lighting

Another group of plants which can be profitably produced in commercial PFALs in the near future are root vegetables such as turnip, radish, and carrot. Growth of turnip was remarkably enhanced by high CO_2 concentration (1000 ppm or higher), a long photoperiod of 24 h per day, and a high PPFD (300 μmol m^{-2} s or higher) compared with that of leaf lettuce (Ikeda et al., 1988). This is attributable to the large sink volume of turnip roots which induced smooth translocation of carbohydrates from leaves to roots. Ikeda et al. (1988) showed that the fresh weight of roots and whole plants and the harvest index (ratio of root fresh weight to whole plant) of turnips were 2.25, 1.54, and 1.22 times greater, respectively, when grown at 1100 ppm CO_2 and 24-hour photoperiod than at 400 ppm CO_2 and 12-hour photoperiod, both at PPFD of 237 μmol m^{-2} s^{-1}.

It can be concluded that the economic profitability of root vegetable production in PFALs can be improved by: (1) enhancing photosynthesis; (2) increasing growth and harvest indexes; (3) selecting a proper cultivar; (4) making the leafy part edible by harvesting early when leaves are still tender (Fig. 23.16 shows baby carrots produced in a PFAL at a planting density of over 1000 plants m^{-2} in a 3×3 cm plant spacing); (5) using the leafy part as feed; and (6) improving the quality and thus economic value per kg of produce.

**Average length: 10 cm, Average weight: 5.9 g each,
9.5 JPY or .093 USD per stick**

FIGURE 23.16 Immature (baby) carrots with edible leaves and a plant height of about 15 cm (The root is about 10 cm long) (Left). Planting density in the cultivation bed in plant factories with artificial lighting is about 1000 plants per square meter. These field-grown baby carrots are added to a beef stake plate (after a few minute boiling) or a salad bowl after chopping, for example. Photo of 21 carrots in a plastic bag shown in the middle and right were sold at a greengrocery as fresh salad sticks with a retail price of 200 JPY (1.9 USD). *The photos were taken in Chiba, Japan in February 2021.*

23.6 Conclusion

Despite a number of outstanding issues, the production of head vegetables in PFALs is set to be commercialized in the near future. Since much is unknown about the ecophysiology of head vegetables in a controlled environment, research and development in this area will open new fields in plant science, engineering, and business. The key research issues to be conducted include: (1) measurement, analysis, and control of 3D distributions of photon spectrum, gas concentrations, and diffusion coefficients of CO_2, H_2O, and others inside the head; (2) nondestructive measurement of 3D structure of inner leaves and their phenotypes; (3) environmental control for enhanced head formation and growth, and enhanced translocation of photoassimilates from outer leaves to the head, (4) breeding; and (5) lighting systems suited to head vegetable production in PFALs. Production of head vegetables as well as root vegetables is considered to contribute to solve global and local issues on food safety, natural resource saving, environmental conservation, and quality of life.

References

Ahmed, H.A., Tong, Y., Yang, Q.C., 2020. Lettuce plant growth and tipburn occurrence as affected by airflow using a multi-fan system in a plant factory with artificial light. J. Therm. Biol. 88 https://doi.org/10.1016/j.jtherbio.2019.102496.

Aloni, B., 2015. Enhancement of leaf tipburn by restricting root growth in Chinese cabbage plants. J. Hortic. Sci. 61 (4), 509–513. https://doi.org/10.1080/14620316.1986.11515733.

Barta, D.J., Tibbitts, T.W., 1991. Calcium localization in lettuce leaves with and without tipburn: comparison of controlled-environment and field-grown plants. J. Am. Soc. Hortic. Sci. 116 (5), 870–875.

Chinese Academy of Agricultural Sciences, 2017. http://www.caas.cn/en/. (Accessed 5 February 2021).

e-Stat, Statistics of Japan, 2019. https://www.e-stat.go.jp/en (Accessed 5 February 2021).

FAOSTAT, 2019. http://www.fao.org/faostat/en/?#data/QC (Accessed 5 February 2021).

Goto, E., Takakura, T., 2003. Prevention of lettuce tipburn by supplying air into inner leaves. Trans. Amer. Soc. Agr. Eng. 35, 641–645.

Hara, T., Sonoda, Y., 1982. Cabbage-head development as affected by nitrogen and temperature. Soil Sci. Plant Nutr. 28 (1), 109–117.

Ikeda, A., Nakayama, S., Kitaya, Y., Yabuki, K., 1988. Basic study on material production in plant factory (2) effects of photoperiod, light intensity, and CO_2 concentration on photosynthesis of turnip. Environ. Control Biol. 26 (3), 113–117 (in Japanese with English abstract and figure/table captions).

Ito, H., Kato, T., 1957. Studies on the head formation of Chinese cabbage – histological and physiological studies of head formation -. Soc. Hort. Sci. 26 (3), 154–162 (written in Japanese with English abstract and captions).

Kato, T., Pak, Y., 1966. Studies on the head formation of Chinese cabbage, II relationships between leaf shape and head formation. Res. Rep. Kochi Univ. Nat. Sci. 30, 79–90.

Kato, T., Sooen, A., 1978. Physiological studies on the head formation in cabbage. 1. Effect of defoliation of wrapper leaves on the head formation posture. Japan. Soc. Hort. Sci. 47 (3), 351–356 (written in Japanese with English abstract and captions).

Kato, T., Sooen, A., 1979. Physiological studies on the head formation in cabbage. 2. Effect of root pruning on the head formation posture. Japan. Soc. Hort. Sci. 48 (1), 26–30 (written in Japanese with English abstract and captions).

Kato, T., Sooen, A., 1980. Physiological studies on the head formation in cabbage. 3. Role of terminal bud in the head formation posture. Soc. Hort. Sci. 48 (4), 426–434 (written in Japanese with English abstract and captions).

Kikuchi, Y., Kanematsu, Y., 2020. Life cycle assessment. In: Plant Factory: An Indoor Vertical Farming System for Efficient Quality Food Production, second ed., pp. 383–395.

Kruk, J., 2005. Occurrence of chlorophyll precursors in leaves of cabbage heads – the case of natural etiolation. J. Photochem. Photobiol. B Biol. 80, 187–194. https://doi.org/10.1016/jphotobiol.2005.04.003.

Lee, Y.J., Ha, J.Y., Oh, J.E., Cho, M.S., 2014. The effect of LED irradiation on the quality of cabbage stored at a low temperature. Food Sci. Biotechnol. 23 (4), 1087–1093.

Nishijima, T., Fukino, N., 2005a. Geometrical analysis of development of erect leaves as a factor in head formation of *Brassica rapa* L. (I) geometrical change of growing leaves in head cultivars. Sci. Hortic. (Amst.) 104, 407–419.

Nishijima, T., Fukino, N., 2005b. Geometrical analysis of development of erect leaves as a factor in head formation of *Brassica rapa* L. (II) comparative analysis of headed and non-headed cultivars. Sci. Hortic. (Amst.) 104, 421–431.

Nishijima, T., Fukino, N., 2006. Autonomous development of erect leaves independent of light irradiation during the early stage of head formation in Chinese cabbage (*Brassica rapa* L. var. Rupr.). J. Jpn. Soc. Hortic. Sci. 75 (1), 59–65.

Parzkill, D.A., Tibbitts, T.W., Williams, P.H., 1976. Enhancement of calcium transport to inner leaves of cabbage for prevention of tipburn. J. Am. Soc. Hortic. Sci. 101 (6), 645–648.

Saengtharatip, S., Goto, N., Kozai, T., Yamori, W., 2020. Green light penetrates inside crisp head lettuce leading to chlorophyll and ascorbic acid content enhancement. Acta Hortic. 1273, 261–269.

Saengtharatip, S., Joshi, J., Zhang, G., Takagaki, M., Kozai, T., Yamori, W., 2021. Optimal light wavelength for a novel cultivation system with a supplemental upward lighting in plant factory with artificial lighting. Environ. Control Biol. 59 (1), 21–27. https://doi.org/10.2525/ecb.59.21.

Sago, Y., 2016. Effects of light intensity and growth rate on tipburn development and leaf calcium concentration in butterhead lettuce. Hortscience 51 (9), 1087–1091. https://doi.org/10.21273/HORTSCI110668-16.

CHAPTER

24

Concluding remarks

Toyoki Kozai[1], Genhua Niu[2] and Joseph Masabni[3]

[1]Japan Plant Factory Association, Kashiwa, Chiba, Japan; [2]Texas A&M AgriLife Research, Texas A&M University, Dallas, TX, United States; [3]Texas A&M AgriLife Extension Service, Dallas, TX, United States

24.1 Introduction

This book was written for readers who are interested in contributing to or willing to contribute to the use of plant factories with artificial lighting (PFALs) to solve global and local issues in (1) food security and safety, (2) shortages in natural resources such as water and arable land, (3) environmental conservation for preventing the loss of ecological biodiversity, global/local climate change, natural and historical landscapes, and water pollution, and (4) quality of life along with advancing the Sustainable Development Goals and the Environment, Society, and Corporate Government. Maximum use of the unique features of PFALs described below and maximum application of recent advanced technologies described in this book are key contributions to solving the above-mentioned four issues.

24.2 Unique features of PFALs

As described throughout this book, the PFAL cultivation room is characterized by unique features which are different from those of greenhouses in the following ways: high airtightness, high thermal insulation, high degree of sanitation, and nontransmission of solar light into the cultivation room. The use of sensors and actuators can facilitate high observability and controllability of resource consumption rates, production rates, the environment inside and outside the cultivation room, and plant phenotype as affected by genotype, environment, and management. Such unique features also enable high predictability and reproducibility of plant growth and energy/mass balances through the use of various simulation models (Fig. 3.7, Chapter 3). Thus, PFALs can potentially achieve optimal yield and quality with minimum consumption of resources and minimum emission of environmental pollutants, thereby achieving high resource use efficiency and high resource and monetary productivities (Chapters 12 and 13).

In reality, however, most existing PFALs are far from achieving these potentials, since the PFAL industry is still an emerging industry, and the development of PFAL technologies is still in its infancy. On the other hand, the number of research papers on plant factories published in 2020 was fourfold the number in 2010 (Fig. 2.4, Chapter 2), and this growing body of knowledge will advance PFAL technologies in the coming years. Optimal exploitation of the above unique features of PFALs will also contribute to the realization of the innovative, creative, and evolutional PFALs described in this book.

24.3 Diversity and commonality

PFALs are also characterized by two additional aspects, namely their commonality and diversity. PFAL commonality (Parts 1 and 2 of this book) is derived from the above unique features which make the energy and material balances of the cultivation room and the plant—environment interrelationships much simpler than those of soil cultivation in a greenhouse. Parts 1 and 2 of this book focus mainly on the commonality of PFALs, while Parts 3 and 4 focus mainly on PFAL diversity which can be reflected in the following:

1) plant species: from algae and moss to fruiting vegetables and fruit trees
2) types of products: from food, medicine, healthcare goods and transplants to industrial products including cosmetics and food/drink additives
3) physical scale of PFALs: from micro- and mini-PFALs for use in restaurants, convenience stores, and homes to large-scale PFALs,
4) structure and types of cultivation systems and degree of automation depending upon the purpose and scale of production and sales,
5) geographical location and/or climate: from extremely cold to hot, and from extremely dry to wet regions/climates,
6) investment resources for business and research: from 0.1 million to 1 billion USD,
7) fields of R&D: from natural and social sciences to interdisciplinary and transdisciplinary fields, and
8) fields of business: from food, medicine, healthcare, cosmetics, manufacturing, and IT industries to education, social welfare, and urban planning.

24.4 Challenges of sustainable PFALs and opportunities

Realizing the goal of a most efficient PFAL poses many challenges and offers many opportunities as described in Section 3.11 of Chapter 3.

With only a history of about 20 years, the technologies used in PFALs with LEDs for commercial plant production are still in their infancy compared to the technologies used in modern agriculture after industrial revolution with a history of around three centuries, and even those used in greenhouse horticulture, which have a history of around one century. PFALs can adopt recent advanced technologies such as AI and IoT relatively easily. On the other hand, the efficient implementation of such advanced technologies to improve the resource and monetary productivities requires a deep understanding of the fundamental characteristics of PFALs as well as an appropriate vision, mission, and goals.

24.5 Conclusion

The PFAL has been developed and created for use on our planet. However, the fundamental concept and core technology of the PFAL is also compatible and common to that of space farming (plant production on spaceships, on the moon, etc.). This commonality is only natural because PFALs and space farms are both designed to produce plants at the highest yield and quality with minimum resource consumption and minimum generation of waste in an airtight or enclosed, thermally insulated clean structure. Resources include electricity for lighting and air conditioning, water, fertilizer, seeds, and a three-dimensional use of production space. There is a need to reduce and/or recycle waste including plant residue, heat energy, and wastewater generated by both PFALs and space farms, so that ultimately there is no waste.

Closed plant production systems such as PFALs are often connected with other biological systems to recycle waste into a resource. For example, CO_2 respired by heterotrophic living organisms is used by plants for photosynthesis, and O_2 produced by plants during photosynthesis is used by heterotrophic living organisms for respiration. People often find growing or living with plants and other living organisms fulfilling or relaxing. In the near future, all the electricity is expected to be generated using natural energy sources. To achieve the above goals, the introduction of recent advanced technologies including AI and IoT is essential.

In conclusion, the PFAL will play an essential role as a new means of plant production in the coming years as the number of people living in limited spaces such as urban areas or in nonfertile wastelands in harsh weather conditions with limited natural resources including water increases.

Index

Note: 'Page numbers followed by "f" indicate figures and "t" indicate tables.'

Printed in the United States
by Baker & Taylor Publisher Services